r. Naimuddin Principal Scientist (Plant Pathology) at ICAR-ndian Institute of Pulses Research, Kanpur has more than 25 years f experience of working on diseases of pulses. He has about fifty esearch papers, many book chapters, popular articles and bulletins. He has co-edited two books and attended international course on 'Biotechnology in Agriculture: Plants and Microorganisms' held at The Hebrew University of Jerusalem, Faculty of Agricultural, Food and Quality Sciences, Rehovot, Israel. He is reviewing research papers of ls and has been in the editorial board of Pulses Newsletter, three NAAS journals and annual reports of the ICAR-IIPR, Kanpur.

Dr. R.K. Mishra Principal Scientist (Plant Pathology) at ICAR-Indian Institute of Pulses Research, Kanpur has 15 years of experience in the field of Plant Pathology. He is involved in characterization and management of major biotic stress of pigeonpea with major focus on wilt and Phytophthora stem blight including host plant resistance and biological control. He has published about 100 research papers/ reviews/conference papers/ abstracts/ books/ bulletins and popular urnals of international/national repute. He is reviewing research articles of ng journals. Dr. Mishra has to his credit five books, many technical bulletins, d extension folders on biotic stresses management in crop plants.

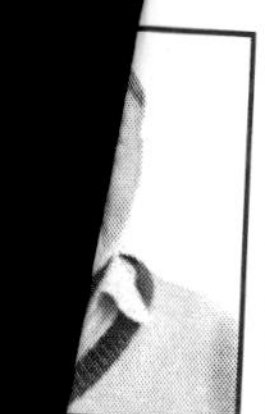

Dr. Mohd. Akram Principal Scientist (Plant Pathology) at Division of Crop Protection, ICAR-Indian Institute of Pulses Research, Kalyanpur, Kanpur has guided 20 M.Sc. students and published 55 research papers, 27 popular articles, 12 book chapters and 5 technical bulletins. He has co-edited 2 books. He is recipient of Young Scientist Award 2003-2004 in Plant Protection by NAAS, New Delhi, Young Scientist Award 2005 by CST, Lucknow, Uttar Pradesh and Lal Bahadur Shastri Young t Biennium Award 2005-2006 in Crop Sciences by ICAR, New Delhi. He received S Norman E. Borlaug Fellowship-2007 to attend training at Cornell University, New York and DST International Travel Fellowship to attend 10th International Pathology Congress, China.

Diseases
their

THE EDITORS

Environmental
different journa
rated research

articles in jo
many leadi
manuals a

Scientis
Indo-U
Ithaca,
Plant

Diseases of Pulse Crops and their Management

– Editors –

Naimuddin

R.K. Mishra

Mohd. Akram

ICAR-Indian Institute of Pulses Research, Kanpur, Uttar Pradesh

2022

Daya Publishing House®

A Division of

Astral International Pvt. Ltd.

New Delhi – 110 002

ISBN: 978-93-5222-134-9 (HB)

Published by : **Daya Publishing House**®
A Division of
Astral International Pvt. Ltd.
– ISO 9001:2015 Certified Company –
4736/23, Ansari Road, Darya Ganj,
New Delhi-110 002
Ph. 011-43549197, 23278134
e-mail: info@astralint.com
Website: www.astralint.com

Digitally Printed at : **Replika Press Pvt. Ltd.**

উত্তরবঙ্গ কৃষি বিশ্ববিদ্যালয়

পুণ্ডিবাড়ী, কোচবিহার, পশ্চিমবঙ্গ-৭৩৬১৬৫

UTTAR BANGA KRISHI VISWAVIDYALAYA

P.O. PUNDIBARI, DIST. COOCH BEHAR, WEST BENGAL- 736165

ড: চিরন্তন চট্টোপাধ্যায়
উপাচার্য
Dr. Chirantan Chattopadhyay
M.Sc (Ag.)(BCKV), Ph.D. (IARI)
ICAR: ARS
FNAASc, FPSI, FPPAI, FISMPP, FISOR
FIMS, FSRMR
Vice-Chancellor

ফোন/Phone: 03582-270141/270013 (O)
ফ্যাক্স/Fax: 03582-270249
মো°/Mob: +91-9434748016
ইমেল/E-mail: vcubkvv@gmail.com
ওয়েবসাইট/ Website: www.ubkv.ac.in

Ref. No. VC/UBKV- 3331

Date: 14 Jan 2021

Foreword

Pulses are a group of crops which are grown not only for their protein rich seeds but also for their ability to fix atmospheric nitrogen and for improving physical, chemical and biological properties of soil. These crops hold the key to the sustainable agriculture. The current trend of growing wheat and paddy in certain parts of the country has led not only to the deterioration of agricultural land but also the over-exploitation of ground water. Pulses by virtue of their ability to fix atmospheric nitrogen and improve soil health are known to help overcome these problems and therefore inclusion of pulses in different cropping systems may help in developing sustainable production systems.

Of the biotic constraints faced by pulse crops diseases caused by fungal and viral pathogens are economically important often rendering the crop uneconomical. Information on the various diseases of these crops and options for their management is essential for the mitigation of production losses thereby adding to the national pulse basket. Bringing information on various diseases and their management at one place is always desirable and helpful to all the stakeholders. Since considerable information on the understanding of pathogen-host interaction and the management of diseases in different pulse crops has been generated, three of my erstwhile colleagues –Drs. Naimuddin, R.K. Mishra and Mohd. Akram, Principal Scientists in Plant Pathology at the Division of Crop Protection, ICAR-Indian Institute of Pulses Research, Kanpur have produced this book "Diseases of Pulse Crops and Their Management". This book contains twenty one chapters written by some experienced as well as young and emerging scientists/teachers working on diseases of different pulse crops in the country. The information provided in the book will definitely help in managing diseases thereby reducing the yield losses.

I like to congratulate all the contributors to the book as also the editors for bringing out this book and hope that the book shall open new vistas in the aspect of disease management of pulse crops.

14 Jan 2021

(CHIRANTAN CHATTOPADHYAY)

Preface

Pulses are group of leguminous crops and one of the important components for sustainable agriculture in many parts of the world. These crops are grown for their nutrient rich grains. Pulses are important source of protein for the poor as well as the vegetarian population of the world. In countries like India, amino acids of pulses protein complement that of cereals as they are generally eaten together. Pulses are important not only in India but also in many other poor countries where these crops are seen as an easy source of nutrition. It is because of this United Nations Organization has declared 2016 as International year of Pulses. Chickpea, lentil, pea, pigeonpea, mungbean, urdbean, lentil, pea and rajma are important grain legumes. Besides there are many arid legumes crops like mothbean, cowpea, horsegram *etc.* grown in arid areas of the world. Diseases of pulses are one of the most important biotic constraints to their production. Their management strategies need to be percolated to the growers so as the losses caused by the diseases is contained and the growers are benefitted economically. Pulses encounter a wide range of diseases caused by a variety of pathogens that are mostly fungal and viral. However, there are many diseases caused by bacteria and nematodes too. The book includes 21 chapters on important diseases like wilts of chickpea, pigeonpea and lentil, collar rot of grain legumes, foliar fungal diseases and viral diseases of these crops, yellow mosaic in pulse crops, sterility of pigeonpea, diseases of mungbean, frenchbean and arid legumes. Diseases caused by *Rhizoctonia, Sclerotinia sclerotiorum* and *Sclerotium rolfsii*, have also been discussed. Chapters on various management options, virus-vector relationship, and role of mycorrhizal fungi in legume crops and culture collection of pathogens have also been included.

Editors

Contents

List of Contributors

Aggarwal, Rashmi
Division of Plant Pathology, ICAR-Indian Agricultural Research Institute, New Delhi – 110012

Akram, Mohd.
ICAR-Indian Indian Institute of Pulses Research, Kanpur – 208024, Uttar Pradesh

Bahadur, Amar
College of Agriculture, Tripura, Lembucherra, Agartala-799210, West Tripura

Bashyal, B.M.
Division of Plant Pathology, ICAR-Indian Agricultural Research Institute, New Delhi – 110012

Biswas, S.K.
Department of Plant Pathology, C.S. Azad University of Agriculture and Technology, Kanpur – 208002, Uttar Pradesh

Choudhary, Damaram
Division of Plant Pathology, Indian Agricultural Research Institute, New Delhi – 110012

Devi, T. Prameela
Division of Plant Pathology, Indian Agricultural Research Institute, New Delhi – 110012

Dubey, S.C.
Division of Plant Quarantine, ICAR-National Bureau of Plant Genomic Research, Pusa, New Delhi – 110012

Dwivedi, P.K.
KVK Raisen, Madhya Pradesh

Gupta, Meenu
Department of Vegetable Science, Dr. Y.S. Parmar University of Horticulture and Forestry, Nauni, Solan – 173230, Himachal Pradesh

Gupta, S.K.
Department of Plant Pathology, Dr. Y.S. Parmar University of Horticulture and Forestry, Nauni, Solan – 173230, Himachal Pradesh

Jaiswal, R.K.
Krishi Vigyan Kendra, Jawaharlal Nehru Krishi Vishwa Vidyalaya, Panna – 488001, Madhya Pradesh

Kamil, Deeba
Division of Plant Pathology, Indian Agricultural Research Institute, New Delhi – 110012

Kharayat, Bhupendra Singh
Division of Plant Pathology, ICAR-Indian Agricultural Research Institute, New Delhi – 110012

Kharwar, R.N.
Department of Botany, BHU, Varanasi – 221005, Uttar Pradesh

Kirar, K.S.
Krishi Vigyan Kendra, Jawaharlal Nehru Krishi Vishwa Vidyalaya, Panna – 488001, Madhya Pradesh

Kumar, Gagan
Institute of Agricultural Sciences, Department of Mycology and Plant Pathology, BHU, Varanasi – 221005, Uttar Pradesh

Kumari, Anupam
Department of Plant Pathology, Bihar Agricultural University, Sabour, Bhagalpur – 813210, Bihar

Kumari, Sunita
Krishi Vigyan Kendra, Kishanganj, BAU, Sabour, Bihar

Mathur, Manish
ICAR-Central Arid Zone Research Institute, Jodhpur, Rajasthan

Mawar, Ritu
ICAR-Central Arid Zone Research Institute, Jodhpur, Rajasthan

Mishra, Monika
ICAR-Indian Indian Institute of Pulses Research, Kanpur – 208024, Uttar Pradesh

Mishra, R.K.
Division of Crop Protection, ICAR-Indian Institute of Pulse Research, Kanpur – 208024, Uttar Pradesh

Naimuddin
ICAR-Indian Indian Institute of Pulses Research, Kanpur – 208024, Uttar Pradesh

Panday, A.K.
Project Co-coordinating Unit (Sesame and Niger), JNKVV, Jabalpur (Madhya Pradesh)

Pandey, M.K.
ICAR-Indian Institute of Vegetable Research (IIVR), Varanasi – 221005, Uttar Pradesh

Pandey, Rakesh
Department of Agricultural Entomology, Banda University of Agriculture and Technology, Banda, Uttar Pradesh

Pandey, Sonika
ICAR-Indian Indian Institute of Pulses Research, Kanpur – 208024, Uttar Pradesh

Pandey, Sumit Kumar
Subject Matter Specialist, Krishi Vigyan Kendra, Sidhi, J.N.K.V.V., Jabalpur, Madhya Pradesh

Pandey, Vibha
Department of Plant Pathology, JNKVV, Jabalpur, Madhya Pradesh

Patel, Jai Singh
Institute of Agricultural Sciences, Department of Mycology and Plant Pathology, BHU, Varanasi – 221005, Uttar Pradesh

Prasad, Bishun D.
Department of Molecular Biology and Genetic Engineering, B.A.C., Sabour, Bihar

Sahni, Sangita
Department of Plant Pathology, DRRPCAU, T.C.A., Dholi, Muzaffarpur, Bihar

Sarkar, Ankita
Institute of Agricultural Sciences, Department of Mycology and Plant Pathology, BHU, Varanasi – 221005, Uttar Pradesh

Sarma, B.K.
Institute of Agricultural Sciences, Department of Mycology and Plant Pathology, BHU, Varanasi – 221005, Uttar Pradesh

Singh H.B.
Institute of Agricultural Sciences, Department of Mycology and Plant Pathology, BHU, Varanasi – 221005, Uttar Pradesh

Singh, A.K.
Division of Plant Pathology, Sher-e-Kashnir University of Agricultural Sciences and Technology-Jammu, J&K

Singh, Jai
Subject Matter Specialist, Krishi Vigyan Kendra, Sidhi, J.N.K.V.V., Jabalpur, Madhya, Pradesh

Singh, R.K.
Institute of Agricultural Sciences, Department of Mycology and Plant Pathology, BHU, Varanasi – 221005, Uttar Pradesh

Singh, Surendra
Department of Botany, BHU, Varanasi – 221005, Uttar Pradesh

Srivastava, J.N.
Department of Plant Pathology, Bihar Agricultural University, Sabour, Bhagalpur – 813210, Bihar

Tiwari, Sashi
Naini Agriculture Institute, SHUATS, Allahabad – 211007, Uttar Pradesh

Tomer, A.S.
ICAR-Central Arid Zone Research Institute, Jodhpur, Rajasthan

Toppo, R. Sudeep
Division of Plant Pathology, Indian Agricultural Research Institute, New Delhi – 110012

Vais, S.S.
Department of Mycology and Plant Pathology, Institute of Agricultural Sciences, Banaras Hindu University, Varanasi – 221005, Uttar Pradesh

Yadav, Sudheer Kumar
Institute of Agricultural Sciences, Department of Mycology and Plant Pathology, BHU, Varanasi – 221005, Uttar Pradesh

Chapter 1

Fusarium Wilt Disease of Chickpea (*Cicer arietinum* L.)

R.K. Jaiswal, B.S. Kirar & R.K. Mishra

Chickpea (*Cicer arietinum* L.) is the world's third most important pulse crop, after dry beans (*Phaseolus vulgaris* L.) and dry peas (*Pisum sativum* L.) (Vishwadhar and Gurha, 1998). India is the leading chickpea producing country with 73.3 per cent of the world acreage and 67.4 per cent of the production. Chickpea is contributing nearly 42 to 47 per cent of total pulse production in India and approximately 90 per cent of the area and production is from six states viz., Madhya Pradesh, Rajasthan, Maharashtra, Uttar Pradesh, Karnataka and Andhra Pradesh.

Chickpea is a vital source of plant derived edible protein especially lysine-rich protein in the diets of the poor, and is particularly important in vegetarian diets (Jukanti *et al.*, 2012). Also, it is being used increasingly as a substitute for animal protein. Chickpea is a good source of zinc, folate and protein. They are also very high in dietary fiber and hence a healthy source of carbohydrates for persons with insulin sensitivity or diabetes. It is also low in fat and most of this is polyunsaturated. One hundred grams of mature boiled chickpea contains 164 calories, 2.6g of fat (of which only 0.27g is saturated), 7.6g of dietary fiber and 8.9g of protein. Chickpea also provide dietary calcium (49–53mg/100g). According to the International Crops Research Institute for the Semi-Arid Tropics (ICRISAT) chickpea seeds contain on an average 23 per cent protein, 64 per cent total carbohydrates (47 per cent starch, 6 per cent soluble sugar), 5 per cent fat, 6 per cent crude fiber and 3 per cent ash. High mineral content has been reported for phosphorus (340 mg/100g), calcium (190 mg/100g), magnesium (140 mg/100g), iron (7 mg/100g) and zinc (3 mg/100g). Recent studies have also shown that they can assist in lowering of cholesterol in the bloodstream (Geervani, 1991, Pittaway *et al.*, 2008). Glandular secretion of the leaves, stems, and pods consists of malic and oxalic acids, giving a sour taste and these acids are supposed to lower the blood cholesterol levels (Duke, 1981). Chickpea also has an advantage in the management of soil fertility, particularly in dry lands and the semi arid tropics.

Chickpea is basically a self-pollinated crop however; cross-pollination occurs to some extent ranges between 0-1 per cent (Singh, 1987; Smithson *et al.,* 1985). Chickpea is a quantitative long-day plant, but flowering occurs in every photoperiod (Smithson *et al.,* 1985). Chickpea is a hardy, deep-rooted, dryland crop (Figures 1.1), which can grow to full maturity in conditions that would be unsuitable for most crops (Singh and Reddy, 1991). The deep-tap root system enhances its capacity to withstand drought conditions. It is usually well suited for cultivation in cooler areas with low rainfall. The yield is maximal when the legume crop is grown in sandy or loam soils possessing an appropriate drainage system because it is very sensitive to excess water availability. Frost, hailstones and excessive rains damage the crop; however, some cultivars can tolerate temperatures as low as -9.5°C in early stages. Relative humidity of 21-41 per cent is optimum for seed setting. Optimum conditions include 18-26° C day and 21-29° C night temperatures and annual rainfall of 600-1000 mm (Duke, 1981). Generally it is grown in the months of September-November and harvested in the months of February- April. Maturity period ranges from 95-110 days after sowing.

Chickpea (*Cicer arietinum* L.) wilt caused by *Fusarium oxysporum* Schlecht and Emnd Snyd. and Hans. f. sp. *ciceris* (Padwick) Snyd. and Hans. was first reported from India by Butler (1918). McKerral (1923) considered this fusarial disease is to

Figure 1.1. Chickpea Field Infected with Fusarium Wilt.

be soil borne. Narasimhan (1929), McRae (1932) and Prasad and Padwick (1939) also reported that it was caused by *Fusarium* species. Padwick (1940) identified *F. orthoceras* var. *ciceris* as the cause of chickpea wilt in India. In the studies of Dastur (1935), *Rhizoctonia bataticola* produced wilted plants and he called the disease 'Rhizoctonia wilt'. Later, the fungus causing typical wilt was named as *F. orthoceras* var. *ciceris* (Padwick, 1940). Association of *Verticillium alboatrum* with chickpea wilt has also been observed (Bhatti *et al.*, 1983; Erwin, 1958a). Erwin (1958b) named the pathogen *F. lateritium* f. sp. ciceris. Snyder and Hansen (1940) renamed *F. orthoceras* var. *ciceris* as *F. oxysporum* f. sp. *ciceris* which is now widely accepted.

Fusarium wilt disease causing severe damage wherever this crop is grown (Rangaswami *et al.*, 1999). It is more prevalent in lower latitudes (0-30°N) where growing season is relatively dryer and warmer than in the higher latitudes (30-40°N). It is one of the major diseases of chickpea and at national level the yield losses encountered was reported to the tune of 60 per cent (Singh *et al.*, 2007). It causes complete loss in grain yield if the disease occurs in the vegetative and reproductive stages of the crop (Haware and Nene, 1980; Haware *et al.*, 1990; Halila and Strange, 1996; Navas *et al.*, 2000). *F. oxysporum* f. sp. *ciceris* infects chickpea at seedling as well as at flowering and pod forming stage (Grewal, 1969), with more incidence at flowering and podding stage if the crop is subjected to sudden temperature rise and water stress (Chaudhry *et al.*, 2007).

The pathogen is mainly soil borne as well as internally seed borne in nature. It can survive in the soil up to six years in the absence of susceptible host (Haware *et al.*, 1978). The chlamydospore like structures occurs in the hilum region of the seed. Infected seed plays an important role in long distance dispersal and in transmitting the disease to new areas. The pathogen can also survive in infected crop residues buried in the soil. Among other legumes, lentil, pea and Pigeonpea are symptomless carriers of the chickpea wilt fungus (Haware and Nene, 1982). Considering the nature of damage and survival ability of the pathogen, use of resistant varieties is the only economical and practical solution for the management of wilt disease. Most of the resistant varieties have been found to be susceptible after some years because of breakdown of their resistance due to evolution of variability in the pathogen. The pathogen with high saprophytic ability can survive in soil for long period during which it may have to go through different environmental stresses and biological competition which may lead to the development of physiological races. Therefore, integrated management strategies are the possible solutions to maintain plant health mainly for soil borne plant pathogens. These strategies include modification of cultural practices, growing of resistant varieties with minimum application of chemicals (Bendre and Barhate, 1998), and encouragement of beneficial microbial population to reduce pathogen inoculums. The present manuscript comprehensively deals with all aspect of Fusarium wilt disease comprising distribution, etiology, pathogen biology, epidemiology and its management strategies.

Distribution of the Disease

Occurrence of chickpea wilt has been reported from almost all over the world but it was first described in India (Butler, 1918). The disease has been reported widely and distributed in near about 32 countries of the world (Nene *et al.*, 1996), however, considered as a serious disease of chickpea in India, Iran, Pakistan, Nepal, Burma, Spain, Tunisia, Mexico, Morocco, Algeria and Syria. In India, its incidence varied from 14 to 32 per cent in different states of India as reported by Dubey *et al.* (2010).

F. oxysporum is a ubiquitous, asexual species complex which can cause vascular wilt or cortical rot diseases in many agricultural crops and have been classified into *Formae specials* based on their host specificity (Nelson *et al.*, 1983). Nene and Reddy (1987) recorded the distribution of *F. oxysporum* f. sp. *ciceris* covering North America, Europe, Middle East, Asia and South East Asia. Two pathotypes have been distinguished based on the distinct yellowing or wilting syndromes. In addition to pathotypes, eight races of the pathogen *viz.* 0, 1A, 1B/C, 2, 3, 4, 5, and 6 have been identified so far which are key elements in the development and management of the disease. Races 2, 3 and 4 have only been described from India (Haware and Nene, 1982), whereas races 0, 1B/C, 5 and 6 are found mainly in the Mediterranean region as well as in the USA (California) (Jiménez-Díaz *et al.*, 1993; Halila and Strange, 1996). Race 1A has been reported in India (Haware and Nene, 1982), California and the Mediterranean region (Jiménez-Díaz *et al.*, 1993).

Economic Importance of the Disease

Haware *et al.* (1978) reported that the fungus *F. oxysporum* f. sp. *ciceris* is a primarily soil borne pathogen, however, it can also be transmitted through seeds. Early wilting causes greater loss than the late wilting. Seeds harvested from wilted plants are lighter than those from healthy plants (Haware and Nene, 1980). Annual chickpea yield losses from Fusarium wilt may vary from 10 to 90 per cent (Haware and Nene, 1980; Jimenez-Diaz *et al.*, 1989; Trapero-casas *et al.*, 1985; Singh and Reddy, 1991; Jalali and Chand, 1992 and Halila and Strange, 1996). It is observed that incidence of wilt in the farmer's fields is increasing considerably every year and its severity is directly related to the increasing density of the pathogen inoculum in the soil (Bhatti and Kraft 1992; Sugha *et al.*, 1994 and Zote *et al.*, 1996). Hanif *et al.* (1999) reported that in Pakistan the area under chickpea cultivation on irrigated land has reduced from 50 to 10 per cent. Chickpea wilt causes complete loss of grain yield if the disease occurs in the vegetative and reproductive stages of the crop (Navas *et al.*, 2000; Halila and Strange, 1996; Haware *et al.*, 1990; Haware and Nene, 1980). An estimated annual loss of 12 million rupees was reported from Pakistan (Sattar *et al.*, 1953).

Symptomatology

The main symptoms of the disease are: yellowing and drying of leaves from base upward, drooping of petioles and rachis, improper branching, withering of plants, browning of vascular bundles and finally wilting of plants. Based on symptoms the pathogen is categorized into two groups. *Fusarium oxysporum* f. sp. *ciceris* (races 1A, 2, 3, 4, 5 and 6) – Wilting pathotype. Flaccidity of leaves and succulent shoots, followed by discoloration and chlorosis of leaves, desiccation and death; vascular (xylem) and pith tissues show discoloration, usually evident in cross sections of stem near the base. *Fusarium oxysporum* f. sp. *ciceris* (races 0 and 1A/B) – Yellowing pathotype. Progressive foliar yellowing from the base upwards; abscission of necrotic leaves; vascular (xylem) and pith tissues show discoloration.

The disease occurs at seedling and flowering stage of plant growth. The symptoms which can be observed are drooping of petioles and rachis, yellowing and drying of leaves from base to upward, browning of vascular bundles, improper branching, withering of plants and finally death of plants (Westerlund *et al.*, 1974; Prasad and Padwick, *et al.*, 1939). Erwin (1958) reported that the foliage of infected plant turns yellow before wilting and the xylem tissue shows light brown discoloration (Figure 1.2). Saxena and Singh (1987) reported that the roots of the wilting plants do not show any external rotting but when split vertically, dark brown discoloration of internal xylem is seen. Nene *et al.* (1991) observed that wilt infected seedlings collapse and lie flat on the ground retaining their dull green colour. Haware (1993) reported that the pathogen gets entry into xylem vessel and invades whole vascular system, inducing symptoms of yellowing and wilting. It is reported that the fungus enters the vascular system of the infected plant via the roots (Brayford 1998; Leslie and Summerell, 2006). It produces enzymes that degrade the cell walls so that gels are formed that block the plant's transport system. Discoloration of the internal tissues progresses from the roots to the aerial parts of the plant, then yellowing and wilting of the foliage occurs, and finally there is necrosis. Singh *et al.* (2007) reported that following infection of host roots, the fungus crosses the cortex, enters the xylem tissues and then spreads rapidly upward direction through the vascular system, becoming systemic in the host tissues, and may directly infect the seed. They also observed that the pathogen is primarily confined to the xylem vessels in which the mycelium branches and produces micro-conidia. The micro-conidia detach and are carried upward in the vascular system until movement is stopped, where they germinate and the mycelium penetrates the wall of the adjacent vessel.

Figure 1.2. Fusarium Wilt Infected Chickpea Plants (*Cicer arietinum* L.).

Isolation of *Fusarium oxysporum* f. sp. ciceris

Fisher *et al.* (1982) reported that highly virulent strain of *F. oxysporum* f. sp. *ciceris* was isolated from infected chickpea plant by using Komada's medium (Komada, 1975). Patel *et al.* (2001) studied the colony characters, pigmentation, sporulation and conidial characters of *F. oxysporum* f. sp. *ciceris*. Honnareddy and Dubey (2007) reported that the isolates of *F. oxysporum* f. sp. *ciceris* had variable pigmentation which varied from normal white to violet, brown, reddish violet, greenish violet, yellowish pink and dark green. Variability and the virulence among various isolates of *F. oxysporum* f. sp. *ciceris* under *in vitro* condition were studied by various workers (Paulkar and Raut, 2004; Singh *et al.*, 2010 and Nikam *et al.*, 2011)

Pathogenicity of *Fusarium oxysporum* f. sp. *ciceris*

Single spore isolation method is used for obtaining pure culture of *F. oxysporum* f. sp. *ciceris* through using Czapekdox agar medium (Barnet and Hunter, 1972; Ahmad, 2010). Potato Dextrose Agar (PDA) slants may be used for sub culturing of *F. oxysporum* f. sp. *ciceris* and incubated at 27 ± 1°C for ten days (Patil, 2006). *Fusarium* species are stored at 4°C till use (Hend *et al.*, 2012). Nikam *et al.* (2011) confirmed Pathogenicity of the *F. oxysporum* f. sp. *ciceris* by sick soil inoculation technique in earthen pots under green house conditions by using susceptible cultivar JG-62.

Characteristics of *Fusarium oxysporum* f. sp. *ciceris*

The pathogen is a common soil inhabitant with taxonomic nomenclature of *F. oxysporum* f. sp. *ciceris* (Padwick) Matuo and Sato (Snyder and Hansen, 1940). The mycelium of *F. oxysporum* is septate, profusely branched growing on potato dextrose agar at 25° C initially white turning light buff or deep brown later, fluffy or submerged (Saxena and Singh 1987). The growth becomes felted or wrinkled in old cultures. Various types of pigmentation (yellow, brown, crimson) can be observed in culture. The macro-conidia are straight to slightly curved, slender, thin walled usually with three or four septa, a foot-shaped basal cell and curved apical cell (Couteaudier and Alabouvette, 1990). They are generally produced from phialides on conidiophores by basipetal division. The micro-conidia are ellipsoidal and either have no septum or a single one. Both are formed from phialides in false heads by basipetal division. They are important in secondary infection. The chlamydospores are globose and have thick walls. They are formed from hyphae or alternatively by the modification of hyphal cells. They are important as endurance organs in soils where they act as inocula in primary infection. The teleomorph or sexual reproductive stage of *F. oxysporum* is unknown (Leslie and Summerell, 2006).

Histopathology and Pathogenesis

Kunwar *et al.* (1989) reported that hyphae were inter and intracellular in pith, xylem and cortex; the epidermis was disintegrated and hypertrophy of cortical and pith cells occurred. A mucilage-like substance was present in the cells of the xylem and cortex. Stevenson *et al.* (1997) observed that hyphae occurred in root xylem of wilted plants. In more severe cases, a large portion of stem xylem vessel was also invaded by up to 5 internodes above the points of seed attachment. Khan *et al.*

(2004) reported that the phytotoxin produced by pathogen causing wilting and leaf burning of chickpea crop. Perez-Artes and Tena (1989) observed correlation between pathogen produced pectate lyase with pathogenicity and virulence of pathogen. Jorge *et al.* (2005) reported that pathogen produced xylanases on medium which hydrolysed xylan to xylobiose.

Disease Cycle

The disease cycle begins with infection of host roots by the fungus then it crosses the cortex and enters into the xylem tissues. It then spreads rapidly up through the vascular system, becoming systemic in the host tissues, and may directly infect the seed. Seed infestation and infection is common. In chickpea, *F. oxysporum* f. sp. *ciceris* can be internally seed-borne and the pathogen is found as chlamydospore-like structures in the hilum region of the seed. The movement of infected seed plays an important role in the long distance dispersal and transmission of Fusarium wilt diseases into new areas. The root tips of healthy plants growing in contaminated soil are penetrated by the germ tube of spores or fungal mycelium. Entry is either direct, through wounds, or opportunistic at the point of formation of lateral roots. The mycelium takes an intercellular path through the cortex, and enters xylem vessels through the pits. The pathogen is primarily confined to the xylem vessels in which the mycelium branches and produces micro-conidia. The micro-conidia detach and are carried upward in the vascular system until movement is stopped, at which point they germinate and the mycelium penetrates the wall of the adjacent vessel. Lateral movement between vessels is through the pits. The water economy of infected plants is eventually severely compromised by blockage of vessels, resulting in stomatal closure, wilting and death of leaves, often followed by death of the whole plant. The fungus then invades all tissues of the plant, to reach the surface where it sporulates profusely. Spores may then be dispersed by wind and water or movement of soil or plant debris. *F. oxysporum* can survive as mycelium and chlamydospores in seed and soil, and also on infected crop residues, roots and stem tissue buried in the soil for more than 6 years (Singh *et al.*, 2007). Chlamydospores can survive in soil either in dormant form or saprophytically for without a suitable host. The disease is favoured by warm and dry soil conditions with an optimal temperature of 22-25°C.

Epidemiology

There are many factors that affect the development of Fusarium wilt diseases in fields. These include: the presence of virulent strains in the soil, susceptibility of the crop varieties, soil type, soil fertility, climatic conditions, irrigated or non-irrigated crops and interactions with other soil borne micro-organisms. Inoculum densities in the soil are also important as disease symptoms may not be apparent when there are low levels of the pathogenic strains in the soil (Davis *et. al.* 2006). The disease can occur at any stage of plant growth, particularly when crops are stressed. Factors such as high plant populations, improper cultivation, other soil-borne pathogens and various herbicides are all known to induce injury of young roots, causing plant stress. These can aggravate Fusarium wilt damage. Early infections can kill seedlings soon after emergence, leaving bare areas in the crop.

Later infection can cause wilting, yellowing and death of adult plants. The effect of Fusarium wilt is most apparent during blossoming and early pod set when the plant and its productivity are more sensitive to stress.

The fungus is seed borne and may survive in plant debris in soil (Brayford, 1998). Haware *et al.* (1978) showed that the fungus present in the hilum of the seed in the form of chlamydospores like structures. Once the inoculum is established in soil, it is difficult to eradicate as the chlamydospores survive in the soil for at least 6 years and under favorable conditions germinate and infect the seedlings through tender roots (Haware *et al.*, 1996). Shakir and Mirza (1994) also studied the location of pathogen in seed and reported it to be present in cotyledons and axis. The primary infection is through chlamydospores or mycelia. The conidia of the fungus are short lived however; the chlamydospores can remain viable up to next crop season. The pathogen survives well in roots and stem, even in apparently healthy looking plants growing among diseased ones harbouring enough fungus (Padwick, 1941). The fungus, however, did not survive in the roots placed on soil surface. Plant species other than chickpea *viz. Vigna radiata, V. mungo, Cajanus cajan, Pisum sativum* and *Lens culinaris* may also serve as symptomless carriers of the disease (Haware and Nene, 1982; Gupta, 1991). The pathogen may also parasitize several weeds such as *Cyperus rotendus, Tribulus terrestris, Convolvulus arvensis* and *Cardiospermum halicacabum* (Nene *et al.*, 1980).

The soil type, reaction, moisture and temperature are known to influence disease development. The greenhouse studies supported the fact that the disease is more severe in light sandy soil than heavy clay ones (Kotasthane *et al.*, 1979; Sugha *et al.*, 1994 a). Chandra *et al.* (1974) attributed higher disease severity in light sandy soil to its low water retention ability. However, Srivastava *et al.* (2002) reported that black soil support highest wilt incidence (75.5 per cent). Wilt incidence in sandy-loam, red and clay soil was found to be 64.4, 59.9 and 46.6 per cent, respectively. The chlamydospores lost organic carbon, ability to germinate and pathogenic aggressiveness when exposed to heat stress (Arora *et al.*, 1996). The soil temperatures in the range of 24.8- 28.50 C and soil moisture above 25 per cent were most conducive condition for severity of chickpea wilt (Sugha *et al.*, 1994). The disease increased with decreasing soil metric potential as Bhatti and Kraft (1992) examined the effect of soil moisture on disease incidence using three soil metric potential regimes, high (-40 to -20 k Pa), medium (-260 to -40 k Pa) and low (- 1060 to -260 k Pa). Patel and Anahosur (2001) also observed that infection was pronounced at 50 and 75 per cent soil moisture. High soil temperature and deficiency of moisture appear to have a definite bearing on the incidence of disease. Soil temperature relation showed that the disease was optimum at 25°C. Chauhan (1962b) showed that the disease intensity increased with the lowering of pH, being considerably low at pH 9.2. In another such study, Chauhan (1963) observed that alkaline soils favoured incidence of wilt. However, Saikh (1974) reported that the pathogen tolerated a wide range of pH, with optimum between 5 and 6.5. Sugha *et al.* (1994 b) also reported a pH of 5.2 to be optimum. The amount of organic matter (Chauhan, 1965) and humus (Chauhan, 1962 c) content of soil were found inversely related to wilt incidence. In the studies of Sugha *et al.* (1994 b) increase in phosphorus and potassium contents did not influence the development of wilt but it was favoured by increase in nitrogen and organic carbon. Laboratory studies of Gupta (1999) showed that zinc inhibited the growth of pathogen. Rao and Krishnappa (1996) established the relationship between environmental factors and *Meloidogyne-Fusarium* wilt complex of chickpea.

Host-pathogen Interaction

A complex interaction between plant and its fungal pathogen is an outcome of expression of both, plant defense genes as well as fungal pathogenesis related genes. The result of such a relationship is projected either as host resistance or disease development in the plant. Plants generally reject attacking phytopathogenic fungi. A majority of plants are not readily colonized and parasitized by most pathogens. Such plants are non-host plants for the pathogen and exhibit immunity against them; since the pathogen is unable to surmount the barriers that prevent plant colonization. These various barriers comprise the basic resistance or basic incompatibility of the plant and are the first level of pathogen defense in a plant. The underlying mechanism for basic incompatibility may depend either on the plant, on the phytopathogenic fungus, or on both. The successful parasitism by the fungus depends on the production of pathogenicity factors which it has acquired during its co-evolution with the plant. During the process of colonization the pathogen withdraws nutrients from its host plant and lives and multiplies at its expense; leading to biochemical changes in the plant causing disease symptoms like yellowing of leaves, wilting, necrosis or distortion, reducing the vitality of plant ultimately causing plant death. Basic compatibility is a highly specific phenomenon referring to only a particular plant species and the appropriate pathogen species or forma speciales (Niks, 1988). In spite of the basic compatibility between plant and its pathogen; the host plant may deploy different defense strategies to limit pathogen attack. Firstly, in the hypersensitive reaction (HR) - the plant cells injured by the invading pathogen die very rapidly causing necrosis of the adjacent tissue; ensuring the pathogen to be cut off from the plant. In the second strategy, the plant cells survive pathogen attack and are induced to build up new defense barriers. This is a non-hypersensitive response often led by HR occurring in adjacent tissues (Klement and Goodman, 1967). Presence of gene for resistance against a pathogen in a particular host plant during a compatible reaction can turn it into an incompatible reaction. Such kind of host resistance comes into action because the infecting pathogen induces defense reaction in the host plant and represents as the second level of pathogen defense. These defense mechanisms are highly specific for the pathogen since only certain races of the pathogen are rejected by the resistant cultivar. Hence this kind of defense is also termed as cultivar specific resistance. These block or restrain growth and reproduction of the pathogen thereby restricting the colonization in the plant. Host resistance presupposes the existence of basic compatibility between the plant and its pathogen (Flor, 1971). In cases where a compatible interaction does not result in the appearance of disease symptoms on the plant, or does not evoke any resistance against the pathogen; the plant is said to exhibit tolerance against the pathogen. Many fungi exhibit biotrophy in the initial phase of the life cycle – obtaining nutrients from the metabolically active plant cells; and saprophytism in the later stages, when they acquire nutrients from the dead plant tissue. Cultivar specific resistance is often exclusively observed against attack by biotrophic pathogens (Prell and Day, 2001). Basic resistance is directed against all pathogens in general, while cultivar specific resistance or host resistance is highly selective against only one pathogen species or forma speciales or race. Two types of host resistance genes are recognized, race-nonspecific resistance (horizontal/ uniform or generalized resistance) and race-specific resistance (vertical/specific or differential resistance). Race-non specific resistance is directed against all members or races of a pathogen species, while race-specific resistance is mounted against only a particular pathogen race. In case of race-specific resistance; a specific mutation in the gene enables specific recognition/discrimination of a particular pathogen race from the other races. Thus, a phytopathogen must first be able to conquer basic resistance and then negate the race specific resistance of its host plant. Virulence is thus the ability of a phytopathogen to overcome race specific resistance and is acquired through gene mutation (Prell and Day, 2001). A study enabling to determine the race specific resistance in a cultivar or cultivar specific resistance in a pathogenic race is essential. Such kind of study would help to understand the mutations necessary in both plant defense and fungal virulence genes for such an association/interaction in the host and its pathogen.

Naseema Beevi *et al.* (2005) tested the *in vitro* compatibility of *T. harzianum* with mancozeb, carbendazim and copper oxychloride and found that carbendazim at 0.1 per cent completely inhibited the mycelial growth while mancozeb and copper oxychloride showed compatibility with the antagonist at 0.2 and 0.1 per cent respectively. Dubey *et al.* (2007) studied the efficacy of *Trichoderma* species in combination with carboxin and found that the integration of *T. harzianum* (106 spores/ml/10 g seed) and carboxin (2 g kg- 1 seed) for seed treatment was the best which enhanced seed germination by 12.0 to 14.0 per cent and grain yields by 42.6 to 72.9 per cent and reduced the wilt incidence (44.1 - 60.3 per cent). Khan and Gangopadhyay (2008) tested the compatibility of *P. fluorescens* with the fungicides and revealed that carboxin and carbendazim were least toxic to *P. fluorescens* strain PFBC-25 whereas captan was most inhibitory.

Shahida *et al.* (2010) tested the common fungicides Bordeaux mixture at 0.5, 1.0, 1.5 per cent concentration, copper oxychloride at 0.1, 0.2, 0.3 per cent concentration, potassium phosphonate at 0.2, 0.3, 0.4 per cent concentration and mancozeb at 0.1, 0.2, 0.3 per cent concentrations for their compatibility with the effective isolates of *T. harzianum* and *T. viride,* and found Bordeaux mixture was not compatible and potassium phosphonate is compatible at all concentrations. Similarly, all these fungicides were also tested for their compatibility with *P. fluorescens* and found potassium phosphonate is compatible at all three different concentrations. Kumar *et al.* (2011) reported that *B. subtilis* was compatible to 1000 ppm of hexaconazole, propiconazole, tebuconazole and validamycin.

Integrated Approaches for the Management of *Fusarium* Wilt of Chickpea

Once introduced and established, pathogenic strains of *F. oxysporum* can survive in soil for extended periods, even in the absence of crop hosts, making eradication a long term process. Hence containment procedures to retard the spread of the pathogen are required. Infected crops should be destroyed by burning and ploughing. This will prevent aerial dispersal of the pathogen via infected crop residues. Chlamydospores may persist within infected soil for up to six years or longer. If the pathogen is detected very early it must be eradicated. It is able to produce spores that are dispersed to short distances via rain splash, infected soil and infected plant trash including seed. The pathogen is also capable of being transported over long distances via the movement of infected seed and contaminated vehicles and machinery. Do not irrigate the affected crop areas and use bunding to divert overland flood flows and surface water run-off around them.

Mechanical Control

Deep ploughing over summer and removal of infected trash can reduce inoculum levels of Fusarium wilt of chickpea (Haware 1998). Soil solarization (covering soil with transparent 100 mm thick polythene sheet for 6-8 weeks from April to May) decreased population of *Fusarium* and plant parasitic nematodes (Chauhan *et al.*, 1998). However, this method of control is not practical for broad-acre farming systems. Since the causal pathogen can survive in soils for long periods of time, there still can be some residual disease inoculum even with good management practices (Lange and McLaren 2002); hence eradication can become a long term process. Since Fusarium wilt of chickpea are seed borne, the use of clean seed from disease-free crops is very important.

Cultural Control

Fusarium wilt can be managed through the use of disease prevention strategies, such as rotation with non-host crop species, control of volunteer pulse crop plants in cereal crops, and control of annual weeds in crop borders and headlands. Soil amendment with organic matter, such as wheat or barley straw, has been found to enhance antagonism by other soil micro-organisms. Delayed sowing can reduce disease incidence, but late sowing can dramatically reduce yield potential and its effect on disease development can differ over locations and seasons (Kannaiyan and Nene 1975). Several studies have suggested that higher disease control and yield are obtained when the planting is delayed until the last week of October (Chandra *et al.*, 1974; Mundkur, 1946; Padwick and Bhagwagar, 1943). The lower disease incidence in late-sown crop is considered to be due to low temperature prevailing during the period of late-sown crop. The studies of Navas-Cortes *et al.* (1998) showed that for each year of experiment epidemic development was related mainly to the date of sowing. Thus, advancing the sowing time can slow down the development of epidemic, delay the epidemic onset and minimize the final amount of disease. Plants spaced at 15-20 cm had much higher disease incidence than those spaced at 7.5 cm; this was attributed to the shallower root system in widely spaced plants which were susceptible to wilt when subjected to moisture stress (Bahl, 1976). Planting of seeds at proper depth (10-12 cm) was helpful in reducing the disease incidence (Singh and Sandhu, 1973), while shallow sown crop seemed to attract more disease (Dahiya *et al.*, 1988; Saikh, 1974; Sugha *et al.*, 1994b). Hanif *et al.* (1999) worked on the effect of various sowing depths on wilt incidence in wilt-sick field in Pakistan and concluded that deep sowing had no effect on reduction of *Fusarium* wilt incidence in susceptible chickpea variety Aug 424. The level of infection was similar at the 10 cm sowing depth as it was at the 30 cm depth. Planting the crop with "Pora" method (Saraf *et al.*, 1973) using lower seed rate helped to minimize disease, whereas broadcast method of planting increased wilt incidence (Bedi and Pracer, 1952). Development of wilt is more prominent under moisture stress conditions (Kausar, 1968); one irrigation before flowering decreases the disease incidence and increases the yield (Saraf *et al.*, 1973; Sekhon, 1950). Intercropping/mixed cropping reduces the wilt incidence and increase yield of chickpea. Mixed cropping of chickpea with wheat and berseem (Saraf *et al.*, 1973) has given measurable disease control. Agrawal *et al.* (2002) worked on the effect of wheat, barley, linseed and mustard as intercrops/mixed crops with chickpea on wilt incidence and reported that lowest wilt incidence obtained with intercropping and mixed cropping with linseed. Mayur *et al.* (2003) reported that wilt incidence was significantly reduced by amending the soil with deoiled mustard cake, groundnut cake and farm yard manure.

Biological Control

Isolation of Antagonists against *Fusarium oxysporum* f. sp. *ciceris*

Isolation of potential antagonists has been made by various workers against *F. oxysporum* f. sp. *ciceris* from rhizosphere soil and root endophytes. Nautiyal (1997) isolated 478 chickpea rhizosphere competent bacteria for suppression of *F. oxysporum* f. sp. *ciceris* and found 386 strains that effectively colonize chickpea roots. Parmer and Dadarwal (1997) isolated *Pseudomonas* and *Bacillus* sp. from rhizosphere and rhizoplane of healthy chickpea plants which inhibited the mycelial growth of *F. oxysporum* f. sp. *ciceris*. Rangeshwaran *et al.* (2002) also isolated twenty-five endophytic bacteria from internal tissues of root and stem portions of chickpea, sunflower, niger, chilli and capsicum plants.

Evaluation of the Efficacy of Antagonists against *Fusarium oxysporum* f. sp. *ciceris*

Assessment of efficacy of various antagonists against *F. oxysporum* f. sp. *ciceris* have made by several workers under *in vitro* condition. Rangeshwaran and Prasad (2000) evaluated antagonistic efficacy of rhizospheric bacteria under *in vitro* condition against *F. oxysporum* f. sp. *ciceris* and found four isolates as potential antagonists *viz.*, *Pseudomonas putida* (PDBCAB 19), *P. fluorescens* (PDBCAB 2), *P. fluorescens* (PDBCAB 29) and *P. fluorescens* (PDBCAB 30). Among these, *P. fluorescens* (PDBCAB 29) was found most effective to inhibit the mycelial growth of *F. oxysporum* f. sp. *ciceris*. Further Rangeshwaran *et al.* (2002) screened twenty five endophytes in dual culture on potato dextrose agar (PDA) and tryptic soya agar (TSA) against *F. oxysporum* f. sp. *ciceris, F. udum, R. solani*, and *Sclerotium rolfsii* and found ten isolates of endophytes more effective which inhibited the mycelial growth of the pathogens.

Mane and Pal (2008) also screened the different antagonists, *i.e. Acrophialophora* sp., fluorescent pseudomonas isolates and *Gliocladium virens, G. catanaltum, G. deliguences, T. hamatam, T. harziamum* and *T. koningii* and found that *T. hamatam* inhibited an average of 26.03 per cent of radial growth of *F. oxysporum* f. sp. *ciceris* followed by fluorescent pseudomonas isolate-4 (25.08 per cent) and fluorescent pseudomonas-3 (25.5 per cent), respectively. Mycelial growth and conidial production of *F. oxysporum* f. sp. *ciceris* was significantly reduced by antagonistic species of *T. atroviride, T. harzianum* and *T. longibrachiatum* under both *in vitro* and *in vivo* condition (Boureghda and Bouznad, 2009).

Biological management of Fusarium wilt of chickpea also has been addressed using bacterial and fungal antagonists in recent years (Anjaiah *et al.,* 2003; Hervás, 1997, 1998; Landa *et al.,* 1997; Landa *et al.,* 2001; Vidhyasekaran, and Muthamilan, 1995). Isolates of *Pseudomonas* spp., *Bacillus* spp., *Paenibacillus* spp., and nonpathogenic (NP) isolates of *F. oxysporum* isolated from the chickpea rhizosphere were effective in suppressing Fusarium wilt of chickpea under controlled conditions (Hervás, 1997, 1998; Landa *et al.,* 2001). Therefore, biological control offers potential for suppression of Fusarium wilt under field conditions, particularly when used in combination with cultivars with partial resistance to the disease and choice of sowing date.

Biological modification of soil microflora has proved successful in reducing disease severity (Mazzola, 2002). Enrichment of soils with the beneficial mycorrhiza fungi (Goswami *et al.,* 2007) or *R. leguminosarum* bacteria (Essalmani and Lahlou, 2003) may provide an effective management strategy against pathogenic fungal species. In addition, prior inoculation of chickpea with non-pathogenic isolates of *F. oxysporum* significantly reduced the disease incidence and severity (Hervas *et al.,* 1995; Kaur 2003; Landa 2005).

Prasad *et al.* (2002) reported that the soil application of *T. viride* and *T. harzianum* one week before sowing was more effective in reducing wilt and wet root rot of chickpea. Seed treatment with *T. harzianum* + carbendazim (1.25 g kg^{-1}) resulted in the maximum seed yield (4.6 g plant-1) and lowest disease incidence (2.5 per cent) as reported by Poddar *et al.* (2004). Steinberg *et al.* (2004) reported that integrated use of spent mushroom compost and cattle manure significantly reduced the disease incidence of both *Fusarium* wilt and dry root rot in chickpea. Harichand and Singh (2005) reported that all the plant extracts, *viz.,* eucalyptus (*Eucalyptus globulus*), jatropha (*Jatropha multifida*), neem (*Azadirachta indica*), garlic (*Allium sativum*) except *Calotropis procera,* were significantly superior in reducing wilt incidence in chickpea. Seed treatment with *T. viride* and *A. sativum* bulb extract are identified as important components of integrated management of chickpea wilt. Kapoor *et al.* (2006) found that amendment with *Lantana camara* (10 t ha- 1) + bioagent Tricoguard at 2.5 kg or 62 kg FYM ha-1 + spray with carbendazim at pre flowering stage was most effective in managing the root rot-wilt complex disease in pea. Nikam *et al.* (2007) reported that combined soil application of *T. viride* and ground nut cake followed by neem cake had given good control against chickpea wilt caused by *F. oxysporum* f. sp. *ciceris*. Jayasekhar *et al.* (2008) found that under field conditions soil application of *P. fluorescens* followed by carbendazim spray (0.2 per cent) after 30 days of *Pseudomonas* application recorded the lowest disease incidence of 3.7 per cent. Singh (2009) reported that the application of two bioagents *viz., T. harzianum, P. fluorescens* and two fungicides carbendazim (0.1 per cent) and mancozeb (0.25 per cent) at 45 DAS was found effective for the control of coriander wilt caused by *F. oxysporum* f. sp. *coriandrii.* Srivastava *et al.* (2010) found that combined use of fluorescent pseudomonads, *T. harzianum* and Arbuscular mycorrhizal fungus significantly reduced the disease incidence and severity of tomato wilt by 74 per cent and 67 per cent in pots and field condition, respectively.

Resistant Cultivars

The cheapest, economical and the most ideal way of managing chickpea wilt, is the use of resistant cultivars. Chemical control of wilt is not feasible and economical because of the soil as well as seed-borne nature of the pathogen. Fungal chlamydospores can survive in soil up to 6 years in the absence of the host plants (Haware *et al.,* 1996). The most practical and cost-efficient method for management of *Fusarium* wilt of chickpea is the use of resistant cultivars (Nene and Haware 1980; Nene and Reddy 1987; Bakhsh *et al.,* 2007., Ahmad *et al.,* 2010). Resistance to Fusarium wilt is required in all chickpea cultivars targeted for Semi Arid Tropics and other Fusarium wilt prone areas of the world. There are many sources with high resistance to Fusarium wilt available in desi type, while resistance sources in kabuli type are limited. A world collection of over 13,500 germplasm accessions from 40 countries was evaluated for race 1 of *Fusarium oxysporum* at ICRISAT-Patancheru. Of the 160 resistant accessions identified, only 10 accessions were of kabuli type (Haware *et al.,* 1992).

Ahmad *et al.,* 2010 evaluated 321 genotypes to identify genetic sources of resistance against Fusarium wilt and reported that at seedling stage 173 genotypes were resistant, 54 were tolerant and 94 were susceptible, whereas at reproductive stage, 102 genotypes were resistant, 36 were tolerant and 183 were susceptible. Eighty two (82) genotypes showed steady resistance at both stages. The genotypes that showed resistance or tolerance at both the stages are most suitable for exploitation in breeding programs or for direct sowing in wilt prone areas. Similarly, various workers have also reported variation in wilt resistance at two stages (Nene *et al.,* 1981; Haware 1996). A high level of variability has been reported between isolates of same race collected from different areas and between isolates of different races (Sivaramakrishnan *et al.,* 2002., Iftikhar *et al.,* 2002).

Iftikhar *et al.* (1997) screened 31 chickpea germplasm lines received from ICARDA and found that all of them were highly resistant to wilt disease. Whereas, Bajwa *et al.* (2000) reported that out of 32 genotypes only one line was resistant, 4 lines were tolerant, and 27 were susceptible to highly susceptible against *Fusarium* wilt. Iqbal *et al.* (2005) also identified 14 chickpea lines to be resistant to wilt at seedling stage but no line found to be resistant at reproductive stage. Chaudhry *et al.* (2006) evaluated 414 germplasm lines for resistance against *Fusarium* wilt and found 35 test lines resistant, 208 intermediate, 77 susceptible and 94 highly susceptible. Further Chaudhry *et al.* (2007) screened 196 chickpea germplasm lines for resistance to wilt disease in a wilt sick plot however, none of the test line was found immune or highly resistant. Arvayo-Ortiz *et al.* (2012) evaluated the new lines of chickpea Hoga-012, Hoga-490-2 and Hoga-508, as well as the commercial cultivars Blanco Sinaloa-92 and Costa-2004 and reported that no variety showed resistance to races 0 and 5 of *Fusarium oxysporum* f. sp. *ciceris,* isolated in chickpea fields of Sonora, Mexico. Thus, breeding programme should be continued in search of resistant genotypes for the control of Fusarium wilt disease.

Evaluation of Fungicides against *Fusarium oxysporum* f. sp. *ciceris*

Several workers have also evaluated various fungicides against *F. oxysporum* f. sp. *ciceris* under *in vitro* condition. Singh and Jha (2003) evaluated seven fungicides, *i.e.* Thiram, Bavistin (carbendazim), Blitox (copper oxychloride), Captaf (captan), Indofil M- 45 (mancozeb+thiophanate-methyl) Ridomil MZ (mancozeb+metalaxyl) and Kitazin (iprobenfos), against chickpea wilt under *in vitro* (each at 0.1 per cent concentration) and *in vivo* (as seed treatment each at 2.5 g kg^{-1} seed, except for Kitazin (1.0 ml kg^{-1}) and as soil drenching each at 0.3 per cent) and found thiram and carbendazim as the most suitable fungicides in inhibiting the mycelial growth of *F. oxysporum* f. sp. *ciceris* under *in vitro* condition.

Compatibility of Potential Antagonists with Fungicides

It is necessary to make study about compatibility of potential antagonists with various fungicides for effective control of wilt disease of chickpea; otherwise the fungicides may inhibit the potential of antagonists in field condition. In this regard, several workers have propounded their reports through conducting *in vitro* studies on compatibility of potential antagonists with various fungicides. Vyas (1987) reported that *Trichoderma* sp. and *B. subtilis* showed high degree of tolerance to thiram and carbendazim. Vidhyasekharan and Muthamilan (1995) reported that thiram and carbendazim were not inhibitory to *P. fluorescens* under *in vitro* conditions. The antagonist *T. virens* was found compatible with carbendazim at 100 ppm concentration however, inhibition occurred at 150 and 1000 ppm under *in vitro* condition (Rajeevpanth, 2001; Girija and Umamaheswaran, 2003). Poddar *et al.* (2004) found that *T. harzianum* in combination with carbendazim (100 ppm) is effective against wilt in chickpea (*C. arietinum* cv. JG 62). The mycelial growth of *T. harzianum* was completely inhibited by carbendazim and hexaconazole while 90 and 41 per cent inhibition occurred with copper oxychloride and mancozeb at 1500 ppm (Tiwari and Singh 2004)). Gupta *et al.* (2005) also reported that carbendazim was incompatible with *T. viride* TV2 isolate while carboxin was compatible for integrated treatment.

Other Eco-friendly Management Approaches

One of the eco-friendly control measures opted to provide resistance against fungal pathogens is using a Proteinase Inhibitor (PI) proteins produced in plant tissues, which, act as a defensive mechanism against microorganisms (Habib and Fazili, 2007). Plant Proteinase Inhibitors rely on inhibition of proteases enzymes secreted by microorganisms, causing a reduction in the availability of amino acids necessary for their growth and development (Zavala *et al.*, 2004). Chickpea seeds are rich in serine proteinase inhibitors (PIs) which show strong inhibitory activity against trypsin, chymotrypsin or other proteolytic enzymes (Srinivasan *et al.*, 2005). Trypsin inhibitor showed anti-metabolic activity against *Helicoverpa armigera* (Nair *et al.*, 2013). Nair and Singh (2013) also analyzed antifungal activity of trypsin inhibitor towards the growth of *F. oxysporum* f. sp. *ciceris*. They found a correlation between trypsin inhibitor and resistance of cultivars against Fusarium wilt as resistant cultivar (JG 2001- 12) showed high trypsin inhibitory activity (99 per cent) in the presence of trypsin enzyme using both natural and synthetic substrates. The crude extracts of JG 2001- 12 showed a decrease in radial growth of *F. oxysporum* f. sp. *ciceris* along with reduction in conidial germination up to 45 to 82 per cent. Thus, trypsin inhibitor may be deployed through transgenic plants for the control of Fusarium wilt of chick pea disease.

Conclusion and Future Prospects

Fusarium wilt disease is a major constraint to chickpea production in most areas of the world. Development of the disease is favored by the long survival of the pathogen in soil and the occurrence of at least eight pathogenic races in its populations. These races differ in virulence on chickpea genotypes as well as in aggressiveness on susceptible cultivars. The virulence/aggressiveness is correlated with the amount of inoculum and environmental conditions required for severe disease. Disease incidençe and severity are driven by pathogen inoculum density and warm temperature in soil and susceptibility of chickpea cultivars. Management of Fusarium wilt of chickpea in sustainable cropping systems should be based on strategies that integrate several control measures; and in that approach sowing date, host resistance and quick diagnosis of the pathogen and its pathogenic races are key measures for a rational management of the disease. Molecular protocols have been developed that would be of much help for that purpose. Use of high-yielding, well-adapted chickpea cultivars resistant to the prevalent pathogen race(s) is the most practical and cost-efficient individual disease control measure for the management of Fusarium wilt. Significant progress has been made in the identification of chickpea germplasm lines, as well as in the development high-yielding cultivars carrying complete resistance to one or more races of the pathogen. Also, significant progress has been made in the unraveling of genetics of race specific resistance. This will allow further progress in pyramiding multiple race-specific resistance in chickpea cultivars that would enhance multi location stability, and potentially to combine that with resistance to other important diseases (*e.g.*, Ascochyta blight, root knot and cyst nematodes) and tolerance to environmental stresses (*e.g.*, drought). Lack of a sufficient level of disease resistance can be compensated to some extent by the use of biocontrol treatments. Combining the use of resistance with other preplanting disease control measures including pathogen-free seed, sanitation to reduce inoculum load in the soil and protection of healthy seeds with fungicides or biocontrol agents would enhance efficiency in the integrated management of Fusarium wilt in chickpea. There is need to give emphasis on preplanting decision-taking process for efficient integrated disease management with skillful assistance to farmers through promotion of extension related activities and involvement of well-trained professional plant pathologists. By following this strategy, chickpeas would maintain their critical role in farming systems as a major source of protein and as a contributing factor in agriculture sustainability through improvement of soil fertility.

REFERENCES

Agarwal S.C., Sharma S. and Prasad K.V.Y. (2002). Efficacy of biological components in wilt management of chickpea. *Indian Journal of Pulse Research*. 15 (2): 177-178.

Ahmad M. A. (2010). Variability in *Fusarium oxysporum* f. sp. *ciceris* for chickpea wilt resistance in Pakistan. *Ph.D. Thesis* submitted to the Quaid-i-Azam University, Islamabad, Pakistan.

Ahmad, M. A., Iqbal, S. M., Ayub, N., Ahmad, Y. and Akram, A. (2010). Identification of resistant sources in chickpea against Fusarium wilt. *Pakistan Journal of Botany*. 42(1): 417-426, 2010.

Anjaiah,V., Cornelis, P., and Koedam, N. (2003). Effect of genotype and root colonization in biological control of Fusarium wilts in pigeonpea and chickpea by *Pseudomonas aureoginosa* PNA1. *Canadian Journal of Microbiology*. 49: 85-91.

Arora, D. K., Pandey A. K. and Srivastva A. K. (1996). Effects of heat stress on loss of C, germination and pathogenicity from chlamydospores of *Fusarium oxysporum* f. sp. *ciceri Soil Biology and Biochemistry*. 28: 399-407.

Arvayo-Ortiz, R. M., Esqueda, M., Acedo-Felix, E., Gonzalez-Rios, H. and Vargas-Rosales, G. (2012). New lines of chickpea against *Fusarium Oxysporum* f. sp. *ciceris* wilt. *American Journal of Applied Sciences*. 9 (5): 686-693.

Bahl, P. N. (1976). Crop stand and its bearing on incidence of wilt in Bengal gram. *Indian Journal of Genetics and Plant Breeding*. 36: 351-352.

Bajwa, K. M., Khan, I.A., Alam, S. S., Ahmad, I. and Gill, M. A. (2000). Screening against phytotoxins for resistance to *Fusarium* wilt. *Pakistan Journal of Phytopathology*. 12: 66-70.

Bakhsh, A., Iqbal, S. M. and Haq, I. K. (2007). Evolution of chickpea germplasm for wilt resistance, *Pakistan Journal of Botany*. 39(2): 583-593.

Barnet, H.L and Hunter, B. (1972). Illustrated genera of imperfect fungi. Burgess publishing company, Minnesota.

Bedi, K.S. and Pracer, C.S. (1952). Gram Wilt disease and its control in Punjab. *Farmer*. 4: 296.

Bendre, N.J and Barhate, B.G. (1998). *A souvenir on Disease Management in chickpea*. M.P.K.V.

Bhatti, M.A and Kraft, J.M. (1992). Effects of inoculum density and temperature on root rot and wilt of chickpea. *Plant Disease*. 76: 50-54.

Boureghda, H and Bouznad, Z. (2009). Biological control of *Fusarium* wilt of chickpea using isolates of *Trichoderma atroviride*, *T. harzianum* and *T. longibrachiatum*. *Acta Phytopathologica et Entomologica Hungarica*. 44 (1): 25-38.

Brayford, D. (1998). *Fusarium oxysporum* f. sp. *ciceris*. *IMI Descriptions of Fungi and Bacteria No.* 1113

Butler, E. J. (1918). Fungi and Disease in Plants: An Introduction to the Diseases of Field and Plantation Crops, especially those of India and the East. Published by Bishen Singh Mahendra Pal Singh, 23-A, Cannaught Place, Dehra Dun-248001(India), 1987. pp. 547.

Chandra, S., Tomer, Y. S. and Malik, B. P. S. (1974). Aspects of wilt disease in gram with special reference to Haryana state. *Indian Journal of Genetics and Plant Breeding*. 34: 257-262.

Chaudhry, M. A., Muhammad, F. and Afzal, M. (2006). Screening of chickpea germplasm against *Fusarium* wilt. *Journal of Agricultural Research.* 44(4).

Chaudhry, M. A, Ilyas, M. B., Muhammad, F. and Ghazanfar, M. U. (2007). Sources of resistance in chickpea germplasm against Fusarium wilt. *Mycopathology*. 5: 17-21.

Chauhan, S. K. (1962 b). Influence of pH in sand cultures on disease intensity and crop loss correlation in Fusarium wilt of gram (*Cicer arietinum* L.). *Journal of Indian Botanical Society*. 41: 220-225.

Chauhan, S. K. (1962c). Fusarium wilt of gram (*Cicer arietinum* Linn.) in relation to organic matter of soil. *Vigyana Parishad Anusandhan Patrika.* 5: 73-76.

Chauhan, S, K. (1963). Influence of different soil temperatures on the incidence of Fusarium wilt of gram (*Cicer arietinum* L.). *Proceedings of Indian Academy Science Bulletin*. 33: 552-554.

Chauhan, S, K. (1965). Interaction of certain soil conditions in relation to the occurrence of Fusarium wilt of gram. *Indian Journal of Agricultural Science.* 35: 52-56.

Chauhan, Y.S., Nene, Y.L., Johansan, C., Haware, M. P., Saxena, N. P., Singh, S., Sharma, S. B., Sahrawat, K. L., Burford, J. R., Pupela, O. P., Kumar, R., J.V.D.K., Sithawanthan, S.(1998). Effect of soil solarization of Pigeonpea and chick pea. *Research Bulletin ICRISAT.* 11: 16.

Chickpea Research Highlights (2009-2010). All India coordinated research project on chickpea.

Couteaudier, Y and Alabouvette, C. (1990). Survival and inoculum potential of conidia and chlamydospores of *Fusarium oxysporum* f. sp. *lini* in soil. *Canadian Journal of Microbiology*. 36: 551-556.

Dahiya, S.S., Sharma, S. and Paroda, A. S. (1988). Effect of date and depth of sowing on the incidence of chick pea wilt in Haryana, India. *International Chickpea Newsletter.* 18: 24-25.

Dastur, J. F. (1935). Gram wilts in the Central Provinces. *Agricultural Live-Sock India.* 5: 615-627.

Davis, R. M, Colyer, P. D., Rothrock, C. S. and Kochman, J. K. (2006) Fusarium wilt of cotton: population diversity and implications for management. *Plant Disease.* 90: 692-703.

Demirci, E., Eken, C and Kantar, F. (1999). Pathogenicity of wilt and root rot pathogens of chickpea cv. Aziziye- 94. *Journal of Turkish Phytopathology*. 28(1): 25-32.

Dubey, S.C., Singh, S.R and Singh, B. (2010). Morphological and pathogenic variability of Indian isolates of *Fusarium oxysporum* f. sp. *ciceris* causing chickpea wilt. *Archives of Phytopathology*. 43: 174- 190.

Dubey, S.C., Suresh, M and Singh, B. (2007). Evaluation of *Trichoderma* species against *Fusarium oxysporum* f. sp. *Ciceris* for integrated management of chickpea wilt. *Biological Control*. 40 (1): 118-127.

Duke, J.A. (1981). Handbook of legumes of world economic importance. Plenum Press, New York. p. 52-57.

Erwin, D. C. (1958 a). *Fusarium lateritium* f. sp. *ciceri*, incitant of Fusarium wilt of *Cicer arietinum*. *Phytopathology*. 48: 498-501.

Erwin, D.C. (1958 b). Verticillium wilt of *Cicer arietinum* in southern California. *Plant Disease Reporter*. 42: 1111.

Essalmani, H. and Lahlou, H. (2003) Bioprotection mechanisms of the lentil plant by *Rhizobium leguminosarum* against *Fusarium oxysporum* f. sp. *lentis*. *Comptes Rendus Biologies*, 326: 1163- 1173.

FAOSTAT (2014). FAOSTAT Production Statistics of Crops. Available: http: // faostat.fao.org/site/567/default.aspx#ancor (accessed 10.08.14.).

Fisher, N.L., Burgess, L.W., Toussoun, T.A and Nelson, P.E. (1982). Carnation leaves as a substrate and or preserving cultures of *Fusarium* species. *Phytopathology*. 72: 151- 153.

Flor, H. H. (1971). Current status of the gene-for-gene concept. *Annual Review of Phytopahtology*. 9: 275-296.

Geervani, P. (1991). Utilization of chickpea in India and scope for novel and alternative uses. In: Uses of Tropical Grain Legumes: Proceedings of Consultants' Meeting, 27-30 March, 1989. ICRISAT Center, Patancheru, Andhra Pradesh, India, p. 47-54.

Girija,V. K and Umamaheswaran. (2003). Basal rot of balsam and its management through bioagents. *Plant Disease Research*. 18 (1): 52- 55.

Goswami, B. K., Pandey, R. K., Goswami, J. and Tewari, D. D. (2007). Management of disease complex caused by root knot nematode and root wilt fungus on Pigeonpea through soil organically enriched with Vesicular Arbuscular Mycorrhiza, karanj (*Pongamia pinnata*) oilseed cake and farmyard manure. *Journal of Environmental Science and Health*. 42: 899-904.

Grewal, J. S (1969). Important fungal disease of *Cicer arietinum* in India. Pulse Improvement Project Seminar Report held at Karaj Agricultural College, University of Tehran and USDA, January 7-9, 1969. pp. 35-40

Gupta, O. (1991). Symptomless carriers of chickpea vascular wilt pathogen (*Fusarium oxysporum* f. sp. *ciceris*). *Legume Research*. 14: 193-194.

Gupta O. (1999P. Effect of micronutrient on root pathogens of chickpea (*Cicer arietinum* L.). *Bhartiya Krishi Anusandhan Patrika*. 14: 25-27.

Gupta, S., Kalha, C.S., Vaid, A and Rizvi, S. E. H. (2005). Integrated management of anthracnose of french bean caused by *Colletotrichum lindemuthianum*. *Journal of Mycology and Plant Pathology*. 35: 432-435.

Habib, H. and Fazili, K. M. (2007). Plant protease inhibitors: A defense strategy in plants. Biotechnology and Mo gen-induced proteins with inhibitory activity toward *Phytophthora infestans*. *Plant Cell*. 3: 619-628.

Halila, M.H and Strange, R.N. (1996). Identification of the causal agent of wilt of chickpea in Tunisia as *F. oxysporum* f. sp. *ciceris* race 0. *Phytopathologia Mediterranea*. 35: 67–74.

Hanif, M., Jamil, F. F and Haq, I. (1999). Effect of various sowing depths on wilt incidence of chickpea in a wilt-sick field in Pakistan. *International Chickpea and Pigeonpea News Letter*. 6: 11.

Harichand and Singh S. (2005). Control of chickpea wilt (*Fusarium oxysporum* f. sp. *ciceris*) using bioagents and plant extracts. *Indian Journal of Agricultural Sciences*. 75 (2): 115-116.

Haware, M. P., Jimenez-Diaz, R. M., Amin, K. S., Phillips, J. C and Halila, H. (1990). *Integrated management of wilt and root rots of chickpea* 129-137. Chickpea in the Nineties: *Proceedings of the second international work shop on chickpea improvement*, Patancheru, India.

Haware, M. P, Nene, Y. L, Pundir, R. P. S and Narayana Rao, J. (1992). Screening of world chickpea germplasm for resistance to fusarium wilt. *Field Crops Research*. 30: 147–154.

Haware, M. P., Nene, Y. L and Rajeswari, R. (1978). Eradication of *Fusarium oxysporum* f. sp. *ciceris* transmitted in chickpea seed. *Phytopathology*. 68: 1364-1368.

Haware, M. P and Nene, Y. L. (1980). Influence of wilt at different stages on the yield loss in chickpea. *Tropical Grain Legume Bulletin*. 19: 38-44.

Haware, M.P and Nene, Y.L. (1982). Races of *Fusarium oxysporum* f. sp. *ciceris*. *Plant Disease*. 66: 809-10.

Haware, M.P. (1993). *Fusarium* diseases of crops of India. *Indian Phytopathology*. 46: 101-109.

Haware, M. P. (1998) Diseases of chickpea. In: Allen DJ and Lenné JM (eds.) The Pathology of Food and Pasture Legumes (pp. 473-516) CAB International and ICRISAT, Wallingford, UK.

Haware, M. P., Nene, Y. L. and Natarajan, M. (1996). Survival of *Fusarium oxysporum* f. sp. *ciceri* in the soil in the absence of chickpea. *Phytopathology Mediterranean*. 35: 9-12.

Hend, A., Alwathnani and Kahkashan, P. (2012). Biological control of *Fusarium* wilt of tomato by antagonist fungi and cyanobacteria. *African Journal of Biotechnology*. 11 (5): 1100-1105.

Honnareddy, N and Dubey, S. C. (2007). Morphological characterization of Indian isolates of *Fusarium oxysporum* f. sp. *ciceris* causing chickpea wilt. *Indian Phytopathology*. 60 (3): 373-376.

Hervás, A, Tapero-Casas, J. L and Jiménez-Díaz R. M. (1995) Induced resistance against Fusarium wilt of chickpea by non-pathogenic races of *Fusarium oxysporum* f. sp. *ciceris* and non-pathogenic isolates of *F. oxysporum*. *Plant Disease.* 79: 1110-1116.

Hervás, A., Landa, B., Datnoff, L. E., and Jiménez-Díaz, R. M. (1998). Effects of commercial and indigenous microorganisms on Fusarium wilt development in chickpea. *Biological Control*. 13: 166-176.

Hervás, A., Landa, B., and Jiménez-Díaz, R. M. (1997). Influence of chickpea genotype and *Bacillus* sp. on protection from Fusarium wilt by seed treatment with nonpathogenic *Fusarium oxysporum*. *European Journal of Plant Pathology*. 103: 631-642.

Iftikhar, A. K., Jabbar, A. and Alam, S. S. (2002). Genetic variation among *Fusarium oxysporum* f.sp. *ciceri* isolates in Pakistan as determined by biological pathotyping and vegetative compatibility. *Pakistan Journal of Botany*. 34(4): 433-440.

Iftikhar, A. K., Kahlid, T. and Ilyas, M. B. (1997). Field screening of chickpea germplasm for the sources of resistance against *Fusarium* wilt. *Pakistan Journal of Phytopathology*. 9(1): 31-33.

Iqbal, S. M., Haq, I. U., Bukhari, A., Ghafoor, A. and Haqqani, A. M. (2005). Screening of chickpea genotypes for resistance against *Fusarium wilt*. *Mycopathology*. 3(1-2): 1-5.

Jalali, B. L and Chand, H. (1992). Chickpea wilt. Singh, U.S., Mukhopadhayay, A.N., Kumar, J and Chaube, H.S. eds. Plant Diseases of International Importance, Vol. I. *Diseases of Cereals and Pulses*. Englewood Cliffs, NY, USA: Prentice Hall. 429-444.

Jayasekhar, M., Manonmani, K and Justin, C.G.L. (2008). Development of integrated biocontrol strategy for the management of stem rot disease (*Fusarium oxysporum* f. sp. *vanillae*) of vanilla. *Agricultural Science Digest*. 28 (2): 109-111.

Jimenez - Diaz, R. M., Trapero-Casas, A. and Cabrera Dela C. J. (1989). Races of *Fusarium oxysporum* f. sp. *ciceris* infecting chickpea in southern spain. *NATO ASI Series*. 28: 515-520.

Jimenez-Diaz, R. M., Hervas, A. A and Trapero-Casas, J. C. (1993). Pathogenic variability and host resistance in *F. oxysporum* f. sp. *ciceris, C. arietinum* pathosystem. Hodowla Rosin. *Aklimatyzacja Nasiennictwo*. 37: 87-94.

Jorge, I., de la Rosa O, Navas-Cortes J. A., Jimenez-Diaz R. M, and Tena, M. (2005). Extracellular xylanases from two pathogenic races of *Fusarium oxysporum* f. sp. *ciceris*: enzyme production in culture and purification and characterization of a major isoform as an alkaline endo-b-(1,4) xylanase. *Antonie van Leeuwenhoek*. 88: 49-59

Juan A, Navas-Cortes J. A, Bernard, H. and Jimènez-Diaz, M. (2000). Yield loss in chickpeas in relation to development of Fusarium wilt epidemics. *Phytopathology*. 90: 1269–1278.

Jukanti, A. K., Gaur P. M., Gowda, C. L. L. and Chibbar R. N. (2012). Nutritional quality and health benefits of chickpea (*Cicer arietinum* L.): a review. *British Journal of Nutrition*. 108: 11-26.

Kannaiyan, J. and Nene, Y. L. (1975). Note on the effect of sowing date on the reaction of 12 lentil varieties to wilt disease. *Madras Agricultural Journal*. 2: 77-79.

Kapoor, A. S., Paul, Y. S. and Singh, A. (2006). Integrated management of white rot and root rot-wilt disease complex of pea. *Indian Phytopathology*. 59 (4): 467-474.

Kaur, R. (2003). Characterization of selected isolates of non pathogenic *Fusarium oxysporum*, fluorescent pseudomonads and their efficacy against chickpea wilt. *Ph.D. Thesis*, Punjab Agricultural University, Ludhiana. pp.150.

Kausar, A.G. (1968). Performance of blight resistant gram selections in wilt nursery and at six locations. *Pakistan Journal of Agricultural Science*. 5: 264-273.

Khan, M. A and Gangopadhyay, S. (2008). Efficacy of *Pseudomonas fluorescens* in controlling root rot of chickpea caused by *Macrophomina phaseolina*. *Indian Journal of Mycology and Plant Pathology*. 38(3): 580-587.

Khan, M. R., Khan, S. M. and Mohiddin F. M. (2004). Biological control of Fusarial wilts of chickpea through seed treatment with the commercial formulations of *Trichoderma harzianum* and/or *Pseudomonas fluorescens*. *Phytopathologia Mediterranea*. 43: 20-25.

Klement, Z. and Goodman, R. N. (1967). Hypersensitive reaction to infection by bacterial plant pathogens. *Annual Review of Phytopathology*. 5: 17-44.

Komada, H. (1975). Development of a selective medium for quantitative isolation of *Fusarium oxysporum* from natural soil. *Review of Plant Protection Research*. 8: 114-125.

Landa, B. B., Hervás, A., Bettiol, W., and Jiménez-Díaz, R. M. (1997). Antagonistic activity of bacteria from chickpea rhizosphere against *Fusarium oxysporum* f. sp. *ciceris*. *Phytoparasitica*. 25: 305-318.

Landa, B. B., Navas-Cortés, J. A., Hervás, A., and Jiménez-Díaz, R. M. (2001). Influence of temperature and inoculum density of *Fusarium oxysporum* f. sp. *ciceris* on suppression of Fusarium wilt of chickpea by rhizosphere bacteria. *Phytopathology*. 91: 807-816.

Landa, B.B., Navas-Cortes, J.A and Jimenez-Diaz, R. M. (2005). Integrated management of *Fusarium* wilt of chickpea with sowing date, host resistance and biological control. *American Phytopathology*. 94(9): 946-960.

Lange, R. and McLaren, D. (2002) Fusarium wilt – a new disease of canola. University of Manitoba website. *www.umanitoba.ca/afs/agronomists_conf/2002/pdf/lange.pdf*

Leslie, J. F and Summerell, B. A. (2006). *The Fusarium Laboratory manual*. (Blackwell Publishing: Iowa, USA).

Mazzola, M. (2002) Mechanisms of natural soil suppressiveness to soilborne diseases. *Antonie Van Leeuwenhoek*. 81: 557-564.

Mane, S. S and Pal, M. (2008). Screening of antagonists and effects of their cultural filtrate on growth and biomass production of *Fusarium oxysporum* f. sp. *ciceris*. *Journal of Plant Disease Sciences*. 3 (1): 74-76.

McKerral, A. (1923). A note on Fusarium wilt of gram in Burma and measures taken to combat it. *Agriculture Journal of India*. 28: 608-613.

McRae, W. (1932). Report of the Imperial Mycologist, Sc. Agric. Res. Institute, Pusa, 1930-31, pp. 78-86.

Mundkur, B. B. (1946). Report of the Imperial Mycologist, Sc. Rep. Agric. Res. Institute, New Delhi, pp. 57-63.

Nair, M. and Singh, S. S. (2013). A Kunitz trypsin inhibitor from chickpea (*Cicer arietinum* L.) that exerts an antimicrobial effect on *Fusarium oxysporum* f.sp. *ciceris*. *Agricultural Sciences*. 4 (11): 585-594 (2013).

Nair, M., Singh, S. S. and Babbar, A. (2013). Purification of trypsin inhibitor (TI) from seeds of Cicer arietinum (L.) and its insecticidal potentiality against *Helicoverpa armigera* (Hübner). *Theoretical and Experimental Plant Physiology*. 25 (2): 141-152.

Narasimhan, R. (1929). A preliminary note on a Fusarium parasitic on Bengal gram (*Cicer arietinum*). *Madras Agriculture Department Year Book*. 1928: 5-11.

Naseema Beevi, S., Ancy Salim, M., Priya Mohan Deepti, S. and Naseema, A. (2005). Compatability of *Trichoderma harzianum Rifai* with botanicals and synthetic pesticides. *Geobios*. 32 (3): 80-82.

Nautiyal, C. S. (1997). Selection of Chickpea rhizosphere competent *Pseudomonas fluorescens* NBRI1303 antagonistic to *Fusarium oxysporum* f. sp. *ciceris, Rhizoctonia bataticola* and *Pythium* sp. *Current Microbiology*. 35: 52-58.

Navas-Cortes, J. A, Hau, B. and Jiménez-Diaz R. M. (1998). Effect of sowing date, host cultivar and race of *Fusarium oxysporum* f. sp. *ciceris* on development of Fusarium wilt of chickpea. *Phytopathology*. 87: 1338-1346.

Navas-Cortes, J. A., Hau, B. and Jimenez- Diaz, R. M. (2000). Yield loss in chickpea in relation to development to *Fusarium* wilt epidemics. *Phytopathology*. 90: 1269-1278.

Nelson, P. E, Toussoun, T. A. and Marasas, W.F.O. (1983) Fusarium Species: An Illustrated Manual for Identification. University Park, USA: Pennsylvania State University Press.

Nene, Y. L. and Haware, M. P. (1980). Screening chickpea for resistance to wilt. *Plant Disease*. 64: 379-380.

Nene, Y. L., Kannaiyan, J., Haware, M. P. and Reddy, M. V. (1980). Review of work done at ICRISAT on soil borne disease of Pigeonpea and chickpea. In: Proceedings of the consultants' group discussion on the resistance to soil borne diseases of legumes, 8-11 January 1979, ICRISAT Patancheru, India. Pp 3-39.

Nene, Y. L., Haware, M. P. and Reddy, M. V. (1981). *Chickpea diseases: Resistance screening techniques*. Information Bulletin No. 10, International Crop Research Institute for the Semi Arid Tropics, Patancheru, 1981, pp. 1-10.

Nene, Y. L. and Reddy, M. V. (1987). *Chickpea diseases and their control*. In: The Chickpea (Eds. Saxena and Singh). CAB International, Wallingford. 233-270.

Nene, Y. L., Reddy, M.V., Haware, M.P., Ghanekar, A.M and Amin, K. S. (1991). Field diagnosis of chickpea diseases and their control. *Information Bulletin* no. 28. ed. By International Crops Research Institute for the Semi Arid Tropics, Patancheru, India.

Nene, Y. L., Shelia, V. K and Sharma, S. B. (1996). *A world list of chickpea and Pigeonpea pathogens* 5th Edn. Patancheru, Andhra Pradesh, India. ICRISAT, p. 27.

Niks, R. E. (1988). Nonhost plant species as donors for resistance to pathogens with narrow host range. II. Concepts and evidence on the genetic basis of nonhost resistance. *Euphytica*. 37: 89-99.

Nikam, P. S., Jagtap, G. P and Sontakke, P. L. (2007). Management of chickpea wilt caused by *Fusarium oxysporium* f. sp. ciceris. *African Journal of Agricultural Research*. 2 (12): 692-697.

Nikam, P. S., Jagtap, G. P and Sontakke, P. L. (2011). Survey, surveillance and cultural characteristics of chickpea wilt caused by *Fusarium oxysporium* f. sp. *Ciceris*. *African Journal of Agricultural Research*. 6 (7): 1913-1917.

Padwick, G.W. (1940). The genus Fusarium III. A critical study of the fungus causing wilt of gram (*Cicer anetinum* L.) and of the related species of the subsection Orthocera, with special relation to the variability of key characteristics. *Indian Journal of Agricultural Sciences*. 10: 241-284.

Padwick, G.W. (1941). Report of the Imperial Mycologist. Sci. Rep. Agric. Res. Institute, New Delhi, pp. 94-101.

Padwick, G. W. and Bhagwagar, P. R. (1943). Wilt of gram in relation 10 date of sowing. *Indian Journal of Agricultural Sciences*. 13: 289-290.

Parmer, N. and Dadarwal, K. R. (1997). Rhizobacteria from the rhizosphere and rhizoplane of chickpea. *Indian Journal of Microbiology*. 37 (4): 205-210.

Patel, S. T and Anahosur, K. H. (2001). Variability among the isolates of *Fusarium oxysporum* f. sp. ciceris and *Fusarium solani* from chickpea. *Madras Agricultural Journal*. 88 (3): 124-126.

Patil S. K. (2006). Studies on *Fusarium oxysporum* Schlecht Fr f. sp. *gladioli* (Massey) Snyd. and Hans. causing wilt of gladiolus. *Ph.D Thesis* submitted to the University of Agricultural Sciences, Dharwad.

Paulkar, P. K and Raut, B. T. (2004). Variability among the isolates of *Fusarium oxysporum* f. sp. *ciceris*. *Journal of Mycology and Plant Pathology*. 34 (1): 20-23.

Perez Arte ìs E. and Tena, M. (1989). Pectic enzymes from two races of *Fusarium oxysporum* f. sp. *ciceris*. Enzyme production in culture and enzymatic activity on isolated chickpea cell walls. *Journal of Phytopathology*. 124: 39–51

Pittaway, J. K., Robertson, I. K., Ball, M. J. (2008). Chickpeas May Influence Fatty Acid and Fiber Intake in an Ad Libitum Diet, Leading to Small Improvements in Serum Lipid Profile and Glycemic Control. *Journal Academy of Nutrition and Dietetics*. 108 (6): 1009-1013.

Poddar, R. K., Singh, D. V and Dubey, S. C. (2004). Integrated application of *Trichoderma harzianum* mutants and carbendazim to manage chickpea wilt (*Fusarium oxysporum* f. sp. *ciceris*). *Indian Journal of Agricultural Sciences*. 74 (6): 346-348.

Prasad, N. and Padwick, G. W. (1939). The genus *Fusarium* 11. A species of *Fusarium* as a cause of wilt of gram (*C. arietinum* L.). *Indian Agricultural Sciences*. 9: 731-735.

Prasad, R. D., Rangeshwaran, R., Anuroop, C. P and Rashmi, H. J. (2002). Biological control of wilt and root rot of chickpea under field conditions. *Annals of Plant Protection Sciences*. 10 (1): 72-75.

Prell, H. H. and Day, P. R. (2001). Plant Fungal Interaction. Springer-Verlag, Berlin.

Rangaswamy, G and Mahadevan, A. (1999). *Diseases of crop plants in India* (4th edition) Prentice Hall of India Pvt. Ltd., New Delhi, pp. 607.

Rangeshwaran, R and Prasad, R. D. (2000). Isolation and evaluation of rhizospheric bacteria for biological control of chickpea wilt pathogens. *Journal of Biological Control*. 14 (1): 9-15.

Rangeshwaran, R., Wasnikar, A. R., Prasad, R. D., Anjula, N. and Sunanda, C. R. (2002). Isolation of endophytic bacteria for biological control of wilt pathogens. *Journal of Biological Control*. 16 (2): 125- 134.

Rao, V.K. and Krishnappa, K. (1996). Relationship between environmental factors and *Meloidogyne-Fusarium* wilt disease complex in chickpea. *Indian Journal of Mycology and Plant Pathology*. 26: 221-223.

Sattar, A., Arif, A. G. and Mohy-ud-din, M. (1953). Effect of soil temperature and moisture on the incidence of gram wilt. *Pakistan Journal of Scientific Research*. 5: 16-21.

Saxena, M. C. and Singh, K. B. (1987). *The chickpea* published by C.A.B. Int. ICARDA. pp. 250-252.

Shahida, K., Surendra, K., Gopal, Sally, K and Mathew (2010). Efficacy of native bio agents against *Phytophthora meadii* causing *Phytophthora* rot in vanilla and its compatibility with fungicides. *SAARC Journal of Agriculture*. 8 (1): 103-111.

Shaikh, M. H. (1974). Studies on wilt of gram (*Cicer arietinum* L.) caused by *Fusarium oxysporum* f. *ciceri* in Marathwada region. M.Sc. (Ag.) thesis, Marathwada Krishi Vidyapeeth, Parbhani, India.

Saraf, C. S., Ahlawat, I. P. S. and Singh, A. (1973). Agronomic Investigation on rabi pulses during 1972-73 at IARI. Report presented at the All India Rabi Pulses Workshop, Calcutta.

Sekhon, G. S. (1950). Effect of time of irrigation on the yield and incidence of wilt attack on the gram crop. *Punjab Farmer*. 3: 191-195.

Singh, K.B. (1987). Chickpea breeding. p.127-162. In: M.C. Saxena and K.B. Singh (eds.), The Chickpea. CAB International, UK.

Singh, D. K and Jha, M. M. (2003). Effect of fungicidal treatment against chickpea wilt caused by *F. oxysporum* f. sp. *ciceris*. *Journal of Applied Biology*. 13 (2): 41-45.

Singh, K. B. and Reddy, M. V., 1991. Advances in disease resistance breeding in chickpea. *Advance Agronomy*. 45: 191-222.

Singh, R. K., Hasan, A. and Chaudhary, R. G. (2010). Variability in *Fusarium oxysporum* f. sp. *ciceris* causing vascular wilt in chickpea. *Archives of Phytopathology and Plant Protection*. 43 (10): 987-995.

Singh, G., Chen, W., Rubiales, D., Moore, K., Sharma, Y. R. and Gan, Y. (2007). Diseases and their management. In *Chickpea Breeding and Management* (Eds Yadav, Redden, Chen and Sharma). CAB International, pp. 497–519.

Singh, K. B. and Sandhu, T. S. (1973). *Cultivation of Gram in Punjab*. Punjab Agriculture University, Ludhiana, India.

Sivaramakrishnan, S., S. Kannan and S.D. Singh. (2002). Genetic variability of *Fusarium* wilt pathogen isolates of chickpea (*Cicer arietinum* L.) assessed by molecular markers. *Mycopathologia*. 155(3): 171-178.

Smithson, J. B., Thompson, J. A. and Summerfield, R. J. (1985). Chickpea (*Cicer arietinum* L.). p. 312-390. In: R.J. Summerfield and E.H. Roberts (eds.), Grain Legume Crops. Collins, London, UK.

Srinivasan, A., Chougule, N. P., Giri, A. P., Gatehouse, J. A. and Gupta, V. S. (2005). Pod borer (*Helicoverpa armigera* Hübn.) does not show specific adaptations in gut proteinases to dietary *Cicer arietinum* Kunitz proteinase inhibitor. *Journal of Insect Physiology*. 51: 1268-1276.

Srivastava, R., Khalid, A. Singh, U. S. and Sharma, A. K. (2010). Evaluation of arbuscular mycorrhizal fungus, fluorescent Pseudomonads and *Trichoderma harzianum* formulation against *Fusarium oxysporum* f. sp. *lycopersici* for the management of tomato wilt. *Biological control*. 53(1): 24-31.

Steinberg, C., Edel-Hermann, V., Guillemaut, C., Perez-Piqueres, A., Singh, P. and Alabouvette, C. (2004). Impact of organic amendments on soil suppressiveness to diseases. *Bulletin OILB/SROP*. 27(1): 259- 266.

Sugha, S. K., Kapoor, S. K and Singh, B. M. (1994). Factors influencing *Fusarium* wilt of chickpea (*Cicer arietinum* L.). *Indian Journal of Mycology and Plant Pathology*. 24: 97-102.

Snyder, W. C. and Hansen, H. N. (1940). The species concept in *Fusarium*. *Journal of Botany*. 27: 64-67.

Tiwari, R. K.S and Singh, A. (2004). Efficacy of fungicides on *Rhizoctonia solani* and *Sclerotium rolfsii* and their effect on *Trichoderma harzianum* and *Rhizobium leguminosarum*. *Journal of Mycology and Plant Pathology*. 30 (2): 134-139.

Trapero-Casas, A. and Jiménez-Díaz, R. M. (1985). Fungal wilt and root rot diseases of chickpea in southern Spain. *Phytopathology*. 75: 1146- 1151.

Vidyasekharan, P and Muthamilan. (1995). Development of formulations of *Pseudomonas fluorescens* for control of chickpea wilt. *Plant Disease*. 79: 782-786.

Vishwadhar G. S. N. (1998). *Integrated Management of chickpea diseases*. Chamola and Dubey, O.P. (eds.) ABH Publishing Co., New Delhi (India). p. 249.

Vyas, S. C. (1987). Effect of seed treatment fungicides thiram and carbendazim on the antagonists of soybean and dry root rot pathogen (*Rhizoctonia bataticola*). *Abstracts of papers presented in workshop on biological control of plant diseases,* TNAU, Coimbatore, India pp. 23.

Westerlund, F. V., Campbell, R. N and Kimble, K. A. (1974). Fungal root rot and wilt of chickpea in California. *Phytopathology*. 64: 432–436.

Zavala, J. A., Patankar, A. G., Gase, K., Hui, D. and Baldwin, I.T. (2004). Manipulation of endogenous trypsin proteinase inhibitor production in Nicotiana attenuata demonstrates their function as antiherbivore defenses. *Plant Physiology*. 134: 1181-1190.

Zote, K. K., Haware, M. P., Jayanthi, S and Narayana Rao, J. (1996). Effects of inoculum density of *Fusarium oxysporum* f. sp. *ciceris* Race 1 and Race 2 on chickpea wilt. *Phytopathology Mediterranea*. 35: 43-47.

Chapter 2

Fungal Foliar Diseases of Chickpea and their Management

Jai Singh Patel, Sudheer Kumar Yadav, Gagan Kumar, Ankita Sarkar, H.B. Singh, Surendra Singh, R.N. Kharwar & B.K. Sarma

The production of chickpea (*Cicer arietinum* L.) is severely hampered due to infestations by various pathogens (Figure 2.1). The disease inciting agents may be fungal, bacterial, viral and nematode origin. Chickpea is tolerant to heat and drought and therefore it is suitable for cultivation even in low-fertility soils. India is the largest producer of chickpea in the world accounting for about 75 per cent of the world chickpea production. Chickpea belongs to the genus *Cicer*, tribe cicereae, family fabaceae, and sub family papilionaceae. However, its place of origin is south eastern Turkey (Ladizinsky, 1975). The best suitable time for its sowing is from mid October to mid November in India.

As far as the diseases of chickpea are concerned, approximately 172 pathogens have been reported to incite diseases in chickpea and some of them are most devastating. For example, ascochyta blight and botrytis can cause upto 100 per cent grain and yield loss under favourable conditions (Nene *et al.*, 1996). Various strategies like fungicides, crop rotation, use of disease free healthy seeds and biological control are in practice for the management and minimization of economic losses due to the diseases occurred in this crop. Out of several practices, biological control is an environmentally safe, economically and ecologically sounds strategy of plant disease management. A number of cases have been reported about the augmentation of natural resistance of plants by the application of beneficial biocontrol agents

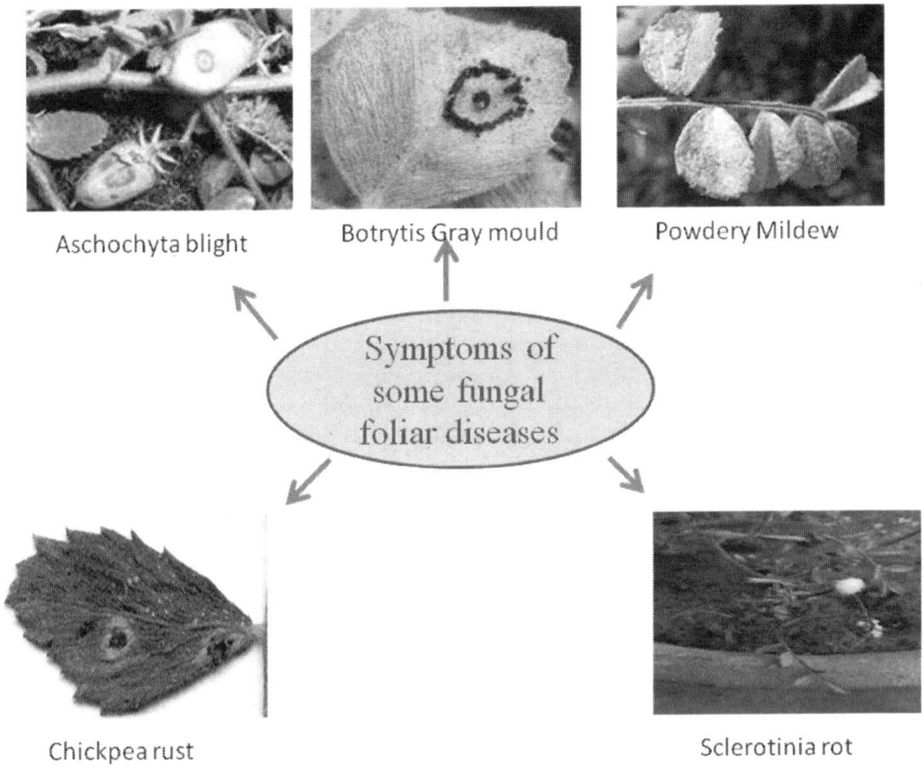

Figure 2.1. Foliar Diseases of Chickpea that Causes Severe Losses.

(BCAs) (bacteria, fungi, mycorrhiza, actinomycetes, non pathogenic fungi, *etc.*), either singly or in combination (Figure 2.2, Table 2.1). A number of studies showed that biochemical changes that occur in plants such as accumulation of phenolic compounds, ROS related enzymes, enzymes of phenylpropanoid pathways, *etc.* after application of BCAs found to confer tolerance against diseases and the role of BCAs were vital in the process. Srivastava *et al.* (2001) showed induction of systemic resistance (ISR) against the charcoal rot pathogen *Macrophomina phaseolina* (Tassi) Goidanich in chickpea when treated with the *Pseudomonas fluorescens* isolate 4-92. PR (pathogenesis related) proteins such as chitinases and glucanases were also found to be accumulated at higher levels in plants pre-treated with *P. fluorescens* isolate 4-92 compared to control plants and 33 per cent increase in disease resistance was observed after application of *P. fluorescens* isolate 4-92. BCAs like *Trichoderma viride* and *P. fluorescens* were analysed for their capability to reduce the advancement of *M. phaseolina* under *in vitro* and field conditions. However, two microorganisms *T. viride* (Tvr) and *P. fluorescens* (Pf4) were found to be most effective. Capabilities of these two organisms for management of disease were compared with some commercial seed treatment chemicals like carbendazim and hexaconazole. The mixture of BCA-augmented farm-yard manure along with seed bio-priming was

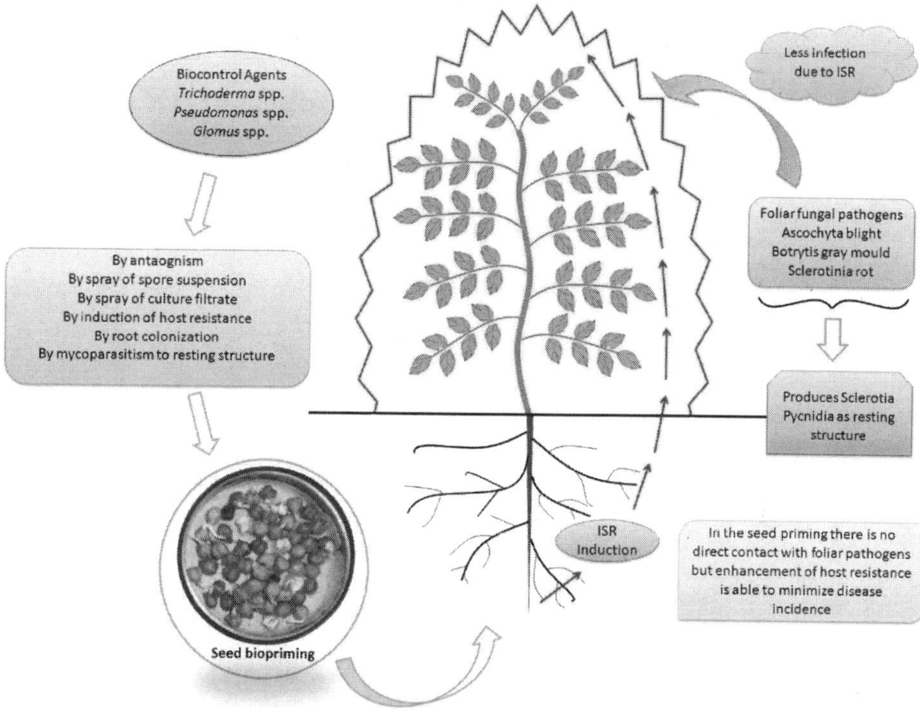

Figure 2.2. Interaction of Bio-agents and Activation of Defence Response against different Fungal Foliar Pathogens.

found to be best in terms of germination, root rot index and yield in contrast to chemical seed treatments and BCAs used individually (Manjunath *et al.*, 2013). Prasada *et al.* (2002) showed that seven days prior to soil application of *Trichoderma harzianum* (PDBCTH 10) and *T. viride* (PDBCTV) was found to be more efficacious as compared to seed treatment in minimizing wilt and wet root rot of chickpea. *T. harzianum* and pathogen treated plots registered 1.2 per cent root rot after 60 days of sowing and 5.1 per cent wilt incidence after 90 days of sowing, whereas only pathogen treated plots registered 17 per cent root rot incidence after 60 days of sowing and 16 per cent wilt incidence at 90 days.

Botrytis gray mold (BGM) is also another devastating disease in chickpea. BCAs such as *T. harzianum* Rifai (Haware 1998) and *T. viride* (Mukherjee and Haware, 1993) were reported to control this disease in the field as well as in green house conditions. Complete parasitisation of *B. cinerea* was observed *in vitro* by *T. viride* and *T. harzianum*. Agarwal and Tripathi (1999) have reported increment in per cent seed germination, root and shoot length, plant vigour along with unaffected nodulation and reduced BGM occurrence, when seeds were treated with *T. viride*. Mukharjee *et al.* (1995, 1997) have also reported that combined use of fungicide vinclozolin and *T. viride* have shown better control of BGM as compared to vinclozolin alone.

Table 2.1: Different Fungal Foliar Diseases and their Reported Bio-control Agents

Sl. No.	Name of the Disease	Causal Organism	Biocontrol Organisms	References
1.	Botrytis Gray Mold	*Botrytis cinerea*	*Bacillus cereus* CRS7, *Pseudomonas fluorescens* CRS 31	Kishore and Pande 2006
2.	Botrytis Gray Mold	*B. cinerea*	*Trichoderma* spp and *Gliocladium* spp.	Davidson *et al.*, 2004
3.	Sclerotinia rot	*Sclerotinia sclerotiorum*	*Streptomyces rochei*	Srivastava *et al.*, 2014
4.	Sclerotinia rot	*S. sclerotiorum*	*Pseudomonas sp. Trichoderma* spp. and *Gliocladium virens, Streptomyces sp.*	Lee *et al.*, 1984, Sharma *et al.*, 1994. Tahtamouni *et al.*, 2006, Sarkar *et al.*, 2014
5.	Ascochyta blight	*Ascochyta rabiei*	*Chaetomium globosum, Trichoderma viride*	Dugan *et al.*, 2009
6.	Alternaria blight	*Alternaria alternata, Alternaria tenuissima*	-	-
7.	Downy mildew	*Peronospora sp.*	-	-
8.	Phoma blight	*Phoma medicaginis*	-	-
9.	Pleospora leaf spot	*Pleospora herbarum*	-	-
10	Powdery mildew	*Erysiphe sp., Leveillula taurica,*	-	Not economically important
11	Rust	*Uromyces ciceris -arietini*	-	-
11.	Scopulariopsis leaf spot	*Scopulariopsis brevicaulis*	-	-
13.	Stemphylium blight	*Stemphylium sarcinifome*	-	-

Burgess *et al.* (1997) showed that protection ability of *T. virens* and *Gliocladium roseum* was similar to the protection provided by the fungicide Thiram. The build up of reactive oxygen (ROS) species and activation of defence related pathways in plants are the immediate indirect effect of bio-agents via induction of plant defence under various stresses. One pathway involved in such response is ISR with the key enzyme phenylalanine ammonium lyase (PAL) which catalyzes the first step of this pathway (Jones, 1984; Bate *et al.*, 1994). PAL is the main enzyme which connects the synthesis of primary metabolites with synthesis of secondary metabolites and involved in synthesis of phytoalexin, phenolic compounds that works against the pathogens (Binutu and Cordell, 2000).

Another disease of chickpea is stem rot caused by *Sclerotinia sclerotiorum.* This pathogen is one of the most devastating diseases in chickpea. However, its host range is very broad. A study by Sarkar *et al.* (2013) showed the use of fluorescent pseudomonads (Cgr and S1) for the control of this disease by accumulation of phenolics via activation of PAL and peroxidase (PO). In another recent study by Srivastava *et al.* (2014) showed the use of actinomycete *Streptomyces rochei* against the fungus *S. sclerotiorum*. Further, they showed the defence responses were activated and regulated through the ethylene mediated signal transduction. The present chapter discuses some important foliar diseases of chickpea and their management practices with special emphasis on the use of beneficial BCAs.

Ascochyta Blight (*Ascochyta rabiei*)

Ascochyta rabiei Labrousse (teleomorph: Didymella rabiei (Kovachevski) Arx.) is the causative agent of Ascochyta blight. This pathogen is one of the important foliar pathogens of chickpea. The major incidence of this disease occurs in the area where growing season of chickpea coincides with cool and moist weather. Ascochyta blight of chickpea is one of the calamitous and ubiquitous diseases in the chickpea growing areas responsible for heavy crop loss. There was a decline in the area of chickpea production in western Canada from over 500000 ha in 2001 to less than 130000 ha in 2006 (Statistics Canada, 2001, 2007) due to continuous occurrence of this disease. The yield loss can be up to 100 per cent when weather conditions are favourable for development of this disease. This disease is a seed and stubble-borne and produces both airborne ascospores and water-splashed conidia during the cropping season (Armstrong *et al.*, 2001; Gossen and Miller, 2004).

Ascochyta blight is the major disease, which can limit production of chickpea worldwide. Its occurrence has been reported in major chickpea growing areas of the world (Nene and Reddy, 1987; ICARDA, 1996; Akem, 1999; Khan *et al.*, 1999; Kaiser *et al.*, 2000; Chongo *et al.*, 2003). Some strategies like crop rotation may contribute for the management of this disease to become epidemic. Many regions of the world namely Australia (Ackland *et al.*, 1998; Knights and Siddique, 2002), Canada (Chongo and Gossen, 2001), Latin America (Kaiser *et al.*, 2000), southern Europe (Trapero- Casas and Jime´nez-Dý´az, 1986), the USA (Kaiser and Muehlbauer, 1989), and West Asia (Akem *et al.*, 2000) are under the effect of this disease and major cause to the economic loss. Infection on petioles, leaflets and on young branches of chickpea includes early symptoms such as epinasty and loss of turgor, followed by water soaking. This pathogen secretes a toxin which maintains the symptom. This disease spreads mainly by the infected seeds produced from infected crops (Weltzien and Kaark, 1981; Kaiser, 1997). Introduction of this disease into Australia (Cother, 1977), Canada (Morrall and McKenzie, 1974), Iran (Kaiser, 1972), and the USA (Kaiser and Muehlbauer, 1984) is due to the use of infected seeds for cropping. The chickpea seeds with light infection by *A. rabiei,* appear usually discoloured with very low seed weight. However, in severe infection, the pathogen appears on the seed coat, cotyledons and embryo axes (Dey and Singh, 1994). On the infected chickpea seed, the mycelium of *A. rabiei* grows on the seed coat (Luthra and Bedi, 1932), on cotyledons (Lukashevich, 1958) and on the emerging seedlings (Kaiser and Hannan, 1988).

Pycnidium is a structure for the survival of *A. rabiei* on the debris lying on the surface of soil (Navas-Corte´s *et al.*, 1995; Galloway and MacLeod, 2003). It can also survive as teleomorph having ability to colonize naturally infested debris and can form viable pseudopothesia on the uninfected parts of the debris. Pycnidium produces conidia in favourable conditions and acts as primary inoculums for the initiation of disease in new crop. However, the teleomorph develops on infected chickpea debris remains in the field and in favourable condition produces ascospores for long distance spread of the disease (Navas-Corte´s *et al.*, 1995; 1998).

Management

Management strategies of Ascochyta blight include application of fungicides, cultural control and improvement of host resistance. Although some cultivars of chickpea are reported to have genetic resistance against this pathogen, but the resistance is only partial and starts breaking down during flowering stage of plant (Nene and Reddy, 1987; Chongo and Gossen, 2001). Number of fungicides have been reported for the effective control of this pathogen but the residual effect of these fungicides can contaminate environment and harmful for the human kinds. Integrated disease management (IDM) is the main key for successful cropping and production of chickpea. Treatment of seeds of chickpea with effective chemical fungicides is one of the methods to reduce the loss caused by Ascochyta blight. From the 1930 to early 1960, number of chemical and physical strategies were employed for the seed treatment and minimization of the disease *e.g.* copper sulphate (Sattar, 1933), malachite green (Zachos, 1951), pimaracin (Zachos *et al.*, 1963), thiram (Khachatryan, 1961) and hot water (Sattar, 1933) treatment. However, these seed treatments generally had low efficacy in controlling seed-borne transmission of the disease.

Due to extensive use of fungicides, the microbial population residing in the plants were also affected very badly. This is well known that legumes have the ability to fix atmospheric nitrogen (N) (Kyei-Boahen *et al.*, 2001; Gan *et al.*, 2005). Symbiotic N-fixing microorganism *Rhizobium* is in direct contact with the applied fungicides in the field. This reduces the survival of *Rhizobium* and affects the establishment of symbiosis between plant root and microorganism (Revellin *et al.*, 1993; Kutcher *et al.*, 2002). The rate of fungicide application on the seed is the key factor affecting viability of *Rhizobium* applied on the seed coat as biofertilizers. Excessive use of fungicides than the recommended for the seed treatment may be the major cause for the decrease in rhizobial population (Kyei- Boahen *et al.*, 2001). Duration of incubation with fungicides may also influence the survival of the *Rhizobium* (Matus *et al.*, 2003; Gan *et al.*, 2005). For IDM there should be decrease in the use of chemical pesticides, which helps in the combined use of pesticides and *Rhizobium* (Welty *et al.*, 1988). Fungicides should be applied prior to *Rhizobium* inoculation. The sequence of application of *Rhizobium* and fungicides should be done in a way so that it can support better survival of *Rhizobium* on seed coat. Use of granular rhizobial inoculants may decrease possibility of physical contact between fungicide which ultimately increases symbiosis (Kyei-Boahen *et al.*, 2002; Gan *et al.*, 2005).

Use of BCAs has become a better alternative to control this disease. A number of researchers are presently working in this area to find out biocontrol agents from the environment having better capabilities against this pathogen. Few fungal antagonists *Chaetomium globosum, Trichoderma viride* and *Acremonium implicatum* were studied under *in vitro* and *in vivo* conditions: for their biological control activity against Ascochyta blight. *A implicatum* isolate-1 inundated the pathogen mycelium, and caused its breakdown. *A. implicatum* isolate-2 produced a clear zone. However, *C. globosum* and *T. viride* covers the mycelium of *A. rabiei* and there is increase in spore production by *T. viride.* On application of culture filtrates of antagonist microorganisms, a remarkable impediment in pycnidiospore germination and colony expansion of *A. rabiei* was observed. However, in the glasshouse condition, the culture filtrates of all the three antagonists were found effective in bringing a curtailment in disease development. *C. globosum* was most effective and causes about 73.12 per cent deduction of disease index. Dugan *et al.* (2009) reported the natural occurrence of *Aureobasidium pullulans* in the post harvest debris of chickpea. Spray of spore suspension of *A. pullulans* to chickpea debris leads to 38 per cent lesser incidence of Ascochyta blight.

Botrytis Grey Mould (*Botrytis cinerea*)

The causal organism of Botrytis grey mould (BGM) is *Botrytis cinerea* Pers. ex. Fr. It is one of the economically important diseases of chickpea. This disease occurs mainly in the area where, there is persistence of cool, cloudy, and humid weather. In the off season, this pathogen survives on the infected crop debris and seeds. The asexual stage of this necrotrophic fungus is dominant in crop chickpea. On potato dextrose agar (PDA) medium it appears as white, cottony, which turns light gray after few days. Young hyphae of *B. cinerea* are thin, hyaline, and 8–16µm wide, which become septate with age. Conidiophores are light brown, septate, and erect ramified pseudodichotomically. They have slightly enlarged tips, bearing small pointed sterigmata which bear 1–2-celled, hyaline, oval, or globose conidia forming clusters. The size of conidia is 4–25×4–18µm and 4–16×4–10µm on infected chickpea plants and PDA, respectively (Nene and Reddy 1987; Pande *et al.*, 2002). Joshi and Singh (1969) have suggested that sporodochia formed by the pathogen on the host surface measure 0.5-5.0µm in diameter and may turn into hard sclerotial masses. Sporulation of *B. cinerea* largely depends upon temperature, nutrients and ecological factors. Shaw and Ajrekar (1915) first reported BGM on chickpea from India followed by Butler and Bisby (1931). However, first epidemic of this disease which causes crop loss of about 95 per cent was reported by Carranza (1965) in Argentina. After that a number of epidemics caused by BGM were reported showing almost complete loss from many chickpea growing countries. The disease has caused serious concerns in India, Bangladesh, Nepal, Pakistan and Australia. BGM is reported on chickpea globally. Numbers of reports from Argentina (Haware and McDonald, 1992, 1993; Bakr *et al.*, 1993; Dhar *et al.*, 1993; Karki *et al.*, 1993; Malik *et al.*, 1993; Haware, 1998; Pande *et al.*, 2002; Davidson *et al.*, 2004) have showed almost complete yield loss (about 100 per cent) under conducive environmental conditions. However, this disease is also reported from Canada, Chile, Colombia, Hungary, Mexico, Myanmar, Spain, Turkey, the USA, and Vietnam (Nene *et al.*, 1984; Pande *et al.*, 2002). During sexual stage, *B. cinerea* fertilised sclerotia germinate by the emergence of apothecia that release sexually produced ascospores (Faretra and Grindle, 1992). Formation of apothecia requires either 2 sexually compatible isolates (MAT 1-1 and MAT 1-2) or a pseudo homothallic isolate (MAT-1/2) (Faretra and Grindle, 1992). Natural occurrence of sexual structure was not known, however, its occurrence is reported in India under laboratory conditions (Singh, 1997).

An antifungal compound produced by *Botrytis* is reported to kill yeast and most often stops fermentation, prior to sufficient accumulation of alcohol in the fermenter. During 1978–79, this disease had become epidemic in India and destroyed about 20000 hectares of chickpea (Grewal and Laha, 1983; Grewal *et al.*, 1992; Haware, 1998). However, in Nepal, this disease occurs almost each year of cropping and average yield loss was about 15 per cent (Joshi1992). The disease was first documented in Bangladesh during 1981 and later on it became devastating, destroying the whole crop in 1988 (Bakr and Ahmed, 1992). Pod yield depends upon the onset of disease, its growth, disease severity, weather conditions and level of inoculums of the pathogen.

Management

Use of disease free seeds can reduce the chances of disease to be epidemic. Manipulation in sowing time, use of erect cultivars and lower density of plants may be helpful in reducing the level of BGM in chickpea (Haware, 1998). Seed treatments with different fungicides such as iprodione, thiabendazole, mancozeb, triadimefon, vinclozolin, triadimenol, thiram, benomyl, captan or carbendazim were found to be effective in reducing seed infection (Cother 1977; Grewal and Laha, 1983; Singh and Bhan, 1986; Singh and Kaur, 1990; Bakr *et al.*, 1993; Haware, 1998; Pande *et al.*, 2002; Davidson *et al.*, 2004). It has been reported that BGM can be effectively controlled by treating the seeds with fungicides (Knights and Siddique, 2002).

B. cinerea strains greatly differ in their genetic variability. There is no clear establishment about the nature of infection processes and genetic basis of pathogen variability. However, extensive investigations on pathological, physiological and molecular characteristics of *B. cinerea* were completed. The lack of information about the disease has forced people to use fungicides resulting deployment of host plant resistance (HPR) as a major option for BGM management. However, predictive models for the application of fungicides may be helpful. IDM of BGM has been proved to be strongest management strategy to control this disease compared to any other individual disease management practices available in India, Nepal, and Bangladesh. However, intensive study about this disease is needed for better control of this disease via IDM.

A study in green house condition by Haware (1998) suggested the use of *T. harzianum* and *T. viride* for the control of BGM. In another study, Mukherjee and Haware (1993) also showed the effectiveness of *T. harzianum* and *T. viride* in the field conditions. Five different isolates of *T. viride* and *T. harzianum* are reported to be highly antagonistic to *B. cinerea* and have ability to parasitize *B. cinerea* completely on PDA medium. There is swelling in hyphal tips of *B. cinerea* due to the presence of *T. viride*. Seed priming with the spores of *T. viride* increases seed germination, also influence positively the growth of shoot and root and ultimately the plant vigour. However, it has no harmful effect on the nodulation process and significantly reduces the disease incidence (Agarwal and Tripathi, 1999). Mukherjee *et al.* (1995) have used a *Trichoderma* sp. strain T15 isolated from the rhizosphere of chickpea and found that is an effective BCA against BGM. Benomyl-tolerant isolates of T15 and vinclozolin-tolerant isolates of *T. viride* were found to be equally effective in the biocontrol of BGM. The fungicide tolerant *T. viride* can be used in integration with the fungicide and the result was more effective in combating BGM than vinclozolin alone (Mukherjee *et al.*, 1995, 1997). Another BCA *Gliocladium roseum* and *T. virens* were found to be effective against *B. cinerea* on seed treatment with spores and artificial inoculation of the BGM (Burgess *et al.*, 1997). The protection level was equivalent to the protection level of fungicide Thiram.

Plant essential oils were found to be effective against *B. cinerea.* A study on efficacy of seven essential oils showed efficacy of Clove oil, cinnamon oil and gerniol by using paper disc method. These compounds at the minimum concentration of 1000 ppm inhibited the *in vitro* conidial germination and reduction in the germ tube length of *B. cinerea* by >90 and 80 per cent, respectively. They have developed a rapid procedure for the detection of antifungal activity of plant extracts and essential oils against *B. cinerea* by using an automatic microtitre plate reader and by using this method they that plant extracts of *Allium* and *Capsicum* spp., and essential oils of *Cymbopogon martini, Thymus zygis, Cinnamomum* were effective against the pathogen. Since, there is no adequate level of genetic resistance available against the BGM and the fungicides become ineffective during conditions of high disease pressure. Hence, there is need of IDM practices against the disease to mitigate yield loss. A number of studies (Agarwal *et al.*, 1999; Haware *et al.*, 1999) showed successful management of the disease by adopting integrated use of *T. viride* with fungicides. Ahmed *et al.* (2002) have reported the use of BGM tolerant genotype ICCL 87322 in combination with Bavistin.

Genetic Engineering in BGM Management

Genetic engineering is an effective tool to transfer new traits such as BGM resistance in chickpea. However, there is requirement of efficient regeneration and transformation systems for gene transfer. In addition there should be knowledge about the genes conferring resistance against *B. cinerea* for transformation. So, integration of genes coding for antifungal metabolites may be a feasible approach for development of resistant crop. Numbers of antifungal proteins such as the fungal cell-wall degrading hydrolytic enzymes like chitinase (Kishimoto *et al.*, 2002) suppress the fungal growth in transgenic tobacco plants and cucumber (Tabei *et al.*, 1998), and *Dendranthema morofolium* (Takatsu *et al.*, 1999). In another study, expression of soybean β-1,3-glucanase gene in kiwifruit reduced the symptoms of *B. cinerea* infection (Nakamura *et al.*, 1999). Deak *et al.* (1999) have reported the protection from the oxidative damage caused during necrosis by *B. cinerea* infection by expression of an iron-binding protein, ferritin, from alfalfa. However, pathogenesis and defence related proteins are potential gene candidates for the transformation but more study is needed for identification of potential genes against BGM.

Sclerotinia Rot

Sclerotinia sclerotiorum (Lib.) de Bary is a cosmopolitan necrotrophic fungal pathogen with a broad range of hosts worldwide. A number of diseases are caused by *S. sclerotiorum* or related sp. such as stem rots, soft rots and wilts of horticultural and agricultural crops including chickpea (Fuhlbohm *et al.*, 2003). It appears as white cotton like growth on the plant. This fungus is a soil-borne fungus that attacks plant in all stages of growth including young seedlings, mature plants and their harvested products distributed worldwide (Agrios, 1997; Purdy, 1979). This fungus may cause almost 100 per cent loss to the crop and millions of dollars are lost annually around the world due to this disease (Purdy, 1979; Subbarao, 1998; Tahtamouni *et al.*, 2006). The survival off season structure of this disease is sclerotia, which can survive for several years and can germinate carpogenically to produce apothecia. After maturity the apothecia release large amount of ascospores that spread over a long distance and serve as primary source of inoculum for infection. Due to these features, the management of this disease is difficult. Many alternatives are being exploited for the control of this disease due to harmful effect of pesticides used to control this disease (Sarma *et al.*, 2007). In India, regular occurrence of the disease has been reported in some districts of Himachal Pradesh, where it causes considerable yield losses (Grewal and Pal, 1986). The yield loss ranges from 21 to 47 per cent at pod bearing stage of the crop (Sharma, 1990).

Management

Due to the limitations of fungicide use and its harmful effect, use of fungal antagonists as BCAs against *S. sclerotiorum* is likely to be the best alternative to conventional chemical control methods. Numbers of studies have reported various biological control measures to control this disease (Adams and Ayers, 1982; El-Tarabily *et al.*, 2000). Some of the chitinase producing microorganisms are studied more extensively against S. *sclerotiorum.* Since chitin is the major constituent of fungal cell wall chitinase producing microorganisms would have a great potential against the disease including *S. sclerotiorum* (Gupta *et al.*, 1995; El-Tarabily *et al.*, 2000; Tahtamouni *et al.*, 2006). Many species in the order actinomycetales are reported to be effective against this disease and most of them are from the genus *Streptomyces* (Saadoun and Al-Momani, 1997; Saadoun *et al.*, 2000; El-Tarabily *et al.*, 2000; Tahtamouni *et al.*, 2006). Srivastava *et al.* (2014) have recently reported the efficacy of *Streptomyces rochei* against *S. sclerotiorum.* Sharma (1994) suggested the use of *T. harzianum* which has the ability to inhibit mycelium of the pathogen by 75 per cent. Other BCAs like *A. cylindrospora* and *T. viride,* while *Gliocladium virens* caused least inhibition of the pathogen. The major reason for the inhibition antagonists is due to secretion of diffusible metabolites, and host parasite relationship revealed hyphal breakage and swelling of S. *sclerotiorum.* Lee and Wu (1984) reported that *G. virens* and *T. viride* produced antibiotics that inhibit the mycelial growth of S. *sclerotiorum* completely and induce swelling and plasmolysis of affected cells. Disease management by using BCAs is major part of integrated disease management and is the most potential management strategy for crop pathogens through reducing pathogen inoculum in soil (Lopez-Lima *et al.*, 2013). The most effective mechanism used by BCAs is the antagonism which is most effective strategy for biological management of crop pathogens. However, numbers of studies have reported the combined use of biocontrol agents against a number of diseases (Singh *et al.*, 2014; Jain *et al.*, 2014). One of the reasons is explained for the improbability of the BCAs

to come in direct contact with the infection causing ascospores of *S. sclerotiorum*. However, BCAs capable of stimulating induced systemic resistance (ISR) responses in hosts have a greater chance to lower the infection level of *S. sclerotiorum* insisted through airborne ascospores. Recently, microbe-induced systemic tolerance was given more emphasis for providing protection to plants. Plant receives appropriate signal from the BCAs, and keep their defence responses in the primed stage. When the pathogen attacks on the plant, it activates a number of robust biochemical changes in host plant against the pathogen attack through ISR mechanisms.
Therefore, identification, characterization and utilization of such microbes for possible alleviation of stresses in plants have always considered important. It has been demonstrated that plant growth-promoting rhizobacteria (PGPR) can increase plant's tolerance to soil-borne pathogens (Singh *et al.*, 2003; Singh *et al.*, 2013a).

Other Fungal Foliar Diseases

Chickpea rust is caused by *Uromyces ciceris-arietini*. This disease has also been reported in Mediterranean region, South-East Europe, South Asia, Australia, East Africa, Mexico and USA (Asthana, 1957; Reddy, *et al.*, 1980; Rubiales and Sillero, 2011; Sillero *et al.*, 2006; Venette and Stack, 1987). This disease has importance only in the area where it occurs. However, regions of central Mexico and Australia or Italy are found to be severely affected with this disease (Ragazzi, 1982; Diaz-Franco and Perez-Garcia, 1995; Jones, 1983). Normally it become epidemic, late in the season, therefore, yield is usually not affected significantly by the infection. However, when plants are attacked during early stages of crop growth it can cause severe reduction in yield. Very few studies are available showing the control measures for this disease in *C. arietinum*. Chemical control with propiconazole has been reported for its control (Dýaz-Franco and Perez-Garcýa, 1995; Gomez and Paredes, 1985). Reports are also available regarding chickpea resistance source and molecular markers for resistance breeding programmes (Madrid *et al.*, 2008; Rubiales *et al.*, 2001).

Chickpea is also attacked by powdery mildew. The first report on chickpea powdery mildew was reported from USA near Pullman, Whitman Co., WA. Disease signs were observed on all the chickpea cultivars in the fields with high incidence (> 80 per cent) (Attanayake *et al.*, 2008). Typical powdery mildew symptoms and signs of the disease were observed on plants in early senescence. Dense white powdery mass on dorsal leaf surface and petiole, with chlorotic areas beneath the fungal growth, then necrotic, followed by defoliation unbranched or branched conidiophores developed typically from internal mycelium and emerged singly or in groups of one to five through stomata. Conidiophores bear hyaline primary and secondary conidia. Primary conidia were lanceolate, (47.5-50)-(66.5-68.5) × (13.5-14-20.5) µm~ Secondary conidia were ellipsoidal to cylindrical, and measured 46.5-62 × 14.5-18.5 µm. The teleomorph was not observed. Morphological features of the anamorph fit the description of *Leveillula taurica* (Lév.) Arnaud (Braun, 1987). This disease does not have any economic impact on the chickpea crop. Therefore, no control measures were suggested. However, further research on the control measures would be needed if the disease occurs in the future, early in the season and causes significant yield losses.

REFERENCES

Ackland, S., Moore, K., Schwinghamer, M., Sykes, J. 1998. Chickpea Foliar Disease Update. *AGNOTE-NSW Agriculture*, Australia, pp. 1–4.

Adams, O. B., Ayers, W. A. 1982. Biological control of *Sclerotinia* of lettuce drop in field by *Sporidesmium sclerotivorum*. *Phytopathology*, 72: 485-488.

Agarwal, A., Tripathi, H.S. 1999. Biological and chemical control of botrytis gray mould of chickpea. *J. Mycol. Plant Pathol.* 29: 52–56.

Ahmed, A.U., Bakr, M.A., Hossain, M.S., Chowdhury, J.A. 2002. Integrated management of Botrytis grey mould disease in chickpea. *Bangladesh Journal of Agricultural Research* 27, 237–242.

Akem, C.N., 1999. Ascochyta blight of chickpea: present status and future priorities. *Int. J. Pest Manage.* 45, 131–137.

Akem, C.N., Malhotra, R.S., Bayaa, B., Mouselli, M.N., Kabbabeh, S., 2000. Taking blight off chickpea. *ICARDA-CARAVN* 2, 18.

Armstrong, C.L., Chongo, G., Gossen, B.D., Duczek, L.J., 2001. Mating type distribution and incidence of the teleomorph of *Ascochyta rabiei* (*Didymella rabiei*) in Canada. *Can. J. Plant Pathol.* 23, 110–113.

Asthana, R.D. 1957. Some observations on the incidence of *Uromyces ciceris-arietini* (Grognon) Jakz. and Boyer on *Cicer arietinum*. *Nagpur Agric. Coll. Mag.* 31: 20A–20B

Attanayake, K.P.R.N. Glawe, D.A. McPhee, K.E. Dugan, F.M. Chen, W. 2008. First report of powdery mildew of Chickpea (*Cicer arietinum*) caused by *Leveillula taurica* in Washington State. http: //www.plantmanagementnetwork.org/pub/php/brief/2008/chickpea/

Bakr, M.A, Ahmed, F. 1992. Botrytis gray mold of chickpea in Bangladesh. In 'Botrytis gray mold of chickpea. Summary Proceedings of the BARI/ICRISAT Working Group Meeting'. (Eds MP Haware, DG Faris, CLL Gowda) pp. 10–12. *(ICRISAT: Patancheru, AP, India)*

Bakr, M.A, Rahman, M.M., Ahmed, F., Kumar, J. 1993. Progress in management of botrytis gray mold of chickpea in Bangladesh. In 'Recent advances in research in botrytis gray mold of chickpea'. (Eds MP Haware, CLL Gowda, D McDonald) pp. 17–19. (*ICRISAT: Patancheru, AP, India*)

Bate, N.J., Orr, J. Ni. W., Meromi, A., Nadler-Hassar, T., Doerner, P.W., Dixon, R.A., Lamb, C.J. and Elkind, Y. 1994. Qualitative relationship between phenylalanine ammonia-lyase levels and phenylpropanoid accumulation in transgenic tobacco identifies a rate-determining step in natural product biosynthesis. *Proc. Natl. Acad. Sci. USA* 91: 7608-7612.

Binutu, O.A. and Cordell, G.A. 2000. Gallic acid derivative from *Mezoneuron benthamianun* leaves. *Pharmacol. Biol.* 38: 284–286.

Braun, U. 1987. A monograph of the erysiphales (powdery mildews). *Beih. Nova Hedwigia* 89: 1700.

Burgess., D.R., Bretag., T., Keane., P.J. 1997. Biocontrol of seedborne *Botrytis cinerea* in chickpea with *Gliocaldium roseum*. *Plant Pathol.* 46: 298–305.

Butler, E.J., Bisby, G.R. 1931. 'The fungi of India. Sci. Monograph No. 1.'pp. XVIII 237. (*ICAR: New Delhi*)

Chongo, G., Buchwaldt, L., Gossen, B.D., Lafond, G.P., May, W.E., Johnson, E.N., Hogg, T., 2003. Foliar fungicide to manage ascochyta blight (*Ascochyta rabiei*) of chickpea in Canada. *Can. J. Plant Pathol.* 25, 135–142.

Chongo, G., Gossen, B.D., 2001. Effect of plant age on resistance to *Ascochyta rabiei* in chickpea. *Can. J. Plant Pathol.* 23, 358–363.

Cother, E.J., 1977. Identification and control of root rot fungi in *Cicer arietinum* (chickpea). *Plant Dis. Rep.* 61, 736–740.

Davidson, J.A., Pande, S., Bretag, T.W., Lindbeck, K.D. and Kishore, G.K. 2004. Biology and management of *Botrytis* spp. in legume crops. In *Botrytis: Biology, Pathology and Control.* (Eds. Y Elad, B Williamson, P Tudzynski, N Delen) pp. 295–318. (The Netherlands: Kluwer Academic Publishers).

Deak, M., Horvath, G.V., Davletova, S., Totok, K., Sass, L., Vass, I., Barna, B., Kiraly, Z., Dudits, D. 1999. Plants ectopically expressing the iron-binding protein, ferritin, are tolerant to oxidative damage and pathogens. *Nature Biotechnol* 17, 192–196. doi: 10.1038/6198

Dey, S.K., Singh, G., 1994. Seedborne infection of *Ascochyta rabiei* in chickpea and its transmission to aerial plant parts. *Phytoparasitica* 22, 31–37.

Dhar, V., Rathi, Y.P.S., Tripathi, H.S., Pal, M., Singh, G., Haware, M.P., Upadhyay, J.P. 1993. Multilocational screening for Botrytis gray mold resistance in India. In 'Recent advances in research on Botrytis gray mold of chickpea'. (Eds MP Haware, CLL Gowda, D McDonald) pp. 14–16. (*ICRISAT: Patancheru, AP, India*)

Dýaz-Franco, A., Perez-Garcýa, P. 1995. Control quýmico de la roya y la rabia del garbanzo y su influencia en el rendimiento de grano. *Rev. Mex. Fitopatol.*13: 123–125.

Dugan, F.M., Akamatsu, H., Shari L. Lupien, S.L., Chen, W., Chilvers, M.L., and Peever, T.L. 2009. Ascochyta blight of chickpea reduced 38 per cent by application of *Aureobasidium pullulans* (anamorphic Dothioraceae, Dothideales) to post-harvest debris. *Biocontrol Sci. Techn.* 19(5): 537-545.

El-Tarabily, K. A., Hardy, G. E., Sivasithamparam, K., Hussein, A. M. and Kurtboke, J. D. 1997. The potential for the biological control of cavity spot disease of carrots caused by *Pythium coloratum* by streptomycete and non-streptomycete actinomycetes in Western Australia. *New Phytologist*, 137: 495-507.

Faretra, F., Grindle, M. 1992. Genetic studies of *Botryotinia fuckeliana* (*Botrytis cinerea*). In 'Recent Advances in Botrytis Research. Proceedings of the 10th International Botrytis Symposium'. Heraklion, Crete, Greece. (Eds K Verhoeff, NE Malathrakis, B Williamson) pp. 7–16. (*Pudoc Scientific Publishers: Wageningen, The Netherlands*)

Fuhlbohm, M.J., Tatnell, J.R., Ryley, M.J. 2003 First report of stem rot and wilt of chickpea caused by Sclerotinia minor in Queensland, Australia. *Australias Plant Pathol* 32: 323–324.

Galloway, J., MacLeod, W.J., 2003. Initiation of ascochyta disease from pulse stubble in the Mediterranean climate of Western Australia. In: *Proceedings of the 8th International Congress of Plant Pathology*, Christchurch, New Zealand, 2–7 February, p. 102.

Gan, Y.T., Selles, F., Hanson, K.G., Zentner, R.P., McConkey, B.G., McDonald, C.L., 2005. Effect of formulation and placement of *Mesorhizobium* inoculants for chickpea in the semiarid Canadian Prairies. *Can. J. Plant Sci.* 85, 555–560.

Gomez, G.R., Paredes, P. (1985). El cultivo del garbanzo blanco en el centro de Sinaloa. Campo experimental Valle de Culiacan, INIFAP, Folleto tecnico 4: 32

Gossen, B.D., Miller, P.R., 2004. Survival of *Ascochyta rabiei* in chickpea residue on the Canadian Prairies. *Can. J. Plant Pathol.* 26, 142–147.

Grewal, J.S., Laha, S.K. 1983. Chemical control of botrytis blight of chickpea. *Indian J Phytopathol* 36, 516–520.

Grewal, J.S., Pal, M., Rewal, N. 1992. Botrytis gray mold of chickpea in India. In 'Botrytis gray mold of chickpea'. (Eds MP Haware, DG Faris, CLL Gowda) pp. 6–8. (*ICRISAT: Patancheru, AP, India*)

Grewal, J.S., Mahendra, Pal. 1986. Fungal disease problems of chickpea. In : *Vistas in Plant Pathology*. (Eds., Varma, A. and Verma, J.P.). *MPH, New Delhi*, pp. 157-170.

Gupta, R., Saxena, R., Chaturvedi, P. and Virdi, J. 1995. Chitinase production by *Streptomyces viridificans*: its potential in fungal cell wall lyses. *J Appl Bacteriol*, 78: 378-383.

Haware, M.P., McDonald, D. 1992. Integrated management of botrytis gray mold of chickpea. In 'Botrytis gray mold of chickpea'. (Eds MP Haware, DG Faris, CLL Gowda) pp. 3–6. (*ICRISAT: Patancheru, AP, India*)

Haware, M.P., Mukherjee, P.K., Lenne, J.M., Jayanthi, S., Tripathi, H.S., Rathi, Y.P.S. 1999. Integrated biological-chemical control of Botrytis gray mold of chickpea. *Indian Phytopathol* 52, 174–176.

Haware, M.P. 1998 Diseases of chickpea. In 'The pathology of food and pasture legumes'. (Eds DJ Allen, JM Lenne) pp. 473–516. (*ICARDA, CAB International: Wallingford, UK*).

ICARDA, 1996. Field Survey of Chickpea Diseases. Legume Annual Report 1996. International Centre for Agricultural Research in the Dry Areas (*ICARDA*), Aleppo, Syria.

Jain, A., Singh, S., Sarma, B.K., Singh, H.B. 2012. Microbial consortium–mediated reprogramming of defence network in pea to enhance tolerance against *Sclerotinia sclerotiorum*. *J Appl Microbiol* 112(3): 537-50.

Jones, D.H. 1984. Phenylalanine ammonia-lyase: regulation of its induction, and its fole in plant development. *Phytochem*. 23: 1349-1360.

Jones, D.R. 1983. Chickpea rust—a disease new to Australia. Queensland Department of Primary Industry Farm Note AGDEX 168/637

Joshi, M.M., Singh, R.S. 1969. A Botrytis gray mould of gram. *Indian Phytopathol* 22, 125–128.

Joshi, S. 1992) Botrytis gray mold of chickpea in Nepal. In 'Botrytis gray mold of chickpea'. (Eds MP Haware, DG Faris, CLL Gowda) pp. 12–13. *(ICRISAT: Patancheru, AP, India)*

Kaiser, W.J., 1972. Occurrence of three fungal diseases of chickpea in Iran. *FAO Plant Prot. Bull.* 20, 73–79.

Kaiser, W.J., Muehlbauer, F.J., 1984. Occurrence of *Ascochyta rabiei* on imported chickpea in eastern Washington. *Phytopathology* 74, 1139 (abstract).

Kaiser, W.J., Ramsey, M.D., Makkouk, K.M., Bretag, T.W., Acikgoz, N., Kumar, J., Nutter, F.W. 2000. Foliar diseases of cool season food legumes and their control. In: 'Linking Research and Marketing Opportunities for Pulses in the 21st Century'. (Ed Knight, R.) pp. 437–455. (*Kluwer Academic Publishers, The Netherlands*)

Kaiser,W.J., Hannan, R.M., 1988. Seed transmission of Ascochyta rabiei in chickpea and its control by seed-treatment fungicides. *Seed Sci. Technol.* 16, 625–638.

Karki, P.B., Joshi, S., Chaudhary, G., Chaudhary, R.N. 1993. Studies on Botrytis gray mold of chickpea in Nepal. In 'Recent advances in research on Botrytis gray mold of chickpea'. (Eds MP Haware, CLL Gowda, D McDonald) pp. 11–13. *(ICRISAT: Patancheru, AP, India)*

Khachatryan, M.S., 1961. Seed transmission of ascochytosis infection in chickpea and its effectiveness of treatment. Sbotnik Nauchnykh Trudov Nauchno-Issledovatel'sk Zemledel Armyarkoi 2, 147–155.

Khan, M.S.A., Ramsey, M.D., Corbiere, R., Infantino, A., Porta-Puglia, A., Bouznad, Z., Scott, E.S., 1999. Ascochyta blight of chickpea in Australia: identification, pathogenicity and mating type. Plant Pathol. 48, 230–234.

Kishimoto, K., Nishizawa, Y., Tabei, Y., Hibi, T., Nakajima, M., Akutsu, K. 2002. Detailed analysis of rice chitinase gene expression in transgenic cucumber plants showing different levels of disease resistance to gray mold (*Botrytis cinerea*). *Plant Science* 162, 655–662. doi: 10.1016/S0168-9452(01)00602-1

Knights, E.J., Siddique, K.H.M, 2002. Chickpea status and production constraints in Australia. In: 'Integrated Management of Botrytis Grey Mould of Chickpea in Bangladesh and Australia'. (Eds Abu Bakr, M., Siddique, K.H.M., Johansen, C.) pp. 33–41. (BARI: Joydebpur, *Bangladesh and CLIMA Crawley, W. Australia*)

Kutcher, H.R., Lafond, G., Johnston, A.M., Miller, P.R., Gill, K.S., May, W.E., Hogg, T., Johnson, E., Biederbeck, V.O., Nybo, B. 2002. *Rhizobium* inoculant and seed-applied fungicide effects on fieldpea production. *Can. J. Plant Sci.* 82, 645–651.

Kyei-Boahen, S., Slinkard, A.E.,Walley, F.L., 2001. Rhizobial survival and nodulation of lentil as influenced by fungicide seed treatment. *Can. J. Microbiol.* 47, 585–589.

Kyei-Boahen, S., Slinkard, A.E.,Walley, F.L., 2002. Evaluation of rhizobial inoculation methods for chickpea. *Agron. J.* 94, 851–859.

Ladizinsky, G. 1975. A new *Cicer* from Turkey. *Notes from the Royal Botanical Gardens.*

Lee, Y.A. and Wu, W.S. 1984. The antagonism of *Trichoderma* spp. and *Gliocladium virens* against *Sclerotinia sclerotiorum. Plant Proto Bull. Taiwan, 28:* 101-109.

Lopez-Lima, D., Sanchez-Nava, P., Carrion, G., Nunez-Sanchez, A.E. 2013 89 per cent reduction of a potato cyst nematode population using biological control and rotation. *Agron. Sustain. Dev.* 33: 425–431.

Lukashevich, A.I., 1958. Control measures against ascochytosis of chickpea. J. Agricult. Sci. 5, 131–135.

Luthra, J.C., Bedi, K.S., 1932. Some preliminary studies on gram blight with reference to the cause and mode of perennation. *Ind. J. Agricult. Sci.* 2, 499–515.

Madrid, E., Rubiales, D., Moral, A., Moreno, M.T., Milla´n, T., Gil, J., Rubio, J. 2008. Mechanism and molecular markers associated with rust resistance in a chickpea interspecific cross (*Cicer arietinum* L. 9 *Cicer reticulatum* Lad.). *Eur. J. Plant Pathol.* 121: 43–53

Malik, B.A., Iqbal SM, Bashir, M. 1993. Present status and Future research outlook for Botrytis gray mold – a potential threat to chickpea in Pakistan. In 'Recent advances in research in botrytis gray mold of chickpea'. (Eds MP Haware, CLL Gowda, D McDonald) pp. *19–20. (ICRISAT: Patancheru, AP, India)*

Manjunatha, S.V., Naik, M.K., Khan M.F.R. and Goswami, R.S. 2013. Evaluation of bio-control agents for management of dry root rot of chickpea caused by *Macrophomina phaseolina. Crop. Prot.* 45: 147-150.

Matus, A., Sadleir, J., Cronkwright, L., McLeod, R., 2003. Effects of crown (carbathiin and thiabendazole), allegiance fl (metalaxyl), vitaflo 280 (carbathiin and thiram), and apron maxx (fluodioxonil and metalaxyl) on N-fixation of chickpea, dry bean, lentil, and pea. In: Proceedings of the Soils and Crops Workshop, 2003. *Saskatoon, SK, Canada, 12–14 February* (CD-ROM).

Morrall, R.A.A., McKenzie, D.L., 1974. A note on the inadvertent introduction to North America of *Ascochyta rabiei*, a destructive pathogen of chickpea. *Plant Dis. Rep.* 58, 342–345.

Mukherjee, P.K. and Haware, M.P. 1993. Biological control of Botrytis gray mold of chickpea. *International Chickpea Newsletter* 28: 14–15.

Mukherjee, P.K., Haware, M.P. and Jayanthi, S. 1995. Preliminary investigations in integrated biocontrol of botrytis gray mold of chickpea. *Indian Phytopathol.* 48: 141–149.

Mukherjee, P.K., Haware, M.P. and Raghu, K. 1997. Induction and evaluation of benomyl-tolerant mutants of *Trichoderma viride* for biological control of Botrytis grey mould of chickpea. *Indian Phytopathol.* 50: 485–489.

Nakamura, Y., Sawada, H., Kobayashi, S., Nakajima, I., Yoshikawa, M. 1999. Expression of soybean β-1,3-endoglucanasecDNAand effect on disease tolerance in kiwifruit plants. *Plant Cell Reporter* 18, 527–532. doi: 10.1007/s002990050616

Navas-Cortes, J.A., Trapero-Casas, A., Jimenez-Dýaz, R.M., 1998. Influence of relative humidity and temperature on development of *Didymella rabiei* in chickpea debris. *Plant Pathol.* 47, 57–66.

Nene YL, Sheila VK, Sharma SB (1984) A world list of chickpea (*Cicer arietinum*) and pigeonpea (*Cajanus cajan* (L.) Millsp.) pathogens. *ICRISAT Pulse Pathology Progress Report* 32, Patancheru, Andhra Pradesh, India.

Nene, Y.L., Reddy, M.V. 1987. Chickpea diseases and their control. In: 'The Chickpea'. (Eds Saxena, M.C., Singh, K.B.) pp. 233–270. (*CAB International, Oxfordshire, UK*)

Nene, Y.L., Shelia, V.K. and Sharma, S.B. 1996. "A World List of Chickpea and Pigenpea Pathogens," 5th Edition, Patancheru, Andhra Pradesh, p. 27.

Pande, S., Singh, G., Rao, J.N., Bakr, M.A., Chaurasia, P.C.P., Joshi, S., Johansen, C., Singh, S.D., Kumar, J., Rahman, M.M., Gowda, C.L.L. 2002. Integrated management of botrytis gray mold of chickpea. Information Bulletin No. 61, *ICRISAT, Andhra Pradesh,India.*

Prasad, R.D., Rangeshwaran, R., Anuroop, C.P., Rashmi, H.J. 2002. Biological control of wilt and root rot of chickpea under field conditions. *Ann. Plant Prot. Sci.* 10 (1): 72-75.

Purdy, L. H. 1979. *Sclerotinia sclerotiorum*: History, diseases and symptomatology, host range, geographic distribution, and impact. *Phytopathol* 69: 875 -880.

Ragazzi, A. 1982. Un grave attacco di ruggine su foglie di cece. [A serious attack of rust on *Cicer arietinum* leaves]. *Inform Fitopatol* 32: 41–43.

Reddy, M.V. and Nene, Y.L. 1979. A case for induced mutation in chickpea for *Ascochyta* blight resistance. Proceedings, Symp. on the Role of Induced Mutations in Crop Improvement, Osmania University, Hyderabad, India, pp 398-408.

Reddy, M.V., Gridley, H.E., Kaack, H.J. 1980. Major disease problems of chickpea in North Africa. *Int. Chickpea Newsl.* 3: 13–14

Revellin, C., Leterne, P., Catroux, G., 1993. Effect of some fungicide seed treatments on the survival of *Bradyrhizobium japonicum* and on the nodulation and yield of soybean [*Glycine max* (L) Merr.] *Biol. Fertil. Soils* 16, 211–214.

Rubiales, D., Moreno, I., Moreno, M.T., Sillero, J.C. 2001. Identification of Partial Resistance to chickpea rust (*Uromyces ciceris-arietini*). In: 'European Association for Grain Legume Research-AEP'. (Eds *Proceedings of the 4th European Conference on Grain Legumes*) pp 194–195. (*Cracow*)

Rubiales, D., Sillero, J.C. 2011. Rust of chickpea. In: 'Compendium of chickpea and lentil diseases and pests'. (Eds Chen, W., Sharma, H.C., Muehlbauer, F.J.) pp 54–55. (American *Phytopathological Society, Minessota*)

Saadoun, I. and Al-Momani, F. 1997. Streptomycetes from Jordan soils active against *Agrobacterium tumefaciens. Actinomycetes,* 8: 30-36.

Saadoun, I., Hameed, K. M., Al-Momani, F., Malkawi, H., Meqdam, M. and Mohammad, M. 2000. Characterization and analysis of antifungal activity of soil streptomycetes isolated from north Jordan. *Egypt.J. Microbiol.*, 35: 463-471.

Sarkar, A., Patel, J.S., Yadav, S.K., Sarma, B.K., Srivastava, J.S. and Singh, H.B. 2014. Studies on rhizosphere-bacteria mediated biotic and abiotic stress tolerance in Chickpea (*Cicer arietinum* L.). *Vegetos*. 27 (1): 158-169.

Sarma, B.K., Basha, S.A., Singh, D.P., Singh, U.P. 2007. Use of on-conventional chemicals as an alternative approach to protect chickpea (*Cicer arietinum*) from Sclerotinia stem rot. *Crop Prot.* 26: 1042–1048.

Sarma, B.K., Singh, D.P., Mehta, S., Singh, H.B., Singh U. P. 2002. Plant growth-promoting rhizobacteria-elicited alterations in phenolic profile of chickpea (*Cicer arietinum*) infected by *Sclerotium rolfsii*. *J. Phytopathol.* 150, 277–282.

Sarma, B.K., Yadav, S.K., Patel, J.S. and Singh, H.B. 2014. Molecular mechanisms of interactions of *Trichoderma* with other fungal species. Open Mycol. J. 8 (Suppl-1, M6): 140-147.

Sattar, A., 1933. On the occurrence, perpetuation and control of gram blight caused by *Ascochyta rabiei* (Pass.) Labrousse, with special reference to Indian conditions. *Ann. Appl. Biol.* 20, 612–632.

Sharma, B.K. 1994. Efficacy of biocontrol agents for the control of chickpea stem rot. *J BioI. Control* 8: 115- 117.

Sharma, B.K. and Singh, B.M. 1990. Biological control of white rot of pea caused by *Sclerotinia sclerotiorum* (Lib) de Bary. *J Bio. Control* 4: 132-134.

Shaw, F.J.F., Ajrekar, S.L. 1915. The genus *Rhizoctonia* in India. *Mem. Dept. Agric. India. Bot. Series* 7, 117.

Siddiqui, Z.A. and Akhtar, M.S. 2007. Biocontrol of a chickpea root-rot disease complex with phosphate-solubilizing microorganisms. *J. Plant Pathol.* 89 (1): 67-77.

Sillero, J.C., Fondevilla, S., Davidson, J., Vaz Patto, M.C., Warkentin, T., Thomas, J., Rubiales, D. 2006. Screening techniques and sources of resistance to rusts and mildews in grain legumes. *Euphytica* 147: 255–272

Singh, A., Sarma, B.K., Upadhyay, R.S., Singh, H.B. 2013a. Compatible rhizosphere microbes mediated alleviation of biotic stress in chickpea through enhanced antioxidant and phenylpropanoid activities. *Microbiol Res.* 168: 33–40.

Singh, G. 1997. Epidemiology of botrytis gray mold of chickpea. In 'Recent advances in research on botrytis gray mold of chickpea'. (Eds MP Haware, JM Lenne, CLL Gowda) pp. 47–50. (*ICRISAT: Patancheru, AP, India*)

Singh, G., Bhan, L.K. 1986. Chemical control of gray mold in chickpea. *International Chickpea Newsletter* 15, 18–20.

Singh, G., Kaur, L. 1990. Chemical control of gray mold of chickpea. *Plant Disease Research* 5, 132–137.

Singh, U.P., Sarma, B.K., Singh, D.P. 2003. Effect of plant growth-promoting rhizobacteria and culture filtrate of *Sclerotium rolfsii* on phenolic and salicylic acid contents in chickpea (*Cicer arietinum*). *Curr Microbiol.* 46: 131–140.

Singh, A., Jain, A., Sarma, B.K., Upadhyay, R.S., Singh H.B. 2013. Rhizosphere microbes facilitate redox homeostasis in *Cicer arietinum* against biotic stress. Ann. *Appl. Biol.* 163 (1) 33-46.

Singh, A., Jain, A., Sarma, B.K., Upadhyay, R.S., Singh H.B. 2014. Rhizosphere competent microbial consortium mediates rapid changes in phenolic profiles in chickpea during *Sclerotium rolfsii* infection. *Microbiol. Res.* 169(5-6): 353-60.

Singh, P.K., Singh, M., Agnihotri, V.K. and Vyas, D. 2013. Arbuscular mycorrhizal fungi: Biocontrol against fusarium wilt of chickpea. *Int. J. Sci. Res. Publ.* 3(1): 1-5. 131-40.

Singh, U.P., Sarma, B.K. and Singh, D.P. 2003. Effect of plant growth-promoting rhizobacteria and culture filtrate of *Sclerotium rolfsii* on phenolic and salicylic acid contents in chickpea (*Cicer arietinum*). *Curr. Microbil.* 46: 131-140.

Srivastava, A.K., Singh, T., Jana, T.K. and Arora, D.K. 2001. Induced resistance and control of charcoal rot in *Cicer arietinum* (chickpea) by *Pseudomonas fluorescens*. *Can. J. Bot.* 79(7): 787-795.

Srivastava, S., Patel, J.S., Singh, H.B., Sinha, A., Sarma, B.K., 2014. *Streptomyces rochei* SM3 induced stress tolerance in chickpea against *Sclerotinia sclerotiorum* and NaCl. *J. Phytopathol.* DOI: 10.1111/jph.12358

Statistics Canada, 2001. Field Crop Reporting Series 80(8). Minister of Industry, Ottawa, Canada, pp. 13–14 (Catalogue no. 22-002-X1B).

Statistics Canada, 2007. Field Crop Reporting Series 86(5). Minister of Industry, Ottawa, Canada, p. 19 (Catalogue no. 22-022-X1E).

Tabei, Y., Kitade, S., Nishizawa, Y., Kikuchi, N., Kayano, T., Hibi, T., Akutsu, K. 1998. Transgenic cucumber plants harboring a rice chitinase gene exhibit enhanced resistance to gray mold (*Botrytis cinerea*). *Plant Cell Reports* 17, 159–164. doi: 10.1007/s002990050371

Tahtamouni, M., Hameed, K. M. and Saadoun, I. 2006. Biological control of *Sclerotinia sclerotiorum* using indigenous chitinolytic actinomycetes in Jordan. *Plant Pathology Journal*, 22: 107-114.

Takatsu, Y., Nishizawa, Y., Hibi, T., Akutsu, K. 1999. Transgenic *Chrysanthemum* (*Dendranthema grandiflorum* (Ramat.) *Kitamura*) expressing a rice chitinase gene shows enhanced resistance to gray mold (*Botrytis cinerea*). *Scientia Horticulturae* 82, 113–123. doi: 10.1016/S0304-4238(99)00034-5

Trapero-Casas, A., Jimenez-Dýaz, R.M. 1986. Influence of sowing date on fusarium wilt and ascochyta blight of chickpea in Southern Spain. In: 'International Food Legume Conference'. (Eds O'Keeffe, L.E., Muehlbauer, F.J.) pp.-11. (*College of Agriculture, University of Idaho, Moscow/USA*)

Venette, J.R., Stack, R.W. 1987. First report of rust (*Uromyces ciceris-arietini*) on garbanzo (chickpea) in the Unites States. *Plant Dis.* 71: 101

Welty, L.E., Prestbye, L.S., Hall, J.A., Mathre, D.E., Ditterline, R.L., 1988. Effect of fungicide seed treatment and rhizobial inoculation on chickpea production. Appl. Agricult. Res. 3, 17–20.

Weltzien, H.C., Kaark, H.J., 1981. Epidemiological aspects of chickpea ascochyta blight. In: Proceedings of the Workshop on Ascochyta Blight and Winter Sowing of Chickpeas, ICARDA, Aleppo, 4–7 May, p. Syria.

Zachos, D.G., 1951. Studies on disinfection on chickpea seeds infected with *Ascochyta rabiei*. *Ann. l'Institut Phytopathologique Benaki* 5, 76–87.

Zachos, D.G., Panagopulos, C.G., Makris, S.A., 1963. Researches on the biology, epidemiology and the control of anthracnose of chickpea. *Ann. l'Institut Phytopathologique Benaki* 5, 167–192.

Chapter 3

Distribution, Symptomatology and Management of Wilt of Pigeonpea in Indian Perspective

R.K. Mishra, Naimuddin,
Monika Mishra & Shashi Tiwari

Pigeonpea [*Cajanus cajan* (L.) Millsp.], an important grain legume crop, is predominantly cultivated in tropical and subtropical regions of the world. India remains the largest producer and consumer of pigeonpea followed by Myanmar, Malawi and Kenya (FAO, 2019). In India, pigeonpea is the second most important legume after chickpea and the country alone contributes 62.6 per cent of world cultivated area with 63.4 per cent of world production (FAO, 2019). Maharashtra, Karnataka, Gujarat, Andhra Pradesh, Madhya Pradesh and Uttar Pradesh are the major pigeonpea growing States. The crop is affected by a variety of plant pathogens including fungi, bacteria, viruses, phytoplasma and nematodes at various stages of crop (Reddy *et al.*, 2012). Among the biotic stresses, Fusarium wilt caused by *Fusarium udum* Butler. is one of the most devastating soil borne diseases of pigeonpea worldwide (Kannaiyan *et al.*, 1984). In India, the disease was first reported by E.J. Butler in 1906.

However, the heavy losses due to this disease warrant for an urgent need to identify suitable and effective management strategies, which paves the way for increasing pigeonpea production in the country. Several methods have been exploited and recommended by the various researchers for management of this important disease. Management of wilt disease through chemicals is not feasible and also it causes several other adverse effect to human, animal and environments. Hence it is imperative to identify stable sources of resistance and exploit them to develop resistant varieties of pigeonpea through breeding approaches. A number of strategies to deal with wilt were identified over the past two decades by screening pigeonpea genotypes under national and international programs. Some of these sources have been effectively used in breeding programs and released as resistant varieties for disease prone areas (Nene and Kannaiyan, 1982; Okiror, 1999; Sharma and Pande, 2011; Sharma *et al.*, 2012a). Use of biocontrol agents offers an effective and environmentally safe strategy for the management of wilt. This needs to be exploited in combinations with other cultural interventions that are known to reduce with inoculums in the fields. The aim of this chapter is to review the symptomatology, epidemiology, variability and characterization of the pathogen and their integrated management through cultural, chemical, host plant resistance, biological and molecular approaches for effective management of wilt of pigeonpea.

Geographical Distribution

In India, the disease was first reported by E.J. Butler in 1906. It also occurs in east Africa, Malawi, Bangladesh, Grenada, Indonesia, Mauritius, Mynmar, Nepal, Nevis, Venezuela, Trinidad, and Tobago (Kannaiyan *et al.*, 1984; Reddy *et al.*, 1993; Marley and Hillocks, 1996). In recent, this pathogen was reported to be spreading in Southern Africa reaching areas in Mozambique (Southern Zambezia province) (Gwata *et al.*, 2006). The disease has also been reported in Zambia and Ghana, although the information on disease incidence and their distribution information is not available (Reddy *et al.*, 1993). In Kenya, the disease was first reported in 1983 when the first released variety (Munaa) broke down with Fusarium wilt and was withdrawn from the farmers (Kimani, 1991). The disease is found in all pigeonpea growing areas in Africa but incidences are high in the eastern areas (Kannaiyan *et al.*, 1984; Hillock and Songa, 1993).

Symptomatology

Wilt is one of the most destructive disease of pigeonpea throughout India caused by fungal pathogen *i.e., Fusarium udum* Butler. The main symptoms are wilting of seedlings and adult plants. The wilting starts gradually showing yellowing and drying of leaves following by wilting of whole infected plant. The affected plants can easily be recognized in patches in the field. Wilt appears on the young seedlings but mainly observed during flowering and podding stage. Disease symptoms resemble plants suffering from drought even with availability of plenty of soil moisture. The characteristic internal symptom of wilt is the browning to blacking of the xylem vessels from the root system to the stems. The xylem gradually develops black streak, and brown or dark purple bands appear on the stem of wilted plants expending upwards from the base of affected plants (Reddy *et al.*, 1993; Pandey *et al.*, 2013). Young infected plants may not show brown or purple bands, but internal browning or discoloration is seen under cross and longitudinal section. When the bark of such bands is peeled off, browning of the wood beneath can be seen. In advance stage of infection, aerial growth of the fungus may appear on basal portion of stem near the ground level as fine, white, cottony mycelia growth, along with slimy masses of pink spores (Reddy *et al.*, 1993). The details symptoms have been also described in detail by several researchers (Butler, 1918; Satyanarayana and Kalyansundaram,

Figure 3.1. Symptoms of Wilt on Pigeonpea Plants and Fields.

1952; Subramanian, 1963; Chaube, 1968; Amin *et al.*, 1976; Upadhyay and Rai, 1989; Nene *et al.*, 1989).

Losses Due to Wilt

Losses in pigeonpea due to wilt disease ranging from 0.2 to 100 per cent in India (Butler 1906, Patil 1984, Kannaiyan *et al.*, 1984 and Gade 2007, Vishwadhar *et al.*, 2005). Even though, the incidence of disease varied from 30 to 60 per cent at flowering and crop maturity stages (Kannaiyan and Nene, 1981), however many times it can also cause yield losses up to 100 per cent in susceptible cultivars (Kannaiyan *et al.*, 1984; Reddy *et al.*, 1990; Okiror, 1999; Dhar *et al.*, 2005, Kiprop *et al.*, 2002, Pandey *et al.*, 2011; Sharma *et al.*, 2006, Karimi *et al.*, 2012). The yield loss due to wilt is also depends upon the stage of the crop at which the plant wilt and it can approach over 50 per cent and even up to 100 per cent when wilt occurs at the pre pod stage (Okiror, 2002). Fusarium wilt causes economic loss in pigeonpea of about 97000 tonnes per year in India and 30,000 t of grain in Africa (Saxena *et al.*, 2010 and Joshi *et al.*, 2001).

Epidemiology and Disease Development

The incidence and severity of the disease depend primarily on soil conditions and the host cultivar. The incidence of the disease is more on vertisols and sandy soil than alfisols (Kannaiyan *et al.*, 1981). Slightly acidic to alkaline soils contains 50 per cent or more sand particles favour the wilt incidence in susceptible cultivars (Singh and Hussain, 1964). Upadhayay and Rai (1981b) made an attempt to correlate the soil physico-chemical properties with fungistatis and the wilt of pigeonpea. They concluded that less incidence of the wilt disease in soils of Southern states of India observed by Kannaiyan *et al.* (1984) might be due to a higher level of soil fungistatics against the pathogen. The disease incidence in a particular soil depends mainly on the saprophytic activity and survival of *F. udum* in soil which are favoured by continued presence of the host substrate. The disease becomes increasingly more severe when susceptible varieties of pigeonpea are grown in infested soils successively (Arora, *et al.*, 1992).

Soil temperature and moisture also play a significant role in occurrence of the wilt disease. Uphadhayay (1979) observed that 20-29°C temperature at 6-16 per cent moisture was most suitable for the disease development. Mundkur (1935), earlier reported a temperature range of 17-29°C favourable for the disease developemnt. Long and medium duration genotypes are more vulnerable than short and extra short duration genotypes probably because the infected plants wilts only after the basal half of the main stem is colonized by the fungus in about 3-4 months. In fact, development of perithecial stage of *Fusarium udum* is favoured by cloudy weather, high humidity and a combination of low and high temperature (Singh, 1983).

Early planting and vigorous plant growth favour the wilt incidence. The disease is not only more prevalent in early planted than late-planted crops, but is also more prevalent in irrigated than rain-fed crops (Chaudhary *et al.*, 2001). It is dispersed through irrigation, rain water and displacement of host debris by termites that feed frequently on dead wilted plants (Upadhayay and Rai, 1982). *Fusarium udum* propagules have been recorded from the bodies of termites which also feed on the wilted host roots (Upadhyay and Rai, 1982). Further, infected seed may be the primary means of spreading *F. udum* over a long distance into new areas where it could become a primary source of inoculum (Haware and Kannaiyan, 1992). Even though wilt incidence is also increased the nematode infestation in pigeonpea (Hasan, 1984, Siddiqui and Mahmood, 1996).

Disease Cycle

The disease is mainly soil borne in nature and sometimes it is also seed borne nature. The fungus is a facultative saprophyte. It can survive in the diseased plant derbies in the absence of the living host plants. Even though plants get infected at an early stage, they seem able to "keep fighting" with the fungus until flowering and podding. Therefore, pigeonpea plants are prone to wilt infection throughout their life cycle, but are most susceptible during the reproductive stage (Rangaswamy and Mahadevan, 1999).

The primary infection takes place through fine lateral branches of the roots and the fungus continues to grow in the xylem vessels. The fungal hypae penetrate the fine rootlets by the hyphae. The secondary infection in case of this disease is rare as a conidia produced on aerial parts of the host seldom succeed in causing infection. The pathogen mainly spread within the field or from one field to another field through irrigation water or through implements (Upadhayay and Rai, 1983). The pathogen may also be introduced into a new area through seed.

The disease cycle was known earlier to occur only through the imperfect state. But after the discovery of the perfect state or the sexual stage (Upadhayay and Rai, 1983), the disease cycle of the wilt may occurs both through imperfect (*F. udum*) and perfect (*G. indica*) stages. However, the asexual cycle is more common since the perfect stage does not frequently develop in nature. Depending on the environmental conditions, chlamydospores are formed both in parasitic and saprophytic phases from the hyphae and conidial cells. These chlamydospores serve as structures of survival during prolonged absence of the host and at the advent of favourable conditions; the resting spores germinate to initiate infection. After death of the infected plant the fungus lives saprophytically on the host substrates for a limited period and then forms chlamydospores to enter the dormant resting stage.

Causal Pathogen

The causal organism of pigeonpea wilt is *Fusarium udum* Butler, is a soil borne facultative parasite, specific to the crop (Upadhyay and Rai, 1989) and was first described by Butler (1910). A detailed description of the pathogen has also been provided by several other researchers' worldwide (Wollenwebber, 1931, Wollenwebber and Reinkings, 1935a and Subramanian, 1955, Booth, 1971, Booth *et al.*, 1978, Gerlach and Nirenberg, 1982, Upadhyay and Rai, 1989). Padwick (1940) also studied the cultural characteristics of *Fusarium udum* and proposed the name *Fusarium udum* Butler. var. *cajani*. However, the name of *Fusarium udum* is commonly accepted as the presence of macroconidia of *Fusarium udum* are distinguished by a prominent hook (Booth, 1971).

Variability, Characterization and Pathogenicity

The pathogen from the same site or diverse geographical origins have been shown to high variable in cultural and morphological characteristics (Reddy and Chaudhary, 1985; Gaur and Sharma, 1989) and their virulence or pathogenicity on pigeonpea genotypes (Soko *et al.*, 1995; Baldev and Amin, 1974; Shit and Gupta, 1978; Nene *et al.*, 1981; Okiror, 1986; Gaur and Sharma, 1989; Okiror and Kimani, 1997; Mayer *et al.*, 1998; Kiprop *et al.*, 2002; Parmita *et al.*, 2005). Variability of fusarium wilt reactions between countries and even sites within the same country is due to existence of different virulent isolate and environmental factors (Songa *et al.*, 1995; Hillock *et al.*, 2000). These high variations in cultural and morphological characteristics of this pathogen could be due to environmental conditions, age of the isolates, sub culturing, method of storage and culturing conditions Kiprop *et al.* (2002). According to Okiror and Kimani (1997) and Kiprop *et al.* (2002) a wide variation in virulence (pathogenecity) in different genotypes of pigeonpea among *F. udum* isolates could be due to environmental conditions and inoculation techniques used. Baldev and Amin (1974) tested 10 isolates of *F. udum* from various sources on 10 pigeonpea lines. Only three genotypes showed resistance to all the isolates. They also characterized these isolates as the races of this pathogen. The existence of different variants/races of *F. udum* is a major drawback for breeding programs for wilt resistance (Okiror and Kimani, 1997, Mishra and Dhar, 2003; Mishra, 2004; Singh *et al.*, 2011; Tiwari and Dhar, 2011). *F. udum* isolates from diverse geographical origins have been shown to exhibit high variability in their virulence (Mishra and Dhar, 2003). Eventhough, the pathogen is highly variable in morphological, cultural and physiological characteristics, production of enzymes and virulence (Upadhyay and Rai, 1989) indicating existence of of physiologic races. There is no relationship between cultural characteristics and aggressiveness. Mishra and Vishwadhar (2003) characterized 17 isolates of *F. udum* on the basis of variation in size and septation of macro-conidia in to three group's *viz.*, smaller conidia, medium conidia and large conidia. They also observed that isolates with large conidia and more septation were most virulent causing 100 per cent mortality; medium conidia were moderately virulent causing 76.5 per cent wilting and smaller conidia were less virulent causing 55.5 per cent wilting in inoculated plants.

There has been conflicting reports on the diversity of the pathogen using molecular markers. Conflicting reports on the relationship between cultural characteristics and the virulence of the isolate are also available. While studies of genetic diversity using isozyme markers revealed low variation in *F. udum* isolates (Shit and Sen Gupta, 1980; Okiror and Kimani, 1997), Kiprop *et al.* (2002) using AFLP reported high variability. The available reports on the differential reactions of certain pigeonpea genotypes to different isolates raises issues on whether isolates qualify to be called races of *F. udum* or not. However, apart from studies aimed at characterizing *F. udum* isolates (Kiprop *et al.*, 2002), there are few reports attempting to characterize the races of *F. udum*. Most of the available reports relate different isolates to different races (Gwata *et al.*, 2006).

Researchers of ICAR-IIPR, Kanpur carried out the extensive survey and collected several diseased samples from pigeonpea growing areas of the country for study of *Fusarium udum* variability present in the country. More than 700 isolates of *Fusarium udum* isolated and well characterized on their cultural, morphological characteristics and relative pathogenicity. On the basis of virulence and geographical locations, 50 most virulent isolates were selected for further grouping into races using a set of 7 pigeonpea genotypes as differentials (BAHAR, C 11, ICP 8863, ICP 7035, BDN 1, KPL 44 and ICP 9174) under the ambit of the National Network Project on wilt of crops (ICAR-Phytofura). Based on reaction of these differential genotypes, these 50 isolates were categorized in to 5 distinct races (Mishra *et al.*, 2017). Prevalence of all these five races has been indicated in major pigeonpea growing areas in the country. However, some isolates could not be grouped into these five races indicating the presence of even more races. This requires the identification of more differentials to further elucidate pathogenic variability of *F. udum*. Findings established that a vast diversity among isolates of *F. udum* representing different parts of the country.

Management Approaches

Cultural Methods

Crop rotation is undoubtedly the best way of elimination soil borne infection. Significant reduction in pigeonpea wilt (25-55 per cent) has been obtained through crop rotation and intercropping/mixed cropping with sorghum (Dhar, 2003). Sorghum, pearl millet, cotton, castor, maize and groundnut inhibit the soil population of *F. udum* if cultivated with pigeonpea as inter- and mixed-cropping systems, because root exudates of such crops acted as barrier for the pathogen. Mixed-cropping with sorghum (Jowar) is most effective and practical solution of this problem. Natarajan *et al.* (1985) have been recorded that, pigeonpea intercropping with sorghum produced a large reduction in wilt incidence in the first year (down to 55 per cent) and thereafter it stabilized at about 20–30 per cent regularly. Similar finding were also found by Singh, 1983, a 4-5-year rotation has been found to be free the field completely by the wilt pathogen.

Soil solarizaion and organic amendments are known to favour native antagonists to suppress the soil borne diseases (Mukhopadhyay, 1994, Katan 1981, 2010, 2017). The mode of action of organic amendments leading to plant disease control and stimulation of microorganisms is complex and dependent on the nature of the amendments (Akhtar and Malik, 2000).

Rai and Singh (1995) tested different oil cakes *viz.*, neem, mustard, mahua, coconut, linseed and sesamum at concentrations of 0.25, 0.5, 1.0 and 2.0 per cent against radial growth of *F. udum* and reported that neem, mustard and mahua cakes were most effective in reducing fungal growth. Further, they also reported that neem oilcake was most effective in controlling wilt incidence. The soil composition and certain cultural practices can affect wilt incidence (Upadhyay and Rai, 1981). Inhibition of radial growth of *F. udum* in castor cake amended soil was also reported (Singh and Singh, 1982).

Host Plant Resistance

Screening of Pigeonpea against Wilt

The pigeonpea genotypes/cultivars must be screen in the wilt sick-field under artificial epiphytotic conditions for identification of resistant sources. Sowing should be done during second fortnight of July. Each genotype was planted in 2 rows of 4-5m length with row to row spacing of 60 cm and plant to plant spacing of 20cm. Bahar variety used as susceptible check even some local check also be used if available. Data on wilt incidence was recorded from each entries/varieties regularly from November onwards just after appearance of the disease. Percentage wilt incidence of each test genotype was calculated by following formula:

$$\text{Disease incidence (per cent)} = \frac{\text{Number of infected plants}}{\text{Total number of plants}} \times 100$$

Data on seedling mortality in each test genotype was recorded and wilt incidence (per cent) was worked out. The genotypes were graded as resistant (0-10 per cent wilt incidence), moderately resistant (10.1-25 per cent), susceptible (25.1-50 per cent), and highly susceptible (>50 per cent wilt) as per the key recommended by All India Co-ordinated Project on Pigeonpea (AICRP).

Identified Resistant Cultivars/Donors

The resistance breeding programme in pigeonpea against wilt was started in 1976 (Nene and Kannaiyan, 1982) and the reported several resistant sources for wilt. The deployment of cultivars with resistance to Fusarium wilt remains the most effective means of wilt disease management. The search for sources of resistance to wilt in pigeonpea began as early as 1905 in India. A large number of wilt resistant/ tolerant cultivars are available in long duration pigeonpea developed by Indian Institute of Pulses Research, Kanpur, ICRISAT and different SAU,s under AICRP (Table 3.1). Some of these genotypes *i.e.*, ICP 8863, AWR 74/2, Banda Palera, DA 11. Varieties C-11, C-28, C-36, F-18, NP-15, NP-38, NP-41, and T-17 had been considered as resistant to wilt across the seasons and locations (Nene and Kannaiyan, 1982,

Amin *et al.*, 1933a, Reddy *et al.*, 1995, Kimani, 1991, 1994; Songa *et al.*, 1995; Silim *et al.*, 2005; Gwata *et al.*, 2006; Karimi *et al.*, 2010, Dhar, and Reddy, 2004, Singh and Rai, 2010).

Table 3.1: Wilt Resistant/Tolerant Varieties of Pigeonpea Released in India

Varieties	*Zone-wise Recommendation*
Amar	Central Zone (CZ), West Zone (WZ), North West Plain Zone (NWPZ)
NP (WR) 15	Central Zone (CZ), West Zone (WZ), North West Plain Zone (NWPZ)
BDN1	Central Zone (CZ), West Zone (WZ)
BDN2	Central Zone (CZ), West Zone (WZ)
Mukta	North West Plain Zone (NWPZ), North East Plain Zone (NWPZ)
Sharda	Central Zone (CZ), North West Plain Zone (NWPZ)
C11	Central Zone (CZ), West Zone (WZ)
TT6	South Zone (SZ)
Maruti (ICP8863)	South Zone (SZ), Central Zone (CZ)
Asha (ICP87119)	South Zone (SZ), Central Zone (CZ)
N P 5	Central Zone (CZ), West Zone (WZ), North West Plain Zone (NWPZ)
IPA 203	North East Plain Zone (NEPZ)
IPA 206	North East Plain Zone (NEPZ)

Besides this, some of the promising lines of IIPR (DPPA 85-1, DPPA 85-2, DPPA 85-3, DPPA 85-7, DPPA 85-8, DPPA 85-12, DPPA 85-13 and DPPA 85-14) have shown resistant reaction against wilt under field conditions and have been identified as potential donors (Singh *et al.*, 1998, Mishra and Vishwadher, 2005, Mishra *et al.*, 2016, 2017).

Chemical Methods

The wilt caused by *Fusarium* spp. are primarily soil borne hence, seed treatment with fungicide is a genuine method to manage disease effectively (Vyas, 1993). Because of the vascular occurrence of *F. udum* it is necessary to find a systemic fungicide which could be absorbed by the host plant and transported via the xylem or phloem system to the site of pathogen activity, where eradication or effective suppression of the fungus could take place (Kulka and Schmeling, 1987). Rohodes and Long (1974) reported that carbendazim is immobile in soil and do not leach or move significantly from the site of application. This favours to exploit the possibility of use of carbendazim as seed treatment even at lower doses. Vyas (1984) reported that *Fusarium* spp. were more sensitive to systemic fungicides. Khan *et al.* (2005) reported 51 per cent reduction in wilt and 20 per cent increase in yield of pigeonpea by the seed application of carbendazim @ 2g/kg seed.

Biological Control Methods

In view of the adverse effect of chemicals to the environment and increasing interest in sustainable agriculture, biological control has been reported as an effective, economical and environmental friendly alternative tool for management of soil borne pathogens (Parker *et al.,* 1985). Biological control of plant diseases using antagonistic microorganisms offers a highly Biological control involves use of microorganisms to reduce the population of harmful microorganisms. It is based on the natural phenomenon, where the nature through its own way keeps the population of each microorganism under control and prevents the disease from getting catastrophic. The mode of action of the antagonistic organisms against various soil-borne plant pathogenic fungi includes biosynthesis of antibiotics, production of hydrolytic enzymes (Velzhahan *et al.,* 1999). Production of siderophore and competition for substrates, 2,4-diacetyl phloroglucinol, which inhibits growth of phyto-pathogenic fungi (Nowak-Thomson *et al.,* 1994) and Induced systemic resistance (Sriram *et al.,* 2009). The use of *Trichoderma* species as biological control agents has been investigated for over 70 years but it is only relatively recently that strains have become commercially available (Jayalakshmi, *et al.,* 2003, Lumsden *et al.,* 1993; Monte, 2001). Many *Trichoderma* species *i.e., T. harzianum, T. asperellum, T. longibrachiatum* and *T. ressei* have been identified from pulses rhizosphere and having biocontrol potential against Fusarium wilt of pigeonpea (Mishra *et al.,* 2015, 2016, 2018). Several researchers working on biocontrol aspect are also reported that the local isolate of *Trichoderma* spp. having the potential ability in inhibiting the mycelial growth of *F. udum* under *in vitro* and *in situ* condition (Jayalakshmi *et al.,* 2003, Chaudahry and Prajapati, 2004, Biswas and Das, 1999, Kumar *et al.,* 2009, 2014 and Chaudhary *et al.,* 2017). Vishwadhar *et al.* (2006) studied the efficacy of the bioagents against ten *F. udum* isolates *in vitro* and reported that after 96 hour of incubation, the per cent inhibition of ten isolates of *F. udum* ranged between 35.5-54.8 by, 36.4 - 54.7 by *T. harzianum* and 36.4 - 57.3 by *T. virens.*

Soil amendment with *Trichoderma harzianum* at all pathogen levels has been reported to give a disease control of 22 per cent -61.5 per cent (Prasad *et al.,* 2002). Studies on antagonism, found that *Aspergillus niger, A. flavus, Micromonospora globosa* and *A. terreus* highly suppressed the population of *F. udum* (Upadhyay and Rai, 1981). Sunitha *et al.* (1995) observed a wide zone of inhibition with the use of *T. harzianum* and *Bacillus subtilis* against *F. udum*. Seed treatment with *T. viride* (4g/kg seed) also reduced wilt incidence (Jayalakshmi, *et al.,* 2003). Prasad *et al.* (2002) found that soil application of *T. harzianum* was more effective than seed treatment for suppression of pigeonpea wilt. They also found that even at the highest pathogen density (log 5.34 cfu/g of soil), soil treatment with *T. harzianum* at 10g/m^2 area of seed gave 30 per cent disease reduction at field level. In seed treatment plots, disease control ranged between 4.36 per cent and 13.7 per cent. Singh (1983) reported the production of 'bulbiformin' by *B. subtilis*, which inhibits the growth of *F. udum.* The combination of carbendazim seed treatment of pigeonpea @ 2g/Kg of seeds + soil application of *P. fluorescens, T. viride* each @ 2.5Kg/ha and FYM @ 50Kg/ha recorded least mean wilt incidence of 7.25 per cent with mean yield of 1203.17Kg/ha (Mahesh *et al.,* 2010).

Molecular Approaches

In recent years, many developments were seen in the area of pulses genomics which have resulted in access to a wide variety of modern genomic tools and technologies for pulses improvement. From the point of resistance breeding, DNA markers have emerged as the promising molecular tools to effectively and precisely address the problem of biotic stresses, in particular disease resistance that severely constrain productivity of these crops. Diverse molecular marker systems have been employed across various pulse crops for molecular mapping of resistance to various diseases. Consequently, DNA markers tightly linked with the disease resistance genes/QTLs were made available for the scientific community (Bohra *et al.*, 2014a, b, Varshney *et al.*, 2013, 2015). Once genomic location of a trait linked gene/QTL/ marker is known, molecular breeding techniques like marker assisted backcrossing (MABC) can be implemented to transfer the gene of interest to different genotypes. MABC is particular suitable for transferring the genomic segments that govern substantial amount of phenotypic variation for the trait (Bohra 2013). In case of transfer of disease resistance from donor species to elite lines, MABC has been very remarkably successful.

Conclusions and Future Strategies

Earlier research mainly focused on resistant sources and chemical control for management of any diseases. Wilt is one of the most devastating soil borne pathogens and it is difficult to manage this disease after appearance in the fields. Now the major emphasis is placed to identify most effective, economically viable management tools and evaluating and integrating location specific components of disease management module. Future the management strategies will also account for effects of climate variability-on pathogen dynamics that is contributing to emergence and resurgence of several other disadvantages. HPR, fungicides, natural plant products, biofungicides, botanicals and agronomic practices will remain the potentially viable options for management. In addition, the future modules will incorporate the latest advancements witnessed towards identification of biocontrol agents and products delivered through biotechnological/breeding interventions such as MAS, genetic engineering, and wide hybridization to develop cultivars with enhanced resistance to diseases. In parallel, a renewed attention is needed on using information technology for disease modeling, developing decision support systems, and utilization of remote sensing to up-scale and disseminates integrated disease management technologies.

REFERENCES

Amin, K. S., Balder, B. and Williams, F. J. (1976). Differentiation of phytopathora stem blight from Fusarium wilt of pigeonpea by field symptoms. *FAO Plant Protection Bulletin*, 24: 123- 124.

Biswas, K.K. and Das, N.D. (1999). Biological control of Pigeonpea wilt caused by Fusarium udum with *Trichoderma* spp. *Ann. Pl. Prot. Sci.* 7: 46-50.

Bohra A (2013) Emerging paradigms in genomics-based crop improvement. Sci World J 585467: 17

Bohra A, Jha UC, Kavi Kishor PB, Pandey S, Singh NP (2014b). Genomics and molecular breeding in lesser explored pulse crops: current trends and future opportunities. *Biotechnol Adv* 32: 1410–1428

Bohra A, Pandey MK, Jha UC, Singh B, Singh IP, Datta D, Chaturvedi SK, Nadarajan N, Varshney RK (2014) Genomics-assisted breeding in the four major pulse crops of developing countries: Present status and prospects. *Theoretical and Applied Genetics*, 127: 1263-91

Booth, C. (1971). The genus Fusarium. Common Wealth Mycological Institute, England pp 137. Booth, C., Mordue, J. E. M. and Gibson, I. A. S. (1978). CMI descriptions of pathogenic fungi and bacteria, Section 58: 571-580.

Butler, E. J. (1906). The wilt disease of pigeonpea and pepper. *Agril. J. India*, 1 : 25-36.

Butler, E. J. (1908). Selection of pigeonpea for wilt disease. *Agril. J. India*, 3 : 182-183.

Butler, E. J. (1910). The wilt disease of pigeonpea and parasitism of Neocosmospora *vasinfecta* Smith. Memories of the Department of Agriculture in India, Botanical Series, 2: 1-64.

Butler, E. J. (1918) (Reprinted 1973) Fungi and Plant Diseases. pp 276.

Chaube, H. S. (1968). Combating diseases of arhar and gram. *Indian Farmers Digest* 1: 26-27.

Chaudhary Balkishan, Kumar Sanjeev and Kushwaha SK (2017) Bio-efficacy of *Trichoderma* species against Pigeonpea wilt pathogen. *Journal of Applied and Natural Science* 9 (4): 2327 – 2331

Chaudhary, R. G. and Prajapati, R. K. (2004). Comparative efficacy of fungal, biological agents *Fusarium udum*. *Annals of Plant Protection Sciences*. 12(1): 75-79

Dhar, V., Datta, S., Chaudhary, R.G., Upadhyay, J.P., Saifulla, M., Mishra, S., Prajapti, R.K. and Shamim, M. (2012). Pathogenic and molecular characterizations of pigeonpea wilt pathogen *Fusarium udum*. *Archives of Phytopathology and Plant Protection*, 45: 423-436

FAOSTAT 2016. FAOSTAT Database. Available online at: http: / /www.fao.org/

Gade, R. M., Zote, K. K. and Mayee, C. D. (2007). Integrated management of pigeonpea wilt using fungicide and bioagents. *Indian Phytopath.*, 60 : 24-30.

Gwata ET, Silim SN, Mgonja M. 2006. Impact of a new source of resistance to Fusarium wilt in pigeonpea. *Journal of Phytopathology* 154, 62-64.

Hasan A (1984). Synergism between *Heterodera cajani* and *Fusarium udum* attacking *Cajanus cajan*. *Nematol. Medit*.12: 159-162.

Haware, M. P. and Kanaiyan, J. (1992). Seed transmission of Fusarium udum in pigeonpea and its control by seed treatment fungicide. Seed Sci. Tech., 20: 597-601

Hillocks RJ and Songa WA. 1993. Root-knot and other nematodes associated with pigeonpea plants infected with Fusarium udum in Kenya. *Afro-Asian Journal of Nematology* 3, 143-147

Jayalakshmi, S. K., Sreeramulu and Benigi, V. I. 2003 Efficacy of Trichoderma spp. against pigeonpea wilt caused by *Fusarium udum* Butler. *Journal of Biological Control*, 17: 75–78

Joshi, PK, Rao P, Gowda, CLL, Jones, RB, Silim, SN, Saxena, KB and Kumar J (2001). The world chickpea and pigeonpea economies: facts, trends, and outlook. ICRISAT, Patancheru, Andhra Pradesh, India: 68 pp.

Kannaiyan, J., Nene, Y. L., Raju, T. N. and heila V.K. 1981. Screening of resistance to Phytophthora blight of pigeonpea. *Plant Disease*. 65: 61-62.

Kannaiyan, J., Nene, Y. L., Reddy, M. V., Ryan, J. G. and Raju, T. N. 1984. Prevalence of pigeonpea diseases and associated crop losses in Asia, Africa and Americas. *Trop. Pest Management* 30: 62-71.

Karimi, James Otieno Owuoche, Said Nandy Silim (2012). Importance and management of fusarium wilt (*Fusarium udum* Butler) of pigeonpea. *International Journal of Agronomy and Agricultural Research*. 2 (1): 1-14.

Katan J. (1981). Solar heating (solarization) of soil for control of soil-borne pests. *Annual Review Phytopathology* 19: 211-236

Katan J. (2010). Cultural approaches for disease management: present status and future prospects. *Journal of Plant Pathology*. 92: S4.7-S4.9

Katan J. (2017). Diseases caused by Soilborne Pathogens: Biology, Management and Challenges. *Journal of Plant Pathology*. 99 (2): 305-315.

Kimani PM. 1991. Pigeonpea improvement research in Kenya: An Overview. Pp. 108-117. In: Singh, Laxman, Silim, S.N., Ariyanagam, R.P., and Reddy, M.V. (eds). Proceedings of the First Eastern and Southern Africa Regional Legumes (Pigeonpea) workshop, 25-27 Jun 1990, Nairobi, Kenya. Eastern Africa Regional Cereals and Legumes (EARCAL) Program, International Crops Research Institute for the Semi-Arid Tropics.

Kiprop, E. K., Baudoin, J. P., Mwang-ombe, A. W., Kimani, P. M., Mergeai, G. and Maquet, A. (2002). Characterization of Kenyan isolates of *Fusarium udum* from pigeonpea [*Cajanus cajan* (L.) Millsp.] by cultural characteristics, aggressiveness and AFLP analysis. *Journal of Phytopath.*, 150 : 517-525.

Kumar, S., Thakur, M. and Rani, A. (2014). Trichoderma: Mass production, formulation, quality control, delivery and its scope in commercialization in India for the management of plant diseases. *Afr. J. Agric. Res.* 9: 3838- 3852.

Kumar, S., Upadhyay, J.P. and Rani, A. (2009). Evaluation of Trichoderma species against *Fusarium udum* Butler causing wilt of Pigeonpea. *J. Biol. Con.* 23: 329-332

Marley PS, Hillocks RJ. 1996. Effect of root-knot nematodes (Meloidogyne spp.) on fusarium wilt in pigeonpea (Cajanus cajan). *Field Crops Research*, 46, 15 - 20.

Mishra R. K, Naimuddin, Monika Mishra and Sandeep Sharma. 2016. *Trichoderma*: Potential biocontrol agents for pulses. *Dalhan Alok* 14: 80-86.

Mishra R.K, Naimuddin, PR Saable, Satheesh Naik, Abhishek Bohra, PG Patil, F. Singh and IP Singh (2016). New donors for wilt in pigeonpea. *Pulses News Letter* 27 (1): 8

Mishra RK, Bohra A, Naimuddin, Kumar Krishna, Sujayanand GK, Saabale PR, Naik SJ Satheesh, Sarma BK, Kumar D., Mishra Monika, Srivastava DK and Singh NP. 2018. Utilization of biopesticides as sustainable solutions for management of pests in legume crops: achievements and prospects. *Egyptian Jr. Biological Pest Control.* DOI 10.1186/s41938-017-0004-1.

Mishra RK, Mishra Monika, Naimuddin and Kumar Krishna (2018). *Trichoderma asperellum:* A potential biocontrol agents against wilt of pigeonpea caused by *Fusarium udum* Butler. *Journal of Food Legumes* 31(1): 50-53

Mishra RK, Naumuddin, Akram Mohd., Sachan DK, Mishra Monika and Kumar Krishna. 2016. Characterization of Indigenous *Trichoderma* spp. through ITS based sequences. *Pulses News Letter.* **27** (2): 5.

Mishra RK, Saabale PR, Naimuddin, Jagadeeswaran R, Mishra O. 2015. Potential *Trichoderma* sp from Pulses Rhizosphere. *Pulses News Letter.* April-June pp: 3.

Mishra, R.K., Naimuddin, P.R.Saabale, Satheesh Naik S. J, Abhishek Bohra, F. Singh and IP Singh (2016). Evaluation of promising lines of pigeonpea for resistance to wilt caused by *Fusarium udum* Butler. *J. Food Legumes.* 29 (1): 64-66

Mishra, R.K., P.R.Saabale, Naimuddin, SatishNaik, Abhisek Bohra, Farindra Singh and I.P. Singh (2016) Identification of New Sources of Resistance for Wilt of Pigeonpea *Pulses News Letter.*

Mundkur BB (1935). Influence of temperature and maturity on the incidence of sunhemp and pigeonpea wilt at PUSA. *Indian J. Agric. Sci.,* 5: 606-618.

Natarajan M, Kannaiyan J, Willey RW and Nene YL (1985). Studies on the effects of cropping system on fusarium wilt of pigeonpea. *Field Crops Research.* 10: 333-346

Nene, Y.L. and Kannnaiyan, J. (1982). Screening pigeonpea for resistance to Fusarium wilt. *Plant Disease,* 66: 306-307.

Nene, Y.L., Reddy, M.V., Beniwal, S.P.S., Mahmood, M., Zote, K.K., Singh, R.N. and Sivaprakasam, K. (1989). Multilocational testing of pigeonpea for broad based resistance. *Indian Phytopathology,* 42: 44-48

Okiror M. A. (1999). Evaluation of pigeonpea (*Cajanus cajan*) germplasm for resistance to Fusarium wilt. *Indian J. Agric. Sci.* 69, 600–601.

Padwick, G. W. (1940). "Genus Fusarium" 5 : *Fusarium udum* Butler, *F. vnsinfectum. Indian J. Agric. Sci.,* 10: 863-866

Patil, P. Y. (1984). Study on pigeonpea wilt. Ph. D. thesis, Mahatma Phule Agricultural University, Rahuri, India, pp. 125-131.

Prasad, R. D., Rangeshwaran, R., Hegde, S. V and Anuroop, C. V. 2002. Effect of seed and soil application of *Trichoderma harzianum* on pigeonpea wilt caused by *Fusarium udum* under field conditions. *Crop Protection,* 21: 243–297.

Reddy MV, Raju TN, Sharma SB, Nene YL, McDonald D. 1993. Handbook of pigeobpea diseases, (In En. Summaries in En. Fr.). Information Bulletin. 14. Patancheru, A.P. 502 324, India: International Crops Research Institute for the Semi-Arid Tropics. p. 64

Reddy, M. V., Sharma, S. B. and Nene, Y. L. (1990). Pigeonpea: Disease management. In: The Pigeonpea (Eds. Y. L. Nene, S. D. Hall and V. K. Sheila), CAB International, Wallingford, U.K., 303-347 pp.

Reddy, MV, Raju, TN, Sharma, SB, Nene, YL, Mcdonald, D, Pande, S and Sharma, M (2012). Handbook of Pigeonpea Diseases (Revised). Inf. Bull. No. 42. Patancheru, A. P. India: ICRISAT. 64 pp.

Satyanarayana, G. and Kalyansundaram, R. (1952). "Soil conditions and root diseases. Symptomology of wilted cotton and redgram". *Proceedings of Indian Academy Science* (B): 36: 54.

Saxena, RK, Saxena, KB, Kumar, RV, Hoisington, DA and Varshney, RK (2010). Simple sequence repeat-based diversity in elite pigeonpea genotypes for developing mapping populations to map resistance to Fusarium wilt and sterility mosaic disease. *Plant Breed.*, 129: 135-141.

Sharma M. and Pande S. (2011). New sources of resistance to Fusarium wilt, sterility mosaic disease and Phytophthora blight in vegetable pigeonpea germplasm. *Indian J. Plant Prot.* 39, 288–293.

Sharma M., Rathore A., Mangala U. N., Ghosh R., Sharma S., Upadhyay H. D. (2012a). New sources of resistance to Fusarium wilt and sterility mosaic disease in a mini-core collection of pigeonpea germplasm. *Eur. J. Plant Pathol.* 133, 707–714.

Siddiqui ZA, Mahmood I (1996). Effects of *Heterodera cajani, Meloidogyne incognita* and *Fusarium udum* on the wilt disease complex of pigeonpea. *Indian J. Nematol.* 26: 102-04

Singh, I.P., Chaudhary, R.G. and Dhar, V. (1998). Inheritance of resistance to Fusarium wilt in pigeonpea. *Indian Journal Agricultural Sciences*, 68: 729-731

Subramanian, C. V. (1955). Studies on south Indian Fusaria IV. 'The Wild Type' in *Fusarium udum* Butler. *J. Indian Bot. Soc.* 34: 29.

Subramanian, S. (1963). Fusarium wilt of pigeonpea I symptomology and Infection Studies. Proceedings of Indian Academy of Science (Section B): 57: 134.

Upadhyay RS and Rai B (1982). Ecology of *Fusarium udum* causing wilt disease of pigeonpea: population dynamics in the root region. *T. Brit. Mycol. Soc.*, 78: 209

Upadhyay RS and Rai B. (1989). Wilt disease of pigeonpea and its causal organism *Fusarium udum*. In: Perspective of Phytopathology (Eds. Agnihotri, V. P., Singh, U.S., Chaube, H. S., Singh, N. and Dwivedi, T. S. Today and Tomorrow's Printers and Publishers. New Delhi. pp. 388.

Upadhyay RS and Rai B. 1981. Effect of cultural practices and soil treatments on incidence of wilt disease of pigeonpea wilt. *Plant and Soil* 62: 309-312

Upadhyay, R. S. (1979). Ecological studies on *Fusarium udum* Butler causing wilt disease of pigeonpea. Ph.D. thesis, Banaras Hindu University, India.

Upadhyay, R. S. and Rai, B. (1981b). Fungistatic activity of different Indian soils against *Fusarium udum* Butler. *Plant and Soil*, 63: 181-187.

Vishwadhar, Mishra S and Chaudhary RG (2006). Potential efficacy of bioagents against *Fusarium udum* isolates. *Indian Phytopathology*, 59: 290–293

Vishwadhar, Reddy, MV and Chaudhary, RG (2005). Major Diseases of Pigeonpea and their Management. In: Advances in Pigeonpea Research, Eds. Masood Ali and Shiv Kumar, Indian Institute of Pulses Research, Kanpur, India, pp. 229-261.

Vishwadher, Reddy MV and Chaudhary RG (2005). Major Diseases of Pigeonpea and their management. In: Advances in Pigeonpea Research, (Eds. Ali, M., Kumar, S.) IIPR, Kanpur, India, pp. 229-261.

Wollenwebber H W and Reinkinn O. A (1935a). Die Fusarien. Ihre Beschreibung, schadwirking und Bckarnpfung. Berlin : Paul parey,1935, pp: 335.

Wollenwebber, H. W. (1931). Fusarium-Monographee Fungi parasitici *etc.* saprophytici. Z, Parasitenkol. 3: 31: 269.

Chapter 4

Wilt Disease of Lentil and Options for Management

Naimuddin, Mohd. Akram, R.K. Mishra & Shameem Ahmad

Lentil, botanically known as *Lens culinaris* Medikus ssp. *culinaris* is one of the oldest legumes known to mankind. It is grown and consumed in many countries of the world. Its grains are rich sources of protein and micronutrients. In many Middle Eastern countries lentil straw is much valued as animal feed. Its botanical name *Lens* is derived from the resemblance of lentil seeds in shape to the lens of eye. As far as introduction of lentil into India is concerned, it seems to have arrived with the Sanskrit speaking people- the Aryans. In Sanskrit, lentil has many names *viz., Masura, Renuka, Mangalaya etc.* In Hindi it is known as *Masur* or *Masoor*. Of the various diseases known to affect lentil, wilt caused by *Fusarium oxysporum* f.sp. *lentis* is the most widely occurring and therefore assumes great economic importance in South Asia, Sub-Saharan Africa and West Asia and North Africa (WANA) region (Kumar *et al.,* 2010). In India, this disease is widely spread in all lentil growing areas and is the single most important biotic factor that constitutes a major threat to lentil production especially in lentil growing areas particularly in central India. The wilt disease is so devastating year after year that the growers in many parts of "Bundelkhand" area of U.P. has no option but to abandon cultivation of lentil.

Of the various options, use of host plant resistance is the best strategy to ameliorate the losses caused by wilt. Most of the recently released varieties of lentil have the potential of 20 q/h; but the productivity of this crop remains as low as 6-7 q/h. Efforts to develop wilt resistant varieties of lentil are continuing and it is heartening to note that these efforts have resulted in the release of many varieties with resistance to wilt. Variability in wilt pathogen has been reported; however, studies needs to be up scaled before existence of races is claimed. At present, use of resistant genotypes and seed treatment with fungicides or bio-control agents are the most advocated means of wilt management in lentil.

Occurrence

Wilt disease, one of the most important diseases of lentil is present in all the lentil growing countries. Wilt disease of lentil was for the first time described from Hungry in the year 1937 (Fleischmann, 1937). There was, however, no evidence of pathogen involved. A wilt disease caused by *F. orthorcerus* was reported from Peru by Carrea (1940). Soon after, Padwick in 1941 observed wilt of lentil in Delhi and Karnal and showed association of *Fusarium* sp. Few years later, a wilt disease was found damaging lentil crop in Kanpur (Dey (1948).

The first authentic and detailed description of wilt disease of lentil was published by Vasudeva and Srinivasan (1952) who noted about 67 per cent wilt incidence in lentil crop in Delhi. This disease has been reported to occur in America (Wilson and Brandsberg, 1965), Czechoslovakia (Ujevic *et al.*, 1965) erstwhile USSR (Kotova 1965), France (Moreau 1978), Myanmar and Pakistan (Bahl *et al.*, 1991), Nepal (Karki, 1991), Syria (Bayaa *et al.*, 1986; El Ahmad and Mouselli, 1987), Egypt (El Morsy *et al.*, 1997; Abou Zeid *et al.*, 1997), Algeria (Belabid *et al.*, 2000; Setti and Bouznad, 1998), Ukraine (Rodigni,1973), Turkey (Soran, 1979), Poland (Pieta and Pastucha, 1997). Wilt disease caused by *Fusarium oxysporum* f.sp. *lentis* has been reported from Italy also (Tosi and Cappelli, 2001). Presently, wilt disease of lentil is considered a serious disease in parts of South America, the Mediterranean basin and south Asia (Erskine and Saxena, 1991). In India, wilt is the main problem in states of Uttar Pradesh, Madhya Pradesh, Himachal Pradesh, Bihar, West Bengal, Assam, Rajasthan, Haryana and Punjab and other area where lentil is grown (Agrawal *et al.*, 1991). In Madhya Pradesh wilt incident as high as 50-78 per cent has been reported in some fields (Khare *et al.*, 1979; Agrawal *et al.*, 1991).

Yield Losses

There seems to be no systematic studies on the extent of losses due to wilt disease of lentil. It is generally considered that the yield loss is equal to the per cent plant mortality *i.e.* 50 per cent plant mortality due to wilt is expected to cause 50 per cent yield loss. This however, may not be practical because some plants killed at advance stage give some yield. In Syria, Erskine *et al.* (1989), however, showed a correlation between wilt incidence and grain yield and found 8.77 per cent yield loss for every 10 per cent increases in wilt incidence. This model has been used for crop loss assessment at formers' field. In India, no such study seems to have been carried out and generally the per cent incidence of wilt is converted to per cent approximate yield loss. Most devastating form of wilt occurs when crop is approaching maturity and the disease leads to mortality of affected plant often before seed formation. In affected plant, if seed had been formed they are few in numbers and may carry seed borne infection.

Symptoms

Wilt disease occurs at both pre emergence as well as post emergence stages. In case post emergence infection, plants are affected at seedling stage (Vasudeva and

Srinivasan, 1952; Claudius and Mehrotra, 1973a; Khare, 1979). Seed germinates but seedling failed to emerge above the soil surface as they die during or immediately after germination. Seed may also rot even before germination process starts. When such seed are split open, browning may be noticed at the micropylar end (Claudius and Mehrotra 1973). In the post emergence stage plant may be affected at all the stage of growth. However, post emergence wilting is more prominent at two stages of plant growth *i.e.* seedling stage and adult stage (flowering and podding). Seedlings are affected at very early stage, often within 4-5 days after germination. Such seedlings show browning of the radicle and reduced number of secondary (smaller) roots. Such seedlings remain stunted, their foliage drop and finally die. Seedling mortality may continue even for about a month or so. In early stage infection, the tap root is damaged at the tip and remains abnormally short. Proliferation of secondary roots occurs in a cluster above the affected level (Vasudeva and Srinivasan 1952; Khare, 1980). Infection at the adult stage leads to the development of the true vascular wilt. In a broadcast crop, wilt occurs in isolated patches more or less circular in outline, which enlarges as the season advance. When the crop is sown in the rows, the disease appear to progress along the lines. Curling of the leaves begins from the lower end and extends upward. Foliage then droops and ultimately the plant dies. The root system of the affected plant is discoloured brown partially or wholly (Vasudeva and Srinivasan, 1952).

Khare (1980) described the symptoms of advanced wilt stage in lentil as - stunting of plants, shrinking and curling of leaves starting from the lower part of the plant and extending upward, epinasty in some cases and finally yellowing and drying of whole plant, conspicuous root discoloration, poor development of root system and of rhizobium nodules, stunting of tap root and its damage at tip. No fungal growth was observed on affected root system. In transverse sections of the diseased roots, fungal mycelium was, however, seen in vascular tissue. The wall of the xylem vessels in wilt-affected roots were discolored brown showing xyloses (Vasudeva and Srinivasan, 1952; Khare, 1980). Shulindin (1950) studied wilt disease of lentil in the erstwhile USSR and described the symptoms of the disease. He observed that the symptoms were inconspicuous in young seedlings but after 12-18 days, the growth of infected seedlings was inhibited, leaves remained smaller and were abnormally dark green and closer to the stem compared to those of healthy plants. Plants did not develop normally and produced few pods with small and undeveloped seeds. Affected plants finally died (Figure 4.1). Conidia were seen in affected cortex of the infected root. In Syria, El Ahmad and Mouselli (1987) differentiated the symptoms of vascular wilt and root rot of lentil caused by *Fusarium oxysporum* f.sp. *lentis* and *Fusarium solanum* respectively. Both diseases were characterized by yellowing and defoliation of leaves and drying of plants. In case of wilt, vascular tissues remained intact while roots rotted completely in the root rot. Generally, symptoms of wilt disease include overall reduction in growth, yellowing and drooping of leaves followed by drying and death of the whole plant.

Casual Organism

Vascular wilt of lentil is caused by *Fusarium oxysporum* f.sp. *lentis* (Figure 4.1). Vasudeva and Srinivasan (1952) were the first to study the fungus causing lentil

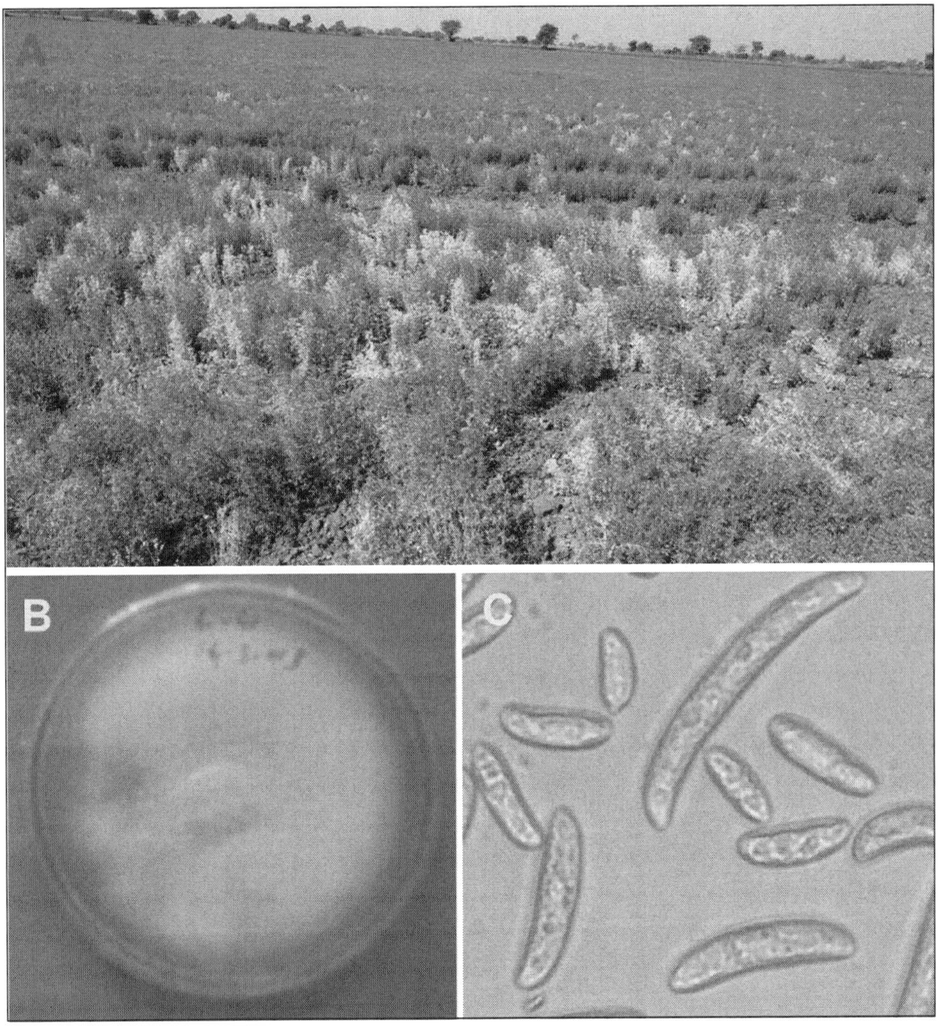

Figure 4.1. A: Wilt symptoms in lentil (a field view), B: Colony and growth pattern of *Fusarium oxysporum* f. sp. *lentis* and C: Typical macro and micro conidia of *Fusarium oxysporum* f. sp. *lentis*.

wilt in detail and following the classification of Wollenweber and Reinking (1935) identified it as *F. orthoceros* var. *lentis*. Consequent upon revision of classification of genus *Fusarium* and placing section Elegans in *F. oxysporum* Schlect by Snyder and Hensen (1940), nomenclature was changed to *Fusarium oxysporum* f.sp. *lentis* by Chattopadhya and Sen Gupta (1967). And it is this name which is widely accepted and is being practiced worldwide. Recently, *Fusarium equiseti* and *Fusarium redolens* have been shown to be associated with a wilt disease of lentil in Pakistan (Rafique *et al.*, 2020a and b). Vasudeva and Srinivasan (1952) described *Fusarium oxysporum* f.sp. *lentis* as – stroma thin, plectenchymatous, not sclerotial, erumpent, pale to purple. Aerial mycelium abundant, wolly, white, sometimes turning pink, collapsing after 3-4 weeks and becoming gelatinous, tough, pinnotes and sporodochia absent. Conidia form freely on mycelium at the end of free conidiophores. Microconidia often agglutinated into false head, non septate, ellepsoid to ovoid, cylindrical, oblong or slightly curved, hyaline. Macroconidia nearly straight to fusiform–falcate, slender with thin delicate walls and indistinct septa, mostly single celled, a few 1-5 septate. Base of conidia papillate or tending to foot cell formation, chlamydospore terminal or inter calary, borne on mycelium rarely on conidia, spherical to pyriform, smooth, hyaline, rich in protoplasmic constituents. One celled, occasionally 2 celled, abundantly formed on certain media and rarely on others.

Many media have been used for culturing *Fusarium oxysporum* f.sp. *lentis* by different workers. *Fusarium oxysporum* f.sp. *lentis* grows well on natural and synthetic media (Kumar *et al.*, 2000). Potato dextrose agar and a temperature range 27-30 °C was found to good for the growth of the fungus (Vasudeva and Srinivasan, 1952). Khare (1980) found PDA and temperature range of 20-25 °C to be best for fungal growth. Sharma and Agnihotri (1971) and Kumar *et al.* (2000) on the other hand reported that fungus grows well on natural as well as synthetic media.; however, Richards' medium was best for growth and sporulation and mannitol and $NaNO_3$ were the best carbon and nitrogen sources, respectively. A C/N ratio of 26:1 was most favourable for growth (Sharma and Agnihotri, 1971). Erskine *et al.* (1990) observed best growth and sporulation of *Fusarium oxysporum* f.sp. *lentis* on lentil dextrose agar medium and at 22.4 °C. Fungus tolerates a wide range of pH, with an optimum best growth at 4.6-5.8 pH (Vasudeva and Srinivasan, 1952; Kumar *et al.*, 2000). On the other hand, Khare (1980) found *Fusarium oxysporum* f.sp. *lentis* to grow well on a wide range of pH between 5.0- 8.5. A wide range of culture media have been used to culture Fusarium *oxysporum* f.sp. *lentis,* Asthana and Hawker liquid medium (Dhingra *et al.*, 1974), lentil dextrose agar medium (Erskine *et al.*, 1990), sand maize medium (Vasudeva and Srinivasan, 1952). Khare (1980) tried different solid media for growth of the fungus *Fusarium oxysporum* f.sp. *lentis* and found PDA to be the best followed by lentil extract agar. Richards' nutrient agar was also good. Minimum growth was observed on Martins' medium. PDA was the best solid medium for culturing *Fusarium oxysporum* f.sp. *lentis*. For pathogenicity test, sand maize medium has been used; whereas for mass culturing of the fungus, lentil seeds and straw are also used.

Host Range

Padwick (1941) reported that lentil wilt pathogen could infected lentil as well as chickpea, but subsequent workers reported the lentil wilt pathogen to be highly host specific in nature (Vasudeva and Srinivasan, 1952; Sharma and Agnihotri, 1972; Khare, 1980; Kannaiyan and Nene, 1978a) infecting only lentil. Other hosts like chickpea, pea, green gram, French bean, paddy, wheat, sorghum, maize, sugar beet, cotton, linseed, mustard and brinjal were not infected.

Variability

Though *Fusarium oxysporum* f.sp. *lentis* is host specific, presence of variability has been reported in it. Sharma and Agnihotri (1972) identified the isolates, 'A',

'B' and 'C' on the basis of their virulence in lentil. Claudius and Mehrotra (1973a) isolated 14 isolates of *Fusarium oxysporum* f.sp. *lentis* at Sagar (M.P.) and found them to differ from each other in regards to their virulence on two verities of lentil. Khare (1980) studied eight isolates of *Fusarium oxysporum* f.sp. *lentis* designated as L1, L2,. L8 for their variability. Isolates differed from each other in number of cultural and morphological characteristics such as colour of mycelium and pigmentation in substrate, growth habit, morphology of micro- and macro-conidia and chlamydospores. The mycelium colour varied from white to pinkish/purple/violet, whereas substrate color in different isolate was whitish yellow, purple with light green at the point of inoculation, whitish yellow, pink to purple becoming dark, pink/purple becoming dark, purple to violet, dark blackish purple. The mycelium was cottony, fluffy, appressed and raised at some points, cottony, submerged, cottony rough granular. Micro conidia size in different isolates varied between 4.99 – 9.99 ? 3.33 - 4.99µ. Isolates varied in macro conidia size (20.54 – 39.96 x 3.33 - 4.99) and septation (0 – 6). Dhingra *et al.* (1974) and Khare, 1980) studied the effect of temperature on the biomass accumulation (mycelium growth) of these eight isolates and found that 25?C was best for growth of these isolates except for L6 isolate that yielded maximum mycelium at 15?C. The strains, however, differed amongst themselves in the amount of the mycelial yield at one or the other temperature level. Using food poisoning techniques Agrawal and Khare (1975b, 1977a) studied effect of nine fungicides, namely Demosan, Dithane M45, Captan, Vitavax, Plantavax, Phaltan, Blitox-50, Difolaton and RH 539 on growth and macroconidia in eight strain of *Fusarium oxysporum* f.sp. *lentis.* All the eight strain of fungus differed from each other in their growth on one or the other fungicides. Some of the fungicides had great influence on macroconidia production by some of the strains. The data on influence of fungicides on growth and macroconidia formation was used to develop a key to differentiate eight isolates from each other.

Belabid and Fortas (2002) studied virulence and vegetative compatibility of 32 Algerian isolates of *Fusarium oxysporum* f.sp. *lentis* causing wilt disease of lentil in different parts of Algeria. All the 32 isolate were found to represent a single race but differ in their aggressiveness on susceptible lines. Isolates were also assessed for vegetative compatibility using nitrate non-utilizing mutants (nit). On basis of their ability to form hetrokaryon, all the lentil pathogenic isolates were grouped in to single VCG 0471 indicating homogeneity of the Algerian population of *Fusarium oxysporum* f.sp. *lentis*. A study involving isolates of *F. oxysporum* f.sp. *lentis* from Iran, Syria and Algeria revealed high molecular variation within regions. Variation among regions also exists, with Iranian populations differing significantly from non-Iranian ones. The clustering of isolates suggested the presence of seven ancestral *F. oxysporum* f.sp. *lentis* lineages, being three present exclusively in Iran indicating that Iran could be a regional centre of origin for *F. oxysporum* f.sp. *lentis* (Pouralibaba *et al.*, 2019). RAPD, SSR and ITS primers revealed varying degree of genetic diversity in 15 selected isolates of FOL from different lentil growing areas of India; and based on the coefficient of similarity, the isolates from north and north east regions grouped separately and formed two clusters (Datta *et al.*, 2011). Cultural, morphological and genetic variability among isolates of *F. oxysporum* f.sp. *lentis* has been reported by other workers also (Hiremani and Dubey, 2019, Saxena *et al.*, 2019). Based on the reactions of 10 genotypes of lentil, 50 Indian isolates of *Fusarium oxysporum* f. sp. *lentis* were grouped in to 8 pathotypes or races (Hiremani and Dubey, 2018). Pouralibaba *et al.* (2016) on the other hand differentiated 52 isolates of *Fusarium oxysporum* f. sp. *lentis* from Iran, Syria and Algeria into 7 pathotypes based on reaction of a differential set of four genotypes of lentil. There was, however, no correlation between the pathotype and the geographical origin of the isolates *Fusarium oxysporum* f. sp. *lentis*.

Host Pathogen Interaction

Most wilt causing Fusaria are reported to produce certain toxins and enzymes that play the important role in inducing wilt in the host plant. These toxins are mostly non-specific in their action. Sharma and Agnihotri (1967) found the lentil wilt pathogen, *Fusarium oxysporum* f.sp. *lentis* to produce a thermostable and non-specific metabolite that could induce wilt symptoms in lentil. Growing fungus longer in the medium resulted in higher production of toxin as indicated by the less time for wilt symptoms to develop in such filtrate. Mehrotra and Claudius (1973a) reported production of hydrolytic enzyme by *Fusarium oxysporum* f.sp. *lentis* which caused same disease of wilting in lentil as caused by the pathogen. These enzyme induced loss of turgor in foliage leading to drooping of top before finally withering. In another study, Claudius and Mehrotra (1973b) reported inhibition of spore germination of *Fusarium oxysporum* f.sp. *lentis* by root exudates of 21 days old lentil. This inhibition was attributed to the glycine and phenylalanine contents of root exudates. This may also partly explain the reason for lower incidence of wilt in lentil plants >3 weeks old inoculated with *Fusarium oxysporum* f.sp. *lentis.* Agrawal and Khare (1975a) also reported production of certain toxin metabolites by *Fusarium oxysporum* f.sp. *lentis*. The toxin reduced vigour of seedling of lentil, which formed very small dark green leaves with loss of turgidity. The leaves finally turned yellow. Agrawal *et al.* (1986) noticed 90 per cent seedling mortality in lentil on inoculation with *Fusarium oxysporum* f.sp. *lentis* culture filtrate indicating presence of a strong toxin in culture filtrate.

Bhalla *et al.* (1992) found that hyphae of *Fusarium oxysporum* f.sp. *lentis* got the entry into the cells of zone of cell elongation at the juncture of epidermal cells within 8 hrs after inoculation. Absence of sheath of the material found on the surface of the inoculated roots and penetrated indicated involvement of chemicals of fungal origin. Fungal hyphae reached cortical cells 10-12 h after inoculation and led to the rupture of cell wall and indicated break down of cell wall components by fungal enzymes. Although extensive proliferation of hyphae occurred throughout the cortex 24 h after inoculation, endodermis and vascular cylinder were free from hyphae for at least 72 h. *Fusarium oxysporum* f.sp. *lentis* is seed borne and may be present in all parts of the seed. It may survive in seed for 9-13 months (Singh *et al.*, 2010).

Disease Cycle and Epidemiology

Association of *Fusarium oxysporum* f.sp. *lentis* with lentil seeds was demonstrated in erstwhile Czechoslovakia by Ujevic *et al.* (1965) who concluded that the pathogen is transmitted by seeds and remains infected plants. In India, Khare *et al.* (1979)

demonstrated the association of fungus with seed of lentil. They isolated the fungus from all plant parts- from roots to seed through stem, branches, pedicle and placenta and concluded that the fungus was systemic in plant and invaded all plant parts. Erskine *et al.* (1990), however, did not find *Fusarium oxysporum* f.sp. *lentis* to be associated with seeds obtained from wilt affected plants. They, therefore, considered seed born nature of the disease related to either on external contamination of seed, possibly following threshing of pod infected saprophytically by fungus while in post harvest heaps in field or via trash such as infected stem, soil *etc.* which could carry the pathogen. Findings of Erskine *et al.* (1990) notwithstanding, seed transmission of the wilt pathogen is extremely important in the transfer of inoculum from infected to non-infected area leading to the introduction of disease in new area (Khare, 1980). But, lentil wilt pathogen survives mainly in soil on infected plant disease. Mehrotra and Claudius (1973c) reported competitive saprophytic ability of *Fusarium oxysporum* f.sp. *lentis* to be very higher in agar plate and straw colonization test. Several biotic and abiotic factors of soil, host *etc.* influence development of wilt. Sharma and Agnihotri (1972) reported that lentil plants are highly susceptible to wilt pathogen at the age of 22 days and susceptibility decreases with advancing age.

Soil Moisture

Vasudeva and Srinivasan (1952) investigated soil factors that favoured wilt in lentil and found maximum wilting (93.3 per cent) at the soil moisture of 25 per cent water holding capacity. Wilt incident was reduced to 16.7 per cent at 5 per cent soil moisture (saturation) and no wilting at 75 per cent or 100 per cent soil moisture (saturation).The disease (wilt) did occur when soil moisture was reduced from 75 per cent or 100 per cent to 25 per cent. Saxena and Khare (1988b) also noticed maximum plant mortality due to wilt at 25 per cent soil moisture and a decrease in mortality with increasing moisture. Saxena and Khare (1994) however, recorded maximum wilting in sandy soils at 30 per cent soil moisture. Moisture content in the soil is also known to influence relative occurrence of the fungi generally associated with wilt complex of lentil (Chaudhary *et al.*, 2009).

Soil Temperature

Wilt development is also influenced by temperature. Vasudeva and Srinivasan (1952) studied effect of temperature ranges on wilt incidence and found 7.5, 96.4 and 41.7 per cent wilt incidence at soil temperature ranges of 13 - 23 ?C, 17 - 31 ?C and 13 - 38 ?C, respectively. Saxena and Khare (1988b) also observed wilt to be more sever at 20-30 ?C soil temperature. In further studies (Saxena and Khare, 1994), sandy loam soils at 30 °C were found to be more conducive for development of wilt. Kaushal and Sharma (1998) reported 24-27 ?C to be the optimum temperature range for development of lentil wilt at Palampur. Cessation of wilt incidence in lentil in December and January under Delhi condition was also attributed to the prevailing sub-optimal temperature during these months. Fresh cases of wilt appeared in February- March were obviously due to the optimal temperatures during these months (Vasudeva and Srinivasan, 1952).

Soil Type

Development of wilt disease is more (100 per cent) in pure sand as compared to only 10 per cent in soil. Also disease appeared 3 - 5 days earlier in pure sand and 3:1 sand-soil mixture compared to combinations with less sand (Vasudeva and Srinivasan, 1952). This was attributed to the effect of aeration on the disease in sand. Sharma *et al.* (1971) studied lentil wilt incidence in sandy loam and clay soils of five district of M.P. and observed that wilt incidence in lentil crops grown in sandy loam soils in Bhind, Jabalpur and Raipur had wilt incident between 52-68 per cent where as in crops grown in clay soils of Jabalpur, Indore and Narsinghpur had lower wilt incidence *i.e.* between 18-40 per cent. Saxena and Khare (1994) also reported maximum wilting in sandy loam soils.

Soil pH

Wilt development is more at pH 7.5 to 8.0 (Sharma *et al.*, 1971, Saxena and Khare, 1988b). In sandy loam soils, wilt was more at pH 8.5 (Saxena and Khare, 1994). On the other hand, Kaushal and Sharma (1998) reported pH range of 5.2-5.8 as favourable to *Fusarium oxysporum* f.sp. *lentis*.

Inoculum Potential

Claudius and Mehrotra (1973a) reported that inoculum levels up to 2 per cent do not produce wilt in lentil and that disease severity increases with an increase in inoculum. Sharma *et al.* (1971) found a spore load of 3 x 10^{6}/g soil as optimum for lentil wilt incidence. Very high inoculum levels increases mortality at pre- and early post- emergence stage.

Sowing Time

Wilt development is influenced by sowing time as the prevailing conditions especially temperature has a great effect on disease development. Maylova (1973) observed seven times more sever wilting in early sown crops then in late ones. In Ukraine, Rodigni (1973) reported 6 times more wilting in lentil crops sown late (May – June) than in those sown in April. In India at Pantnagar Kannaiyan and Nene (1975a) observed least wilting in lentil crop sown on 15th December, while advance stage wilt was less in crops sown on 15th November. Bhatt and Sharma (1992) reported more seedling wilt in lentil crop sown up to the third week of October and lesser in later sowings. Izquierdo and Morse (1975) in Uruguay observed more wilt in early sown lentil compared to late sown crop. In Syria, late sowing (February) of lentil increases wilt incidence compared to normal sowing in December (ICARDA, 1987). Erskine *et al.* (1990) attributed it to the fact that in delayed sowing a large function of the growth period of the crop coincides with optimum or near optimum temperature for the pathogen compared to normal sowing.

Soil Fertility

Application of chemical fertilizer to the soil influences activities of wilt pathogen in the soil thereby influences wilt incidence. Khare *et al.* (1979) reported considerable reduction in wilt incidence due to application of high doses of P_2O_5. Kaur and Sharma

(1993b) reported that application of Zn at 10 kg/ha effectively controlled wilt disease of lentil. Crop density has also been reported to influence wilt incidence. Ujevic *et al.* (1965) observed more wilting in densely grown crop as compared to normal population. Myalova (1973) however reported the reverse trend.

Soil Biota

Association of nematode species like *Ditylenchus dipsaci*, *Aphelenchoides* spp., *Aphelenchus* spp., *Helicotylenchus* spp., *Heterodera* spp., *Meloidogyne* spp., *Pratylenchus* spp. and *Tylenchorhynchus* spp. with lentil wilt in the field has been reported. However, *Ditylenchus dipsaci* was observed more frequently and its population increased with wilt incidence in lentil (Erskine *et al.*, 1990).

Host Resistance

A number of workers have reported wilt resistant genotypes of lentil. Fleischmnn (1937) for the first time suggested development of a wilt resistant population of lentil by the mass selection method. He also noticed superiority of small seeded lentil over the large types. Less wilt in small seeded varieties than in medium and large seeded varieties was also reported in the erstwhile USSR by Shulindin (1950) who also observed differences in wilt incidence among lentil varieties. In erstwhile Czechoslovakia, Kovacikova and Suchanek (1973, 1974) found breeding lines MDPa, MDPb, MDPe and variety Afghanistan to be resistant and lines Lentille Blonbe 386/6 and 385/5, MDPc, MDPd and varieties Slovenska Modra and Souk et Theta de Safi as moderately resistant to wilt disease.

In India during eighties large number of lentil genotype were screened against wilt at Jabalpur and Pantnagar (Kanniayan and Nene, 1976; Khare *et al.*, 1976, Khare *et al.*, 1979). Kanniayan and Nene (1976) screened 158 lines of lentil against wilt under greenhouse and found 32 of them to be immune from *Fusarium oxysporum* f.sp. *lentis*. Of these, resistant lines - JL 80, Pusa 3, Pant 234, showed high degree of tolerance to not only wilt but also to wet root rot and collar rot and were recommended for cultivation. Lentil lines LWS15 and LWS 21 had resistance to 5 strains of the wilt pathogen (Kanniayan and Nene, 1978b). Khare *et al.* (1976) at Jabalpur screened 2000 lines of lentil and found L 642, P 338, Pant 638, Pusa 1, Pusa 7, LG 7, LP 6 and many others to be resistant to wilt. Tiwari and Singh (1980) also evaluated lentil germplasm against wilt and identified JL500 as a source of resistance to wilt for breeding wilt resistant variety of lentil. Kannaiyan and Nene (1978b) tested lentil genotypes against wilt pathogen at Pantnagar and identified LSW 21 as immune against 5 of the seven strains of *Fusarium oxysporum* f.sp. *lentis*.

Pandey *et al.* (1988) screened 49 lines and three varieties of lentil under greenhouse and found identified 8 germplasm lines resistant to wilt. Kaur and Sharma (1993a) screened 58 cultivars of lentil and found none of them to be immune to wilt but there were, however, 4 cultivars with wilt incidence between 1-10 per cent and graded them as resistant and 23 other cultivars were graded as moderately resistant (11-30 per cent wilt incidence). Pradhanang *et al.* (1993) screened 19 lines of lentil pot as well field conditions in Nepal and found none of them to be resistant to wilt. Singh and Singh (1994) evaluated 58 cultivars of lentil and found six lines

(SL 420, SL 896, SL 914, LC 202, LC 270, LC 271) to be resistant to wilt. ICARDA, Syria released six wilt resistant lines *viz.*, ILL 422, ILL 813, ILL 1220, ILL 1462, ILL 2313 and ILL 2684) for use by researchers in wilt resistance breeding programmes (Sarker *et al.*, 2001). None of the 48 lentil cultivars screened against wilt caused by *Fusarium oxysporum* f.sp. *lentis* in 1992-93 and 1993-94 in Bihar was found immune; the cultivars BR 25, Pant L 234 and Pant L 639 however showed resistance to wilt (Kumar *et al.*, 2003). In Bulgaria, Mikhov *et al.* (1987) found lentil varieties Nsalda, Zhana, Anisiya and Tadzhikskya 95 resistant to *Fusarium oxysporum* f.sp. *lentis*. Stoilova and Chavdarov (2006) tested 32 lines of lentil during 2003 and 2004 and found none of them to be resistant to wilt caused by *Fusarium oxysporum* f.sp. *lentis*.

Pandya *et al.* (1980) developed a wilt resistant variety L 406 by selection from an indigenous stock 'P 495'. Another wilt resistant variety of lentil Pant L4 was developed by Singh *et al.*(1994). This variety was recommended for cultivation in North West plain zone of India. Sharma *et al.* (1993) developed a lentil variety 'VL 4' with good level of resistance to wilt. Lentil germplasm PI 592998, known to possess resistance to wilt and released in Lebanon as 'Talia 2' in 1990 was registered as ILL 5588 (Erskine *et al.*, 1996). At Kumarganj, Faizabad, a wilt resistant and high yielding variety 'Narendra Lentil-1' was developed by Singh *et al.* (1997). In Syria, ICARDA has released wilt resistant lentil varieties named 'Idlib-2', 'Idlib-3', 'Idlib-4', 'Idlib-5' (El- Ashkar *et al.*, 2003, 2004a and b, ICARDA, 2015). Lentil genotypes PL 156, PL 168, LL 1136, LL 1197, RVL 11-4, JL 1 and JL 3 showed resistant reaction against wilt at Pantnagar (Singh and Kushwaha, 2015). In Jharkhand, lentil genotypes VL 133 (small seeded), LL 985 (large seeded) and VL 515 (extra large seeded) were identified as resistant to Fusarium wilt (Akhtar *et al.*, 2012). In Pakistan, Ali *et al.* (2018) screened 466 genotypes of lentil against wilt in Pakistan and found one genotype (LPP 12089) to be resistant. Under AICRP (MULLaRP) screening of germplasm and breeding lines against wilt is a regular feature and as a result of testing for last four decades or so, about two dozen lines that consistently show resistance against wilt have been identified for cultivation.

Inheritance of Resistance

Kamboj *et al.* (1990) studied inheritance of resistance to Fusarium wilt in lentil using crosses involving three letil genotypes resistant to resistant to *Fusarium oxysporum* f.sp. *lentis* (viz., Pant L 234, JL 446, LP 286) and two susceptible *viz.* JL 641, LP 286). Segregating pattern for wilt reaction in F2 BC(P_1),BC(P_2) and F2 generations in field and glasshouse conditions indicated that resistance to fusarium wilt is under the control of two dominant duplicate genes in Pant L234 and two independent dominant genes with complimentary effects in JL 446 and L 286. Allelic tests suggested that presence of five independently segregating genes for the resistance to wilt. The symbols were designated in term of fuw (fusarium wilt) with *Fuw* for dominant and *fuw* for recessive counterpart. The symbols designated are 'Fuw_1' and 'Fuw_2' for the two dominant duplicate genes in Pant L 234, 'Fuw_3' and 'Fuw_4' for two complementary genes in LP 286 and JL 446, and 'Fuw_5' for the third complementary gene present in JL 641 and L 9-12. A third dominant gene complementary to the dominant genes in JL 446 and LP 286 is present is present I two susceptible line. Eujayl *et al.* (1998) studied inheritance of resistance to wilt in a cross between resistant (ILL5588) and susceptible (L692-16-1(s) lines of *Lens culinaris*. Assessment of RIL populations revealed that resistance to wilt was determined by a single dominant gene (Fw) that was linked to RAPD marker OPK-15900 at 10.8cM and was associated with three more RAPD markers (OP-B17800 and OP-D15500).

Morphological, anatomical and biochemical bases of wilt resistance in lentil have not been studied much. Saxena and Khare (1988a) reported that lentil varieties with shorter shoots or lesser number of secondary shoots exhibited less wilt resistance. Saxena *et. al* (1992) found that such variety possessed a compact dark cambium and narrow meta xylem. Saxena and Khare (1988b) worked on the biochemical parameters of wilt resistant (Pant L 234 and Pant L 639), moderately resistant (JPL 355 and JPL 1014) and susceptible (JL 1 and JPL 18) and concluded that a higher amount of amino acids, sugar, phenols, orthohydric phenols, phosphorus and potassium and less nitrogen contents and permeability in roots are responsible for resistant to wilt. These reports notwithstanding, studies involving more number of varieties having different grades of disease resistance are required to establish any such correlation.

Management

Since the pathogen is soil and seed borne, all the management strategies are aimed at either reducing the seed borne or soil borne inoculums. Seed borne inoculums is mainly responsible for introducing the disease/strain in newer areas. Many approaches which are followed to mitigate the losses caused by wilt of lentil are:

Cultural

Manipulation of sowing date may be used to overcome the losses due to wilt, but its efficiency would differ from place to place, as it is the prevailing weather of an area that influence development of wilt disease. Reduced wilt incident in lentil due to delayed sowing was also observed by Mylova (1973), where as in Syria, delayed sowing increased wilt incidence in lentil (Erskine *et al.*, 1990). In Egypt, sowing on first December resulted in less incidence of wilt than in crop sown on first November though the yield was more in latter case. Also sowing at 2 cm depth was most appropriate resulting in least wilt and sowing at depth lesser or more that 2cm depth resulted in increased wilt infection (Sallam and Abdel Moneim (2012). Kannaiyan and Nene (1975a) observed least wilt in 15 December sown lentil crop at Pantnagar. Advanced stage wilt was however, less in the crop sown on 15 November. More seedling wilt is seen in crop sown early *i.e.* up to third week of October (Bhatt and Sharma 1992). Ghatak *et al.* (2015) advocated delayed (November sowing) under Bhagalpur, Bihar conditions for mitigating the losses caused by wilt disease in lentil. Generally, under north Indian conditions delayed sowing results in less disease (wilt) development but it adversely affects grain yield.

Preceding crop may also influence the incidence of wilt in the lentil crop. Kannaiyan and Nene (1979) observed that cultivation of rice or sorghum in the *kharif* season reduced wilt in succeeding crop of lentil in *rabi*. But incidence of collar rot increases if the stubbles of these crops are incorporated in the soil (Agrawal *et al.*, 1976). Intercropping with linseed is also known to reduce wilt disease in lentil and is traditionally practiced in Bundelkhand area, the lentil bowl of the country. Soil fertility also influence wilt disease. High doses of P_2O_2 were found to reduce population of wilt fungus (*Fusarium oxysporum* f.sp. *lentis*) resulting is reduced wilt incidence. (Khare *et al.*, 1979). Zinc @ 10 kg/ha effectively reduced incidence of wilt disease in lentil (Kaur and Sharma, 1993b). Seed bed configuration is also reported to influence incidence of lentil wilt. Lentil crop grown under raised bed conditions had lower wilt incidence as compared to flat bed in central highlands of Ethiopia (Bedasa and Zewdie, 2019).

Chemical

Several fungicidal chemicals have been tested against *Fusarium oxysporum* f.sp. *lentis* by many workers. Sharma *et al.* (1971) found benlate, ceresan wet, mercuric chloride, thiride and mixture of captan, thiram and PCNB to be complete inhibitors of growth of *Fusarium oxysporum* f.sp. *lentis.* Myalova (1973) in the erstwhile USSR found seed treatment with 0.25 ammonium molybdate, 0.15 per cent boric acid or 0.15 per cent potassium permanganate decreased *Fusarium oxysporum* f.sp. *lentis* infection and markedly increased seed yield. Seed treatment with zinc sulphate decreased wilt infection but had little effect on yield. Agrawal *et al.* (1974) evaluated 12 fungicides against the lentil wilt pathogen, of which thiram, benlate, rhizotol and ceresan checked the growth completely. Spore production was however not correlated to the growth inhibition as demosan and vitavax both inhibited the growth equally (46.3 per cent) but sporulation was heavy in demosan and poor in vitavax. The germinability of the spores on poisoned media also differed to a greater extent from fungicide to fungicide. Agarawal and Khare (1977b) reported organomurcurials (ceresan wet) and arsenic fungicides such as rhizoctol to be fungicidal while captan, thiram, folpet, difolatan, benomyl and oxycarboxin were fungistatic in their reaction on *Fusarium oxysporum* f.sp. *lentis.* Saxena and Mukhopadhyaya (1987) reported bavistin to be inhibitory to *Fusarium oxysporum* f.sp. *lentis. Fusarium oxysporum* f.sp. *lentis* has been reported to the seed born and to survive in soil. The chemical control measures are directed at seed and soil.

Kovacikova (1970) reported protection of *Fusarium oxysporum* f.sp. *lentis* inoculated lentil seeds when treated with captan, thiram and phaltan at 2 g/kg seed. Sharma *et al.* (1971) reported best control of lentil wilt by seed treatment with benlate @ 0.25 per cent (w/w) in addition to its soil application @ 20 mg/kg soil. Mehrotra and Claudius (1973b) reported effective control of lentil wilt by applying Mn and Zn @ 80 ppm. They observed reduction in wilt by presoaking of seeds in 80 ppm Mn or Zn and foliar spray with same chemical. Based on result of seed treatment trials for 4 consecutive years, Agrawal *et al.* (1975) found thiram + brassicol or thiram + bavistin (1:1) at 0.25 per cent w/w to be effective seed dressing fungicidal mixtures against lentil wilt. Soil application or fungicides such as roger 5G, diptrex 5G or solvirex 5G (@15 kg/ha) or azodrin 5G @15 kg/ha were effective in reducing the wilt of lentil (Kannaiyan and Nene,1975b). Kannaiyan and Nene (1974) studied effect of fungicidal seed treatment on wilt disease of lentil and found seed treatment with benlate (0.3 per cent), BAS 3050 F (0.2 per cent), BAS 3302 F (0.2 per cent), and EL-273 (0.2 per cent) reduced the incidence of wilt. Seed treatment with

Benlate (0.3 per cent) also resulted in significantly higher grain yield. Kannaiyan *et al.* (1975) found that benlate, a systemic fungicide got degraded in root and shoot of plant from treated seeds of lentil. Of the two degradation product, MBC was detected up to 45 days of sowing and the other (unidentified) only up to 15 days. This indicates that seed treatment with fungicides may provide protection against wilt only up to 45 days or so.

Seed treatment with a combination of two fungicides has also been found effective in reducing the wilt caused by *Fusarium oxysporum* f.sp. *lentis*. Saxena and Saxena (1993) noticed seed treatment with thiram + carbendazim (2:1, 2.5g/kg seed) and thiram + quintozene (1:1, 2.5 g/kg) or benomyl (1 g/kg) caused reduction in plant mortality in lentil due to wilt. Seed treatment with benomyl has also been reported to be effective against lentil wilt in Pakistan (Rafiq *et al.*, 2016). De and Chaudhary (1999) found seed treatment with carbendazim gave maximum wilt control (65 per cent) in lentil variety Sehore 74-3. Seed treatment and soil drenching with carbendazim is also reported to effectively reduce wilt incidence and increase grain yield (Maheshwari *et al.*, 2008).

Biological

Biological control agents against lentil wilt pathogen, *Fusarium oxysporum* f.sp. *lentis* are known and have been used to manage wilt disease of lentil. Ujevic *et al.* (1970) found isolates of Pseudomonas, *Ervinia*, *Rhizobium*, *Penicillium expansum* and *Trichoderma lignorum* to be antagonistic to *Fusarium oxysporum* f.sp. *lentis* on lentil. The efficiency of these antagonists was more pronounced in sterilized soil. In Syria, Erskine *et al.* (1990) found a Pseudomonas sp. Isolated from soil to possess antagonistic activity against *Fusarium oxysporum* f.sp. *lentis*. In India, Mehrotra and Claudius (1972) were the first to report disease (wilt) suppression by *Trichoderma viride*, *Streptomyces gougeroti*, two other isolates of *Streptomyces* and two bacterial isolates in lentil. Antagonism of *T. harzianum* and *T. koningii* against *Fusarium oxysporum* f.sp. *lentis* was demonstrated by Saxena and Mukhopadhyay (1987) and Mukhopadhyay *et al.* (1989). A simple and low cost technique for mass multiplication of *Trichoderma* sp. on wheat bran-saw dust medium in autoclvable plastic bags that can be practiced at farmers' field has been developed (Mukhopadhyay, 1987). Shrestha and Mukhopadhyay (1992) found that seed treatment with bioagent, *Gliocladium virens* and vitavax @ 0.5g/kg seed resulted in 140 per cent increase in plant stand and 188 per cent increase in yield of lentil. They found application of antagonist to the soil at 200kg/ha to be more effective.

Seed treatment with bio-agent like *Trichoderma* sp. and *Gliocladium* sp. alone often fail to protect germinating seeds because establishment of bio-agent in the rhizosphere takes time and hence germinating seeds need to be protected at early stage. Once germinating seeds are protected in the early stage, bio-agent will multiply in sufficient number and will provide better protection against wilt pathogen. With this view in mind Mukhopadhyay *et al.* (1992) developed a new treatment strategy involving both biological and fungicidal seed treatment for the management of wilt disease of lentil. This method involved seed treatment of lentil seeds first with *Gliocladium virens* (10^7 spores/ml) and then with 0.1 per cent carboxin.

They carried out large scale trials on efficacy of this new seed treatment (biological + fungicidal) against wilt of lentil and found it to be very effective in controlling not only the wilt disease but other soil borne diseases of lentil at farmers' fields in and around Pantnagar. De and Chaudhary (1999) found a combination of *Bacillus subtilis*, carboxin and *Trichoderma harzianum* or *T. viride* or *Gliocladium virens* reduced wilt by 79 per cent in lentil variety PDL 2. Combination of *Pseudomonas fluorescens*, *T. viride* and carboxin decreased wilt by 65 per cent. Singh and Mukhopadhyay (2002) investigated the potential of gamma radiation induced mutants of *G. virens* (M 3 and M 7) along with wild type in the control of wilt complex of lentil. Soil application of *G. virens* isolates @100 and 200 kg/ha) and seed treatment with 10^7 spores/ml (for wild type and M 3) and 10^8 chlamydospores/ml (for M 7) all resulted in significant increase in seedling emergence, plant stand and yield. Based on two years results, the average increase in yield in treatments with different *G. virens* isolates ranged from 72.3 to 111.35 per cent. They concluded that the treatments with *G. virens* gave control equally effective to chemical control (vitavax) under field conditions. De *et al.* (2003) found that seed treatment with carbendazim + thiram, and with *Gliocladium virens* + *Pseudomonas fluorescens* + carboxin controlled respectively, 48.8 and 44.2 per cent wilt in lentil variety Sehore 74-3. Both the treatments were nearly equally effective in controlling the wilt of lentil.

El Hassan *et al.* (1997) screened a number of isolates of *Bacillus* sp. and found 35 of them to be antagonistic to *Fusarium oxysporum* f.sp. *lentis*. They tested six of them and found that seed coating with bacterial antagonist had no effect on seed germination. The antagonist did not affect the growth of three lentil specific rhizobia strains *in vitro* and their growth was not affected by benomyl or thiabendazole seed treatments. It was concluded that selected *Bacillus* strains could be exploited for the control of lentil vascular wilt. El Hassan and Gowen (2002) studied the dynamics between the wilt pathogen of lentil, *Fusarium oxysporum* f.sp. *lentis* and *Bacillus subtilis*, a biological control agent in glasshouse experiment and found that *B. subtilis*, alone or in presence of wilt pathogen, reached a population density greater than 10^9 (cfu/g) in soil 2 weeks after application of bacterium in soil. *B. subtilis* was effective in reducing wilt incidence by 70 per cent. El Hassan and Gowen (2003) studied the interaction between *Trichoderma hamatum* and *Fusarium oxysporum* f.sp. *lentis* both *in vitro* and *in vivo*. *T. hamatum* was found to be a mycoparasite of *F. oxysporum* f.sp. *lentis*. *In vivo*, soil treatment with *T. hamatum* significantly increased the population of *T. hamatum* to be effective and reduced the population of *F. oxysporum* f.sp. *lentis* indicating that *T. hamatum* was a strong competitor in the rhizosphere with *F. oxysporum* f.sp. *lentis* in loamy soil. Their results suggest that mycoparasitism and competition are some of the mechanisms involved in the inhibitory interaction of *F. oxysporum* f.sp. *lentis* with *T. hamatum*. Combined seed treatment with *Trichoderma harizanum* + *Pseudomonas fluorescens* significantly reduced lentil wilt and increased grain yield (Garkoti *et al.*, 2014).

Reports on efficacy of extracts of certain plant species against wilt pathogen of lentil, *Fusarium oxysporum* f.sp. *lentis* are also available. Singh and Tripathi (1992) evaluated leaf extract of higher plant species against lentil wilt pathogen, *Fusarium oxysporum* f.sp. *lentis* and found extract from *Adenocalynma allicea* and *Artabotrys hexapetalus* to be 100 per cent fungitoxic. In another study, extract of *Mentha spicata* completely inhibited the growth of *Fusarium oxysporum* f.sp. *lentis* and remained active even after autoclaving for 30 min at 1.05 kg/cm^2 (Singh and Tripathi,1994). Leaf extract of *Ranunculus sceleratus* were also found to be fungitoxic to *Fusarium oxysporum* f.sp. *lentis* (Singh and Tripathi, 1995a). Of the aqueous extracts of leaves from 40 species of higher plants screened against *Fusarium oxysporum* f.sp. *lentis*, extracts from *Impatiens balsamiana* and *lawsonia inermis* completely inhibited mycelial growth (Singh and Tripathi, 1995b). These findings do not appear to have been evaluated at field level and thus remain to be exploited further.

Conclusion

Lentil is exposed to the wilt pathogen from seed to maturity. Seed is exposed to the wilt pathogen as soon as it is sown and may be colonized by the latter resulting in seed rot. Crop suffers due to both pre- and post- emergence mortality till maturity and this necessitates the need to protect seed and to target the soil borne inoculums of wilt pathogen. All the management interventions for reducing the wilt disease in fact help not only reduction of soil borne inoculums of the wilt pathogen but enhance population of antagonistic microbial population in soil. An intervention that enhances or supports soil microbial community consisting of antagonistic fungi and bacteria and also PGPR appears to be the most desired strategy for sustainable management of wilt disease. Biological control using seed treatment or application of biological control agents to the soil needs to be aimed at increasing the suppressiveness of the soil. Such studies should be planned and monitored on long term basis and aimed at increasing the resident antagonists that in turn will impart suppressiveness to the soil. Use of host plant resistance, sowing as per the recommendation for a particular area and application of biological control agents against the wilt pathogen should form the essential components of an integrated disease management approach for lentil wilt. Seed treatment with bio-formulations should find more focus after the recent ban on use of most of the fungicides which were being recommended for seed treatment against soil borne diseases including wilt.

REFERENCES

Abou Zeid, N.M., El-Morsy, G.A., Hassanein, A.M. and Arafa, M.K. (1997). Major organisms causing root-rot/wilt and their relative importance on faba bean, lentil and chickpea. *Egyptian Journal of Agricultural Research* 75 (3): 529-542.

Agrawal, R.K., Kotasthane, S.R., Gupta, Om and Khare, M.N. (1986). Influence of culture and culture filtrate of three soil-borne fungi on the mortality of lentil seedlings. *Indian Journal of Mycology and Plant Pathology* 16: 299-300.

Agrawal, S.C. and Khare, M.N. (1975a). Toxin production in culture medium by *Fusarium oxysporum* f. sp. *lentis*. *Science and Culture* 41: 239-240.

Agrawal, S.C. and Khare, M.N. (1975b). Effect of fungicides on macroconidia formation of *Fusarium oxysporum* f. sp. *lentis*. *Science and Culture* 41: 280-281.

Agrawal, S.C. and Khare, M.N. (1977a). Role of fungicides in differentiating eight strains of *Fusarium oxysporum* f. sp. *lentis*. *Pesticides* 11: 39-40.

Agrawal, S.C. and Khare, M.N. (1977b). Action of some systemic fungicides against *Fusarium oxysporum* f. sp. *lentis*. *Pesticides* 2: 51-52.

Agrawal, S.C., Khare, M.N. and Joshi, L.K. (1975). Effect of pre-sowing treatments of seeds on pre- and post-emergence losses in lentil. Second Annual Session Madhya Pradesh Vigyan Academy (Abstr.).

Agrawal, S.C., Khare, M.N. and Kushwaha, L.S. (1974). *In vitro* evaluation of fungicides against *Fusarium oxysporum* f. sp. *lentis*. *Indian Phytopathology* 27: 419-421.

Agrawal, S.C., Khare, M.N. and Tomar, K.S. (1976). Effect of some kharif crops on the incidence of collar rot of lentil caused by *Sclerotium rolfsii* Sacc. *Food Farming and Agriculture* 8: 4.

Agrawal, S.C., Singh, K. and Lal, S.S. (1991). Plant protection of lentil in India. Paper presented in ICAR-ICARDA seminar 'Lentil in South Asia', New Delhi, 11-15 March 1991, pp. 37.

Akhtar, J., Murmu, H.P., Kumar, Y. and Singh, P.K. (2012). Genotypic response and soil sickness for identification of resistant donors against fusarium wilt of lentil. *Journal of Food Legumes* 25(1): 81-82

Ali, S., Rizwan M., Idrees M, Abbas, G., Akhtar, K.P.,Iqbal M., Khan, M.E., Saleem, U., Rehmam, A.U., Rafiq M., Asghar, M.J. and Aslam, M. (2018). Identification of multiple sources of *resistance* in *lentil* against some potential fungal diseases. *Pakistan Journal of Agricultural Sciences* 55 (4): 875-880.

Bahl P.M., Lal, S. and Sharma, B.M. (1991). An overview of the production and problems of lentil in South Asia. Paper presented in ICAR- ICARDA seminar 'Lentil in South Asia', New Delhi 11-15 March 1991, 10 pp.

Bayaa, B., Erskine, W. and Khoury, L. (1986). Survey of wilt damage on lentils in northern Syria. *Arab Journal of Plant Protection* 4 (2): 118-119.

Bedasa, T. and Zewdie, A. (2019). Evaluation of lentil varieties and seedbed types for the management of lentil fusarium wilt disease (*Fusarium oxysporum* f. sp. *lentis*) in Central Highlands of Ethiopia. *African Journal of Agricultural Research* 14(24): 1012-1019.

Belabid, L. and Fortas, Z. (2002). Virulence and vegetative compatibility of Algerian isolates of *Fusarium oxysporum* f. sp. *lentis*. *Phytopathologia Mediterranea* 41 (3): 179-187.

Belabid, L., Fortas, Z., Dalli, D., Khiare, M. and Amdjad, D. (2000). Wilt and root rot of lentils in north-western Algeria. Cahiers Agricultures 9 (6): 515-518.

Bhalla, M.K., Nozolillo, C. and Schneider, E.F. (1992). Observations on the responses of lentil root cells to hyphae of *Fusarium oxysporum*. *Journal of Phytopathology* 135: 335-341.

Bhatt, J.C. and Sharma, A.K. (1992). Effect of sowing dates on disease incidence in lentil in Kumaon hills. *Indian Phytopathology* 45 (Suppl.) LXIX (Abstr.).

Carrera, C.M. (1940). The genus *Fusarium* in the Argentine Republic: study and systematic classification (second contribution). *Revista Argentina de* Agronomía 7: 277-296.

Chattopadhyay, S.B. and Sengupta, P. (1967). Studies on wilt diseases of pulses, 1.I, Variation and taxonomy of *Fusarium* spp. associated with wilt diseases of pulses. *Indian Journal of Mycological Research* 5: 45-53.

Chaudhary, R.G., Vishwa Dha and Singh R. K. (2009). Association of fungi with wilt complex of lentil at different crop growth stages and moisture regimes. *Archives of Phytopathology and Plant Protection* 42 (4): 340-343.

Claudius, G.R. and Mehrotra, A.S. (1973a). Wilt and root rot disease of lentil (*Lens esculenta*) at Sagar. *Indian Phytopathology* 26: 268-273.

Claudius, G.R. and Mehrotra, A.S. (1973b). Root exudates from lentil seedlings in relation to wilt diseases. *Plant and Soil* 38: 315-320.

Datta, S., Chaudhary, R.G., Md Shamim, Singh, R.K and Vishwa Dhar. (2011). Molecular diversity in Indian isolates of *Fusarium oxysporum* f.sp. *lentis* inciting wilt disease in lentil (*Lens culinaris* Medik.). *African Journal of Biotechnology* 10 (38): 7314-7323.

De, R.K. and Chaudhary, R.G. (1999). Biological and chemical seed treatment against lentil wilt. Lens Newsletter 26 (1-2): 28-31.

De, R.K., Dwivedi, R.P. and Narain, U. (2003). Biological control of lentil wilt caused by *Fusarium oxysporum* f.sp. *lentis*. *Annals of Plant Protection Sciences* 11(1): 46-52.

Dey, P.K. (1948). Plant Pathology-Administrative Report, Agriculture Department, Uttar Pradesh. 1945-46, pp. 43-46.

Dhingra, O.D., Agrawal, S.C., Khare, M. N. and Kushwaha, L.S. (1974). Temperature requirements of eight strains of *Fusarium oxysporum* f.sp. *lentis* causing wilt of lentil. *Indian Phytopathology* 27: 408-410.

El-Ahrned, M. and Mouselli, N. (1986). Lentil wilt in south Syria. *Arab Journal of Plant Protection* 4 (1): 30 (Abstr.).

El-Ahrned, M. and Mouselli, N. (1987). Wilt and root rot of lentils (*Lens culinaris*). LENS Newsletter 14: 10-14.

El-Ashkar, F., Sarker, A., Haddad, N. I., Bayaa, B., El-Hassan, H. and Erskine, W. (2003). Registration of " Idlib-2" lentil. *Crop Science* 43 (2): 728-729.

El-Ashkar, F., Sarker, A., Erskine, W., Bayaa, B., El-Hassan, H., Kadah, N. and Karim, B.A. (2004a). Registration of 'Idlib-3' Lentil. *Crop Science* 44: 2261.

El-Ashkar, F., Sarker, A., Erskine, W., Bayaa, B., El-Hassan, H., Kadah, N. and Karim, B.A. (2004b). Registration of 'Idlib-4' Lentil. *Crop Science* 44: 2261-2262.

El-Hassan, S., Bayaa, B., Erskine, W. and Akem, C. (1997). Antagonistic effects of *Bacillus* spp. against the causal organism of lentil Fusarium wilt (*Fusarium oxysporum* f.sp. *lentis*). *Arab Journal of Plant Protection* 15 (2): 65-72.

El-Hassan, S.A. and Gowen, S.R. (2002). Studies on population dynamics of *Bacillus subtilis* and *Fusarium oxysporum* f.sp. *lentis*, the causal organism of lentil vascular wilt. The BCPC Conference: Pests and diseases, Volumes 1 and 2. Proceedings of an international conference held at the Brighton Hilton Metropole Hotel, Brighton, UK, 18-21 November 2002, 753-756pp.

El-Hassan, S.A. and Gowen, S.R. (2003). Mechanisms in the biological control of lentil vascular wilt (*Fusarium oxysporum* f.sp. *lentis*) by *Trichoderma hamatum*. The BCPC International Congress: Crop Science and Technology, Volumes 1 and 2 Proceedings of an international congress held at the SECC, Glasgow, Scotland, UK, 10-12 November, 2003. pp. 651-654.

El-Morsy, G.A., Abou Zeid, N.M. and Hassanein, A.M. (1997). Identification of Fusarium wilt caused by *Fusarium oxysporum* and pathogen variability in faba bean, lentil, and chickpea crops in Egypt. *Egyptian Journal of Agricultural Research* 75 (3): 551-564.

Erskine, W. and Saxena, M.C. (1991). Breeding lentil at ICARDA for southern latitudes. Paper presented in ICAR-ICARDA seminar 'Lentil in south Asia', New Delhi, 11-15 March 1991, pp. 19.

Erskine, W., Bayaa, B. and Dholli, M. (1990). Effect of temperature and some media and biotic factors on the growth of *Fusarium oxysporum* f.sp. *lentis* and its mode of seed transmission. *Arab Journal of Plant Protection* 8 (1): 34-37.

Erskine, W., Bayaa, B., Ibrahim, H., Malhotra, A.S., Fares, A. and Hanati, G. (1989). Lentil improvement: Screening for vascular wilt resistance, pp. 125-128. In: Food Legume Improvement Programme, Annual Report 1989, ICARDA, Syria.

Erskine,W., Bayaa, B. and Saxena, M.C. (1996). Registration of ILL 5588 lentil germplasm resistant to vascular wilt and ascochyta blight. *Crop Science* 36 (4): 1080.

Eujayl, I., Erskine, W., Bayaa, B. and Baum, M. (1998). The inheritance and DNA-marker linkage analysis of *Fusarium* vascular wilt and frost injury in lentil. 3rd European conference on grain legumes. Opportunities for high quality, healthy and added-value crops to meet European demands. Valladolid, Spain, 14-19 November 1998. 1998, pp. 66-67.

Fleischmann, A. (1937). Observations on lentil wilt. Pflanzenbau 14: 49-56.

Garkoti, A., Kumar, V. and Tripathi, H. S. (2014). Control of wilt disease of lentil through bio control agents and organic amendments in tarai region of Uttarakhand, *India. J. Environ. Biol.* 35(6): 1067-70.

Ghatak, A., Shukla, N., Ansar, M., Balodi, R. and Kumar, J. (2015). Effect of sowing time, soil temperature and inoculum density on suppression of Fusarium wilt in lentil (*Lens culinaris*). *International Journal of Bio-resource and Stress Management*. 6 (2): 268-273.

Hiremani, H.S. and Dubey, S.C (2018). Race profiling of *Fusarium oxysporum* f. sp. *lentis* causing wilt in lentil. *Crop Protection* 108: 23-30.

Hiremani, H.S. and Dubey, S.C. (2019). Genetic diversity of *Fusarium oxysporum* f. sp. *lentis* populations causing wilt of lentil in India. *Indian Phytopathology* 72(4): 657–663.

ICARDA. (1987). Annual report of Food Legumes Improvement Programme. International Centre for Agriculture Research in Dry Areas, Aleppo, Syria.

ICARDA. (2015). A new high-yielding lentil variety 'IDLIB 5' released in Syria. October 4, 2015. ICARDA Communication Team.

Izquierdo, J.A. and Morse, R. (1975). First selection for wilt resistance in lentil (*Lens culinaris* Medic) in Uruguay. LENS Newsletter 2: 20-28.

Kamboj, A.K., Pandey, M.P. and Chaube, H.S. (1990). Inheritance of resistance to *Fusarium* wilt in Indian lentil germplasrn (*Lens culinaris* Medic.). Euphytica 50 (2): 113-117.

Kannaiyan, J. and Nene, Y. L (1974). Effect of fungicidal seed treatment on wilt disease of lentil. *Indian Journal of Plant Protection* 2: 80-85.

Kannaiyan, J. and Nene, Y. L (1975b). Effect of soil application of insecticide- granules on the incidence of lentil wilt disease. *Indian Journal of Plant Protection* 2: 77-79.

Kannaiyan, J. and Nene, Y. L. (1975a). Note on the effect of sowing dates on the reaction of 12 lentil varieties to wilt disease. *Madras Agricultural Journal* 62: 240-242.

Kannaiyan, J. and Nene, Y. L. (1978a). Host range of *Fusarium oxysporum* f.sp. *lentis*, *Sclerotium rolfsii* and *Rhizoctonia solani* isolated from lentil. LENS Newsletter 5: 11-12.

Kannaiyan, J. and Nene, Y. L. (1978b). Strains of *Fusarium oxysporum* f. sp. *lentis* and their pathogenicity on some lentil lines. LENS Newsletter 5: 8-10.

Kannaiyan, J. and Nene, Y.L. (1976). Reaction of lentil germplasm and cultivars against three root pathogens. *Indian Journal of Agricultural Sciences* 46: 165-167.

Kannaiyan, J. and Nene, Y.L. (1979). Effect of crop rotation and planting time on the incidence of wilt in lentil. *Indian Journal of Plant Protection* 7: 114-118.

Kannaiyan, J., Tripathi, R.K. and Nene, Y. L. (1975). Degradation and persistence of benomyl in lentil. *Indian Phytopathology* 28: 305-306.

Karki, P.B. (1991) Present status and future prospects of plant protection research in Nepal. Paper presented in ICAR- ICARDA seminar 'Lentil in South Asia', New Delhi 11-15 March 1991, 13 pp.

Kaur, A. and Sharma, L.C. (1993a). Screening of lentil varieties against *Fusarium* wilt. *Indian Journal of Mycology and Plant Pathology* 23: 172-173.

Kaur, A. and Sharma, L.C. (1993b). Effect of *Rhizobium*, inoculation and amendments in soil infested with lentil wilt pathogen. *Indian Journal of Pulses Research* 6: 115-117.

Kaushal, R.P. and Sharma, C. (1998). Effect of temperature, soil pH and phosphorus level on lentil wilt development. LENS Newsletter 25 (1-2): 65-67.

Khare, M. N. (1980). Wilt of lentil. Final Technical Report: Project PL-480, Grant No. FG-IN-430. JNKVV, Jabalpur. 155 pp.

Khare, M.N., Agrawal, S.C. and Jam, A.C. (1976). Studies on diseases of lentil. Proc. XVI Annual Workshop on Rabi Pulses, BHU, Varanasi, 27-30 September 1976. pp. 1-4.

Khare, M.N., Agrawal, S.C. and Jam, A.C. (1979). Lentil diseases and their control- Technical Bull, JNKVV, Jabalpur M.P., India, 29 pp.

Khare, M.N., Agrawal, S.C., Dhingra, O.D. and Kushwaha, L.S. (1975a). Variation in the growth of eight strains of *Fusarium oxysporum* f. sp. *lentis* on different solid media. *Indian Phytopathology* 28: 126-128.

Kotova, V.V., Khaleeva, J.N. and Shikuhova, *E.G.* (1965). Fungus diseases of legumes. In: Distribution of Pests and Diseases of Crops in the USSR in 1964. Trudy Vses. Inst. Zaschita Rastlin 25: 159-164.

Kovacikova, E. (1970). Seed treatment of lentil and pea against some fungal diseases. Ochrana Rostlin 6: 117-126.

Kovacikova, E. and Suchanek, A. (1973). A method of lentil inoculation with *Fusarium oxysporum* Schlecht. Zentralblatt fur Bakteriologie, Parasitenkunde Infektionskrankheiten und Hygiene 2, 128 (1-2): 12-22.

Kovacikova, E. and Suchanek, A. (1974). The resistance or the lentil collections to *Fusarium* wilt in conditions of natural infections. Ochrana Rostlin 10: 59-68.

Kumar, R. and Jha, D.K. (2003). Influence of root on Fusarium wilt of lentil. *Journal of Research*, Birsa Agricultural University 15 (2): 249 250.

Kumar, R., Jha, D.K., and Dubey, S.C. (2000). Influence of nutrition and pH on growth and sporulation of *Fusarium oxysporum*. *Journal of Research*, Birsa Agricultural University 12 (1): 61-65.

Kumar, S., Kumar, J., Singh, S., Ahmad, S., Chaudhary, R.G. and Sarker, A. (2010). Vascular wilt disease of lentil: a review. *Journal of Lentil Research* 4: 1-14.

Maheshwari, S.K., Bhat, N.A., Masoodi, S.D. and Beig, M.A. (2008). Chemical control of lentil wilt caused by *Fusarium oxysporum* f. sp. *lentis*. *Annals of Plant Protection Sciences* 16 (2): 419-421.

Mehrotra, R.S. and Claudius, G.R. (1972). Biological control of root rot and wilt diseases of *Lens culinaris* Medic. *Plant and Soil* 36: 657-664.

Mehrotra, R.S. and Claudius, G.R. (1973a). Role of metabolites and enzymes in the root rot and wilt diseases of *Lens culinaris*. *Indian Journal of Mycology and Plant Pathology* 3: 8-16.

Mehrotra, R.S. and Claudius, G.R. (1973b). Effect of chemical amendments and foliar application on lentil wilt. *Plant and Soil* 39: 695-698.

Mehrotra, R.S. and Claudius, G.R. (1973c). Competitive saprophytic ability of *Fusarium oxysporum* f.sp. *lentis* and *Sclerotium rolfsii* responsible for wilt and root rot diseases of *Lens culinaris* at Sagar. *Journal of Indian Botanical Society* 52: 191-198.

Mikhov, M.I., Stoeva, I. and Ivanov, P. (1987). Biological and economic characters of regionally released and promising lentil varieties in Bulgaria. Rasteniev"dni-Nauki 24 (4): 45-51.

Moreau, B. (1978). Diseases and pests of lentil. Institute National de Vulgarisation pour les Fruits, Legumes et Champignons, September 1978, pp. 95-100.

Mukhopadhyay, A.N. (1987). Biological control of soil-borne plant pathogens by *Trichoderma* spp. *Indian J. Mycol. Plant Pathol.,* 17, i-x.

Mukhopadhyay, A.N., Kaur, N. and Saxena, H.O. (1989). Biological control of soil-borne diseases of chickpea and lentil. *Indian Phytopathology* 42: 315 (Abstr.).

Mukhopadhyay, A.N., Shrestha, S.M. and Mukherjee, P.K (1992). Biological seed treatment for control of soil-borne plant pathogens. *FAO Plant Protection Bulletin* 40 (1-2): 21-30.

Myalova, L.A. (1973). *Fusarium* wilt of lentil. Zashchita Rostenii 4: 53.

Padwick, O.W. (1941). Report of the Imperial Mycologist, Scientific Report, Agriculture Research Institute. New Delhi, 1939-40. pp., 94-101.

Pandey, B.K., Shukla, A. and Pandya, B.P. (1988). Variability of resistance to root rot/wilt complex in lentil germplasm. *Indian Phytopathology* 41: 359-362.

Pandya, B.P., Pandey, M.P. and Singh, J.P. (1980). Development of Pant L 406 lentil. resistant to rust and wilt. LENS Newsletter 7: 34-37.

Pieta, D. and Pastucha, A. (1997). *Fusarium oxysporum* Schl. as a pathogen to some leguminous plants. Annales Universitatis Mariae Curie Sklodowska Sectio EEE, Horticultura 5: 227-236.

Pouralibaba, H.R., Rubiales, D. And Fondevilla, S. (2016). Identification of Pathotypes in *Fusarium oxysporum* f.sp. *lentis*. *European Journal of Plant Pathology* 144(3): 539-549.

Pouralibaba, H.R., Šatoviæ, Z., Cobos ma J., Rubiales, D. and Fondevilla, S. (2019). Genetic Diversity and Structure of *Fusarium oxysporum* f.sp. *lentis* Isolates from Iran, Syria and Algeria. *European Journal of Plant Pathology* 153 (4): DOI : 10.1007/s10658-018-01617-7

Pradhanang, P.M., Pandey, R.R. and Ghimire, S.R. (1993). Response of different lentil varieties to wilt (*Fusarium* spp.). LENS Newsletter 20 (1): 62-64.

Rafiq *et al.,* 2016 Rafique, K., Chaudhary A.R., Naz, F. and Shabbir, G. (2016). Management of vascular wilt of lentil through host plant resistance, biological control agents and chemicals. *Pakistan Journal of Botany* 48 (5): 2085-2092.

Rafique, K., Kang, S., Aziz-ud-Din, Mahmood, T., Imran, Ullah, I. and Mahmood, H. (2020b). First report of vascular wilt on lentil (*Lens culinaris*) caused by *Fusarium redolens* in Pakistan. New Disease Report, Published July 2020. https: //doi. org/10.1094/PDIS-11-19-2519-PDN

Rafique, K., Rauf, C.A. and Kang, S. (2020a). First report of *Fusarium equiseti* causing wilt on lentils (*Lens culinaris* Medik.) in Pakistan. *J Plant Pathol* 102: 571.

Rodigni,V.M.(1973). Fusarium wilt of lentil in the Ukraine. Zashchita Rastenii No. 8: 51.

Sallam, N.M.A. and Abdel-Monaim M.F. (2012). Influence of some agricultural practices on suppression of lentil wilt disease. *Plant Pathology Journal* 11 (1): 32-37.

Sarker, A., Bayaa, B. and Erskine, W. (2001). Registration of six lentil germplasm lines with resistance to vascular wilt. *Crop Science* 41 (5): 1655.

Saxena D.R., Saxena M. and Tiwari, N. (2019). Morphological and cultural variability in *Fusarium oxysporum* f.sp. *lentis* causing wilt of lentil. *Indian Phytopathology* 72 (4): 10.1007/s42360-018-0087-y

Saxena, D.R. and Khare, M.N. (1988a). Root morphology and anatomy in relation to vascular wilt incidence in lentil. *Indian Phytopathology* 41: 280 (Abstr.).

Saxena, D.R. and Khare, M.N. (1988b). Factors influencing vascular wilt of lentil. *Indian Phytopathology* 41: 69-74.

Saxena, D.R. and Khare, M.N. (1994). Influence of soil environment on mortality of lentil due to soil-borne pathogens. *Indian Phytopathology* 47 (1): 60-64.

Saxena, D.R. and Saxena, M. (1993). Evaluation of chemicals against biological drying of lentil. Bhartiya Krishi Anusandhan Patrika 8 (91): 25-29.

Saxena, D.R., Saxena, M., Saxena, B.R. and Khare, M.N. (1992). Root morphology and anatomy in relation to wilt incidence in lentil. LENS Newsletter19 (2): 45-49.

Saxena, H.C. and Mukhopadhyay, A.N. (1987). Biological control of wilt complex in lentil. *Indian Journal of Mycology and Plant Pathology* 17: 123 (Abstr.).

Setti, B. and Bouznad, Z. (1998). *Fusarium* root rot and wilt of lentil in western Algeria. 3rd European conference on grain legumes. Opportunities for high quality, healthy and added-value crops to meet European demands. Valladolid, Spain, 14-19 November 1998.

Sharma, H.C. Deshkar, M.V. and Khare, M.N. (1971). Factors affecting *Fusarium* wilt of lentil (*Lens esculenta* M.). In: Proc Pathological Wilting of Plants, University of Madras, Madras, India, 42 pp.

Sharma, K.S. and Agnihotri, J.P. (1967). Production of toxic metabolites by *Fusarium orthoceras* f.sp. *lentis* Vas. and Sri. Labdev *Journal of Science and Technology* 5: 297-299.

Sharma, K.S. and Agnihotri, J.P. (1971). Influence of nutrition and pH on growth and sporulation of *Fusarium orthoceras* var. *lentis*. *Indian Phytopathology* 24: 553-556.

Sharma, K.S. and Agnihotri, J.P. (1972). Pathogenecity and variability of lentil wilt *Fusarium*. *Indian Journal of Mycology and Plant Pathology* 2: 170.

Sharma, R.K., Koranne, K.D., Joseph, J.K., Singh, P. and Prakash, V. (1993). 'VL Masoor 4': a new lentil variety for the hills of Uttar Pradesh. Indian Farming 42 (10): 19-20.

Shrestha, S.M. and Mukhopadhyay, A.N. (1992). Integrated approach: A successful method of controlling lentil wilt complex. *Indian Phytopathology* 45 (Suppl.): LXV (Abstr.).

Shulindin, A. F. (1950). *Fusarium* infection of lentils sown in the spring and summer. Agrobiology 2: 144.

Singh, B.B., Singh, D.P., Maurya, D.M., Rai, O.P, and Gupta, R.P. (1997). Development of high-yielding, bold-seeded, rust- and wilt-resistant lentil cultivar Narendra Masoor-1 for Uttar Pradesh, India. LENS Newsletter 24 (1-2): 52-53.

Singh, G. and Mukhopadhyay, A.N. (2000). Biocontrol potential of mutants of *Gliocladium virens* for wilt complex of lentil. Legume Research 23 (2): 133-135.

Singh, I.S., Singh, J.P., Singh, A.K. and Chauhan, M.R (1994). Pant lentil 4: A high-yielding, rust, wilt and blight resistant variety for the north-western plains of India. LENS Newsletter 21(2): 8-9.

Singh, J. and Singh, I.S. (1994). Evaluation of lentil germplasm for plant type, initial flowering and disease resistance. LENS Newsletter 21 (1): 5-7.

Singh, J. and Tripathi, N.N. (1992). Efficacy of some angiospermic plants against *Fusarium oxysporum* f.sp. *lentis* Schl. National Academy Science Letters 15 (9): 277-278.

Singh, J. and Tripathi, N.N. (1995a). Effect of growth stages of *Ranunculus sceleratus* L. on its fungitoxicity. *Indian Phytopathology* 48 (3): 363-364.

Singh, J. and Tripathi, N.N. (1995b). Fungitoxicity of some green plants against *Fusarium oxysporum* causing wilt disease in lentil. *Journal of Living World* 2 (1): 7-9.

Singh, V. and Kushwaha K.P.S. (2015). Screening of lentil germplasms against fusarium wilt. *Trends in Biosciences* 8(23): 6491-6494.

Singh, V.K., Naresh, P., Biswas, S.K. and Singh, G.P. (2010). Detection, location and survibility of *Fusarium oxysporum* f.sp. *lentis* Gordan in seeds of lentil. *Annals of Plant Protection Sciences* 18 (2): 464-466.

Snyder, W. and Hansen, R.N. (1940). The species concept of *Fusarium*. *American Journal of Botany* 27: 64-67.

Soran, H. (1979). Detection of causal agents of root rot of lentil in outskirt of Ankara. *J. Turkish Pytopathol.* 8(1): 1-8.

Stoilova, T. and Chavdarov, P. (2006). Evaluation of lentil germplasm for disease resistance to *Fusarium* wilt (*Fusarium oxysporum* f.sp. *lentis*). *Journal of Central European Agriculture* 7 (1): 121-125.

Tiwari, A.S. and Singh, B.R. (1980). Evaluation of lentil germplasm. LENS Newsletter 7: 20-22.

Tosi, L. and Cappelli, C. (2001). First report of *Fusarium oxysporum* f. sp. *lentis* of lentil in Italy. *Plant Disease* 85 (5): 562.

Ujevic, I., Kovacikova, E. and Urosevic, B. (1965). *Botrytis parasitica* and *Fusarium oxysporum* as dangerous parasites of lentil. Polnohospodarstvo 11(8): 584-594.

Ujevic, I., Kovacikova, E. and Urosevic, B. (1970). Interrelation between some parasitic fungi, *Rhizobium* and other microorganisms of lentil (*Lens esculenta*) and the possibilities of biological control. Zentbl. Bakt Parasit Kde, Abt., 1252: 394-408.

Vasudeva, R.S. and Srinivasan, K.V. (1952). Studies on the wilt disease of lentil (*Lens esculenta* Moench.). *Indian Phylopathology* 5: 23-32.

Wilson, V.E. and Brandsberg, J. (1965). Fungi isolated from diseased lentil seedlings in 1963-64. *Plant Disease Reporter* 49: 660-662.

Wollenweber, H.W. and Reinking, O.A. (1935). *Die Fusarien*, ihre *Beschreibung, Schadewirkung und Bekanpfung*. Berlin, 355 pp.

Chapter 5

Sterility Mosaic Disease of Pigeonpea: A Potential Threat to the Cultivation of Pigeonpea

J.N. Srivastava & Anupam Kumari

Pigeonpea [*Cajanus cajan* (L.)] Millspaugh, is one of the major pulse crops of the tropics and subtropics also popularly known as Red gram, Tuar, or Arhar. It is a primary source of protein for millions in India (Bressani *et al.*, 1986).

India, accounts for nearly 65 per cent of the world's pigeonpea production with an acreage of 3.57 m, ha and annual production of 2.36 m, tons (Muller, *et al.*, 1990). Pigeonpea in India is mainly grown in the semi-arid regions of the states of Andhra Pradesh, Karnataka, Maharashtra, Madhya Pradesh, Tamil Nadu and Uttar Pradesh because of its drought resistance. Development of many high yielding short duration cultivars that can fit into diverse cropping systems, including cereal based ones, lead to. Increase in area of pigeonpea cultivation. The simultaneous use of pigeonpea for food, fodder and fuel, its ability to ameliorate soils and its use as a hardy crop on marginal soils fitting into many intercropping systems makes pigeonpea an important crop in the semi-arid tropics (Agrawal, 2003).

Pigeonpea is attacked by a large number of diseases and therefore, the risk involved in its cultivation is quite high, over 63 fungal, 3 bacterial, 19 viral or alike pathogens and 10 nematodes are reported to infect the crop at various growth stages (Nene *et al.*, 1996). Of these, sterility mosaic is very important and widespread.

Sterility mosaic disease caused by Pigeonpea sterility mosaic virus (PPSMV) is widespread and economically important. Pigeonpea sterility mosaic virus transmitted by the eriophyid mite, *A. cajani* (ChannaBasavanna), but it is not transmitted through seed pollen or so (Reddy *et al.*, 1998) and is recognized only in Pigeonpea growing countries of Asia. Sterility mosaic (SMD), caused by pigeonpea sterility mosaic virus (PPSMV) is most serious disease of Pigeonpea because of it's wide spread and destructive nature, causes serious loss of pigeonpea production. The production of pigeonpea in the Indian sub-continent and other countries in Asia is severely affected by sterility mosaic disease **(SMD)** (Ghaneker *et al.*, 1992). Although SMD has become a potential threat to the cultivation of Pigeonpea in Indian subcontinent (Singh, *et al.*, 1999) and causes an estimate annual loss of 300,000 tonnes of grains (more than US $ 150 million) in India alone (Kumar, 2000), because control measures are still not well understood except resistant cultivars. Numerous studies have been attempted to determine the nature of pathogen, symptomatology, epidemiology, disease resistance and control measures since it's recognition in India in 1931 (Mitra, 1931). This review aims to consolidate the existing knowledge about SM that may be useful to combat this "green plague" of pigeonpea.

Distribution and Economic Importance of Sterility Mosaic Disease

Pigeonpea sterility mosaic disease was first reported from Pusa, Bihar (Mitra, 1931) and a similar disease was also reported from Bihar in 1931 (Alam, 1931, Mc Rae, 1932) and Coimbatore in 1938 (Anon. 1938). Earlier disease was confined only to Indian subcontinent (Reddy, *et al.*, 1990) but now it also been reported from Bangladesh, Nepal and Thailand (Nene, *et al.*, 1989), Myanmar (Su, 1931), Srilanka (Newton, 1953) and China (Chaohong, *et al.*, 2001). Sterility mosaic was more severe in Uttar Pradesh, Bihar, Gujarat, Karnataka, and Tamil Nadu (Kannaiyan *et al.*, 1984). Recently, Singh *et al.* (1999) reported it's occurrence from Chhattisgarh.

Yield losses susceptible genotypes due to SMD occurring early in the season can reach over 90 per cent with an annual loss exceeding cause yield loss of 205,000 tons annually in India (valued at USS >I50 million, of grain in India alone (Kanniyan *et a1.*, 1984). According to Kannaiyan *et al.* (1981) disease was more severe in Bihar (21.4 per cent) followed by Uttar Pradesh (15.4 per cent), Tamil Nadu (12.8 per cent), Gujarat (12.2 per cent), Karnataka (9.8 per cent) and Andhra Pradesh (1.6 per cent), respectively, but now it has become a major constraint in pigeonpea production in South Indian States also. Disease incidence varied in main as well as in ratoon pigeonpea and it appeared in severe outbreak in Maharashtra in 1991 (Muniyappa *et al.*, 1980; Zote *et al.*, 1991). SM causes serious and often total yield loss (Singh *et al.*, 1989) and a susceptible cultivars infected in early stage (< 45 days) of the crop growth showed complete sterility and yield loss upto 100 per cent but as the plant grew older (>45 days), their susceptibility to SM decreased and such plant showed partial sterility (Reddy *et al.*, 1981). Significant reduction, due to SM infection, in plant height and number of branches (Dahal, and Neupane, 1991, Singh, and Rathi, 1994), flowers per plant and pod length (Dahal, and Neupane, 1991) and 1000 grains weight (Singh and Rathi, 1994) has also been reported.

Symptomatology

SMD infected plant, shows a conspicuous mosaic symptoms and become sterility. Therefore the disease is referred to as "sterility mosaic." Disease plant easily can noticed in the field from a distance, because yellowish green leaves, reduced leaves, a bushy appearance and accompanied by sterility/without flowers or pods (Alam, 1933), pale green colour and upright professed vegetative growth (Capoor, 1952). Similar symptoms have been described by other workers (Kandaswamy and Ramakrishnan, 1960, Prasad, 1965, Seth, 1962). Nene (1972) reported severe stunting, reduction in leaf size, an increase in number of secondary and tertiary branches arising from the leaf axles and complete or partial sterility of reproductive structures.

Figure 5.1: Infected Plant with Yellowish Green Leaves, a Bushy Appearance and without Flowers or Pods (Reddy *et al.*, 2012).

An initial symptom of SM was vain-clearing in younger leaves in seedlings but primary leaves did not produce any symptoms. Young leaves are show vein clearing symptoms and older laves showed mosaic symptoms but primary leaves no any symptoms (Capoor, 1950, Reddy and Nene, 1979).

Major three types of symptoms (Ramakrishnan, and Kandaswamy, 1972) has been described as

(a) Severe mosaic on leaflets: plants do not produce flowers and pods (complete sterility)

Figure 5.2: Mosaic Symptoms on Leaf (Reddy *et al.*, 2012).

Figure 5.3: infected Leaf with Ring Spot Symptom (Reddy *et al.*, 2012).

(b) Mild mosaic: plants produce flowers and pods to some extent (partial sterility) and

(c) Ring spot: symptoms are characterized by Green Island surrounding by a chlorotic halo on leaflets and this symptoms disappear as the plants mature (no sterility).

During multilocational screening trials variation in symptom expression by some genotypes were noted (Reddy *et al.*, 1998). A variety showing resistant (no symptoms) or susceptible reaction (mosaic or chlorotic ring spots) at one place may show a different kind of reaction at another location.

Pigeonpea is a cross-pollinated crop. In addition to environmental factors, genotypic variability induced as a result of cross-pollination also likely to play an important role in symptomatology. Thus, variability in the pathogen, the plant genotype, the environment, and the mixed infection with other pathogens, may all influence the symptom expression (Nambiar, and Ramakrishnan, 1968). Symptoms are suppressed during the hot summer months, but with monsoon rains they reappear on the new growth. Shade and humidity encourage mite multiplication, especially in hot summer weather. Disease incidence is high when pigeonpeas are inter- or mixed cropped with sorghum or millets.

Biochemical Changes

Biochemical alteration in SM diseased plants, studied by several workers, have allocated

(A) Absence of gluamic, alanin, maleic and citric acid and reduced chlorophyll, carotene, xanthophylls, sugar synthesis and it's translocation, choloroplastic protein, C: N ratio, peroxidise activities, organic and ascorbic acid contents.

(B) Increased chlorophyllase, catalase, nitrate reductase and proteolytic enzymes activities and presence of unidentified amino acid (Narayanswamy, and Ramakrishnan, 1965a, Narayanswamy, and Ramakrishnan, 1965b, Narayanswamy, and Ramakrishnan, 1965c) and (C) reduction in total carbohydrates, starch, resin, Ca, K, Na, Mn, photosynthesis rate of hill reaction (Singh and Mal, 1976), reducing sugar contents, RNA and DNA level, total Nitrogen and free amino acids in disease plants (Nambiar, and Ramakrishnan, 1968, Nambiar, and Ramakrishnan, 1969a, Nambiar, and Ramakrishnan, 1969b). Lower in dry weight, chlorophyll and photosynthetic rate (Singh and Mall, 1976, 1978) and presence of specific peroxidise enzymes and protein (Rathi *et al.*, 1986) in diseased plants has been reported.

Histopathological Changes

In addition to physiological and biochemical changes, reduction in leaf thickness, epidermal and palisade cell also was observed in SM infected pigeonpea plants. In transverse section of diseased mid-vein, fibre was found absent completely or present, partially and the phloem cells were hypertrophied with larger cellular spaces (Singh, and Rathi, 1994a, Singh and Rathi, 1996).

Causal Organism

In very recent studies, using a new purification method SMD causal agent was identified as a novel virus provisionally named Pigeonpea sterility mosaic virus (PPSMV) (Kumar *et a1.*, 1999 a and b; 2000). The causal agent of the disease is PPSMV, a virus with a segmented, negative-sense, single-stranded RNA genome, transmitted in a semi-persistent manner by an eriophyid mite *Aceria cajani* - Channabassavanna (Acari: Arthropoda). Both the virus and vector are highly specific to pigeonpea and a few of its wild relatives, such as *C. scarabaeoides* and *C. cajanifolius*. The virus particles are highly flexuous and of branched filamentous (Kumar *et al.*, 2002; Kumar *et al.*, 2001: Kumar *et al.*, 2000).

In a separate study, mite populations obtained from SMD-affected plants from various SMD-endemic locations from the lndian subcontinent were analysed for biodiversity using PCR-based nuclear ribosomal DNA fingerprinting technique morphological studies using scanning electron microscopy (Kumar *et al.*, 2001b).

It was apparent that there was no other *Aceria* mite species or biotypes of *A. cajani* that differed in their ability to transmit PPSMV It was therefore concluded that resistance breakdown at some locations is due to the existence of PPSMV strains and this was confirmed (Reddy *et al.*, 1998).

Pigeonpea Sterility Mosaic Virus (PPSMV)

Taxonomy

PPSMV was recently classified as a species in the genus *Emaravirus*, a genus whose members show features resembling those of members of the genera *Tospovirus* (Family: *Bunyaviridae*) and *Tenuivirus*, both of which comprise single-stranded RNA viruses that encode proteins by an ambience strategy.

Genome Organization and Gene Functions

The virus particle of asymmetric morphology measuring 3-8 mm in diameter with 32 kDa coat protein and 4-5 segmented RNA of size 0.8-3.5 kilo-base is responsible for this sterility mosaic disease (Kumar *et al.*, 1998). The PPSMV genome contains five segments of single-stranded RNA that are predicted to encode proteins in negative sense. The ribonucleo protein complex is encased in quasi-spherical, membrane-bound virus particles of 100-150 nm (Kumar *et al.*, 1999, 2001 b). The largest segment, RNA-1, is 7022 nucleotides in length and codes for RNA-dependent RNA polymerase (2295 amino acids); RNA-2, with a sequence length of 2223 nucleotides, codes for glycoproteins (649 amino acids); RNA-3, with a sequence length of 1442 nucleotides, codes for nucleocapsid protein (309 amino acids); RNA-4, with a sequence length of 1563 nucleotides, codes for a putative movement protein p4 (362 amino acids); and RNA-5, with a sequence length of 1689 nucleotides, codes for p5 (474 amino acids), a protein with unknown function(Kumar *et al.*, 2002; Kumar *et al.*, 2001: Kumar *et al.*, 2000).

Detection of Virus in Mite Vectors

For virus detection in mites, electron microscope (Paliwal and Slykhuls, 1967; Takahashi and Orlob, 1969; Stein-Margolina *et al.*, 1969; and Paliwal, 1980), and serological methods such as ELISA, immunofluorescent microscopy, Western blotting and dot-immunobinding assay were used (Sherwood, 1987). ELISA- and RT-PCR-based diagnostic tools for the precise and sensitive detection of virus in infected plants were developed (Kumar *et al.*, 2001a). The purified PPSMV preparations were not infective to plants but PPSMV In crude plant extracts was transmitted experimentally to herbaceous hosts (*Nicotiana baenthamiana and N clevelandili*) by mechanical inoculations (Kumar *et al.*, 2002)

Transmission

Transmission of PPSMV has been reported through graft (Capoor, 1952; Janarthan *et. al*, 1972; Mali *et al.*, 1977; Nene, 1972) sap (Capoor, 1952) and phytopathogenic nematode (Narayanswamy *et al.*, 1963), however, phytopathogenic nematode and sap transmission could not be established (Mali *et. al*, 1977; Mc Rae, 1932; Nene, 1972; Ramakrishnan and Kandaswamy, 1972). It is also not transmitted through seeds (Anon, 1979, Capoor, 1952; Nene, 1972) and dodder (Nene, 1972; Seth, 1965). Seth (1962) first showed that under natural conditions, PPSMV is transmitted through an eriophyid mite (Aceria cajani Channa Basavanna). This was confirmed by (Seth, 1965; Nene, 1972; Nene and Reddy, 1976a; Kulkarni *et al.*, 2002; Ramakrishnan and Kandaswamy, 1972; Reddy *et al.*, 1989). In nature, *A. cajani* populations were almost exclusively observed on SMD-infected pigeonpea, but not on healthy plants, indicating a strong communalistic relationship between the virus-infected plants and the vector (Kulkarni *et al.*, 2002).

Three methods are being used for experimental transmission of PPSMV *viz.*, the 'leaf stapling technique' (Nene and Reddy, 1976a), 'infector-hedge technique'' (Nene *et al.*, 1981a) and 'spreader row' inoculation technique (Nene *et al.*, 1981 b). All these methods use mites to transmit virus from source leaf to the healthy plant.

Vector

Eriophyid Mite

Aceria cajani Channabasavanna (1966)

Distribution

Bangladesh, India, Myanmar (Burma), Nepal, and Sri Lanka.

Economic Importance

These mites are widespread and common in pigeonpea throughout South Asia. A serious problem in Nepal and India where it is estimated to cause annual pigeonpea grain losses worth US$282 million.

Taxonomy

It belong to Acari: Arthropoda, order- Acarina in family-Eriophyidae

Aceria cajani, as the vector of the pigeonpea sterility mosaic virus causing the most damaging sterility mosaic disease of this crop in India. It is a common and widely distributed insect. It is uniformly distributed on the leaves with major concentration in the younger leaves. More than 90 per cent population of *A. cajani* occur on the lower surface of leaves in the diseased plants. The mites are very small but can be seen easily under a stereo binocular (40x) microscope. This mite is highly host-specific and is largely confined to pigeonpea and its wild relatives *viz., Cajanus scarabaeoids* and *C. cajanifolius* growing as wild plants that is commonly found on wastelands and field bunds. They feed on the lower surface of the leaf with short chellceral stylets. The short stylets (- 2.03pm) of these mites allow penetration of epidermal cells and the mite feeding cause no obvious damage to pigeonpea (Sheila *et al.*, 1988)

Morphology and Habitat

These mites are small, pink in colour, spindle shaped and normally feed on the underside of leaflets. The mites are difficult to see with the naked eye. Vector (*A. cajani*) is worm like eriophyid mite about 200-250 um in length and has two pair of legs attached one after another. Adults measure 200 to 250 μm and have a life cycle of about 2 weeks comprising egg (30 x 40 µm) and two nymphal stages. It lays milky white eggs on the vegetative terminals. At room temperature eggs hatch in 3-5 days and the adult emerges from the final nymphal stage about a week later (Oldfield *et al.*, 1981).

Figure 5.4: Eriophyid Mite (Magnified) (Ranga Rao and Shanower, 1999).

Many of nymphs are found on young folded leaflets. *A. cajani* is light-shy and seen to feed on lower side of the leafltets, preferably at the terminal of pigeonpea plants and found totally or partially buried in thick mass of hair of pigeonpea leaves (Ghanekar *et al.*, 1992). The wind dispersal of infective mites may cause plant to plant infestation and its spread to adjoining crops. They can be wind borne up to 2 km from the source of inoculum. Perennial and volunteer pigeonpeas, and the ratooned growth of harvested plants provide reservoirs of the mite vector and the pathogen. Disease incidence is high when pigeonpeas are inter- or mixed cropped with sorghum or millets.

As a result of the transmission of the sterility mosaic virus, infected plants develop light green or chlorotic leaves which have mosaic patterns. Most infected plants do not bear flowers, hence the name of the disease. Vegetative buds harbour significantly larger population of mites than leaves.

Pathogen Vector Relationship

Single mite is sufficient to transmit the PPSMV successfully but 10 viruliferous mites are required for cent per cent infection and nymph and adults both are equally effective in PPSMV transmission (Ramakrishan and kandaswamy, 1972 Janarthan, 1972, 1973). Incubation period may be 18-43 days depending upon the number of mites present on leaves (Ramakrishan and kandaswamy, 1972). Rathi, 1983 reported incubation period may be 18- 23 days. The minimum acquisition access period was found to be 30 min (Nene, 1995). Vector/SM pathogen may have more than one biotype/strain at the same location (Reddy, *et al.*, 1991) or at geographically separated location (Reddy, *et al.*, 1993). SMD involves the virus, mite vector, cultivar and environmental conditions. Infected perennial and volunteer plants serve as a source for both the virus and its vector mites, and play an important role in the disease cycle (Ahmed, and Benigno, 1985).

Epidemiology

Survival and Spread of PPSMV and its Vector in Nature

PPSMV and its vector survive on the pigeonpea (Seth, 1965), *Oxalis corniculata* Rathi,(1983) and off-season and ratooned pigeonpea (Reddy, *et al.*, 1990, Singh and Rathi, 1995). Ghanekar *et al.* (1992) observed that *A. cajani* and mild mosaic symptoms on *Atylosia scarabaeoides* but failed to transmit the pathogen to healthy pigeonpea. Later, Reddy *et al.* (1987) confirmed (Cajanus) *Atylosia scarabaeoides* as an alternative host of PPSMV and its vector. Singh and Rathi, (1995) observed that *A. cajani* on the *Cannabis sativa* (Bhang) but its role in disease cycle is not confirmed. Vector is carried alive through wind against the wind direction and as for as 2 km in wind direction (Reddy, *et al.*, 1990) from the source of inoculums. Very recently, PPSMV was transmitted by mechanical sap inoculation onto *Nicotiana benthamiana* and *N. clevelandii* (Kurnar *et al.*, 2002).

The epidemiology of SMD involves the virus, mite vector, cultivar and environmental conditions. Infected perennial and volunteer plants serve as a source for both the virus and its vector mites, and play an important role in the disease cycle.

Meteorological Factors and Vector Population

Seasonal variation in vector population has been reported by Reddy and Raju (1993) High humidity prevailing after summer rain did not favour mite multiplication (Ramakrishan and kandaswamy, 1972) but disease was highest up to 95 per cent wherever the crop was irrigated or grown near the other irrigated crops, possibly due to increased population of vector. Highest summer rainfall (149.1 mm), lowest maximum (37.7°C) and minimum (21.9°C) temperature highest maximum (69 per cent) and minimum (29.3 per cent) relative humidity were associated climatic factors during summer that favour an outbreak of SM in following crop season at ICRICAT and temperature ranging from 10 to 25 °C (minimum) and 25 -35°C (maximum) was congenial for the build up mite population (Dhar, *et al.*, 1995). Mite population was positively correlated with rainfall and relative humidity and negatively with temperature (Reddy and Raju, 1993) wherever, Singh and Rathi, (1997) reported that mite population positively correlated with maximum and minimum temperature negatively correlated with maximum relative humidity.

Interaction of PPSMV with other Pathogens and Pests

In addition to sterility mosaic, several diseases caused by fungi, bacteria, viruses and phytoplasmas also affect pigeonpea, Infection with one pathogen either Increases the severity of another disease or protects the plants. PPSMV infection protects pigeonpea from the severity of *Fusarium udum* (Anon 1964; Chadha and Raychaudhary, 1966) and Mungbean yellow mosaic virus (Rathi, 1977), but it predisposed infected plants to powdery mildew infection and red spider mites (*Schizotetranychus cajani*) (Anon. 1979; Reddy *et al.*, 1989; Sithanantham *et al.*, 1989)

Management Approaches

The best option for prevent mite (*A. cajani*) and also sterility mosaic disease by the use of resistant/tolerant varieties of pigeonpea.

Cultural and Sanitary Methods

The various crop management methods to reduce mite infestation include: the destruction of mite-infested plants at an early stage of the crop growth, sowing new pigeonpea (*Cajanus cajan*) crops away from perennial pigeonpea, the rotation of pigeonpea with other crops to reduce vector populations, and the removal of perennial and volunteer plants of pigeonpea (Seth, 1965, Raychaudhary and Nariani, 1977). Ramakrishnan and Kandaswamy (1972) recorded higher disease incidence in the crop sown in February and March than those in other months whereas, Rathi (1977) and singh and Rathi (1996a) found no significant difference in the disease incidence in various dates of sowing. SMD incidence was lower at 50cm spacing than 150cm and higher at low plant dencity. However plant density Ramakrishnan and Kandaswamy (1972), spacing (25cm, 50cm, and 75cm). Inter-cropping with sorghum (Bhatnagar, *et al.*, 1984), and pearl millets (Siddappaji *et al.*, 1979) or both (Zote, *et al.*, 1988), and border and inter-cropping with jwar and sunhemp (Singh, 1992, 1995) had no effect on disease incidence.

Host-Plant Resistance

The use of sterility mosaic disease (SMD)-resistant varieties of pigeonpea can reduce mite infestations (Amin *et al.*, 1993) as mites prefer SMD-susceptible plants for multiplication (Muniyappa and Nagia, 1982; Reddy, *et al.*, 1989). Alam (1931) first reported pigeonpea cultivar, Sabour 2E as resistant to SMD. ICRISAT early maturing lines ICPL-151 is tolerant to SM and others as ICPL-146, 269, 366, 8327, DA-12, DA-15, DA-51, MA-17, Sehore-367, DPPA-84-61-3, DPPA-84-8-4, Pant A-104, Bhawanisagar 1 and NPPR 1 all are resistant to SM. Besides these, few more lines with multiple disease resistance as ICT- 7198, 8024,8860, 8862, 9142, 10960, PR-5149, ICPL-83-227 (wilt and SM), ICPL 11302-11304 (Wilt, SM and Phythophthora blight) (Das and Gupta, 1992) and ICP-8861, 8862 and 10960 (Wilt, SM and Alternaria blight) has also been reported. Recently, Kulkarni *et al.* (2003) observed that, out of 62 accessions of *Cajanus scarabaeoides,* 21 accessions (ICP15684, 15688, 15692, 15695, 15697, 15699, 15700, 15701, 15702, 15703, 15707, 15712, 15725, 15726, 15728, 15734, 15736, 15737, 15739, 15740 and 15741) were resistant to both the mild and severe strain of PPSMV.

ICPL-87119, a multiple disease resistant genotype, is being extensively grown by farmers of Andhra Pradesh. Muniyappa and Nangia (1982) and Reddy and Nene (1980,1981) reported ring spot symptomson certain genotype (NPWR 15) and considered it as tolerant reaction because such genotypes showed normal flowering and podding inspite of infection. Simultaneously, they also observed that these tolerant lines did not permit prolonged multiplication of vector or not preferred by the vector.

Chemical Control

A number of organo-phosphorus chemicals are available for seed dressings for systemic protection, and foliar sprays (*e.g.* oxythioquinox, kelthane, dinocap, and tetradifon) are available to control *A. cajani* populations and to minimize the spread (Ghanekar *et al.*, 1992). The timing of applications and dosage are critical for effective control.

Seed Treatment

A large number of granular insecticides *viz.*; Temik, Thimet and Disyton, as basal and top dressing, were not effective Ramakrishnan and Kandaswamy (1972) but test plants from the seeds treated with Temik 10G exhibited no symptoms after inoculation in glass house. Seed dressing with Furadan 3G (Carbofuran) (25 per cent) was effective up to 45 days (Ghanekar *et al.*, 1992).

Foliar Spray

Spraying of Tedion, Morestan (Oxythioquinox) and Dicofol (Kelthane) (0.1 per cent) killed the mites 90 per cent (Ghanekar *et al.*, 1992). Three sprays of Metasystox (0.05 per cent), Nuvacran (Monocrotophos) (0.04 per cent) and Karathane (Dinocap) (0.5 per cent), starting from the first symptoms appearance and after 20 days intervals, failed to provide protection under field conditions (Singh, 1992). Nematicides also proved to be ineffective against SMD while spray of acaricides, *viz.*; Morocide, Dicofol (Kelthane), Sulphur, Moreston and Golecron, at 10 to 15 days interval, reduced the disease incidence up to 50 per cent Ramakrishnan and Kandaswamy (1972). Disease incidence was reduced by 43.0 per cent in Demecron (0.05 per cent) (Janarthan, 1973) and Metasystox (Oxydemeton-methyl) (0.1 per cent) and Quenching oil (2 per cent) spray (Rathi, 1977), as compared to unsprayed plots.

Soil Application

Soil application of Furadon 3G (40 kg/ha) and Temik (15kg/ha) was effective up to 75 days after sowing (Ghanekar *et al.*, 1992).

Field Monitoring/Economic Threshold Levels

The migration of *A. cajani* is difficult to monitor. The presence of *A. cajani* is noticed as a result of the symptoms on pigeonpea plants due to pigeonpea sterility mosaic virus (PPSMV) infection. The monitoring of mite populations on the leaves in crops may aid in decision-making about the necessity and timing of acaricidal applications to prevent secondary spread in the field [more than five mites per leaf are required for the successful spread of *A. cajani* in the field, Reddy *et al.* (1989).

IPM Approaches

Seed treatment with systemic acaricides will prevent early mite infestations. Foliar sprays to check the secondary spread in the field and the cultivation of sterility mosaic disease (SMD)-resistant pigeonpea cultivars can limit the perpetuation of *A. cajani* in the field.

Phytosanitary Measures

A. cajani is only found in Asia, it is important to restrict the movement of SMD-infected pigeonpea plant material to the regions where this disease is not known to occur (*e.g.* Australia, Africa and the Americas). In addition, PPSMV, which is vectored by A. cajani, occurs as various isolates in several geographic regions within Asia. Therefore, the movement of infected plant material should be restricted even within Asia, to avoid the introduction of new PPSMV isolates into different agro-ecological regions.

Early Warning Systems

No specific early warning systems are available to indicate outbreaks of A. cajani. However, the summer rain (in March to April) contributes to a high mite infestation and an increased incidence of SMD on new pigeonpea crops (Reddy and Raju, 1993). This is because the summer rain supports the survival of leftover SMD-infected pigeonpea plants in the fields. Such plants harbour *A. cajani* and allow a high multiplication of the pest during the off-season. The plants act as a source of A. cajani spread and contribute to an increased mite infestation on the newly sown crop.

REFERENCES

Agrawal, S. C. 2003. Diseases of Pigeonpea. Concept Publishing Company, New Delhi. 352 pp. Devraj B. and Jain R. 2011. Pulse Expert: An expert system for the diagnosis and control of diseases in pulse crops. Expert Systems with Applications 38 (9): 11463-11471.

Ahmed, K. M. and Benigno, D. A, 1985. Virus-vector relationship in mosaic disease of garlic. *Indian Phytopath.*, 38: 121 -1 25.

Alam, M., 1931. Admistrative report of Botanical Section 1930-31. Rept. Expt. Agric. Bihar and Orissa, India pp. 42-65.

Alam, M., 1933. Arhar sterility. Proc. of the 20th Annual Meeting of the Indian Science Congress, Poona, Section: Agriculture 43: 15-1 6.

Amin, K. S., Reddy, M. V., Nene, Y. L, Raju, T. N., Pratibha Shukla, Zote, K. K., Arjunan, G., Bendre, J. N., Rathi, Y. P. S. Sinha, B. K., Gupta, R. P., Anikumar, T. B., Chauhan, V. B., Gurdeep Singh, Jha, D. K. And Gangadharan, K., 1993. Multlocatronal evaluation of pigeonpea for broad based resistance to sterility mosaic disease in India. *Indian J.of Agric. Sci*, 60: 542-546.

Anonymous, 1938. New Plant disease recorded in India. *Intern. Bull. Pl. Prot.* 12: 122-123.

Anonymous, 1964. Agric. *Res.* 4: 209-222.

Anonymous, 1979. ICRISAT. Ann. Rept. 1975-76. Hyderabad, India, 87-111.

Bhatnagar, V. S., Nene, Y. L. and Jadhav, D. R., 1984. Sterility mosaic disease of Pigeonpea in a sorghum based cropping system. *Int. Pigeonpea Newsl.*, 3: 37-38.

Bressani, R, Gómez-Brenes, R. A. and Elías, L.G. 1986. "Nutritional quality of Pigeonpea protein, immature and ripe, and its supplementary value for cereals". *Arch Latinoam Nutr.*, 36 (1): 108–16. PMID 363219.

Capoor, S. P., 1950. Report on Pigeonpea sterility mosaic with symptoms and transmission *Proc. Indian Sci. Cong.* 37: 56-59.

Capoor, S. P., 1952, Observation on the sterility disease of pigeonpea in Bombay. *Indian J. Agric. Sci.*, 22: 27 1.

Chadha, K.C. and Raychaudhary, S. P., 1966. Interaction between sterility virus and *Fusarium udum* Butl. In pigeonpea. *Indian J Agric. Sci.*, 36: 133-139.

Chaohong, Z., Saxena, K. B. Zheng, L. Xuxiao, Z. And Shiying, Y. 2001. Pigeonpea diaeaes in China. *Chickpea Pigeonpea Newl.* 8: 39.

Dahal, G. And Neupane, R. K. 1991. Incidence and effect of sterility mosaic disease of Pigeonpea in Nepal. *Inter. Pigeonpea Newl.* 13: 23-24.

Das, N. D. Gupta, S. K. 1992. Evaluation of some pigeonpea lines for resistance to wilt and sterility mosaic. *Inter. Pigeonpea Newl.* 15: 23-24.

Dash P. K. 2012. Pigeonpea hybrid Pushkal. *Agbiotech Digest*, 3(2): 1-2.

Dhar, V. Rathore, Y. S. And De, R. K. 1995. Studies on population dynamics of *Aceria cajani* vector of Pigeonpea sterility mosaic disease. *J. Mycol. Pl. Pathol.* 25 (1 and 2): 72-73.

Ghanekar, A. M., Sheila, V. K., Beniwal, S. P. S., Reddy, M. V., and Nene, Y. L. 1992. Sterility mosaic of pigeonpea. Pages 415-428 in: Plant Diseases of International Importance: Diseases of Cereals and Pulses, Vol. I. (U. S. Singh, A. N. Mukhooadhyay, J. Kumar, and H. S. Chaubey, eds.) New Jersey, USA, Prentice Hall.

Janarthan, R, 1973. Ann Rept 1972-73 T N A U Colmabatore, India 138 pp.

Janarthan, R, Navneethan, G. Subramanian, K. S. And Samual, G. S. 1972. Sme observation of transmission of Sterility mosaic of pigeonpea. Cur. Sci. 41: 646-647.

Kandaswamy, T. K. and Ramakrishnan, K., 1960, An epiphytotic of sterility mosaic of pigeonpea at Coimbattorc. Madras Agric. *J.*, 47: 440-441.

Kannaiyan J, Nene Y. L, Reddy M. V, and Raju T. N. 1981. Prevalence of wilt and sterility mosaic in India (1975-80) *Inter. Pigeonpea Newl.* 1: 24-26.

Kannaiyan J, Reddy M. V, Nene Y. L, Ryan J. G. and Raju T. N. 1984. P Pigeonpea disease and associated crop losses in Asia, Africa and Americas. *Trop. Pest Manag.*, 30: 62-71.

Kulkarni, N. K. Kumar, P. L. Muniyappa, V., Jones, A. T. and Reddy, D. V. R. 2002. Transmission of Sterility mosaic virus by eriophyid mite, Aceria cajani (Acari: Arthopode. *Plant Disease*. 86: 44-48.

Kulkarni, N. K., Reddy, A. S., Kumar, P. L., Vijaynarasimha, J., Rangasawamy, T. K., Muniyappa, V., Reddy, L. J. Saxena, K. B., Jones, A. T. and Reddy, D. V. R.

2003. Broad based pigeonpea Sterility mosaic disease in accessions of *Cajanus scarabaeoides* (L) Benth. *Indian J. Pl. Prot.* 31(1) : 6-11

Kumar, P. L, 2000. A novel virus associated with pigeonpea sterility mosaic disease in India. In Association of Applied Biologist Conference on Advances in Plant Virology. September *20-22,* 2000. Dundee University, Dundee, Scotland.

Kumar, P. L., Fenton, B., Duncan, G., Jones, A. T., Sreenivasulu, P., and Reddy, D. V. R. 2001. Assessment of variation in *Aceria cajani* (Acari: Eriophyidae) using analysis of nuclear rDNA ITS regions and scanning electron microscopy: Implications for the variability observed in host plant resistance to pigeonpea sterility mosaic disease. *Ann. Appl. Biol.* 139: 61-73.

Kumar, P. L., Jones, A. T. and Reddy, D. V. R., 2002, Mechanical transmission of Pigeonpea sterility mosaic virus. J. *Mycol. PI. Pathol., 32:* pp15.

Kumar, P. L., Jones, A. T. Sreenivasulu, P. and Reddy, D.V.R., 1998. Isolation of a virus associated with sterility mosaic disease of pigeonpea (*Cajanus cajan*). *Indian J. Plant Prot.,* 26: 164-166.

Kumar, P. L., Jones, A. T., and Reddy, D. V. R. 2002. Methods for the detection of Pigeonpea sterility mosaic virus and screening for SMD resistance - Training course manual. ICRISAT Publication, Patancheru, India.

Kumar, P. L., Jones, A. T., Duccan, G., Roberts, I. M. and Reddy, D. V. R., 2001a Characterization of a novel mite transmitted virus associated with pigeonpea sterility mosaic disease. *Phytopathology,* 91 : 51.

Kumar, P. L., Jones, A. T., Fenton, B., Sreenivasulu, P. and Reddy, D. V. R., 2001 b, Characterization of a vlrus from Pigeonpea with affinities to species in the genus *Aureuvirus,* Family *Tombusviridae. Plant Disease,* 85: 208- 215.

Kumar, P. L., Jones, A. T., Sreenivasalu, P. and Reddy, D. V. R. 2000. Breakthrough in the identification of the causal agent of pigeonpea sterility maosaic disease. J. Mycol. Plant Pathol. 30: 249.

Kumar, P. L., Jones, A. T., Sreenivasulu, P. and Reddy. D V. R, 1999a. Unfolding the mystery of the causal agent of pigeonpea sterility mosaic disease. In *Abstracts, lnternational Conference on Life Sciences in Next Milleniurn* University of Hyderbad, India, pp-65-66.

Kurnar, P. L., Jones, A. T., Sreenivasulu, P. and Reddy, D. V. R, 1999b, Characterization of a virus associated with sterility mosaic disease of pigeonpea. In: Abstracts, *xth International Congress of Virology,* 9-13 August, 1999, Sydney. Pp90.

Mali, V. R., Shirashta, A. M. and Godbole, G. M., 1977. Occurrence of PSMV In Marathawada, *Res. Bull. Marathawada Agnc Univ.* 1: 148-149.

Mc Rae, W. 1932, Report of Imperial mycologist. Sci. Rept. Imp. Instt. *Agric.* Res Pusa 1930-31, pp. 73-81.

Mitra, M, 1931. Report of Imperial Mycologist. Sci. Rept. Imp. Inst. Agric. Res. Pusa 1929-30, 45-71.

Mitra, M. 1931. Report of Imperial Mycologists. *Scientific Reports of the Agricultural Research Institute,* Pusa, 1929-30. Pusa, India, pp.58-71.

Muller, R. A. E., Parthasarathy Rao, P and Subba Rao, K. V. 1990. Pigeonpea: markets and outlook. In The Pigeonpea (eds. Y. L. Nene, S. D. Hall and V. K. Shiela) CAB international ICRISA T, pp 457-479.

Mumyappa, V. and Chandrashekharaiah, S. C. 1980. Prevalence of sterility mosaic Disease in Karnataka. *Curr. Sci.,* 9: 196.

Muniyappa,V. and Nangia, N., 1982. Pigeonpea cultivars and selection for resistance to sterility mosaic in relation to prevalence of eriophyid mite (*Acena cajani* Channabasavanna). *Trop. Gram. Legume Bull.,* 25: 28-30.

Nambiar, K. K. N. and Ramakrishnan, K., 1968, Studies on Pigeonpea sterility mosaic disease VIII. Effect of disease on carbohydrate. Proceedings of Indian Academy of Sciences. Section B, 68: 259-300.

Nambiar, K. K. N. and Ramakrishnan, K., 1969a. Studies on Pigeonpea sterility mosaic disease VIII. Effect on mineral metabolism. Proceedings of Indian Academy of Sciences. Section B, 70: 37-41.

Nambiar, K. K. N. and Ramakrishnan, K., 1969b. Studies on Pigeonpea sterility mosaic disease IX. Effect on nitrogen metabolism. In Proceedings of lndian Academy of Sciences. Section B, 70: 200-207.

Narayanaswamy, P. and Ramakrishanan, K. 1965a. Studies on the sterility mosaic disease on pigeonpea. I- Transmission of the disease. In: Proceedings of lndian Academy of Sciences, Section B, 61 : *73-76.*

Narayanaswamy, P. and Ramakrishanan, K. 1965b. Studies on the sterility mosaic disease on pigeonpea. II- Carbohydrate metabolism of infected plant. In: Proceedings of lndian Academy of Sciences. Section B, 62: 130-1 39.

Narayanaswamy, P. and Ramakrishanan, K. 1965c. Studies on the sterility mosaic disease on pigeonpea.lV- Organic acid metabolism and respiration of infected plant. In: Proceedings of lndian Academy of Sciences. Section B, 64: 135-152.

Narayanaswamy, P., Sheshdri, A.R. and Ramakrishnan, K., 1963. Perliminary notes on the suspected nematode transmission of redgram sterility mosaic. *Madras Agric. J.,* 50: 109-110.

Nene, Y. L. 1995. Sterility mosaic disease on pigeonpea: Challenge continues Presidential Adreess. J. *Mycol. PI. Pathol.,* 25(1 and 2): 1-11.

Nene, Y. L, and Reddy, M. V 1976a. Screening for resistance to sterility mosaic of pigeonpea. *Plant Dis. Reptr.,* 60: 1034-1036.

Nene, Y. L. 1972. A survey of viral diseases of the pulse crops in U.P. Res. *Bull.* 4, Pantanagar, U.P., India, GBPUA and T Pantanagar.

Nene, Y. L., Kanniyan, J. and Reddy, M. V. 1981a. Pigeonpea diseases: resistance-screening techniques. *Information Bull. No.9.* Patancheru 502 324, Andhra Pradesh, India: International Crops Research Institute for the Semi-Arid Tropics. 19pp.

Nene, Y. L., Kanniyan, J. and Reddy, M. V., 1981 b, Resistance to major pigeonpea diseases. In: *Proceedings of International Workshop on Pigeonpea,* Dec. 15-16, 1980, ICRISAT center, Patancheru, India, pp 121-128.

Nene, Y. L., Reddy, M. V. Beniwal, S. P.S., Mahmood, M., Zote, K. K., Singh, R. N and Sivaprakasam, K., 1989. Multilocational testing of pigeonpea for broad based resistance, *Indian Phytopathol.,* 42: 44-48.

Nene, Y.L, Sheil,a V. K. and Sharma, S. B. 1989. A World List of Chickpea (*Cicer arietinum* L.) and Pigeonpea (*Cajanus cajan* L.) Millsp.) Pathogens. Legume Pathology Progress Report-7.Patancheru, A.P., India: ICRISAT, 23.

Nene, Y.L., Sheila, V.K. and Sharma, S.B., 1996, A world list of Chichpea and Pigeonpea pathogens, stheedition. ICRISAT publication, pp.25.

Newton, W. and Peiris, J.W.L. 1953. Virus diseases of plants in Ceylon.*FAO Plant Prot. Bull.,* 2: 17-21.

Oldfield, G.N., Reddy, M.V., Nene, Y.L. and Reed, W., 1981. Preliminary studies of the eriophyid vector of sterility mosaic. Int. *Pigeonpea Newsl.,* 1. 25-27.

Paliwal,Y.C. 1980. Relationship of Wheat streak mosaic and barley stripe mosaic viruses to vector and non-vector Eriophyid mites. *Archives of Virology,* 63: 123-132.

Paliwal,Y. C. and Slykhuis, J.T. 1967. Localization of wheat streak mosaic virus in the alimentary canal tract of its vector, *Aceria tulipae* K. *Virology,* 32: 344-353. Prameela, H. A,, Viswanath, S. and Anilkumar, T. B. 1990. The predesposing effect of sterility mosaic to powdery mildew infection in pigeonpea. *Zentralblatt fur Mikrobiologie,* 144: 111-114.

Prasad, S. N., 1965. Studies on the sterility mosaic disease of Arhar (*Cajanus cajan*). *Allahabad Farmer,* 39: 235-237.

Ramakrishnan, K. and Kandaswarny, T. K, 1972. Investigation on virus diseases of pulse crops in Tamil Nadu. *Final Technical Report,* T.N.A.U. Coimbattore, Tamil Nadu, India. pp.53.

Ranga Rao, G.V., and Shanower, T.G. 1999. Identification and Management of Pigeonpea and Chickpea Insect Pests in Asia. Information Bulletin no. 57. (In En. Summaries in En, Fr.) Patancheru, 502 324, A.P., India: International Crops Research Institute for the Semi- Arid Tropics. ISBN 92–9066–412–6

Rathi, Y. P. S., 1977. Ann. Rept. Presented at *All India Pulses Workshop* 1976-77, held at T.N.A.U., Coimbattore, India, pp.13.

Rathi, Y. P. S., 1983. Studies on sterility mosaic disease of pigeonpea. *Final Technical Report, GBPU 8 AT,* Pantanagar, U P., India. pp. 23.

Rathi, Y. P. S., Bhat, A. and Singh, U. S. 1986. Biochemical changes In: pigeonpea (*Cajanus cajan*) leaves in relation to resistance against sterility mosaic disease. J. *Bioscience,* 10: 467-474.

Raychaudhary, S. P. and Narayani, T. K., 1977. Sterility mosair: of pigeonoea, In Virus and Mycoplasma diseases of plants in India, Oxford and IBH Publishing Co., pp. 19-20.

Reddy MV and Nene YL (1979) Ring spot symptoms: A genotypic expression of Pigeonpea sterility mosaic. *Trop. Grain Legume Bull.*, 15: 27-29.

Reddy MV, Sharma SB and Nene YL (1990) Pigeonpea: Disease management. in: The Pigeonpea. (Y. L. Nene, S. D. Hall, and V. K. Sheila, eds.) CAB International, Wallingford, UK. pp. 303-347.

Reddy, M. V., Raju, T. N., Sharma, S. B., Nene, Y. L., McDonald, D., Pande, S. and Sharma, M. 2012. Handbook of Pigeonpea Diseases (Revised). Information Bulletin No. 42. Patancheru, AP 502 324, India: International Crops Research Institute for the Semi-Arid Tropics. pp. 30-33.

Reddy, M. V, and Nene, Y. L., 1980. Influence of sterility mosaic resistant pigeonpea on multiplication the mite vector. *Indian Phytopathology*, 33: 61-63.

Reddy, M. V. and Raju T. N. 1993. Some clue to increased incidence and seasonal variation of Pigeonpea sterility mosaic in peninsular India. *Inter. Pigeonpea Newsl.*, 18: 22-24, ICRISAT Centre, Patancheru.

Reddy, M. V. and Nene, Y. L., 1981. Estimation of yield loss due to sterility mosaic. In: Proceedings of International Workshop on Pigeonpea 15-19, Dec. 1980. ICRISAT centre, India Vol: 2, Patancheru, A.P., pp. 405-312.

Reddy, M. V., 1987, Diseases of Pigeonpea and Chickpea and Their management. In Plant Protection in Field Crops. (Eds. M Veerabhdra Rao and Sithanantham) Plant Protection Association of India, Rajendranagar, Hyderabad, India. 175-1 84.

Reddy, M. V., Beniwal, S. P. S., Sheila, V. K., Sithananthan, S and Nene. Y. L. 1989. Role of eriophyid mite (Aceria cajani) in transmission and spread of sterility mosaic of pigeonpea. Pp121-127 in Progress in Acarology (Channbasavanna, G. P. and Virakthmath, C.A. eds), Oxford and IBH Publishing Co. New Dellhi, India.

Reddy, M. V., Raju, T. N. and Lenne, J. M., 1998, Diseases of pigeonpea. In: The Pathology of food and pasture legumes (Eds. D.J. Allen and J. M. Lenne) CAB International ICRISAT, Pp 51 7-558.

Reddy, M. V., Raju, T. N. and Nene, Y. L., 1991, Appearance of a new strain of pigeonpea sterility mosaic pathogen. *International Pigeonpea News Letter*, 14: 22-23.

Reddy, M. V., Sharma, S. B. and Nene, Y. L., 1990. Pigeonpea: Disease management. In The Pigeonpea (Eds., Y. L. Nene, S. D. Hall and V. K. Sheila). CAB International-ICRISA T, pp. 349-374.

Reddy, M. V., Sheila, V. K., Murthy, A. K. and Padma, N, 1995. Mechanism of resistance to Aceria cajani in pigeonpea. Int. *J.* Trop. Plant Dis., 13: 51-57.

Reddy, M.V., Raju, T.N., Sharma, S.B., Nene, Y.L. and McDonald, D. 1993. Handbook of pigeonpea diseases. Information Bulletin No. 42. International Crops Research Institute for the Semi-Arid Tropics. ISBN: 92-9066-277-8.

Reddy, S. V., Reddy, M. V., Ghanekar, A. M., Nene, Y. L. and Amin, K.S. 1988. Annual recurrence of of pigeonpea sterility mosaic in Eastern U.P., India, *Int. Pigeonpea Newsl.*, 7: 30-33.

Seth, M. L. 1962, Transmission of pigeonpea sterility mosaic by an eriophyid mite *Indian Phytopathol.*, 15. 225-227.

Seth, M. L. 1965. Further observation and studies on pigeonpea sterility mosaic. *Indian Phytopathol*, 18: 317-318.

Sheila, V.K., Manohar, D.K. and Nene, Y.L., 1988. Biology and morphology of *Aceria cajani. Int. Pigeonpea Newsl.*, 7: 28-29.

Shewood, J. L., 1987. Comparison of a Filter paper lmmunobinding assay, Western blotting and an Enzyme Linked lmmunosorbant Assay for the detection of Wheat streak mosaic virus. *J. Phytopathol.*, 11 3: 68-75

Siddappaji, C., Raghumurthy, M., and ChannaBasavanna, G. P., 1979. Effect of intercropping on the incidence of pigeonpea sterility mosaic transmitted by *Aceria cajani* (Acari-Eriophyidae). (ed. G. P. Channbasavanna) In: Contributions to Acarology in India, *In Proceedings of All India Symposium in Acarology*, Bangalore, pp. 80-84.

Singh AK and Rathi YPS (1997) Epidemiology of vector of Pigeonpea sterility mosaic virus. *Indian J. Virol.*, 13: 143-145.

Singh, A. K. and Rathi, Y. P. S., 1994a. Histopathology, cytopathology and fluorescent microscopy of infected pigeonpea. *New Agriculturist. 5* : 225.

Singh, A. K. and Rathi, Y. P. S., 1994b. Nature of losses caused by sterility mosaic in pigeonpea. *J. Appl. Biol.*, 4: 35-37.

Singh, A. K. and Rathi, Y. P. S., 1995, Survival of pigeonpea sterility mosaic virus (PSMV) and its vector (Aceria cajani). New Approaches Agric. Tech. 3: 595-598.

Singh, A. K. and Rathi, Y. P. S., 1996a, Nuclear inclusions produced in the phloem cells of pigeonpea diseased mid-vein infected with pigeonpea sterility mosaic virus. *lndian J. Virol.*, 12: 147-149.

Singh, A. K. and Rathi, Y. P. S., 1996b, Nuclear inclusions produced in the phloem cells of pigeonpea diseased midvein infected with pigeonpea sterility mosaic virus. *lndian J. Virol.*, 12: 147-149.

Singh, A. K., 1992. Sterility mosaic of pigeonpea: Etiology, Epidemiology and management. Ph. D. Thesis, GBPU and AT, Pantanagar, U. P., India, 151pp.

Singh, A. K., Rathi, Y. P. S. and Agrawal, K. C. 1999. Sterility mosaic of Pigeonpea A challenge of 20th Century. *lndian J. Virol.*, 15: 85-92.

Singh, O., Shah, R. P., Deewan, B. B. Sexana, K. B., and Bharti, M. P. 1989. ICPL-366, A long duration pigeonpea cultivar found promising in Nepal. *Intern. Pigeonpea Newl.* 11: 14.

Singh, R, and Mall, T. P., 1976. Effect of infection by arhar sterility mosaic virus on the hill-reaction and primary productivity of pigeonpea leaves. *Indian J. Experi. Biol.,* 14: 376-377.

Singh, R, and Mall, T. P., 1978. Effect of pigeonpea sterility mosaic virus on yield and reproductive capacity of pigeonpea cultivars. *Indiain J. Mycol. Pl. Pathol.,* 8: 178-1 82.

Sithanantham, S., Reddy, M. V. and Rameshwar Rao 1989. Increased damage by the spider mite, *Schizotetranychus cajani* in pigeonpea plants affected by sterility mosaic. Pages 11-14 (In) Progress in Acarology, Vol 2. G. P Channabasavnna and C. A. Viraktamath, eds. Oxford and IBH Publishing Co., India.

Stem-Margolina, V., Cherni, N. E. and Razvyazkina, G. M., 1969, Phytopathogenic viruses In plant cells and in the mite vector- electron microscopic investigation. *lzvestia Akad Nauk. Sr. Biology,* 1: *62-68.*

Su. U. 1931. Plant disease in Burma. *Inter. Bull. Plant Prot.,* 5: 141-142.

Takahashi, Y. and Orlob, G. B. 1969. Distribution of wheat streak mosaic virus like particles in *Aceria tulipae. Virology, 38:* 230-240.

Zote, K. K., Mali, V. R., Mayee, C. D., Kulkarni, S. V. and Mote, T. S., 1991. Outbreak of sterility mosaic disease of pigeaonpea in Marathwada region. *Int. Pigeonpea Newsl., 14*: 19-21.

Zote. K. K., Dandanaik, B. P., Khalikar, P. V, and Raut, K. G., 1988. Effect of intercropping with sorghum and pearl millet on incidence of pigeonpea sterility mosaic. *Int. Pigeonpea Newsl.,* 14: 19-21.

Chapter 6

Diseases of Mungbean and their Management

Jai Singh, R.K. Singh & Sumit Kumar Pandey

In India pulses have been important constituents of food preparations since the vedic times (Achaya, 1998). Agri-archaeological researches have revealed that pulses have been grown in the Indian sub-continent for more than 6000 years. We find references to most of the pulses in the scriptures of ancient India, these are the same pulses that we cultivate today (Nene, 2000).

Pulses constitute a very important source of dietary protein for majority of the Indian people, as they provide a perfect mix of vegetable protein component of high biological value when supplemented with cereals. It is also an excellent feed and fodder for livestock. These crops are also endowed with the ability to improve nutrient supply in soils through addition of organic matter (leaf drop) and nitrogen, by fixing atmospheric nitrogen, carbon sequestration, soil amelioration. It also have ability for lower water requirement, capability to withstand harsh climate, pulses have remained an integral component of sustainable crop production system in the dry or rainfed areas. They offer good scope for crop diversification and intensification. This is more significant in view of the impending danger of non-availability of nitrogen as fertilizer from petroleum sources.

Mungbean (*Vigna radiata* (L.) Wilczek) commonly known as green gram or golden gram or sona or mung is an important pulse crop being a crop of tropical and subtropical region mungbean crop has ability to tolerate warm temperature as high as 40°C. It is considered as a "poor men's protein (Mian, 1976). Apart from 26 per cent protein, it also contains 51 per cent carbohydrate, 10 per cent moisture, 4 per cent minerals and 3 per cent vitamins (Khan 1981). It is widely cultivated throughout Asia including India, Pakistan, Bangladesh, Sri Lanka, Thailand, Combodia, Vietnam, Indonesia, Malaysia and South China (Yadav *et al.*, 2014). It is also grown in parts of East and Central Africa, Australia, United State of America and West Indies. In India it is grown in almost all parts of the country (Table-1) during both summer and rainy (kharif) season. India is the largest producer of mungbean and it is the third most important pulse crop with an area of approximately 3.43 million hectares (about 15 per cent of the national pulse crop area), production 1.71 million tones of grain with productivity of 498 kg/ha(Anonymous, 2012).The world annual production area of mungbean is about 5.5 million ha^{-1} and contributes about 75 per cent of the world production of mungbean (Taunk *et al.*, 2012). To meet global demand, it is imperative to improve the current average global productivity (~400kg ha^{-1}) as well as to expand the crop into new regions (Nair *et al.*, 2012). The yield of mungbean is affected by several biotic and abiotic factors (Grewal, 1988). Among the biotic factors, Powdery mildew, yellow mosaic disease, Cercospora leaf spot (CLS), Anthracnose, Dry root rot, Leaf crinkle disease, Web blight and Rust are important diseases and are being discussed here.

Powdery Mildew

Powdery mildew is a major problem to mungbean cultivation. Recently its become a limiting factor in rabi sown mungbean crop in Odisha, Andhra Pradesh, Karnataka,and Taminlnadu. The disease has been reported from Australia, India, Phillipines, Sri Lanka, Taiwan, Colombia, Ethopia, Thailand, Korea and USA. Disease causes 9-50 per cent yield losses at farmers field (AVRDC, 1982).

Symptoms

The disease appears on all above ground part of plant. Initially white powdery patches appear on leaves and other green parts which later become dull colored. These patches gradually increase in size and become circular covering the lower surface also. When the infection is severe, both the surfaces of the leaves are completely covered by whitish powdery growth. The powdery mass consisting of mycelium and conidia evenly turn dirty white. Severely affected parts get shriveled and distorted. In severe infections, foliage becomes yellow causing premature defoliation. The disease also creates forced maturity of the infected plants which results in heavy yield losses.

Pathogen

The powdery mildew incited by obligate parasite *Erysiphae polygoni*, is ectophytic, spreading on the surface of the leaf, sending haustoria into the epidermal cells to obtain nutrients. Conidiophores arise vertically from the leaf surface, bearing conidia in short chains. Conidia are hyaline, thin walled, elliptical or barrel shaped or cylindrical and single celled. Later in the season, cleistothecia appear as minute, black, globose structures with myceloid appendages. Each cleistothecium contains 4-8 asci and each ascus contains 3-8 ascospores which are elliptical, hyaline and single celled.

Favourable Conditions

The pathogen favored by dry weather, temperature range between 22 to 26 °C and relative humidity of 80-85 per cent. Disease incidence is higher on 60 to 80 days old plants than the 40 days old plants. The disease is more severe in closely planted crop during late July than in those with widely spaced late sown crop *i.e.* mid August.

Disease Cycle

The fungus is an obligate parasite and survives as cleistothecia in the infected plant debris. Primary infection is usually from ascospores which develop in cleistothecia. The pathogen has a wide host range and survives in oidial form on various hosts *viz. Vigna mungo, Vicia faba, Lathyrus sativa, Vigna ungulate etc.,* in offseason which may also act as reservoirs of the initial inoculum. The secondary spread is carried out by the air-borne conidia. Rain splash also helps in the spread of the disease.

Management

Destruction of plant debris can minimize the survival habitat of pathogen. The growing powdery mildew resistant varieties/genotype has been considered to be the most effective method of reducing disease incidence, such as TARM-1, TARM-2, TARM-18, COG-4, Pusa-105, Kamdev ML-131, ML-1299 and LGG-460 (Akhtar *et al.*, 2014 and Yadav *et al.*, 2014). Disease incidence can be minimized by early sowing with the wider spacing. Two foliar spray of Carbendazim(12 per cent) + Mancozeb (63 per cent) -75WP @ 0.2 per cent (Yadav *et al.*, 2014) or Karathane (0.1 per cent) (Suryawanshi *et al.*, 2009) or Wettable sulphur 1.5 kg or Benlet (0.05 per cent) or Tridemorph @ 0.05 per cent or difencoazole @ 0.1 per cent (Rakhonde *et al.*, 2011) or propiconazole @ 0.1 per cent (Akhtar *et al.*, 2014) at the initiation of disease and repeat 15 days later, has been found effective to control the disease under field condition. Spray with NSKE@ 50 g/lit., or Neem oil 3000ppm @ 20 ml/lit., twice at 10 days interval from early onset of disease.

Anthracnose

Anthracnose caused by *Colletotricum truncatum* (Schw.) is the most important seed-borne fungal disease of mungbean causes heavy reduction in seed quality and yield every year in varying intensity. The disease causes qualitative as well as quantitative losses (Sharma *et al.*, 1971). Losses in yield due to anthracnose have been estimated to be in the range of 24 to 67 per cent (Deeksha and Tripathi, 2002 and Shukla *et al.*, 2014). The yield losses caused by mungbean anthracnose are proportional to the disease severity and vary remarkably depending on the stage of infection, genotypes and environmental conditions.

Symptoms

The disease appears on all aerial part of the plant and at any stage of plant growth. They may occurs more often on the under surface of leaf than on the upper surface, and also on petioles, stem and pods. Initially water soaked lesions seems on pods and leaves later becoming circular, black, sunken spots with dark center and bright red orange margins on leaves and pods. In severe infections, the affected parts wither off. Seedlings get blighted due to infection soon after seed germination.

Pathogen

Five different species of *Collectotrichum* [viz. *Collectotrichum lindemuthianum* Sacc and Marga) Bri and Cav (perfect state *Glomerella lindemuthionum* (Sacc and Margn) Shear); *C. dematium* (Pres. ex Fr.) Grov., *C. trucatum* (Sehw.) Andrus and Moor and *C. graminicola* (Ces. Wilson) have been found associated with theinfected seeds causing anthracnose disease. Of these, *C. linemuthionum* and *C.capsici* are wide spread and causes severe infections under favorable environmental conditions.

Mycelium of the pathogen is septate, hyaline and branched. Conidia are produced in acervuli, arise from the stroma beneath the epidermis and later rupture to become erumpent. A few dark coloured, septate setae are seen in the acervulus. The conidiophores are hyaline and short and bear oblong or cylindrical, hyaline, thin walled, single celled conidia with oil globules. The perfect stage of the fungus produces perithecia with limited number of asci, which contain typically eight ascospores which are one or two celled with a central oil globule.

Favourable Conditions

The disease incidence was correlated positively with relative humidity and rainfall, and negatively with maximum and minimum temperature. The disease score were highest at 91-100 per cent RH and 26-29°C, with low rainfall (2.5-53.5 mm). Early sowing (23 June) were more vulnerable than the late sowing (Thakur and Khare, 1992). Intermediate rains at frequent intervals favour the epidemic development of the disease.

Disease Cycle

The fungus is seed-borne and cause primary infection. It also lives in the infected plant debris in soil. The secondary spread by air borne conidia produced on infected plant parts. Spattering rains splash associated with wind current are the main source for the local dissemination of the pathogen. The pathogen infects several leguminous hosts which may also act as reservoirs of the primary inoculum.

Management

Remove the crop debris from the field after harvest of the crop. Hot water treatment at 58°C for 15 minutes proved effective to minimize the seed borne infection and increase proportion of seed germination. Mungbean genotypes/ cultivars *viz.*, TM-92-2, TARM-18, MLTG-9 were showed highly resistance reaction against anthracnose disease (Rathaiah and Sharma, 2004 and Kulkarni *et al.*, 2009). Seed treatment play an important role in reducing disease to minimize early establishment of the pathogens. Carbendazim 50 WP @ 2 g/kg seed or Thiram+ Carbendazim (2:1) @ 3g/kg seed of were effective in eliminating the seed borne infection. Shukla *et al.* (2014) proved that, two sprays of carbendazim(0.1 per cent) are sufficient to manage the disease and realize the economic yields. Foliar sprayof

Mancozeb (@ 0.2 per cent) soon as the symptoms of disease appeared and second spray after 15 days of the first spray is also effective to reduce disease incidence.

Cercospora Leaf Spot

The occurrence of Cercospora leaf spot was first time reported from Delhi, India (Munjal *et al.*, 1960).The disease prevalent in all part of the country and causes severe leaf spotting and defoliation. Since 1973, it has been considered to be one of the most serious diseases by the Asian Vegetable Research and Development Centre, Taiwan (Mew *et al.*, 1975). Due to this disease yield losses up to 50-70 per cent has been observed (Lal *et al.*, 2001 and Chand *et al.*, 2012). This disease has been reported from Indonesia, India, Taiwan, Philippines, Thailand, Bangladesh and Malaysia (Pandey, *et al.*, 2009). In India disease occurs in most of the mungbean growing stats particularly, Punjab, Haryana, Uttar Pradesh, Madhya Pradesh, Rajasthan, Himachal Pradesh, Odisha, Assam and Maharastra.

Symptoms

The disease appears initially on the lower leaves as water soaked spot of the infected plants and then spread all over the plant. The Characteristic symptoms appear as leaf spots produced are small, numerous in number with pale brown centre and reddish brown margin. Similar spots also occur on branches, petioles and pods. Under favourable environmental conditions the spots increase in size and at the time of flowering and pod formation lead to defoliation. Sometimes the leaves may become malformed and wrinkled. Maturity is delayed in the diseased plants resulting poor pod formation. On the pods symptoms range from few spot to solid blacking and killing of the entire pod and seeds within the infected pod become shriveled, small, immature and dark in colour (Poehlman, 1991). Five species of *Cercospora* infect mungbean with slight variation in their symptoms. *C. cruenta* infects munbean and producing typical spots around flowering stage. The spots on leaves are usually angular being limited by veins, large in size (5-7mm) than produced by *C. cenescens* and have tan to grey centre surrounded by prominent reddish purple border. *C. canescens* produced numerous, small (1-3 mm), circular to semicircular pale brown spots on leaves with reddish brown border. With the advancing age and the spore formation, the spots assumes silvery to grey colour in centre and the spots are mainly form on leaves. *C. kikuchii* produced sub circular to irregular, pale brown with tan to grey center and measure 3-6 mm in diameter. The spots are elliptical on stem and pods. *C. dolichi* is prominent on rabi mungbean crops and develops reddish brown blotches. *C. corocollae* has been found attaching on mungbean.

Pathogen

Among these five species of *Cercospora viz., C. cruenta* Sacc (Perfect state: *Mycospharrella cruenta* Latham), *C. canescens* Ell. and Martin, *C. kikuchii* Matsumato and Tomoyassun, *C. dolichi* Ell and Evt., and *C. corocollae* (Speg.) former two species are frequently observed on mungbean. The fungus produces clusters of dark brown septate conidiophores. The conidia are linear, hyaline, thin walled and 5-6 septate.

The conidiophore of *C. canescens* is long, multiseptate, usually straight or slightly curved, light to dark brown, 20-250 x 4.5 – 5.0 mm; conidia hyline, straight or curved, multiseptate 50-225 x 3.5- 5 mm. The conidiophores of *C. cruenta* are short, aseptate, light brown, straight and rarely branched 10-77 x 3-5 mm in size, conidia subhyline or pale olivaceous, multiseptate, slightly curved, 30-150 x 3-5 mm. Conidiophores are pale olivaceous brown, sparingly septate, straight to curved, 10-45 x 4.5 -5.5 mm in C. dolichi; conidia olive brown, obclavate, cylindrical, multiseptate, 40-100 x 3.7-4.5 mm. In C. kikuchii, conidiophores occur in dense fascicles, medium dark brown, multiseptate, 45-220 x 4-6 mm, conidia hyaline, acicular, multiseptate, 150-300 x 3.7 – 4.5 mm.

Favourable Conditions

The Cercospora leaf spot is most prevalent in rainy season. CLS can cause heavy defoliation in severe conditions on mungbean especially at optimum temperature 25-30°C with RH 90-100 per cent. Symptoms of this disease appear mainly on leaves but also on stem, branches and pods. Hossain *et al.* (2011) described that the disease was mainly observed when crop was at flowering, pod formation and ripening stage. The severity of the disease depends on the humidity, rainfall and temperature in addition to causal agent and host

Disease Cycle

The fungus survives on diseased plant debris and on seeds. Infected seed produce seedlings with cotyledonary lesions. Sporulation on the cotyledonary lesions provides primary inoculum for infecting young leaves. In humid weather, sporulation is profuse in the resulting lesions conidia disseminated by wind and splash of rain to other leaves and stem initiate secondary infections that produce conidia which again infect other leaves, stems and pods during warm and wet conditions. Peak dispersal period for conidia occur at dew off in the morning at the onset of rainfall.

Management

Different approaches such as sanitation, tillage operations, crop rotations, destruction of infected crop debris, removal of collectoral hosts in the vicinity of the crop, host plant resistance, fungicides and different plant extracts are used to managed Cercospora leaf spot of mungbean (Hanson and Panella, 2003; Khan and Smith, 2005; Khan, 2008; Dubey and Singh, 2013). Mungbean entries ML-5, ML-65, ML-131, ML-392, ML-325, ML-406, Ml-408, ML-53, LGG-460 and ML-1299 showed resistance response to Cercospora leaf spot (Mathur *et al.*, 1990; Yadav *et al.*, 2014 and Akhtar *et al.*, 2014). Similarly, Haque *et al.* (1997) reported that twelve genotypes (NH-98, 98-cmg-003, c2/94-4-42, NM-1, NM-2, 98cmg-018, BRM-188, Co-3, Basanti, PDM-11, BARI mung-2 and VC-3960-88) were found highly resistant to Cercospora leaf spot. The resistance in mungbean is governed by a single dominant gene (Chauhan and Gupta, 2004). Agronomic practices *i.e.* intercrop being mungbean with tall growing cereals and millets as well as mulching reduces the disease incidence resulting in increase of yield. Two foliar spray of Difenconazole

(0.0125 per cent) or Carbendazim (12 per cent) + Mancozeb (63 per cent)-75WP (0.2 per cent) or Hexaconazole (5 per cent) + Captan (70 per cent) -75 WP (0.05 per cent) 0r Ridomil Gold (0.1 per cent) or Propiconazole (0.1 per cent) are effective for minimizing Cercospora leaf spot incidence, severity and increasing yield of mungbean (Kapadiya and Dhruj, 1999; Akhtar *et al.*, 2014, Shahbaz *et al.*, 2014, and Yadav *et al.*, 2014). It has been observed that neem leaf extract resulted significant reduction of Cercospora leaf spot of mungbean (Uddin *et al.*, 2013).

Rust

The disease is occurs on mungbean and may pose a threat to the successful cultivation of crop under favorable development of the disease. It has been reported from India and Indonesia, and usually occur in the rainy season.

Symptoms

Initially the disease symptom appears as circular, reddish brown pustules which appear more common on the underside of the leaves, less abundant on pods and sparingly on stems. When leaves are severely infected, both the surfaces are fully covered by rust pustules. Shriveling followed by defoliation results in yield losses.

Pathogen

The disease is caused by *Uromyces phaseoli typica* (Syn: *U. appendiculatus*). It is autoecious, long cycle rust and all the spore stages occur on the same host. The uredospores are unicellular, globose or ellipsoid, yellowish brown with echinulations. The teliospores are globose or elliptical, unicellular, pedicellate, chestnut brown in colour with warty papillae at the top. Yellow coloured pycnia appear on the upper surface of leaves. Orange coloured cupulate aecia develop later on the lower surface of leaves. The aeciospores are unicellular and elliptical.

Favourable Conditions

The temperature range of 21-26°C, availability of moisture at leaves surface, cloudy humid weather, high relative humidity and night with heavy dews favour infection and subsequent disease development.

Disease Cycle

The pathogen is an obligate parasite and being autoecious in nature. It produces all the five stages *i.e.* pycnidial, aecial, uredial and tlial on the same host. The uredial stage repeated several times and it is often noticed on species of *Dolichos* and *Vigna*. The urediospores build up in masses and spread by wind to causes secondary infections. The fungi can perpetuates through the teliospores in the soil and as well as uredosporesin crop debries. The teliospore germinates and produces basidiospores which in turn produce pycnidia, aecia, and uredia. Pycnia appears in yellowish spots on the upper surface of leaves which Orange coloured aecia are formed on the upper surface of the leaf around the pycnia on the opposite side. Primary infection is by the sporidia developed from teliospores. Secondary spread is by windborne uredospores. The fungus also survives on other legume hosts.

Management

Removal and destruction of plant debris and manipulation in dates of sowing may help in reducing the disease incidence. Spray of Wettable Sulphur at the rate of 2 kg/ha or Carbendazime at the rate of 500g/ha or Propiconazole @ 1.0liter/ ha, immediatly on the initiation of disease and repeating it after 15 days. Check the disease spread, Foliar spray of Captan -70 per cent + Hexaconazole -5 per cent (Taqat-75WP) is highly effective to control rust (Adinarayana *et al.*, 2013).

Dry Root Rot

Dry root rot is an important disease of mungbean incited by *Macrophomina phaseolina* (Tassi) Goid is a serious problem and hampering the production of the mungbean in all growing areas. It was the first reported from Jabalpur (M.P.) India.

Symptoms

Pathogen attack on all parts of plants *i.e.* root, stem, branches, petioles, leaves, pods and seeds. In pre emergence stage, the fungus causes seed rot and mortality of germinating seedlings. In post emergence stage, seedling gets blighted due to soil or seed born infection. The fungus attacks the stem at ground level on one month old crop and localized, raised white canker at the base of stem. They gradually enlarge in size and turn brown streaks spreading upwards. The leaves of infected plants are dark green, mottled and reduced in size. The leaves are dropped suddenly and dry, resulting heavy reduction in flowering and podding. When affected plants are splitt vertically from the collar downwards, reddish discolouration of the internal tissues is clearly visible while the internal root tissues appear white. The affected plants can be easily pulled out leaving dried, rotten root portions in the ground. The rotten tissues of stem and root contain a large number of black minute sclerotia.

Pathogen

The disease is caused by *Macrophomina phaseolina* (Tassi) Goid [*Rhizoctonia bataticola* (Taub) Butler]. The pathogen invades the mungbean plants both inter and intra-cellularly. It grows fast covering large area of the host tissues and eventually killing them. The fungus produces dark brown, septate mycelium with constrictions at hyphal branches. It also produces numerous sclerotial bodies on the host tissues. Sclerotia are minute, dark to jet black colour, round and smooth to oblong or irregular and abundance in nature. The fungus also produces dark brown, globose ostiolated pycnidia immersed in the host tissues, are erumpent at maturity. The pycnidiospores are thin walled, hyaline, single celled and elliptical some time curved or irregularly contoured and variable in size, 10-42x 6-10mm.

Favourable Conditions

The fungus reported as a high temperature pathogen. The disease develop rapidly and become sever at high temperature (Day temperature of 30°C) and prolonged water stress (15 per cent RH) followed by irrigation. The disease severity is increases with increasing temperature.

Disease Cycle

The fungus survives in the infected debris and also as facultative parasite in upper layer of the soil. The primary spread is through seed-borne and soil-borne sclerotia which either do not germinate or produce seedlings that may die soon after emergence. Since the fungus attacks a wide range of hosts, it perpetuates freely and becomes virulent when the optimum pre-disposing conditions in the host exist. The fungus being soil and seed borne in nature, it spread in the field through irrigation water, farm implements, and cultural practices. The secondary spread is through air-borne pycnidiospores and sclerotia and causes further spread of the pathogen.

Management

Follow up it 2-3 years crop rotation for reducing the inoculum load from soil. Expose the soil through deep summer ploughing is effective method for weed control as well as disease. Destruction of infected plant debris by burning and burying in soil. Seed dressing with Carbendazim @ 2g/kg seed or *Trichoderma harzianum* or *T. viride* or *G. virense* @ 4 gram per kg seed were effective in controlling the seed borne infections (Kumar *et al.*, 2012, Deshmukh and Raut, 1992 and Manczinger *et al.*, 2002). Foliar spray of Carbendazim and Benomyl @ 500 ppm at 15 days interval reduced Macrophomina leaf blight. Meanwhile, Kumari *et al.* (2012) reported that integrated management approach showed that vermicompost and bavistine in combination was more effective in reducing the root rot incidence in pod conditions.

Yellow Mosaic Disease

Yellow mosaic disease is the most serious limiting factor in mungbean cultivation. The occurrence of Mungbean yellow mosaic virus was first time reported from Delhi, India. Since then, it has been reported from all over India and other countries of the Indian subcontinent (Verma and Malathi, 2003), but it is widely prevalent in Uttar Pradesh, Bihar, Chhattisgarh, Delhi, Punjab, Haryana, Himanchal Pradesh, Rajasthan, Odisha, Andhra Pradesh, Tamil Nadu, Karnataka and Madhya Pradesh. Yield losses due to this disease vary from 5 per cent to 100 per cent depending upon disease severity, susceptibility of cultivars and population of whitefly (Nene 1972 and Rathi 2002). The infection not only drastically reduces yield but also delayed in pod maturity (Singh *et al.*, 1982) and reduce number of pod per plant. In mungbean, Yellow mosaic disease incidence in farmers fields might be as high as 100 per cent (Rishi, 2009).

Symptoms

Initially small yellow patches or spots appear on green lamina of young leaves and soon it develops into a characteristics bright yellow mosaic or golden yellow mosaic symptom. Yellow discoloration slowly increases and leaves turn completely yellow. Maturity is delayed in the diseased plants and flower and pod formation are severely reduced. Seeds that develop on severely infected plants are small and immature. In addition to yellow mosaic to the yellow mottle symptoms a necrotic centre develops in the yellow spots in some of the varieties. The pods are small and distorted. Early infection causes death of the plant before seed set.

Pathogen

It is caused by *Mungbean yellow mosaic India virus* (MYMIV) in Northen and Central region and *Mungbean yellow mosaic virus* (MYMV) in western and southern regions. Germinate virus particles, are isometric, paired, 15x 30µ min size (Honda *et al.*, 1981)., having ssDNA, bipartite genome with two gemonic components DNA-A and DNA-B. These type of DNA were identified as DNA-A and DNA-B by nuclic acid hybridization tests (Verma *et al.*, 1991) The virus particles are limited to phloem associated elements in infected plants (Srivastava, *et al.*, 1985).

Favourable Conditions

The incidence and spread of yellow mosaic disease are generally high in summer crops, as compared with rainy or winter season crops. Weed serves as reservoir of the virus, and are a source of primary inoculum. The virus starts spreading with the onset of source of the mansoon in India (Grewal, 1988), reaching a high incidence in areas where the temperature ranges between 31 to 35 °C and relative humidity is about 70 per cent (Sharma *et al.*, 1993). These conditions favour the disease development and multiplication of the vector *Bemisia tabaci.*

Disease Cycle

It is transmitted by whitefly, *Bemisia tabaci* under favourable conditions in a circulative, and non propogative manner. It is not seed or soil borne or sap transmissible. Disease spreads by feeding of plants by viruliferous whiteflies. Summer sown crops are highly susceptible. Weed hosts *viz., Croton sparsiflorus, Acalypha indica, Eclipta alba* and other legume hosts serve as reservoir for inoculums.

Management

The most effective way to minimize the impact of yellow mosaic disease is to grow resistant varieties like Pant Mung-3, Pant Mung-4, Pant Mung-5, Pant Mung-6, Pusa Vishal, Basanti, ML-267, ML-337, PDM-54, LGG-407, LGG-460, Narendra mung-5, HUM-1, TM-99-37, HUM-16, TARM-1, Pusa-95-31, Sweta and Samrat. Manipulation in sowing dates, plant density and intercropping with two rows of maize (60x30 cm) or Sorghum (45x 15 cm) for every 15 rows of greengram inhibit the congenial atmosphere for vector and reduced the movement of vector. The incidence of MYMV in mungbean was the lowest in crop raised from the seeds treated with Thiamethoxam @ 4 g/kg seed (Ganapathy and Karuppiah, 2004). Thiamethoxam as seed treatment (4g/kg seed) providing production to healthy seedling from MYMV, and its subsequent sprays (0.02 per cent) (Dubey and Singh, 2013) or seed treatment with Imidachloprid 70WS @ 4g/kg seed and two subsequent spray at 21 and 35 days protect the crop up to the pod formation stage.

Leaf Crinkle Disease

Leaf crinkle, caused by urdbean leaf crinkle virus, is a limiting factor for successful cultivation of mungbean (*Vigna radiata* (L.) Wilczek) in India. Urdbean leaf crinkle virus (ULCV) has been reported in natural infections of mungbean in India by Singh *et al.*, 1979. ULCV has the potential to cause heavy yield losses in mungbean. In early infected plants seed production may fail completely. A field experiment conducted over 3 years showed losses in grain yield ranging from 2 to 95 per cent in both mungbean and urdbean, depending on the plant age at symptom onset after 55 days and before 24 days old respectively (Kadian, 1982).The virus also reduced yield components such as plant height, root length, nodulation, pods/plant, pod length, seeds/pod and weight of 100 seeds, but it increases the leaf area and dry weight of roots and shoots (Singh, 1980).

Symptoms

The initial symptoms appear on youngest leaves as chlorosis around some lateral veins and its branches near the margin. The leaves show curling of margin downwards. Some of the leaves show twisting. The veins show reddish brown discolouration on the under surface which also extended to petioles. Plants showing symptoms within 5 weeks after sowing invariably remain stunted and majority of these die due to top nerosis within a week or two. Plants infected in late stages of growth do not show severe curling and twisting of the leaves but show conspicuous veinal chlorosis anywhere on the leaf lamina, petioles as well as internodes are shortened; infected plant gives a stunted and bushy appearance, flowering is delayed, inflorescence, if formed, are malformed with small size flower buds and fail to open. Pollon fertility and pod formation is severely reduced on infected plants which decrease drastically.

Pathogen

The disease is caused by virus which is not yet completely characterized. The virus particles are ungrouped spherical and approximately 25-30 nm in size. Dubey *et al.* (1983) reported that ULCV particle purified from urdbean were 32 nm in diameter. In this sections of urdbean cells, particles measured 25 to 30 nm in diameter(Bhaktavatsalam *et al.*, 1983). The A_{260}/A_{280} ratio was 1.51 (Bhaktavatsalam *et al.*, 1983). No serological relationships were found between ULCV and black gram mottle Carmovirus, broadbean mottle bromovirus or bean pod mottle comovirus (Dubey *et al.*, 1983).

Transmission

The crinkle disease is transmitted through sap, grafting, insect vectors and seeds. Beniwal *et al.* (1980) reported it has been seed transmitted to an extant of 15 per cent in mungbean. Many workers have reported its transmission by various insects *viz.*, Aphid, *Aphis craccivora* and *A. gossypii* (Bhardwaj and Dubey, 1986; Dhingra, 1975), beetle, *Henosepilachna dodecastigma* (Beniwal and Bharathan, 1980) and White fly, *Bemisia tabaci* (Narayanasami and Jaganathan, 1974). The virus is also sap transmissible.

Favourable Conditions

The presence of weed hosts like *Aristolochia bracteata* and *Digera arvensis.* Kharif season crop is highly susceptible, continuous cropping of other legume crops also harbor the virus. Environmental factors showed positive interaction in case of maximum and minimum temperature but in case of relative humidity the interaction was negative. With an increase the temperature, the infection rate increased and with increase in relative humidity the infection rate decreased (Binyamin *et al.*, 2011). Ashfaq (2007) reported that maximum disease incidence was developed at maximum temperature of 35- 42 °C.

Disease Cycle

The virus is seed-borne and primary infection occurs through infected seeds and presence of weed hosts like *Aristolochia bracteata* and *Digera arvensis.* Kharif season crop and continuous cropping of other legumes *i.e.* Urdbean (*Vigna mungo*), Cowpea (*Vigna unguiculata*), Pigeonpea (*Cajanus cajan*), Taparybean (*Vigna aconitifolicus*) (Kolte and Nene, 1975; Beniwal, 1983) serve as source of inoculum. White fly, aphids and beetle helps in the secondary spread of virus.

Management

To avoid early infection into the crop, planting of disease free seeds are important measures of management of disease. Removed collateral weed hosts around the field and rogue out the diseased plants could reduce the chance of further spread and reduced production of more infected seed. Hot water seed treatment at 52 °C for 20-30 min reduced seed borne infection. Cultivation of resistant mungbean entries also decreased diseased incidence. Foliar spray of Thiomethaxam 0.02 per cent at 15 days after sowing; decreased whitefly population and incidence of leaf crinkle (Ganapathy and Karuppiah, 2004). Binyamin *et al.* (2011) reported that 2 per cent neem and its extract reduced the severity of ULCV infection, while Zeshan *et al.* (2012) have reported that nitrogen, phosphorus, potash, zinc, boron and naphthalene acetic acid (NAA) were effectively reducing ULCV infection in different varieties of mungbean by the same order from 65 to 58.57 per cent.

Web Blight

Web blight is one of the major constraints in the production of many pulses in warm humid tropic zones of the world. On Mungbean Rhizoctonia blight was reported for the first time from Philippines (Nacien, *et al.*, 1924) in 1924. In India first report of its occurrence on mungbean was given by Dwivedi and Saksena in 1974 from Kanpur, Uttar Pradesh subsequently this disease has also been reported from Assam (Saikia, 1976), Punjab (Bains *et al.*, 1988), Madhya Pradesh (Tiwari and Khare, 1998), Bihar, Rajsthan, Haryana, Himanchal Pradesh and Jammu and Kashmir (Anonymous, 2004). The disease was observed to reduce 33 to 40 per cent grain yield and 28.6 per cent in 1000 grain weight at different level of disease severity and in different variety of mungbean (Gupta *et al.*, 2010 and Singh *et al.*, 2012). Since then the web blight of mungbean has become one of the most serious problems of this crop in Northern India causing extensive damage to mungbean.

Symptoms

The symptoms of web blight on mungbean occur on roots, stems, petioles and pods but the disease is the most destructive on foliage. The pathogen is soil borne, seed borne and air borne in nature. During second to third week of plants growth, it causes seedling mortality. Seed decay, pre and post emergence mortality occurs. The first symptoms appear as small circular brown spots on the primary leaves. These spots enlarge, often show concentric banding and surrounded by irregular water soaked areas. On young pods the spots are light tan and irregular in shape but generally on mature pods, they are dark brown and sunken. The lesion expands and coalesces and white mycelia fungal growth can be seen under surface of infected leaves and young branches. The mycelium on infected leaves appears as spider web thus suggested the name web blight disease ((Dwivedi and Saksena, 1974 and Singh, 2006).

Pathogen

The pathogen has been reported on mungbean under the name *Thanatephorus cucumeris* (frank) Donk (Dwivedi and Saksena, 1974 and Singh, 2006). Web blight pathogen *R. solani* has been characterized as: Colony colour pale to dark brown:

- ✰ Rapidly growing mycelium of relatively large diameter (> 10 µm);
- ✰ Branching near the distal septum of hyphal cell, often nearly at right angles in older hyphae.
- ✰ Production of moniloid cells (Barrel shaped cells or chlamydospores) in chain or aggregates;
- ✰ Production of sclerotia nearly uniform texture, varying in size (Â 1mm to several cm) and shape (round to thin crust) and

The perfect state *T. cucumeris* develops on healthy tissue adjacent to the lesion on green gram leaves as well as *in vitro* (Tiwari and Khare, 1998). The sexual stage is induced on the host under warm and humid conditions with heavy dew formation during the night. Fructifications are resupinate, creamy to greyish white, loosely attached to the substratum, composed of arachnoid, repent hyphae giving rise to thin hypochnoid or sub membranous fertile patches. Cluster of basidia are produced terminally in a discontinuous hymenium. Basidia are mostly 9-25x 5-12.4 µm in size but some times short, wider, barrel shaped to sub cylindrical without a median constrction. Sterigmata are straight, attenuate towards the apex usually bearing 4 basidium. Basiospore are hyaline smooth thin walled, oblong to ellipsoid and dorsally fattened or broad ovoid and commonly widest at the distal end form secondary basidiospores.

Favourable Conditions

Environmental factors play a vital role in the development of web blight disease caused by *R. solani*. Higher aerial temperature (26 to 32 °C) relative humidity near 100 per cent and soil temperature 30-33°C favoured the development of high disease severity. Rainfall (91-97 mm) had a significant role in severe development of web blight during early stage of crop (Sharma and Thripathi, 2001; Gupta *et.,al.* 2010). The maximum basidiospores production and their discharge occurs during mid night and early morning before sunrise (Saksena and Dwivedi, 1973). The spore production is favoured by night temperature below 24 °C, RH above 95 per cent and rate of evaporation below 1.5 mm per day. The basidiospores germinate in two hours and penetrate leaf through the intact surface by formation of infection cushion or direct penetration through stomatal openings. Host susceptibility is also important in disease incidence and severity.

Disease Cycle

The pathogen is seed, soil and airborne in nature. The epidemiology of *Rhizoctonia* blight may be divided into two phases, one before and the after canopy closures. The first phase is soil borne and the second is leaf borne (Yang, 1990). The pathogen is primarily soil borne and can survive for many years by producing sclerotia in soil and on plant tissue/seedcoat. *R. solani* also survive as mycelium by colonizing soil organic matter as saprophytes. The highest inoculum potential is noted in the top 10 cm soil and no inoculum is found below 40 cm (Kaiser *et al.*, 1970). It germinates to produce vegetative threads (hyphae) of the fungus that can attack mungbean on the leaves during heavy rains. The collateral weed hosts play an important role in initiation and early spread of the disease to the main host (Saksena, 1985), because of their proximity to the soil carrying primary inoculum and their special microclimate. The weed hosts are first to be infected and facilitate the production of the basidiospores of the pathogen, there after the pathogen become air borne to causes leaf, stem or pod infection and produces fresh crops of basidiospores on the main host. In *R. solani* secondary spread via sclerotic through wind and rainsplash is also reported. Cob web like mycelial growth of pathogen also gets transmitted from diseased to healthy leaves in the web blight of legumes (Saksena, 1979; Verma and Thapliyal, 1976).

Management

Plant resistance is one of the most attractive approaches for suppressing plant disease. Singh *et al.* (2007) reported that out of 85 mungbean entries, only 7 *viz.*, NDM-92-2, Ml-406, EC-27130, EC-1243-1, EC-5551, IC- 73362 and IC-39338 showed resistance reaction against web blight disease. Disease injury can be reduced by crop rotation, adjusting time of sowing (early sowing) to avoid periods of low temperature with wider row spacing (50 cm) and draining wet areas of the field. Seed dressing with Carbendazim and Thiophonate methyl were found effective in eliminating the seed borne infection. *Trichoderma virens* and *T. viride* were found effective against *R. solani* (Dubey, 2003). Seed treatment *T. viride*/*T. virens* + Carboxin+ Rhizobium gave maximum seed germination, plant vigour, number of root nodules and grain yield with minimum seedling mortality and disease intensity (Dubey, 2003). Integration of soil application of *P. glabra* cake (200 kg/ha), seed treatment with *T. viride* (2 g/kg seed) + Carboxin (1 g/kg seed) + *Rhizobium* species (25 g/kg seed) and foliar spray of *P. glabra* leaf extract (10 per cent) suppressed disease severity to a significant extent (92.7 per cent). This also increased seed germination, improved plant vigour and enhanced production. The same combination excluding carboxin was also effective and could be an option for organic production of mungbean (Dubey, 2006).

Bacterial Leaf Blight

Bacterial leaf spot was first time reported from China (Fang *et al.*, 1964). In India Patel and Jindal (1972) reported two new disease on mungbean *i.e.*, Bacterial leaf spot and Halo blight during 1968 and 1969. The disease has been reported from

Punjab, Haryana, Rajasthan, Uttar Pradesh, Madhya Pradesh, Maharasthra, Andhra Pradesh and Tamil Nadu.

Symptoms

The disease is characterized by brown, circular to irregular, dry, raised spots on leaves and stem. Initially it appears as superficial eruptions and gradually invades the tissue giving corky or rough appearance. The lesions on petiole are brown, flat or raised and occur as long streaks without defoliations. In severe infections dark brown longitudinal split or canker develops on stem. When the disease is sever, the leaves become yellow and fall off prematurely.

Pathogen

Bacterial leaf blight is incited by *Xanthomonas campestris* pv. *vignaeradiate* [(Sabet, Ishaq and Khalil) Dye]. The bacterium is a gram negative, rod shaped and having single polar flagellum. The pathogen can also infect *Phaseolus vulgaris, P.hontus, P. braclantus, Dolichos lablab* and *Lens culineris*.

Favourable Conditions

The disease develops rapidly under warm and wet weathers. The optimum temperature for the growth of bacterium is lies between 30-35 °C.

Disease Cycle

The bacterium survived on seeds, diseased plant debris as well as other host plants. In seed borne condition, the bacteria develops in seed coat of the germinating seeds and infects the surface of expended cotyledon, penetrating through rifts in the cuticle and grow intercellularly to reached vascular system. It may then proceed as vascular pathogen throughout the large xylem vessels of the plant. Both leaf lesions and stem cankers may arise from systemic infections. Primary local infection may also take place through stomata and produced symptoms on cotyledons. Secondary spread of inoculum from primary lesions is carried out by splashing water or windblown rain, implements, men and animals.

Management

Disease transmission through seeds could be reduced by treatment with Streptomycin sulphate(@ 500 ppm), bleaching powder (0.025 per cent), Vitavax (0.2 per cent) Captan (0.3 per cent). Three protective sprays of Streptomycin (100ppm), Zineb (0.3 per cent), Blitox -50 (0.25 per cent), or Benomyl (0.25 per cent) were found quite effective to reduced secondary infections. reported that application of *Eriwinia herbicola* 48 hrs. before inoculation and bacterization of seeds of Pusa Baisakhi with Bacillus sp. reduced seedling infection. Number of cultivars *i.e.* Jalgaon 781, ML-1, ML-2, ML-3, ML-8, ML-9, ML-10, Yellow mung and P70-68 showed resistance (Chand *et al.*, 1977).

REFERENCES

AVRDC, (1982). Asian Vegetable Research and Development Centre, Shanhua Taiwan, *Center Point* Vol 2 No1.

Achaya, K.T. (1998). Indian food – A historical companion. Oxford University Press, Delhi, India, 322 pp.

Adinarayana, M.; MahaLakshmi, M. S. and Rao, Y. K. (2013). Field efficacy of new fungicide Taqat-75 WP against foliar fungal disease of blackgram. *J. Biopesticides*, 6(1): 46-48.

Akhtar, J.; Lal, H.C.; Kumar, Y.; Singh, P.K.; Ghosh, J.; Khan, Z. and Gautam, N.K. (2014). Multiple Disease resistance in green gram and Blackgram germplasm and management through chemicals under rain fed conditions. *Legume Res.*, 37 (1): 101-109.

Ali, M. and Gupta, S. (2012). Carrying capacity of Indian Agriculture: pulse crops. *Current Sci.* 102 : 874-881.

Anonymous (2004). Annual Report (Kharif) All India Co-ordinated Research Project on MullaRP (ICAR) IIPR, Kanpur, 112 pp.

Anonymous (2012). Project Co-ordinator's report AICRP on MULLaRP crops, IIPR, Kanpur.

Ashfaq, M. (2007). Characterization of epidemiological and biochemical factors in relation to resistance against urdbean leaf crinkle virus (ULCV) and its management, *Ph.D. Thesis*, P: 37, University of Agriculture, Faisalabad, Pakistan.

Bains, S. S.; Dhaliwal, H. S. and Basandrai, A. K. (1988). A new blight of mung and mash in Punjab. *Ann. Biol. Ludhiana*, 4(1-2): 113-114.

Beniwal, S. P. S.; Chaubey, S.N. and Bharathan, N. (1980). Presence of Urdbean leaf crinkle virus in seeds of mungbean germplasm. *Indian Phytopathol*. 33: 360.

Beniwal,S. P. S. (1983). Urdbean leaf crinkle virus, pp.85-86, Viruses of Legumes. 1983. Eds. K.F. Boswell and A.J. Gibbs. Descriptions and keys from VIDE. *Austral. Natl.Univ.*, Canberra.

Bharadwaj, S.V. and Dubey, G.S. (1986). Studies on the relationship of Urdbean leaf crinkle virus and its vectors, *Aphis craccivora* and *Acyrthosiphom pisum*. *J. Phytopathol*. 83: 115

Binyamin,R.; Khan, M.A.; Ahmad, N.; and Ali, S. (2011). Relationship of epidemiological factors with Urdbean leaf crinkle virus disease and its management using plant extracts. *Int. J. Agric. Biol.*, 13: 411-414.

Bora, L.C.; Gangopadhyay, G. and Chand, J.N. (1993). Biological control of bacterial leaf spot of mungbean with phylloplane antagonists. *Indian J. Mycol. Pl. Pathol*, 23: 162-168.

Borah, P.K.; Jindal, J.K. and Verma J.P. (2000). Biological management of bacterial leaf spot of mungbean caused by *Xanthomonas axopodis* pv. *vignae radiate*. *Indian Phytopath*. 53: 384-394.

Chand, J. N.; Yadav, O. P. and Yadav, H. C. (1977). Reaction of genetic stock of mungbean to bacterial blight. *Indian Phytopath*.30: 565.

Chand, R.; Singh, V.; Pal, C.; Kumar, P. and Kumar, M. (2012). First report of new pathogenic variant of *Cercospora canescens* on mungbean (*Vigna radiata*) from India. *New Disease Reports*. 26: p. 6.

Chauhan, M. P. and Gupta, R. P. (2004). Genetics of Cercospora leaf spot disease resistance in mungbean [*Vigna radiata* (L.)Wilczek] *Legume Res*. 27: 155-156.

Deeksha, J. and Thripathi, H. S. (2002). Cultural, Biological and chemical control of Anthracnose of Urdbean. *J.Mycol. Pl. Path.* 32: 52-55.

Deshmukh, P. P. and Raut, J. G. (1992). Antagonism by *Trichoderma* sp. on the five plant pathogenic fungi. *New Agriculturist*.3: 127-130.

Dhingra, K. L. (1975). Transmission of Urdbean Leaf Crinkle Virus by two aphid species. *Indian Phytopathol.* 28: 80

Dubey, G.S.; Sharma, I and Prakash, N (1983). Some properties of Urdbean Leaf Crinkle Virus. *Indian Phytopathol*.36: 762.

Dubey, S. C. (2003). Integrated management of web blight of urd and mungbean. *Indian Phytopath.* 56: 413-417.

Dubey, S.C. (2006). Integrating bioagent with plant extract, oil cake and fungicide in various mode of application for better management of web blight of urd bean. *Archives of Phytopathology and Plant Protection.* 39: 341-351.

Dubey, S. C. and Singh, B. (2013). Integrated management of major diseases of mungbean by seed treatment and foliar application of insecticides, fungicides and bioagent. *Crop Protection*. 47: 55-60.

Dwivedi, R.P. and Saksena, H.K. (1974). Occurrence of web blight caused by *Thanatephorus cucumeris* on mungbean. *Indian J. Farm Sci.*, 2: 100.

Fang, C. T.; Chen, H. Y. and Chu, C. I. (1964). A comparative study of the species of the genus *Xanthonomas* from leguminous plants. *Acta Phytopath Simica*, 7: 21-31.

Ganapathy, T. and Karuppiah, R. (2004). Evaluation of new insecticides for the management of white fly (*Bemisia tabaci* Genn), mungbean yellow mosaic virus (MYMV) and Urdbean leaf crinkle virus (ULCV) diseased in mungbean (*Vigna radiata* (L.) wilezek). *Indian J. Pl. Prot.* 32: 35-38.

Grewal, J. S. (1988). Diseases of pulse crops An overview. *Indian Phytopath.* 41: 1-14.

Gupta, R. P., Singh, S. K. and Singh R.V.(2010). Assessment of losses due to web blight and weather effects on disease development in mungbean. *Indian Phytopath.* 63(1): 108-109.

Gupta, R. P.; Singh, S. K. and Singh R.V.(2010). Assessment of losses due to web blight and weather effects on disease development in mungbean. *Indian Phytopath.* 63(1): 108-109.

Gupta, S. (2011) Indian Institute of Pulses Research (IIPR). Vision2030. In: Indian Institute of Pulses Research (ICAR) (ed.). Kanpur, India. 42 p.

Hanson, L. E. and Panella, L. (2003). Disease control in sugerbeet. Int.Suger.J.105: 60-68.

Haque, M. F.; Mukherjee, A. K.; Mahto, R. N.; Jha, D. K.; Chakraborty, M.; Srivastava, G.P. and Prasad, D. (1997). Birsa-Urd-1 new variety for Chotanagpur region of Bihar. *J.Res.*, 9 : 177-178.

Honda, Y. M.; Twaki, Y.; Saito, P.; Thangmeearkom, P.; Kittisak, K. and Deema, N. (1981). Mechanical transmission, purification and some properties of whitefly transmitted mungbean yellow mosaic virus in Thailand. *Plant Disease* 65: 801-04.

Hossain A. M. M. Z.; D. D. Chatterjee, B. Howlader, K. Kanti and Jalauddin. (2011). Incidence and severities of Cercospora leaf spot on different strain of growing mungbean. *Bangladesh Res. Pub.* J. 5: 1-6.

Kadian, O. P. (1980). Studies on leaf crinkle disease of urdbean (Vigna mungo(L.) Hepper) mungbean (*V. radiata* (L.) Wilczek) and its control, Ph.D. Thesis, P: 177. Department of Plant Pathology, Haryana Agriculture University, Hisar, India.

Kadian, O. P. (1982). Screening for resistance to leaf crinkle in mungbean. *Tropical Pest Management;* 28: 436.

Kaiser, W. J.; Mossahebi, G.H. and Okhorat, M. (1970). Occurence, pathogenicity and distribution in soil of *Rhizoctonia solani* inciting a stem canker disease of mungbean (*Phaseolus aureus*) in Iran. *Iranian Journal of Plant Pathology*, 6: 17-25.

Kapadiya, H. J. and Dhruj, I. U. (1999). Management of mungbean Cercospora leaf spot through fungicides. *Indian Phytopath.*, 32 (1): 96-97.

Khan M. R. I. (1981). Nutritional quality characters in pulses.pp. 199-206.In: Proceedings of National workshop on pulses, BARI, Ghazipur.

Khan, M. F. (2008). Evaluating strategies for managing Cercospora leaf spot of sugerbeet. *Phytopathol.*98: p.80.

Khan, M. F. R. and Smith, L. J. (2005). Evaluating fungicides for controlling Cercospora leaf spot on sugar beet. *Crop Prot.* 24: 79-86.

Kolte, S. J. and Nene, Y.L. (1975). Host range and properties of urdbean leaf crinkle virus. *Indian Phytopathol.* 28: 430-431.

Kulkarni, S.; Benagi, V. I.; Patil, P. V.; Hegde, Y.; Konda, C. R. and Deshpande, V. K. (2009). Sources of resistance to anthracnose in greengram and biochemical parameters for resistance. *Karanataka J. Agric. Sci.*, 22(5); 1123-1125.

Kumari, R.; Shekhawat, K. S.; Gupta, R. and Khokhar, M. K. (2012). Integrated management against root rot of mungbean[*Vigna radiata* (L.)Wilczek] incited by *Macrophomina phaseolina*. *J. Plant Pathol. and Microb.*3: 136 doi: 10.4172/2157-7471.1000136.

Lakshmanan, P., Mohan, S. and Jeyarajan, R. (1990). Antifungal Properties of some plant extracts against *Thanatephorus cucumeris*, the causal agent of color rot disease of *Phaseolus aureum*. *Madras Agric.* J., 77(1): 1-4.

Lal, G.; Kim, D.; Shanmugasundaram, S. and T. Kalb. (2001). Mungbean production by AVRDC. World Vegetable Center, Taiwan, Shanhua: AVRDC-The World Vegetable Center p. 6.

Manczinger, L.; Antal, Z. and Kredics, L. (2002). Ecophysiology and Breeding of Mycoparasitic *Trichoderma* strains (A Review). *Acta Microbiol Immunol Hung.* 49: 1-14.

Mathur, A. K.; Chitale, K.; Gaur, V. K. and Tyagi, R. N. S.(1990). Response of mungbean cultivars to cercospora leaf spot in Rajasthan. *Indian J. Mycol.Pl. Pathol.* 19: 213-214.

Nacien, C. C. (1924). Studies on Rhizoctonia blight of beans. *Philippine Agriculturist,* 8: 315-321.

Mew, I.; Pin, C.; Wang, T. C. and Mew, T. W. (1975). Inoculum production and evaluation of mungbean varieties for resistance to *Cercospora canescens. Plant Dis. Rep.* 59: 397-401.

Mian, A. L. (1976). Grow more pulse to keep your pulse well: An essay of Bangladesh Pulse, Department of Agronomy, Bangladesh Agriculture University, Mymen singh. pp. 11-15.

Munjal, R. I.; Lal, G. and Chona, B. L.(1960). Some Cercospora species from India –IV. *Indian Phytopathol.*13: 144-149.

Nacien, C. C. (1924). Studies on *Rhizoctonia* blight of beans. *Philippine Agriculturist,* 8: 315-321.

Nair. R.M.; Schafleitner, R.; Kenyon, L.; Srinivasan, R.; Easdown, W.; Ebert, A. and Hanson, P. (2012). Genetic improvement of mungbean productivity. In: Proc. Of the 12th SABRAO Congress on Plant Breeding towards 2025: Challenges in a Rapidly changing World, 13-18 Jaunary2012,Chiang Mai, Thailand. P.27-28.

Narayansami, P. and Jaganathan, T. (1974). Characterization of blackgram leaf crinkle virus. *Madras Agric.*J.16: 979.

Nene, Y. L. (1972). A survey of viral diseases of pulse crops in Uttar Pradesh. *Final Tech. Report. Res. Bull.* No. 4, U.P. Agricultural University, Pantnagar, India, pp. 1-91.

Nene, Y. L. (2000). A new vision of pulse pathology research in India. *Indian J. Pulses Res.,* 13 (1): 1-10.

Nene, Y. L.; Naresh, J. S. and Nair, N.G. (1971). Additional hosts of mungbean yellow mosaic virus. *Indian Phytopathol.* 24: 415-417.

Nene, Y. L., Naresh, J. S. and Nair, N. G. (1971). Additional hosts of mungbean yellow mosaic virus. *Indian Phytopathology.*24: 415-417.

Pandey, S.; Sharma, M.; Kumari, S.; Gaur, P. M.; Chen, W.; Kaur, I.; Macleod, W.; Basandari, A. K.; Basandari, D.; Bakr, A.; Sandhu, J. S.; Tripathi, H. S., and Gowda, C. I. I. (2009). Integrated foliar disease management of legumes. In: Grain Legumes: Genetic improvement, Management and Trade, eds. By Masood Ali *et al.*, pp143-161. Indian Society of Pulses Research and Development, Indian Institute of Pulse Research, Kanpur, India.

Patel, P. N. and Jindal, J. K (1972). Bacterial leaf spot and halo blight diseases of mungbean and other legumes in India. *Indian Phytopath.*25: 517-525.

Poehlman, J. M. (1991). The mungbean. Westview Press. BOULDER. pp. 169-274.

Rakhonde, P. N.; Koche, M. D. and Harne, A. D. (2011). Management of powdery mildew of green gram. *J. food Legumes.*, 24 (2): 120-122.

Rathi, Y. P. S. (2002). Epidemiology, yield losses and management of major diseases of Kharif pulses in India. In: *Plant Pathology and Asian Congress of Mycology and Plant Pathology*, Oct.-1-4, 2002. University of Mysore, Mysore,India.

Rathaiah, Y. and Sharma, S. K. (2004). A new leaf spot disease on mungbean caused by *Colletotrichum truncatum*, *J. Mycol.Pl. Path.*, 34: 176-178.

Rishi, N. (2009). Significant plant virus diseases in India and a glimpse of modern disease management technology. *J.Gen. Plant Pathol.*75: 1-18.

Saikia, U. N. (1976). Blight of mung caused by *Corticium sasakii* a new disease recorded from Assam. *Indian Phytopath.*, 29 : 61-62.

Saksena, H. K. (1973). Banded blight disease of paddy in north India, Proceedings, International Rice Research Conference, International IRRI, Los Banos, Philippines.

Saksena, H. K. (1979). Epidemiology of diseases caused by *Rhizoctonia* spp. In proceedings of consultants group. Discussion on the resistance to soil borne diseases of legumes ICRISAT. Hyderabad, 59-64 pp.

Saksena, H. K. (1985). Relationship of environment and Rhizoctonia aerial blight. *Indian Phytopathology*, 38: 584.

Saksena, H. K. and Dwivedi, R. P. (1973). Web blight of black gram caused by *Thanatephorus cucumeris*. *Indian Journal of farm Sciences*, 1(1): 58-61.

Shahbaz, M. U.; Iqbal, M. A.; Rafiq, M.; Batool, A. and Kamran, M. (2014). Efficacy of different protective fungicides against Cercospora leaf spot of mungbean [*Vigna radiata* (L.) Wilczek]. *Pak. J. Phytopathol.*26: 187-191.

Sharma, J and Tripathi, H. S. (2001). Influence of environmental factors on web blight development in urdbean[*Vigna mungo*(L.)Hepper]J. *Mycol.Pl.Pathol.*31: 54-58.

Sharma, S.; Yadav, O. P.; Thareja, R. K. and Kaushik, J.C. (1993). Nature and extent of losses due to mungbean yellow mosaic virus and its epidemiology. *HAU J. Res.* 23: 51-53.

Sharma, H. C.; Khare, M. N.; Joshi, L. K. and Kumar, S.M. (1971). Efficacy of fungicides in the control of diseases of kharif pulses mung and urd. All India Workshop on kharif Pulses. p. 2.

Sharma, L.; Goswami, S. and Narrale, D. T. (2013). Cultural and Physiological varaiability in *Rhizoctonia solani*; responsible for foliar and lesion on aerial part of soybean. *J. Applied and Natural Sci.* 5: 41-46.

Shukla, V.; Bhagel, S.; Maravi, K. and Singh S.K. (2014). Yield loss assessment in mungbean [*Vigna radiata* (L.) WilczekI] caused by anthracnose [*Colletotrichum truncatum* (Schw.) Andrus and Moore]. *The Bioscan* 9 (3): 1233-1235.

Singh, B. R.; Singh, M.; Yadav, M. D. and Dingra, S. M.(1982). Yield loss in mungbean due to yellow mosaic. *Sci. and Culture* 48: 435-436.

Singh, J. (2006). Occurrence, Variability in an management of web blight pathogen (*Rhizoctonia solani* K{hn) of mungbean [*Vigna radiata* (L.)Wilczek]. Ph.D Thesis, B.H.U., Varanasi (U.P.).

Singh, R.; Singh, H. and Dodan, D. S. (1995). Sensitivity of *Rhizoctonia solani* isolates causing seedling mortality of mung to different fungicides. *Plant Dis. Res.,* 10: 41-43.

Singh, J.; Singh, R. B. and Balai, L. P.(2012a). Grain Yield Loss in Mungbean due to Web Blight. *Trends in Biosciences,* 5(2): 147-148.

Singh, J.; Singh, R. B. and Balai, L. P. (2012 b). Manipulation of sowing Dates and inter-Rows Spacing and Eco Friendly Management(*Vigna radiata*) Web Blight. *Trends in Biosciences,* 5(2): 152-153.

Singh, J. P.; Kadian, O. P. and Verma, J. P. (1979). Survey of virus diseases of Mungbean and Urdbean in Haryana. Haryana Agric. Univ.J.Res. 9: 345.

Singh, R.K. and Dwivedi, R.S. (1987). Fungitoxicity of different plants. National Academy Sci. 10(3) 89-91.

Srivastav, K. M.; Raizada, R. K. and Singh, B.P. (1985). Gemini viruses. Perspective in Plant Virology, Print House (India) Lucknow, 1: 153-188.

Suryawanshi, A. P.; Wadje, A. G.; Gawade, D. B. Kadam, T. S. and Pawar, A. K. (2009). Field evaluation of fungicides and botanicals against powdery mildew of mungbean. *Agric. Sci. Digest.,* 29(3): 209-211.

Tariq, V. N.and Magee, A. C. (1990). Effect of volatiles from garlic bulb extracts on *Fusarium oxysporum* fsp. *lycopersici*. *Mycological Res.* 94(5): 617-620.

Taunk, J., Bharti, A. B., Yadav, N. R.Yadav, R. C. and Yadav, R. K. (2012). Genetic diversity among greengram [*Vigna radiat*a (L.)Wilczek] genotypes varying in micronutrient (Fe and Zn) content using RAPD markers *Indian Journal of Biotechnology*. 11: 48-53.

Thakur, M. P. and Khare, M. N. (1992). Epidemiology of mungbean Anthracnose. *Indian J. Pulse Res.*5: 49-52.

Tiwari, A. and Khare, M. N. (1998). Variability among isolates of *Rhizoctonia solani* infecting mung bean. *Indian Phytopath.* 51: 334-337.

Uddin, N. M., Bakr, M. A., Islam, M. R., Hossain, M. I. and Hossain, A. (2013). Bioefficacy of plant extracts to control Cercospora leaf spot of mungbean (*Vigna radiata*). *Int. J. Agril. Res. Innov. and Tech.,* 3(1): 60-65.

Verma, A., Dhar, A. K. and Malathi, V. C. (1991). Cloning and restriction analysis of mungbean yellow mosaic virus. In: International Conf. *Virology in the Tropics Lucknow*, India, pp-14.

Verma, H. S. and Thapliyal, P. N. (1976). *Rhizoctonia* aerial blight of soybean. *Indian Phytopath.,* 29 : 389-391.

Yadav, D. L.; Pratik J. and Pandey, R. N. (2014). Identification of source of resistance in mungbean genotypes and influence of fungicidal application to powdery mildew epidemics. *International Journal of Current Microbiology and Applied Sciences*, 3(2): 513-519.

Yadav, D. L.; Pandey, R. N.; Jaisani, P. and Gohel, N. M.(2014). Sources of resistance in mungbean genotypes to Cercospora leaf spot disease and its management. African *J. Agril.Res.*, 9(41): 3111-3114.

Yang, X. B.; Snow, J. P. and Berggren, G. T. (1990). Analysis of epidemics of Rhizoctonia aerial blight of soybean in Louisiana. *Phytopathology*, 80: 386-392.

Zeshan, A. M.; Ali, S.; Khan, M. A. and Shahi, S. T. (2012). Role of nutrients and Naphthalene acetic acid in the management of Urdbean leaf crinkle virus. *Pak. J. Phytopathol.*24: 79-81.

Chapter 7

French Bean Diseases and their Management

S.K. Gupta & Meenu Gupta

French bean (*Phaseolus vulgaris*) is an important cool season legume vegetable grown for its tender pods, shelled green beans and dry beans (Rajmah beans). It is an herbaceous annual plant belonging to family Fabaceae and is grown throughout the world. Brazil and India are the largest producers of dry beans, while China produces, by far, the largest quantity of green beans. French beans have a soft velvety pod with fleshy seeds and are smaller than common green beans. During cultivation, this crop is affected by several diseases of fungal, bacterial and viral nature which not only reduce the quality but also quantity of the crop.

FUNGAL DISEASES

Root Rot and Web Blight

Root rot and web blight is a very serious disease of this crop throughout the world in warm and wet conditions. Losses from root rot vary greatly in different localities and from year to year. Several workers have reported varying degree of losses due to web blight. Rajnauth (1987) reported 25-100 per cent yield losses in Trinidad. In India, yield losses of green pods at different stages of plant growth varied from 8.4 to 64.6 per cent while incidence of root rot was found up to 15 per cent in fields around Bangalore (Sharma and Sohi, 1980; Sharma and Sohi, 1981).

Host Range

The fungus is polyphagous in nature and has been recorded on 287 genera belonging to dicotyledonous and monocotyledonous families of Angiosperms and also three each of Gymnosperms and Pteridophytes from various parts of the world (Ahuja, 1976).

Disease Occurrence and Distribution

Root rot of French bean was first recorded during the year 1924 from New Jersey (Cook, 1924) while, web blight was first described as destructive wilt of beans and cowpeas in Puerto Rico (Matz, 1921). Thereafter, these have been reported from Egypt, England, Ceylon, Uganda, Tanganyika, Rhodesia, Brazil, Manitoba, Medellin Columbia, Iran, Malaysia, Japan, Venezuella, Spain, Cuba, North Minas Gerais, Turkey, France, Sierre Leone, Florida, Texas, Brazil, Burma, Ceylon, Japan, Philippines, Mexico, Costa Rica, Central America and Trinidad (Gupta and Thind, 2006).

In India, root rot and web blight of French bean were recorded for the first time from Bangalore and Sikkim, respectively (Sharma and Sohi, 1981; Srivastava and Gupta, 1989). Paul and Sharma (1990) reported the occurrence of root rot from the Solan area of Himachal Pradesh while, web blight was recorded during the 1994 crop season (Mathew and Gupta, 1996) from the same locality.

Diagnostic Symptoms

The disease occurs on roots, stem, leaves, pods and seed (Mathew and Gupta, 1996). Sanchez and Cardenaz (1988) reported that root rot affects plant populations from emergence to the first 30 days of crop growth. The distinguishing symptoms of the disease occur in two phases and are described below:

Symptoms appear on roots and stem above and below the surface of the soil as reddish brown cankers. The lesions enlarge rapidly and engirdle the stem at the collar region, extending longitudinally downward to the roots. Roots are affected in advanced stages of infection leading to partial or complete rotting of the root system (Figure 7.1a).

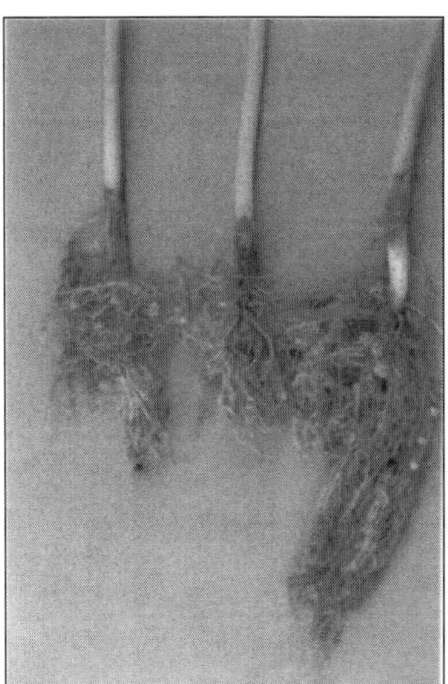

Figure 7.1a. Symptoms of Root Rot

Figure 7.1b. Symptoms of Web Blight.

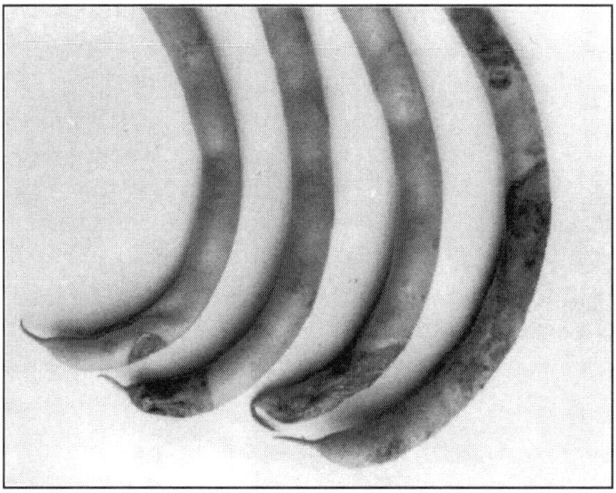

Figure 7.c. Symptoms on Pods.

The web blight symptoms on leaves appear as small, circular, water-soaked spots. These spots increase rapidly in size and cover extensive areas of the leaf blade. Initially, the spots are much lighter in colour than the surrounding unaffected area. At later stages, they become tan brown and surrounded by dark borders. The spots coalesce to form larger areas on the leaf blade (Figure 7.1b). Light tan hyphae develop on both surfaces of the infected leaf and at later stages are studded with minute sclerotia of the fungus. Defoliation occurs in advanced stages of infection. The pods are attacked at all stages of growth. On young pods, circular slimy water-soaked spots appear which are light tan in colour. On mature pods, the spots are dark brown, more or less circular, slightly zonate and definitely sunken (Figure 7.1c). Spots coalesce to cover the entire pod. The distal end of the pods is the most affected. In favourable weather, runner hyphae of the fungus are seen spreading from the point of infection to healthy parts, which later become studded with sclerotia. Tan brown to reddish brown discoloration is observed on infected seed, located immediately below the spots on the affected pods.

Biology

The disease is caused by *Rhizoctonia solani* Kuhn. The fungus consists of septate mycelium. The colour of the hyphae varies from hyaline when young to brown on maturity. The mycelium is branched near the distal septum of the mother hyphal cells at right or acute angles and the branches are constricted at or near the point of origin of the septum. The size of hyphae varies from 117.7 - 173.4 x 6.19 - 9.29 µm. In culture, sclerotia develop within 5-6 days. The size of sclerotia ranges from 1 to 3 mm in diameter (Butler and Bracker, 1970). The sclerotia consist of brown barrel shaped monilioid cells measuring 20-35 x 10-15 µm. The fungus could be seen growing as white crust on infected foliage and as white mycelial web at the distal ends of severely infected pods. In advanced stages, smooth sclerotia measuring 1.0 to 2.5 mm in diameter are seen adhering to infected plant parts. The basidial stage of the fungus (*Thanatephorus cucumeris*) also occurs but very rarely.

Variability in *R. solani* has been studied with respect to cultural, morphological, pathological, physiological, anastomosis grouping and isoenzyme analysis (Ogoshi, 1987). In Himachal Pradesh, Upmanyu *et al.* (2003) also carried out these studies with 18 isolates. The colony colour of all the isolates varied from white to brown and produced varied patterens of sclerotial formation except two (isolate I_7 and I_{18}), which did not produce sclerotia. Isolates also varied significantly in aggressiveness within each cultivar while cultivars showed significant variation in horizontal resistance. Two isolates (I_8 and I_9) were the least sensitive to Bavistin while I_{10}, I_{13} and I_{17} exhibited maximum sensitivity. The isolates were placed in AG-1-1A, AG-1-1B and AG-4 anastomosis groups on the basis of their behaviour. Isolates produced different banding patterns for esterase as well as NADH dehydrogenase and were clustered into different groups on the basis of similarity coefficients. Banding pattern of esterase clearly indicated the relationship with anastomosis groups of *R. solani* isolates. Isolates of AG-4 varied very little while AG-1-1B and AG-1-1A were slightly more variable. Sharma *et al.* (2005) also studied the morphological and molecular variability among 24 isolates of *R. solani* collected from different hosts and locations of north India and classified the isolates into four groups. However, French bean isolates were grouped into two separate clusters indicating the high genetic variability in the pathogen population prevalent in different epidemiological regions of Himachal Pradesh. All the isolates of *Rhizoctonia solani* exhibited slight variation in their physiological requirements. Most of the isolates preferred Czapek's dox agar as solid medium while potato dextrose, broth supported the maximum growth of the isolates. All the isolates grew rapidly within a temperature range of 25 to 30°C and RH above 95 per cent. pH ranging between 5.0 and 6.0 supported the maximum growth of the isolates. However, isolates preferred slightly higher pH as compared to lower ones. Glucose among carbon and ammonium tartrate as well as potassium nitrate among nitrogen sources yielded maximum mycelial growth of most of the pathogen isolates (Upmanyu and Gupta, 2009).

Disease Cycle

The pathogen perennates as mycelium or sclerotia in the soil on plant debris. The fungus is present in most of the soils and once established in a field remains there indefinitely. The germination of sclerotia gives rise to infective propagules, which are sufficient to cause infection. The fungus spreads with rain, irrigation or floodwater, tools and with infected or contaminated propagation material. The association of *R. solani* with bean seed has been reported (Nitshe and Cafati, 1985). Seed transmission upto 46 per cent in cultivars Kentucky Wonder and Contender (Gupta *et al.*, 2000a) have also been reported. Shama and Shama (1989) indicated that seed penetration of *R. solani* occurs directly through intact seed coat or through the hilum tracheids and that the pathogen is transferred from infected seed to the soil. The infected seeds after germination either rots in the soil itself or give rise to plants that later damped off. The soil borne inoculum is splashed by rain to the foliage that causes web blight. On some plants or plant parts the pathogen can enter only through wounds or through openings in the epidermis while in others, it thrives in the soil or in mud splashed on to the plants until it can enter through healthy tissue.

In favourable weather, runner hyphae of the pathogen are seen spreading from the point of infection to healthy parts, which later become studded with sclerotia.

Epidemiology

Root rot is favoured by high soil temperature and soil moisture while, web blight is more prevalent in the wet season. *Rhizoctonia* diseases of roots are most severe under moist soil conditions (Tu *et al.*, 1992). Bruggen *et al.* (1986) reported positive correlation between lesion size and soil moisture. If the moisture is sufficient enough or there is frequent drizzle, pronounced blight symptoms will appear. During humid periods fan shaped whitish hyphae spread from the point of infection and produce the conspicuous and destructive phase of the disease. Diaz Plaza *et al.* (1991) reported that disease incidence and severity increased after rain. Upmanyu and Gupta (2005) studied the role of abiotic factors on the development of this disease and reported that root rot development was favoured at high (80 per cent) soil moisture and 25°C temperature while 100 per cent relative humidity coupled with 25°C temperature favoured web blight. Continuous lesf wetness for atleast 6h was essential for disease initiation and increase in leaf wetness durations from 6 to 12 h showed a corresponding decrease in incubation period from 36 to 26 h whereas further increase in leaf wetness did not exert any effect on incubation period.High rainfall and soil moisture coupled with high relative humidity and soil temperature (23-25°C) were found to be favourable for web blight (Mathew and Gupta, 1996). A high soil moisture (80 per cent) and a temperature of 25°C were the most favourable for root rot development, while web blight development was optimum at >85 per cent relative humidity coupled with 25°C temperature (Upmanyu and Gupta, 2005).

Management

The causal fungus persists in the soil, therefore, an integrated approach of disease management combining the use of cultural, biological and chemical methods is required.

Cultural Practices

Cultural practices *viz.*, planting disease free seed, mulching, adjustment of planting date, medium (30-60 x 15 cm) plant spacing, crop rotation and soil solarization have been recommended to keep the disease under check (Sartorato *et al.*, 1988; Tu, 1992, Sharma and Gupta, 2003). Deep tillage (30 cm subsoiling) can reduce soil compaction and thereby, root rot severity in addition to increased yield (Tan and Tu, 1995). Voland and Epstein (1994) reported the efficacy of fresh manure over composted manure in reducing bean hypocotyl lesions while soil amendment with cotton, mustard and neem cakes was found effective in reducing root rot incidence and increased yields (Upmanyu *et al.*, 2002). The effectiveness of these amendments may be due to enhanced populations of non-parasitic microorganisms exhibiting antagoni-stic effects on the pathogen thereby inhibiting the mycelial growth and sclerotia production of the pathogen. Soil amendments with low soil C: N ratio increased incidence of *R. solani* whereas high C: N ratio amendment did not interfere with the incidence while incidence was independent of soil moisture conditions (Fenille and Souzo 1999). Poultry manure, composted *Urtica* sp. and composted *Lantana camara* treatments were superior to other composts in suppressing Rhizoctonia root rot (Joshi *et al.*, 2009).

Host Resistance

Different varieties/accessions/lines have been evaluated worldover and few were found to possess varying degree of resistance. Dark seeded PI accessions of French bean *viz.* 109859, 163583, 174908, 226895 and 300665 were resistant to *R. solani* while most of the white seeded cultivars were susceptible to both seed infection and pre-emergence damping off (Prasad and Weigle, 1976). However, Sharma and Sohi (1981) reported all the three French bean cultivars *viz.*, "Contender", "Premier" and "Giant Stringless" to be equally susceptible irrespective of their colour. Bush type cultivars/lines are more susceptible to the disease than pole type ones. The cultivars/lines *viz.*, Wisconsin RRR 46, BAT 477, BAT 332, BAT 1753, RIZ 30, EMP 81, A 300, ICA Pijao and Jackson Wonder have been reported to be highly resistant (Hagedorn and Rand, 1980; Tu and Park 1993; Muyolo *et al.*, 1993) while, PLB-43, EC-18834, IC-18593, PI-249554, Phagu Chitra, Pole wall-I, Pole wall-II, and K-13522 were found moderately susceptible to the disease (Gupta *et al.*, 1997). Among 93 cvs./lines, only ET 8396 was found resistant both under natural epiphytotic and in vitro conditions that can be used in breeding programmes as source of resistance (Upmanyu *et al.*, 2004).

Biological Control

Exploitation of the populations of bioagents may prove useful to combat the ravages of soil borne diseases. Biological treatment of soil under glass house conditions with *Trichoderma viride* showed results similar to chemical control (Dumitras and Sesan, 1990). Elad *et al.* (1980) found that in naturally infested soils, wheat bran preparations of *Trichoderma harzianum* significantly decreased the *Rhizoctonia* disease of beans. Tu and Vaartaja (1981) reported that *Gliocladium virens*, a hyperparasite of *R. solani*, inhibited production of sclerotia. Management of this disease by seed/soil treatment and foliar sprays with *T. harzianum*, *T. virens* and *T.viride* has also been reported (Mathew and Gupta, 1998; Hazarika and Das, 1998; Upmanyu *et al.*, 2002). Inhibition of mycelial growth and sclerotial germination of this pathogen by *T. harzianum* and *T. viride* culture filterate has also been observed (Hazarika and Das, 1998). Roberti *et al.* (1993) indicated that *T. virde* produced non-volatile compounds while *T. harzianum* produced coils and hook pincer shaped structures, which prevented the growth of *R. solani*. The efficacy of potential bacterial antagonists *viz.*, fluorescent pseudomonads and *Bacillus* spp. against root rot has also been reported (Sanchez *et al.*, 1994; Wolk and Sarkar, 1994, 1995). An endophytic *Pseudomonas florescens* carrying *Serratia marcescens* Chi A gene either on the plasmid or integrated into the chromosome controlled effectively the phytopathogenic fungus *Rhizoctonia solani* on bean seedlings under plant growth chamber conditions (Downing and Thomson, 2000). Coinoculation of seed with *Bacillus subtilis* MB1600 (Epic) and *Rhizobium tropici* significantly reduced root rot severity and enhanced yield (Estevez-de-Jensen *et al.*, 2002). Efficacy of seed treatment and soil application of *B. subtilis* in the reduction of pre- and post-emergence root rot have also been reported (Sharma and Gupta, 2003). The water extracts obtained from green parts of potato, tomato and rape and *Trichoderma* PBG-1 showed marked antagonistic activity against *Rhizoctonia solani* (Smolinska and Kowalska, 2006). Out of sixty isolates of *Trichoderma*, 7 isolates significantly reduced the incidence of root rot compared to the control. *T. harzianum* isolates Tr-34 and Tr-14 were the most effective, resulting in only 20.8 and 23.4 per cent root-rot incidence, respectively (Joshi *et al.*, 2007).

Chemical Control

Different fungicides have been reported to be effective for the management of this disease. Seed treatment with fungicides *viz.*, benomyl, captan, thiram, thiophanate-methyl, carbendazim, carboxin, terrachlor, pencycuron, iprodione + thiram (Raffat, 1992; Maringoni *et al.*, 1992) gave maximum protection against pre and post emergence mortality and also significantly increased the yield of green pods. However, seed treatment followed by foliar sprays with carbendazim or bitertanol gave best results against web blight (Thakur *et al.*, 1991). Gupta *et al.* (1999) reported good efficacy of Bavistin (0.2 per cent), Celest (0.1 per cent), Topsin-M (0.2 per cent) and Raxil (0.2 per cent) against pre-emergence damping off of beans. Good control of web blight phase of this disease was achieved by seed treatment with Celest (0.2 per cent) or Raxil (0.2 per cent) or Bavistin (0.2 per cent) followed by foliar sprays of Bavistin (0.05 per cent) or tebuconazole (0.05 per cent) (Mathew and Gupta, 1996; Upmanyu *et al.*, 2002).

Integrated Approach

Integration of seed treatment with bioagents and fungicides has been found to be more effective to manage soil borne pathogens. Tu and Zheng (1993) studied the compatibility of seed treatment with *Bacillus subtilis*, *Pseudomonas fluorescence* or *Gliocladium virens* and subsequently dressed with diazinon + captan + thiophanate-methyl (DCT). They observed that fungicide treated seed showed inhibition of bioagents, which was transient, and the growth and sporulation of the bioagents resumed as the effect of the fungicides diminished. They further suggested that bioagents such as *B. subtilis* and *P. fluorescence* are more mobile in the rhizosphere of seedlings grown from treated seed and hence, their incorporation together with fungicides can extend the effect of the chemical to give protection against root rot. Upmanyu *et al.* (2002) also reported the tolerance in *T. viride* to carboxin, tebuconazole and carbendazim. The combined seed treatment of French bean with Contaf (0.025 per cent) and *Gliocladiumvirens* (0.1 per cent) resulted in higher seed germination, lower incidence of the disease and higher grain yield (Mukherjee *et al.*, 2001). Combined treatment of seed (soaking in *Trichoderma harzianum*, 1 x 10^7 conidia/ml) and soil (20 and 100 kg/calcium and micronized sulfur (40 per cent micron sulfur), respectively) before sowing reduced the root rot incidence and increased green pod yield (Ziedan and Mahamoud, 2002). Soil amendment with mustard cake alongwith carbendazim seed treatment and foliar sprays were found highly effective in reducing the pre-emergence and post-emergence root rot and web blight severity (Upmanyu *et al.*, 2002). A combination treatment including soil solarization (SS) for 50 days + soil amendment (mustard cake) + combination of *Allium sativum* ethanol extract and *Bacillus subtilis* was found effective in reducing pre- and post-emergence damping-off while web blight was best contained by SS

+ soil amendment + *A. sativum* (ST + FS) and increased yields (Sharma and Gupta, 2003).

Angular Leaf Spot

It is an important disease of French bean particularly in mid hill conditions where moderate temperatures with high humidity conditions favour the development of this disease. Up to 50 per cent yield losses under epiphytotic condition has been reported from USA whereas in Mexico, it has been considered as a principal fungal disease where up to 90 per cent incidence has been recorded causing significant losses (Campos and Fucikovsky, 1981). In India, the total loss including damaged and unmarketable pods has been estimated to be about 40-70 per cent due to this disease (Singh and Saini, 1980) while Ponappa *et al.* (1976) reported 80 per cent yield loss and premature death of the plants from Bangalore area of Karnataka.

Host Range

The pathogen has a very narrow host range restricted only to the family leguminosae. In 1931, Abramoff reported *I. griseola* to cause angular brown spot on soyabean in Russia. Out of six species of legumes, only *P. lunatus* L. was attacked by this pathogen (Cardona-Alvarez and Walker, 1956). In India, among thirteen leguminous plants tested, only *Phaseolus multiflorus, P. vulgaris* and *P. lunatus* were found susceptible (Verma, 1979).

Disease Occurrence and Distribution

It was reported in the year 1878 from Italy for the first time (Saccardo, 1878). Since then, it has been reported from different continents of the world like U.S.A, Europe, Africa and Asia (Gupta and Shyam, 2003). In India, the disease was reported for the first time from Nilgiri hills in South India (Srinivasan, 1953). Subsequently, the disease has been appearing regularly in varying intensities in all bean-growing areas.

Diagnostic Symptoms

The pathogen infects all green parts of the plants *viz.*, leaves, petioles, stems, pods and seed, producing varied symptoms. The detailed symptoms are described below:

On Leaves

On cotyledonary leaves, the symptoms appear as circular spots while on true leaves, 3-5 angled spots appear in between veins and veinlets (Figure 7.2a). The spots vary from 0.5 to 8.5 mm in size and are dark grayish on upper surface and light gray on the lower surface of the leaves but with the passage of time, change to reddish brown and finally attain dark brown colour. On the lower surface, light reddish discolouration appears on the veins and veinlets, which delimit such spots. Gray moldy coating covers the spots on the lower surface of the leaves. On close observation, the spots reveal the presence of coremia bearing large number of spores. Many of such spots coalesce covering a large portion of the leaf giving it dirty brown appearance, whereas the remaining part turns yellowish. Such severely infected leaves show upward curling and defoliate prematurely.

On Petioles and Stems

Elongated dark brown spots appear both on petioles and stems. These are concentrated towards the point of leaf attachment on the petiole while on the stem, at the point of origin of branches. Under severe conditions, complete defoliation occurs resulting in pre-mature death of the plants. Coremia formation is rare on these parts under natural conditions.

On Pods and Seed

Pods are infected at all the developmental stages. Lesions on the pods are smooth, usually circular and rarely elongated (Figure 7.2b). At first, these spots are superficial having reddish brown center with well-defined ashy black borders but with the passage of time, deeper tissues are involved. In the central portion reddish brown, coremia and spores are produced. Severely infected pods either bear no seeds or produce only shrivelled seeds. In late infections *i.e.* at maturity stage, the seeds underneath the pod lesions show yellowish brown discolouration which is more prominent on hilum region but can occur any where on the seed surface.

Figure 7.2a. Symptoms of Angular Leaf Spot on Leaves.

Figure 7.2b. Symptoms on Pods.

Biology

The disease is caused by fungus *Phaeoisariopsis griseola* (Sacc.) Ferr. On the host tissue, the fungus produces a stromatic structure in the substomatal cavity. Conidiophores are laterally joined together in the form of synnemata or coremia, which are darker at the base and lighter towards the tips, and either curved flexuous or septate. The coremia measure 128-680 µm in length and 28-69 µm in thickness. Conidia are light grey to hyaline, cylindrical to spindle shaped and straight to slightly curved. Conidia are borne singly on the tips of columnar hyphae, which are mostly 1-3 (rarely 4-6) septate, with dimensions ranging from 50-60 x 7-8µm. Karanja *et al.* (1994) also reported cylindrical to spindle shaped brown coloured conidia of average dimensions ranging from 17-95 x 7-24 µm and 2-3 septate. No sexual stage of the fungus has yet been observed.

Villegas (1959) working with 30 isolates of *P. griseola* collected from different bean production sites in Colombia reported the existence of 13 pathogenic races. In India, existence of a new strain of *P. griseola* from Himachal Pradesh, India on the basis of morphological characters *viz.*, length and width of synnemata, conidiophores, conidia and conidial septation has been reported (Sharma and Sohi, 1978). Similarly, Lacerda *et al.* (1994) iden-tified five races (viz., G1, G2, G3, G4 and G5) from Pennambuco state, Brazil of which G1 was the most common and G3 most virulent. Earlier Hocking (1967) recovered a Tanzanian isolate of *P. griseola*, which produced circular lesions and speculated the role of single mutation within natural isolates. Using isozyme analysis Correa-Victoria (1987) divided 55 isolates of *P. griseola* from Latin America, the United States and Africa into two groups. All African isolates displayed the same electrophoretic pattern and were associated with large seeded bean cultivars originating in the Andean region of South America, while a second electrophoretic pattern was associated with small seeded bean cultivars from Central and North America. Similarly, Guzman *et al.* (1995) using RAPD markers differentiated 62 isolates of *P. griseola* from Malawi, USA and Brazil into two major groups (Andean and Mesoamerican). Corrales and Jara (1995) further suggested that these races have evolved with large seeded Andean cultivars while Mesoamerican races have evolved with small or medium seeded varieties from Central America.

Disease Cycle

In areas where the crop is grown regularly year after year in the same field, the fungus overwinters in the infected plant debris. Seed borne nature of the pathogen has also been reported. In India, the pathogen has been reported to be both seed and soil borne (Sohi and Sharma, 1974; Sindhan and Bose, 1980). Infected seed used for sowing serve as the source of primary infection. The infected seed after germination gives rise to a plant with circular brown spots on the cotyledonary leaves. These spots on the under surface bear numerous dot like structures known as coremia which bear condiophores and conidia of the fungus. The conidia may be disseminated from the debris by rain splashes or along with wind blown soil particles and from sporulating lesions by wind currents. The pathogen infects host tissue by entering through stomata and advances intercellularly in the mesophyll and palisade parenchyma. After 7-10 days of infection the fungus develops intracellularly throughout necrotic lesions, stromata develop in the substomatal cavity and sporulation may occur during periods (24 to 48 h) of continuous moisture. The conidia cause secondary infection and the cycle continues. In severe infections, the leaves turn yellow and fall down prematurely. Infection from pods also goes to the developing seed.

Epidemiology

The pathogen (*P. griseola*) overwinters both on plant debris and in infected seed (Sohi and Sharma, 1974; Sindhan and Bose, 1980). The viability of conidia on plant debris is 6 and 8 months in laboratory and field, respectively (Sindhan and Bose, 1980), thus it serves as a source of primary inoculum. Other source of primary inoculum is infected seed. Though the infestation site has been identified in the hilum of seed but in some cultivars both hilum and seed coat seem to be infected (Saettler and Corea, 1988). While disease can initially result from the planting of infected seed causing spots on cotyledonary leaves, the primary route in most bean growing areas involves the rain splashed conidia from plant debris or wind blown conidia with soil particles attacking leaves, pods and stems. Infected leaves become inoculum source for secondary spread of the disease. However, the rate and degree to which secondary spread occurs is dependent on the interactions and influences of all factors involved with the disease and its spread such as climate, pathogen and the host. Moderate temperatures and moist conditions favour the arrival of viable inoculum and successful field infections. Optimum temperature for growth and sporulation, spore germination and development of the disease are 21-24°C, 18-24°C and 24°C (Sindhan and Bose, 1980; Sarotorato, 1988a) while optimum RH for sporulation, spore germination and development of symptoms has been reported as >90 per cent. Mathew *et al.* (1998) studied the effect of leaf wetness durations at optimum temperature and reported that leaf wetness of at least 12 h duration coupled with moderate temperature (25°C) exhibited minimum incubation period of 9 days. Gupta *et al.* (2005) reported that increase in relative humidity from 85.7 to 89.9 per cent coupled with moderate temperatures of 25°C resulted in sharp increase in disease development. Diurnal periodicity of conidial release of *P. griseola* revealed a distinct rhythm at different times of the day with maximum conidial density being observed between mid night and early morning hours. Conidial formation and release was favoured by temperature (23-24°C), high relative humidity (>90 per cent), prevalence of dew (20.00 to 8.00 h) and dark period.

These workers also studied the role of environmental factors on disease development and reported the consistent effect of rainfall, soil moisture and relative humidity on the disease development. Simple correlation coefficients between disease severity and rainfall (0.7596), soil moisture (0.7070) and relative humidity (0.5253) were found to be highly significant and positive. But Chhetry *et al.* (1994) reported non-significant linear correlation and regression coefficients of disease incidence and severity with microclimatic factors. The inoculum concentration from $1x10^4$ to $5x10^4$ conidia of *Phaeoisariopsis griseola*, leaf wetness durations from 3 to 9 h showed a progressive decline of incubation period of angular leaf spot on French bean. Intermittent wetness up to 3 cycles (12 h dry and 12 h wet) was found effective to increase number of lesions per leaf. Higher relative humidity (>90 per cent) and temperature (25°C) favoured disease development (Gupta *et al.*, 2005).

Management

Cultural Practices

Various cultural practices *viz.*, use of disease free seed and removal of diseased leaves and seedlings (Trutmann and Kayitare, 1991); 3-4 year crop rotation (Correa-Victoria and Saetler, 1987); destruction of diseased plant material; deep ploughing (Hocking, 1967); wider spacing (Navarro *et al.*, 1981; Gupta *et al.*, 2000e) and altering of sowing dates (Bhardwaj *et al.*, 1994, Gupta *et al.*, 2000e) and staking of pole type varieties (Gupta *et al.*, 2000e) have been recommended from time to time for the control of this disease. Moreno (1977) studied the effect of different cropping systems (bean grown alone or in combination with maize, sweet potato or Cassava) on the severity of angular leaf spot and reported that severity was maximum when maize was included and lowest with sweet potato and Cassava. However, during a multilocational trial in Tanzania, Kikoka *et al.* (1989) found intercropping with maize effective to reduce the severity of the disease. Boudreau (1993) and Gupta *et al.* (2000e) reported the similar observations from Kenya and India, respectively. Poultry manure, composted *Urtica* sp. and composted *Lantana camara* treatments were superior to other composts in suppressing angular leaf spot (Joshi *et al.*, 2009).

Host Resistance

French bean cultivars are known to possess varying degree of resistance against the pathogen (Buruchara *et al.*, 1988). Various cultivars/lines like Albama No.1, Café, California Small White, Case Knife, Epicure, Mexico blank, McKastan, Navy bean, Negro Costa Rica, Scotia Rojo Chico, Mexico 11, Mexico 12 and Cauca 27, Carpoata (Santos Filho *et al.*, 1978); Negro Tacana (Lopez *et al.*, 1997) and Manteiga Maravilha (Paula *et al.*, 1998) are reported as resistant. Araiyo *et al.* (1989) reported that non-black cv. Ouro and black coloured cvs. Rio Tibagi and Milionario 1732 possess resistance to *P. griseola*.

In India, cvs./lines *viz.*, EC 38921, EC 44621, PLB 148, Kentucky Wonder, Canadian Round, Canadian Long Red, IC 47651-6, EP 146 (BAT-482), HPR-54, HPR-63, HPR-92, HPR-93, HPR-111, HPR-232, HPR-260, HPR-299, HPR-332, HPR-326, Hans, Nagar Local, SVM-1, JK-8, Him-12, IIHR 42-2, IIHR 901, EC 97830, EC 44758 and NIC 14402 were reported to be resistant against this disease (Singh and Sharma, 1975; Bhardwaj *et al.*, 1992; Srivastava *et al.*, 1995; Mohan *et al.*, 1995; Gupta *et al.*, 2000d). The effectiveness of supplementing local bean mixture with 25 per cent or more resistant lines (BAT 76 and A 285) was evaluated and it resulted in significant reduction of angular leaf spot severity in Africa (Pyndji and Trutmann, 1992; Trutman and Pyndji, 1994). Of the 68 French bean genotypes/lines screened for resistance to *Phaeoisariopsis griseola* during 2002 and 2003 at two locations namely, Shalimar and Wadoora in Kashmir, India, 9 genotypes/lines, *viz.* SL-4-PL-1, HIM-1, Bountiful, JKRO-3, L-27, L-28, L-8 and Local Big Beans, exhibited highly resistant reaction on pods. On leaves, 6 genotypes, *viz.* Uri-Red, EC-285559, Medur Tral, Local Red, Local Big Beans and SAW/GH/TA/41A at Wadoora location, and 4 genotypes, *viz.* Gurez Local, Big Bean Black, EC39855 and Local Big Beans at Shalimar location showed highly resistant reaction, while the genotype Shalimar French Bean-1 gave highly susceptible response on leaves at both locations (Bashir *et al.*, 2006).

Chemical Control

Various fungicides have been reported to give good control of this disease in India and other countries. However, during continuous rains which are essential both for good growth of bean plants and rapid development and spread of the disease, routine application of fungicides is sometimes delayed which often results in heavy build up of the disease. Proper timing of initial sprays is of prime importance to achieve desirable level of disease control. Attempts have been made to control the disease by both seed treatment with mercurial and other fungicides and foliar sprays with various fungicides. Sprays of fungicides like wettable sulphur, mancozeb, Bordeaux mixture, zineb, carbendazim, benomyl, triforine, triademorph, (Fortugno, 1977; Sindhan, 1984; Issa *et al.*, 1982; Hidalgo and Araya, 1993; Singh *et al.*, 1995), chlorothalonil, TPTH, TPTA (Castro *et al.*, 1991; Vierra *et al.*, 1998) and mancozeb (Issa *et al.*, 1982; Castro *et al.*, 1991; Vierra *et al.*, 1998), have been reported effective to reduce disease intensity and increase yield. Oliveira *et al.* (1992) reported stannic triphenyl (fentin) acetate, bitertanol and tebuconazole as more effective when compared to Bavistin while, Srivastava and Gupta (1994) reported captan, thiophanate methyl, carbendazim, ziram and captafol to be approximately equally superior to mancozeb, copper oxychloride and tridemorph. Combi fungicide sprays like benomyl + captafol, chlorothalonil + copper oxychloride (Issa *et al.*, 1982) and thiophanate methyl + thiram (Rodrigues *et al.*, 1987) have also been found effective. Combination treatments of carbendazim (seed treatment + foliar spray) and bitertanol (seed treatment + foliar spray) were also found effective in reducing the severity and increasing the green pod yields than seed treatment or foliar spray alone (Mathew *et al.*, 1998).

Various EBI fungicides have longer post-infection and eradicant activity being systemic in nature. Mathew *et al.* (1998) studied the mode of action of EBI's like hexaconazole, penconazole and bitertanol along with mancozeb and carbendazim by providing artificial weather conditions for infection. Carbendazim provided hundred per cent control of the disease up to 120 h while mancozeb upto 96 h. EBI's exhibited poor protective activity only up to 72 h. On the other hand, EBI fungicides demonstrated post-infection activity upto 72 h while it was only 24 h in case of carbendazim and mancozeb. When applied as pre-symptom spray mancozeb had little effect and produced sporulating spots while EBI's and carbendazim produced reddish brown non-sporulating spots when applied 7th and 8th day after inoculation. As post-symptom treatment carbendazim, hexaconazole, flusilazole and bitertanol had high level of antisporulant activity and their residue persisted for more than 10 days. These properties of EBI fungicides can be utilized in planning different spray programmes for effective control under different situations. Of late, spray schedules involving the strobilurine (azoxystrobin), tria-zoles (tebuconazole, propiconazole) and benzimidazole chemical groups were found effective in controlling angular leaf spot (Oliviera and Oliveira, 2003).

Integrated Approach

Fungicidal treatments coupled with careful selection of varieties and sowing dates proved highly effective to reduce disease severity and increased the yield of the crop (Bhardwaj *et al.*, 1994). Effect of two mulches (Pine needles and *Eucalyptus* leaves) alone and in combination with carbendazim (0.1 per cent) sprays on mulch and foliage on the disease severity and pod yields has also been studied by Mathew *et al.* (1998). Sprays of carbendazim on mulches and foliage were highly effective in reducing the disease severity and increasing green pod yields. Sprays of carbendazim on mulches alone were also found quite effective and reduced the severity and increased green pod yields significantly with additional advantage of residue free pods.

Anthracnose

It is an important disease of common bean throughout the world but more severe in tropical and subtropical regions (Pastor-Corrales *et al.*, 1994). Though this disease occurs worldover including tropical and subtropical regions but it causes greater losses in the temperate zones than it does in tropics. Losses from this disease can be up to 100 per cent when highly contaminated seed is planted under favourable disease development conditions. Yield losses of 90-100 per cent have been reported in many countries throughout tropical America and Africa (Pastor-Corrales and Tu, 1989; Lenne, 1992). In Himachal Pradesh, India, the incidence of this disease has been reported to range from 5.0 to 65.0 per cent in different localities leading to considerable yield losses in certain years (Sharma *et al.*, 1994; Padder and Sharma, 2010).

Host Range

Besides *Phaseolus vulgaris*, the host range of *C. lindemuthianum* includes *Vigna radiata, V. sinensis, P. multiflorus, P. lunatus, Macrotyloma uniflorum, Dolichos lablab, D. biflorus* and *Canavalia ensiformis.*

Disease Occurrence and Distribution

The pathogen was actually collected by Mycologists as early as 1843 in France (Desmazieres, 1843), but the first description of bean anthracnose in 1875, was based on the specimens collected from Lindemuth, Bonn, Germany (Zaumeyer and Thomas, 1957). Subsequently, it was reported from England, Italy, France, Germany, North America, Japan, Taiwan, East and West Indies, Belgian Congo, Uganda, Brazil, Guatemala, Venezuela, Canada, Cuba, Australia, New Zealand and erstwhile USSR (Gupta and Thind, 2006). In India, its occurrence was noticed for the first time in Nilgiri hills in 1915 (Hutchinson and Ram Ayyar, 1915). Since then the disease has been recorded in the entire bean growing areas of the country, which have cool and moist weather during growing season.

Diagnostic Symptoms

The disease may appear on any above ground plant part depending upon time of infection and source of inoculum but most striking symptoms appear on immature pods. Initial symptoms may, infact, appear on the cotyledonary leaves as small, dark brown to black lesions. Conidia and hyphae then may be disseminated by rain or dew to the developing hypocotyl where rust coloured specks develop. The specks gradually enlarge lengthwise along, and partially around the hypocotyl and young stems, forming a sunken lesion. On leaves, the infection may occur on both sides but early sign of infection usually appear on the under leaf surface as blackened dead portions of the veins. These may extend to a limited adjoining areas. Later, such spots also appear on the upper leaf surface. Lesions are also formed on petioles and stems. On pods, black sunken spots with lighter or grey central area are seen (Figure 7.3). The central portion of the spots shows pinkish masses of spores of the fungus, especially in wet weather. Later, the sides of these spots appear raised. The seeds obtained from heavily infected pods show brown to light chocolate coloured sunken cankers on the seed coats.

Figure 7.3. Symptoms of Anthracnose.

Biology

The fungus, *Colletotrichum lindemuthianum* (Sacc. and Magn.) Briosi and Cav. is responsible for this disease. The name is still widely used in scientific literature despite the fact that the perfect stage (*Glomerella cingulata* (Stonem) Spauld and Shrenk) of the fungus has been described and occurs very rarely (Kimati and Galli, 1970). The mycelium is branched, septate and hyaline at first becoming dark coloured with age. On culture media, conidia may form singly at the ends of hyphal threads. When stromatoid masses form, acervuli are produced upon them and on the host they form stroma below the cuticle, rupturing the latter if it is still present on the lesion. Conidia, which are borne acrogenously on short conidiophores, develop successively under favourable conditions until pinkish spore masses appear on the surface. Setae occur sparingly. Conidia (4.4-5.3 x 13 to 22 µm) are hyaline, cylindrical with rounded ends or somewhat pointed at one end; a clear vacuole like body often occurs near the centre. One to four, usually two germ tubes arise from a conidium. Appresoria are formed at the tips of germ tubes, especially when they come into contact with host (Figure 7.4).

The perfect stage of the fungus was originally called *Glomerella lindemuthiana* Shear (Shear and Wood, 1913), but now it has been renamed as *G. cingulata* (Kimati and Galli, 1970). The fungus produces perithecia with a diameter of 120 to 210 µm and rostrum 30 to 80 µm in length. Perithecia contain hyaline and filiform periphyses until 27 days of age and asci which measure 8 x 48- 68 µm disappear after 27- 30 days. Each ascus contains eight ascospores, which may be alantoid (605 x 20 µm) in shape. Ascospores are ejected from the tip of ascus. Meyer *et al.* (2001) reported that during attack of this disease, CO_2 assimilation, stomatal conductance to water vapour, and photosynthetic electron transport rate decreased whereas in dark respiration increased.

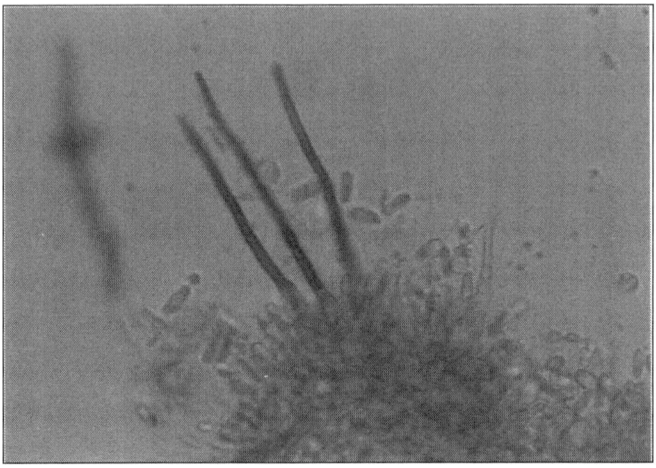

Figure 7.4. Acervulus of *Colletotrichum* Bearing Conidia and Conidiophores.

Barrus (1911) was the first to differentiate two distinct pathogenic races with in a morphologic species of *C. lindemuthianum* on the basis of reaction of host varieties and designated them as Alpha and Beta. Since then many pathotypes of this pathogen has been described in Europe (Fouilloux, 1979), South and Central America (Menezes and Dianese, 1988), Canada (Tu *et al.*, 1984) and Uganda (Pastor-Corrales and Tu, 1989) *etc.* Unfortunately, differential bean cultivars were not identical which made it difficult to compare results and to exchange data or resistant germplasm (Pastor-Corrales and Tu, 1989). To overcome this difficulty, a set of twelve bean differential cultivars has been defined with binary values (Pasto-Corrales, 1991). A number designates a race, which is the sum of the binary values assigned to differentials susceptible to the isolate (Pastor-Corrales, 1991).

Using this system, atleast 130 different *C. lindemuthianum* races have been identified. Race characterization has been carried out on isolates from South America, Central America, USA, Canada, Brazil, India and Spain. In many African countries, even though anthracnose is severe, little or no information exists on diversity of *C. lindemuthianum*. Only 56 isolates have been characterized using the binary system in Africa, and a total of 36 races have been identified up till now. Races 401 and 485 have been recorded for the first time in Burundi, Central Africa. Occurrence of Delta mutant (Race 7), a new race in North of Spain was recorded (Gupta and Thind, 2006).

In India, 85 isolates of *C. lindemuthianum* collected from different bean growing areas of Himachal Pradesh were characterized on the basis of their reaction types on International and CIAT differentials (Drijfhout and Davis, 1989) and 12 and 19 races were identified, respectively (Sharma *et al.*, 1999). On International differentials, specifically Alpha-Brazil, Beta, Gamma and IND-I to IND- IX were characterized. The races designated as IND I to IND IX were different from those identified in Europe and USA, thus forming a new race group from the Indian subcontinent. On CIAT differential set, only races 65 and 73 resembled the North American races. Out of 90 isolates, 12 races resembled American and European virulences, whereas 17 races (101, 103, 115, 119, 131, 195, 537, 551, 581, 598, 613, 615, 631, 639, 707, 775 and 935) were identified as new, thereby enhancing the total race number to more than 140 globally. RAPDdata analysis with nine 10-mer random primers producing polymorphic bands, grouped 90 isolates into two major clusters accommodating 36 and 54 isolates, respectively. There was no congruence between the RAPD pattern and virulence phenogram (Sharma *et al.*, 2007).

Padder *et al.* (2007) analyzed five *Colletotrichum lindemuthianum* populations causing anthracnose in common bean to know about genetic diversity, migration and the probable rate of spread of races capable of overcoming resistance present in elite cultivars. Diversity within each population (HS) was high with values ranging from 0.26 to 0.31. Average differentiation among populations (GST) was 0.12 and populations were isolated by distance between the two populations. This indicates that less short distance gene flow occurs in Himachal Pradesh. Ribosomal DNA analysis of fifteen isolates representing 15 races of *Colletotrichumlindemuthianum* restricted with 2 restriction enzymes (*Msp I* and *Hea* III), fivehaplotypes were observed in the dendrogram thereby revealing low diversity (Padder *et al.*, 2008).

Disease Cycle

The pathogen overwinters in infected seeds and plant debris. Ravi *et al.* (1999) reported that in seed, the pathogen was mostly present in seed coats and cotyledons and rarely in embryonic axes. In infected seed, conidia can remain viable for many years (Tu, 1983) and infect young seedlings. The conidia may germinate in 6 to 9 h under favourable environmental conditions and produce one to four germtubes, which in turn form appressorium and get attached to the host cuticle by a gelatinous layer. The later produces infective hyphae, which penetrate the cuticle and epidermis by mechanical pressure, applied by appressoria. Infective hyphae initially produces biotrophic primary hyphae (100-130 µm) that are wide in diameter and entirely intracellular, followed by necrotrophic secondary hyphae (25-40 µm) that are narrow and either intercellular or intracellular (Perfect *et al.*, 2001). The hyphae enlarge and grow between the cell wall and protoplast for 2 to 4 days without apparent damage to the host cells. Several days later, the cell walls are degraded and the protoplast dies, leading to the appearance of water -soaked lesions. Mycelium then may aggregate within the lesion site and form an acervulus, which ruptures the host cuticle. The acervulus contains a layer of 3 to 50 conidiophores depending on the size of lesion. The conidia are dispersed by rain splashes to nearby plants and the cycle can be repeated several times during the growing season. The perfect stage of the fungus is very rarely encountered in nature. Perithecia are formed in the infected tissue and asci and ascospores are produced within the perithecia. The role of ascospores in disease cycle has not been investigated because their occurrence is very rare.

Epidemiology

Conidia and/or dormant mycelia in the infected seeds and/or infected plant debris germinate and infect the young seedlings. It can remain viable in infected seed for several years (Tu, 1983). When infected seed germinates, lesions appearing on cotyledons serve as the source of secondary inoculum. The spores are almost entirely water-borne. Primary leaves and the hypocotyl are foci of secondary infections. Borucka and Marcinkowska (2001) reported the importance of wet seasons at early growing period besides availability of a source of inoculum for the initiation and occurrence of this disease. The pathogen is very sensitive to changes in temperature and humidity. It develops most abundantly in cool, wet weather and largely disappears under hot and dry conditions. A relative humidity of 92 per cent and above is essential for infection, the optimum being close to 100 per cent. The fungus requires about 10 mm of rain to establish initial infection (Tu, 1981). The optimum temperature for disease development ranges from 18 to 27°C with maximum intensity at 21°C (Sindhan, 1983) and is markedly reduced at 13°C. Heavy and frequent rains with moderate temperatures (19- 25°C) and high relative humidity (>70 per cent) favoured the progress of the disease in terms of vertical and horizontal spread of the disease (Kumar *et al.*, 1999). Moderate rainfalls at frequent intervals are also essential for the local dissemination of conidia present in water-soluble gelatinous matrix and the development of severe anthracnose epidemics. The movement of insects, animals and man may spread conidia particularly when foliage is moist.

Management

Various management strategies have been attempted to reduce yield losses due to bean anthracnose, which are based on use of fungicides, resistant cultivars, cultural practices and biocontrol of seed borne infection.

Cultural Practices

Use of healthy seed, clean cultivation and three-year crop rotation has been recommended for the management of this disease (Sharma and Sohi, 1989). Disease free seed can be produced either by surface irrigation in semi-arid regions where conditions of high temperature and low humidity prevent infection of this fungus, or under a pedigreed seed programme, in which seed plots were isolated and subjected to strict inspection for disease free seed. Best sowing time of French bean in hill regions is in between mid April to mid May for maximum yield and minimum anthracnose incidence.

Host Resistance

The most promising strategy to control this disease is the use of resistant cultivars. However, the diversity and the high variability of the fungus with the continuous emergence of new races pose problems in breeding programme. Anderson *et al.* (1963) reported that the resistance in cv. Charlevoix against alpha and beta races of *C. lindemuthianum* is conferred by single independent dominant gene. A glycoprotein inhibitor in the cell wall of resistant varieties imparts resistance

(Sharma *et al.*, 1986). Due to the presence of much pathogenic variability in the pathogen, non-race specific resistance has been used in breeding for anthracnose. Cultivar IAPAR-Rio Negro-8 is resistant to all races of the pathogen under field conditions (Alberini *et al.*, 1987) while Menezes and Dianese (1988) have identified varieties (Aroana 80, Ayso, Carioca 80, Iguaco, Moruna 80, Riuo Piqiuri, Rio Vermelho, Mulatao, Olho-de-Pombo and Vermelhao) which were resistant to atleast 7 of the 9 races of the pathogen and no race was virulent to all the 70 cvs. screened in Brazil. Chakrabarty and Shyam (1990) demonstrated resistance in cvs./lines VL 60, VL 63, Jawala, HPR 33, B 4 B 6 and P 48 against eight isolates of the pathogen. However, Jawala later exhibited highly susceptible reaction, which may be due to the development of new pathotypes. Kumar *et al.* (1997) screened sixty cultivars/ lines of kidney bean against ten races *viz.*, beta, gamma, Ind I, Ind II, Ind II, Ind IV, Ind V, Ind VI (alpha-Brazil), Ind VII and Ind VIII of *C. lindemuthianum* under field and artificial epiphytotic conditions and found two acceessions (AB-136 and G 2333) as highly resistant to all the above races prevalent in Himachal Pradesh, India, while cultivars Cornell 49-242, EC 43036, EC 57080, KRC 5, PI 207-262 and Widusa exhibited resistance to more than five races whereas rest were susceptible and these can be utilized in breeding of race specific resistant varieties. Later Gonalves-Vidigal *et al.* (2001) reported that a single dominant gene controls the resistance of cv. AB 136 to both races 31 and 69 and the symbol Co-6 was assigned to the gene. In addition, linkage analysis using random amplified polymorphic DNA marker indicated that Co-6 also controls the resistance of this cultivar to other races of the pathogen, or that different genes are present in the same linkage block.

In the recent years few cvs./lines/land races *viz.*, SRC 74, SRC 89A, SRC 90, SRC 95, SRC 201, PDR 14 (Sud and Sood, 2002) and Morden 003 (Mundel *et al.*, 2004) have been reported resistant to this disease. Some exotic accessions like G 2333, Cornell 49242, PI 207262, Mexique 222, TO, Perry Marrow, Kaboon and Widusa were resistant to more than five Indian races, whereas two Indian accessions KRC-5 and Hans showed resistance to six and four races, respectively (Pathania *et al.*, 2006).

Biocontrol

Although research in biocontrol has advanced considerably in many major plant diseases in recent years but this is not the case with bean anthracnose. Limited work has been done like *in vitro* evaluation of the antagonists against this pathogen (Gupta *et al.*, 1991) or the phenomenon of cross protection by inoculating a less virulent strain of the fungus to reduce the severity from severe strains (Sutton, 1979). Infected seed when soaked in 10 per cent culture filterate or treatred with 0.4 per cent talc formulation of *Trichoderma viride* recorded minimum seed infection and maximum seed germination (Ravi *et al.*, 1999). Treatment with *P. chlororaphis* PCL1391 resulted in best biocontrol of anthracnose, while *P. fluorescens* WCS365 showed no significant difference compared to the positive control (Bardas *et al.*, 2009). Padder *et al.* (2010) evaluated three bioagents (*Trichoderma viride, T. harzianum* and *Gliocladum virens*) and five biopesticides (Achook, Neemgold, Wannis, Spictaf and Neemazal) under *in vitro* and *in vivo* conditions against *Colletotrichum lindemuthianum*. All the three antagonistic fungi caused significant inhibition of mycelial growth, maximum being with *T. viride* (69.21 per cent) followed by *T. harzianum* (64.20 per cent). Among the biopesticides tested at four concentrations, Wanis applied @ 1000 µl/ml caused maximum inhibition of 82.12 per cent followed by Spictaf (52.85 per cent). *T. viride* and Wanis @ 1000 µl/ml were most effective in reducing the seed borne infection.

Under greenhouse conditions, the extracts of *M. argyrophylla* and *O. vulgare* caused the greatest reductions (41.82 per cent and 37.65 per cent, respectively) in disease severity when a local effect assay was carried out (Pinto *et al.*, 2010). In this study, it was evaluated the antimicrobial activity of cascalote phenolics against phytopathogenic fungus *Colletotrichum lindemuthianum* which causes anthracnose disease in common beans. The results indicate that cascalote phenolics have *in vitro* fungistatic activity against *C. lindemuthianum* races R-0 and R-1472. Moreover, these compounds provoked on the whole, *in vitro* inhibition of spores germination and cellulase and polygalacturonase activities in these fungi. *In vivo* assays with cascalote phenolics under greenhouse conditions using susceptible cv. PI 206272 of common bean resulted in a good protection against anthracnose severity especially as a preventive treatment (Garcia *et al.*, 2010).

Chemical Control

Various systemic and non-systemic fungicides such as captan, thiram, carbendazim *etc.* have been used by various workers to reduce seed infections (Sindhan and Bose, 1981; Trutman *et al.*, 1992). Foliar sprays of carbendazim, benomyl, zineb, maneb and mancozeb have been recommended for its control (Chakrabarty and Shyam, 1988). Benlate and Bavistin persisted in the plant upto 15-20 days after seed treatment (Sindhan and Bose, 1981). EBI fungicides like triadimefon and triforine were effective against the pathogen under *in vitro* conditions but were phytotoxic to *P. vulgaris* seedlings when used as seed treatment (Chakrabarty and Shyam, 1988). Carbendazim alone or in combination with thiram as seed treatment followed by two or three sprays of mancozeb at 45, 60 and 75 days after sowing effectively managed this malady (Sharma and Sugha, 1995). A combination of seed treatment with tebuconazole (0.1 per cent) and foliar sprays of tricyclazole (0.03 per cent) reduced the disease severity both on leaves and pods while maximum seed yield was recorded in case of seed treatment and foliar sprays of carbendazim (Gupta *et al.*, 2000c).

Continuous use of systemic fungicides has been shown to be associated with the development of resistant biotypes (Tu and McNaughton, 1980). Maringoni *et al.* (2002) reported that all isolates of *C. lindemuthianum* collected from *P. vulgaris* fields located at Paranpanema Valley, Brazil were insensitive to benomyl, carbendazim and thiophanate methyl showing the occurrence of cross resistance to different benzimidazole fungicides while all isolates were sensitive to chlorothalonil. Recently, Oliveira and Oliveira (2003) reported that strobilurin fungicides like azoxystrobin and trifloxystrobin efficiently controlled this disease while chemicals like carbendazim and fluquinconazole applied without rotation were least efficient treatments. However, combination of systemic fungicides with non-systemics and rotation of fungicides can be an effective approach for management of this disease instead of constant use of one systemic fungicide.

Floury Leaf Spot

Floury leaf spot also known as Farinose spot is an important disease of beans in relatively high altitudes in the tropics where moderate temperatures along with plentiful moisture in the form of rain or dew is prevalent. Srivastava and Gupta (1989a) reported about 65 per cent infection of lower leaves, which were defoliated prematurely and reduced yield considerably.

Disease Occurrence and Distribution

The occurrence of this disease was first reported from Ecuador in 1950 (Petrak, 1950). Subsequently it was reported from Brazil, Colombia, Nicaragua, Zambia Uganda, Tanzania, Kenya, Ethiopia, North West India, Malaya, New Guinea, Venezuela, Honduras, Panama, Guatemala and the Dominican Republic (Gupta and Thind, 2006). In India, this disease appears in an epidemic form in the northwestern Himalayan region and its occurrence was first reported from Solan during the year 1964 (Sohi *et al.*, 1965).

Diagnostic Symptoms

The disease appears on lower older leaves as yellowish spots and progresses upward. In severe infection, leaves show yellow mottling and in older infection, the spots become brown with yellow margins. These spots are generally 2-3 mm in diameter, but several spots merge together to affect most of the area. On the under surface, white fluffy growth of the fungus conidiophores can be seen which give to the leaf the appearance of having been sprinkled with coarse flour, hence the disease is named as floury leaf spot. Infected leaves turn brown and finally shed prematurely. Pre-mature defoliation causes heavy yield losses.

Biology

The fungus responsible for this disease is *Mycovellosiella phaseoli* (Drummond) Deighton (Syn.: *Ramularia phaseolina* Petrak). On the under surface of host, colorless conidiophores emerge in tufts from the stomata and interwine in clusters around the leaf hairs. Erect, more or less divergent, somewhat tapering towards the tip, simple or irregularly dichotomous, a few short, erect, single conidiophores are also found scattered over the leaf surface. Conidia are variable in shape and size, mostly ellipsoid or ovoid, tips acute or obtuse, base papillate at the point of attachment, straight hyaline, commonly aseptate, rarely one septate and measure 5.24-21.4 x 3.5- 5.3 μm.

Disease Cycle

Very little information is available in literature regarding the disease cycle of this pathogen. The pathogen cannot be cultured and no perfect stage is known. The fungus is not a seed borne and perpetuates in infected plant debris from one season to another (Sharma and Kaushal, 1990). Conidia loose germination with the passage of time and germinate poorly after one year of dry storage but when supplemented with sugars (0.1 per cent) and amino acids like DL-methionine, L-leucine, valine and DL-threonine, good germination of one-year-old conidia was obtained (Sharma *et al.*, 1998). This indicates the possibility that conidia can remain viable in crop debris. The initial occurrence of lesions on lower most leaves on the plants grown in disease prone areas indicates that soil or debris may act as source of primary inoculum. Overwintering conidia may be splashed by rain to lower leaves, which in the presence of free water germinate on host surface and cause infection. Conidia produced on primary spots are blown by wind and cause secondary infections.

Epidemiology

The disease is more common in altitudes of 1500 to 2000 m with heavy rainfall. Sharma *et al.* (1998) studied the effect of temperature and different humidity levels on the germination of *M. phaseoli* and reported that conidia start germination within 6 h of incubation at 20 and 25°C.

Management

Since the pathogen overwinters in infected plant debris, the disease can be managed by using cultural, chemical and resistant varieties.

Cultural Practices

Cultural practices like collection and destruction of infected fallen leaves, two-year crop rotation with non-leguminous crops and wider plant spacings (Gupta and Mathew, 1999) along with weed free cultivation can be useful in reducing the disease.

Host Resistance

French bean cultivars/lines *viz.*, PIB 326, EC 30020, Mantulan, 12-A, Sel.-P-2-3 and EC 10037 have been reported resistant (Singh *et al.*, 1974; Kapoor *et al.*, 1988) which can be successfully utilized in breeding programmes for developing resistant varieties. Sugha *et al.* (1998) reported that White Local developed minimum disease as compared to other genotypes and observed that all the genotypes irrespective of their being pole or bush types were equally susceptible to the disease which are contrary to the observations of Kapoor *et al.* (1988) who reported pole types as resistant and bush types as susceptible to this disease. Two hybrids of *Phaseolus vulgaris* (contender) x *P. acutifolius* (G 40015 and G 40020) and five of *P. vulgaris* x *P. coccineus* (Pc1 and Pc2) were free of floury leaf spot lesions (Kumar *et al.*, 2005).

Chemical Control

Sprays of various fungicides like benomyl, carbendazim and cercobin have been reported effective in controlling this disease (Singh and Sharma, 1976).

Integrated Approach

Two foliar sprays of *Pseudomonas fluorescens* strain RPB 14, 1st at seedling (10^{-7} cfu ml^{-1}) and 2nd at flowering (10^{6} cfu ml^{-1}) stage, followed by single spray of carbendazim (0.05 per cent) at pod developmental stage was found effective in the management of this disease (Mondal, 2004).

Bean Rust

Rust being a polycyclic disease can cause serious economic losses under environmental conditions favourable to its development by premature defoliation of bean plants, decrease in number of pods/plant and their reduced weight. Various workers have reported varying degree of losses due to this disease in different parts of the world. Issa *et al.* (1980), while assessing yield losses in 5 cultivars of bean (*Phaseolus vulgaris*), encountered 29-76 per cent losses solely due to *Uromyces appendiculatus*, while 36.7 per cent yield losses were observed by Singh and Musyimi (1981) in plants, which had 79 per cent infection. The losses in green pod yield have been reported from 4.7 to 69. 0 per cent depending on the cultivar, age of the crop at the time of infection and per cent disease intensity (Sharma, 1989). Gonzalez and Garcia (1996a) concluded from their studies that this disease could cause yield losses ranging from 28 to 50, 8 to 33 and 13 to 29 per cent in different bean cultivars like ICA- Pijao, Velasco Largo and Bonte 11, respectively. In Himachal Pradesh, severity of rust was observed to be 15-80 per cent in different areas of Solan district (Gupta *et al.*, 2008).

Host Range

Besides French bean (*Phaseolus vulgaris*), the host range of rust pathogen includes *P. lunatus, Dolichos lablab, D. biflorus, Vigna chinensis etc.*

Disease Occurrence and Distribution

The disease has been reported from the entire bean growing areas of the world. This disease was described for the first time from Germany in 1795 (Persoon. 1795). It has been reported from USA, North America, Brazil, Portugal, Ethiopia, France, Uganda, Guatemala, Philippines, Bulgaria, Greece, Tanganyika, Chiles, Hawaii, Peru and Turkey. Butler (1918) was the first to report the occurrence of this disease from India. Now this disease is prevalent in Karnataka and Tarai areas of the country. Recently the occurrence of this disease has been reported on pencil beans in Solan area of Himachal Pradesh (Gupta *et al.*, 2008).

Diagnostic Symptoms

Leaves and pods are the principal plant parts attacked by the disease but occasionally, tender parts of stem and branches are also infected. On leaves, the disease may appear on both the surfaces, but is more common on the abaxial surface. Pustules of the disease usually appear first as small and slightly raised spots that are almost white in colour. These pustules enlarge by forming reddish brown sori, up to 2 mm in diameter, containing the urediniospores. After coalescing, they may occupy larger areas. A ring of secondary sori may develop around the original infection on susceptible cultivars. Telial stage is dark brown to black coloured and linear. The affected leaves may turn yellow and dry or fall prematurely.

Biology

The disease is caused by *Uromyces phaseoli* var. *typica* Arth. and *U. appendiculatus* being autoecious rust, all its spore stages are produced on bean plant but pycnial and aecial stages are not commonly encountered in nature. The urediniospores are brownish, globoid or ellipsoid, one celled echinulate-measuring 23-33 X 16- 23 µm. Spore wall is golden brown 1-1.5 um thick having two equatorial or superequatorial pores. Teliospores are one celled, pedicellate, globoid or broadly ellipsoid. These measure 24-32 X 20-26 µm and may be smooth or possessing a few verucose marks on the wall. The teliospores wall is chestnut brown in colour, 3- 3.5 µm thick with a hyaline papilla over the pore. Telia arise from uredinia and appear as dark brown or black in colour. Pycnia appear as yellowish spots on upper surface of the leaf. Orange coloured aecia are cupulate 0.25- 0.30 mm in diameter. Aeciospores are ellipsoid, one celled and measure 20- 26 X 10- 20 µm.

Variability

Variation amongst different isolates of *U. appendiculatus* had been observed as early as in 1939 (Dundas and Scott (1939). Twenty physiologic races of this pathogen have been distinguished on a set of differential hosts under greenhouse conditions (Harter and Zaumeyer, 1941). Races 1, 4, 7 and 12of the fungus have also been reported by De Menzes (1952). Chirsten and Echandi (1967) identified more common races as 3 and 15 in Alajuela, 3 and 25 in Turrialba, 8 and 22 in San Isdro del General, 10 and 32 in Cartago and 8 in Los Angeles de Tilaren (regions of Costa Rica) from their greenhouse and field studies. In Brazil, B1 and B2 were the most common races identified on the set of six differential hosts (Dalas and De Costa, 1968). Races 3, 10, 18, 19, 32, 33, 34, 36 and 37 have been reported from Cauca Valley where race 33 was encountered most frequently throughout the valley (Zuniga de Rodriguez and Victoria, 1975). Stavely *et al.* (1989) described 58- 70 as the new races of the fungus in north America of which race 67 is the first to produce moderate size urediniospores on a previously resistant Cv. Compuesto Negro Chimaltenango. Occurrence of 13 different races among 15 monopustular isolates of *U. appendiculatus* obtained from 4 bean growing areas of the state of Minas Gerais, Brazil have been reported (Faleiro *et al.*, 1999). From 80 monopustular isolates, Moran *et al.* (1992) reported that 53 races could be distinguished with a set of only eight differential hosts (Kentucky Wonder 814, Early Gallatin, 51051, NEP-2, Ecuador 299, Olathe, Mexico 309 and Redlands Pioneer) instead of 20 differential varieties. Faleiro *et al.* (1999) identified 13 different races among 15 monopustular isolates of *U. appendiculatus* obtained from four *P. vulgaris* growing regions of Minas Gerain, Brazil.

Disease Cycle

It is autoceious rust *i.e.* it has its life cycle confined to a single host. The aecia are rare in nature but have been observed in bean fields in Oregon (Eastman, 1943). The Urediniospores are produced in great numbers in sori on the upper and lower surfaces of the leaf. They spread the disease over great distances and cause severe epidemics.The urediniospores will germinate as soon as they are mature. One or two germ tubes developed from germinated urediniospores (Xin *et al.*, 2000). The penetration of the germtube occur mainly through the stomata but occassionally penetrated directly through the epidermal cell. A well defined appressorium is formed after penetration by the pathogen. The fungus form a sub-stomatal vesicle when it penetrate through the stomata and a round inflated body usually occur. One or two primary hyphe then develop from the inflated body. Once within the bean leaf, the fungus grow intercellularly and the penetration to the leaf cell occur after a haustorium mother cell is formed at the hyphal tip in contact with a host cell. The branches emerge from behind the site of the primary hyphe where a haustorium mother cell develops. The secondary hyphae spread intercellularly. Under favourable environmental conditions, urediniospores produce another generation of same kind of spores in about 10-15 days and produce several generations in a single season (Zaumeyer and Thomas, 1957). When climate becomes unfavourable, teliospores are produced in the same sorus. Teliospores, which require a rest period and retain their viability for many months, are largely instrumental in carrying over the organism from one season to another.

Epidemiology

Obligate parasitism and autoecious nature of rust pathogen helps it to complete entire life cycle on bean plants. It is not seed borne. In cooler regions, survival is ensured through plant debris in the form of teliospores while in other areas through continuous growing of crop including other hosts. Cloudy and humid days, allowing dew to last on the leaves for some time in the morning favour germination of spores and subsequent infection. Harter *et al.* (1935) have reported optimum temperatures for urediniospores germination and infection as 14.5 and 17° C, respectively. Davison and Vaughan (1963) suggested that rust overwnters in or as urediniospores in trash and on trellis poles because 16 per cent urediniospores of *U. phaseoli* race 33 germinated even after storage at –18° C for >600 days and produced abundant infection on bean leaves. Daily mean temperature of 34° C prevents disease development. Sharma (1977) observed that irrigation given at 3 days intervals favoured disease development than with 18-day intervals. Disease intensity is not influenced by varying levels of N, P and K. Farm implements, insects, humus and animals help in local dissemination of the rust fungus (Sumatrini, 1998). Sharma (1998) analyzed the meteorological data for eight years (1986-1993) and showed that excessive precipitation is deleterious to disease development, which explains the occurrence of severe infection during periods of low rainfall.

Management

The disease can be managed by cultural, chemical, biological methods and host resistance.

Cultural Practices

Moreno and Mora (1984) have suggested minimum tillage or no tillage to prevent spread of the disease and intercropping of bean with maize, since, severity and rate of infection was more in monoculture. Mixed cropping reduced rust incidence level by 51 and 25 per cent with sole cropping and row intercropping, respectively (Fininsa, 1996). Plant density and disease incidence are directly related (Flores Revilla *et al.*, 1994). Mishra *et al.* (1998) evaluated seven dates of sowing under Orissa conditions and Arka Komal variety was sown in between 12 October and 23 November, 9th November was found as the optimum date of sowing because the rust incidence was minimum. Sharma and Sohi (1989) also recommended some cultural practices like long crop rotations, collection and destruction of plant debris, wider plant spacings and removal of weeds to permit aeration and drying which check rapid build-up of inoculum and spread of the disease under field conditions.

Host Resistance

More than 35 physiologic races of the rust fungus (*U. appendiculatus*) are present in Australia, Brazil, Hawaii and the USA. Extensive cultivation of the new cultivars may make them susceptible to new virulence. Golden Gate Wax and Brown Kentucky Wonder 298 have been used in breeding for resistance to four races (Dundas, 1942). During rust epidemic of 1946 in Mississippi, only two lines, Virginia 119 and No.780 remained free from infection (Foster, 1947). Frazier and Hendrix (1949) developed a new rust resistant pole French bean named Hawaiian Wonder from the cross Bountiful X Lualualei. Seminole is a bush bean variety having multiple disease resistance including rust (Anonymous, 1953). Cultivars/lines like Redlands Greenleaf and Redlands Pioneer (Anonymous, 1966), Jackson Wonder and Strikton (Makraw *et al.*, 1973), Brazil K-12973 (Sokol *et al.*, 1977), Catu and Carioca (Issa, 1985a) Serva Negra, Ag497 and Agrorrico (Azevedo and Kushalppa, 1986), Gresham (Anonymous, 1989), PV 136, PV 88, PV 89 and PV 3 (Prakasam and Thumburaj, 1991), white seeded Enganador, Chevere and CC 25-9B (Faure *et al.*, 1992), Negro Inifap (Villar-Sanchez and Lopez-Salinas, 1993), Negro Cotaxtle 91 (Becerra Leor *et al.*, 1994),IBRN-6,IBRN-11,Hawaiian Wonder, Bush Bean S-9, Bush Blue Lake and Contender (Ghimire *et al.*, 1995), lines 83-10 and 83-11 with stringless pods (Guu,1995),Ouro Negro (Faleiro *et al.*, 1996), IIHR 220, stringless bean (Mohan *et al.*, 1997), Mimoso Rasteiro (Paula *et al.*, 1998), SVM-1, Hans (Sharma, 1998) and IC 47643, PI 201489 and PI 247761 (Bhat *et al.*, 1999) have been found to be resistant in different parts of the world. Cv. Ouro Negro showed immunity to 3 out of 4 races tested and PI 181395 was the best source of resistance to rust, as it was immune to all the four races of *U. appendiculatus* (Faleiro *et al.*, 1998). Inheritance studies indicated that resistance to rust was controlled by a single, dominant gene (Aghora *et al.*, 2007). Line 'Arka Anoop' developed by breeding programme is resistance to rust and has high pod yield and good pod quality. Germplasm lines/cvs. Like Alapa Trey, Sing Tamey and Local Collection were found resistant to this disease under Solan conditions (Sharma *et al.*, 2014).

Biological Control

Verticillium lecanii has been suggested to be a potential biological agent for the management of rust as it penetrates hyphae and invades urediniospores of the fungus (Allen, 1982). Under field conditions *Verticillium lecanii* reduced the disease incidence and severity to 56 and 17 per cent as compared to 98 and 85 per cent, respectively, in agrochemical package while seed yield was slightly less in biocontrol than agrochemical package which may be due to the use of fertilizer in latter package (Carrion *et al.*, 1999). Different isolates of *Bacillus subtilis* like APPL-1 (Baker *et al.*, 1983 and Baker *et al.*, 1985) and FF-1, FF-5 and FF-6 (Mizubuti *et al.*, 1995) have been selected as the potential bioconrol agents. In culture medium *B. subtilis* isolates released a thermo stable substance(s) antagonistic to the rust fungus as pustule number was reduced by >95 per cent (Centurion *et al.*, 1994). Isolate B 206 showed the greatest efficacy, when cells of the antagonist were automized on to the plant.

Chemical Control

Different workers suggested various chemicals from time to time for keeping the disease under check. Straib (1943) suggested the disinfection of stakes with formaline (0.1 per cent) before reuse. Preventive application of zineb (5 per cent) with a mist blower or as a spray (0.35 per cent) proved effective (Cheo, 1955). Systemic fungicides are more effective than the non-systemic owing to the difficulty of obtaining under leaf coverage with the latter. Efficacy of various systemic fungicides like oxycarboxin, carboxin, triforine, bitertanol, triadimefon, tridemorph, hexaconazole and HWG 1608 has been reported against this disease (Jaffer, 1971; Rolim *et al.*, 1981; Singh and Musymi, 1979a; 1984; Prakasam and Thamburaj, 1991;Gonzalez and Garcia, 1996; Sumartini, 1998) and possibly eradicated the already established infection of *U. phaseoli* in bean plants after three sprays without any sign of phytotoxicity (Singh and Musymi, 1979a; 1984; Gonzalez and Garcia, 1996b).

Fungicides like hexaconazole, maneb and tebuconazole when applied twice before flowering reduced rust infection drastically and increased seed yield by > 200 kg/ha (Becerra Leor *et al.*, 1994). Three sprays of hexaconazole (0.1 per cent) effectively managed this disease in Sikkim (Bhat, 2002). Thiophanate methyl or chlorothalonil alone or in combination increased pod yields and decreased severity on leaves (Balardin, 1994).Mancozeb have been proved to be the most effective fungicide in controlling the disease. The extent of avoidable losses due to this disease was 60.8 per cent during the epidemic year (Sharma, 1998). Certain antibiotics like distamycin (100ppm), griseofulvin (30ppm) and chloramphenicol have also been reported effective in managing rust of beans. Addition of glycerine (1 per cent) and IAA (1ppm) increased the efficacy of griseofulvin, which was the only material displaying systemic activity (Baldacci and Betto, 1958).

BACTERIAL DISEASES

Common Blight

Common blight can cause significant yield losses if conditions are favourable for disease development. Warm and humid weather favours severe outbreaks of the disease, thus the disease is more serious in the tropics (Saettler, 1989). Epidemics of common blight occurred in Ontario, Canada in early 1960's and in 1962, when warm weather along with high humidity occurred (Sutton and Wallen, 1970). In Colombia, yield losses were estimated upto 22 and 45 per cent in natural and artificial infected fields, respectively (Yoshi *et al.*, 1976). It has also been estimated that common blight is responsible for an average 17 per cent yield losses in the Cauca Valley of Colombia (Saettler, 1989). In India, upto 84 per cent yield losses due to this disease in French bean cvs. Contender and Arka Komal have been reported at different crop growth stages (Kishun, 1988).

Host Range

The host range of this pathogen includes both cultivated and weed host species *viz., Dolichos lablab, Emilia sonchifolia, Glycine max, Ipomoea crassicaulis, Lupinus polyphyllus, Phaseolus acutifolius, P. aureus, P. coccineus, Pisum sativum, Strophostyles helvola*, *Vigna mungo* and *V. radiata* var. *radiata*, (Saettler, 1989; Rodriguez *et al.*, 1991). The bacterium was also isolated from diseased leaves of peas, cowpeas, *Ipomoea purpure*, *Malva sysiaca* and *Vigna luteola* which may play significant role in the harbouring and spread of the pathogen in the neighbouring bean fields (Khlaif and Qadous, 1995).

Disease Occurrence and Distribution

The disease was first reported from New York, U.S.A. (Beach, 1892) and subsequently, reported from Australia, Bulgaria, Canada, China, France, Hungary, Japan, Madagaskar, Norway, Phillipines, Russia, Spain, Turkey, South Africa, Uruguay, Yugoslavia, Argentina, Brazil, Colombia, Paraguay, Venezuela, Guatemala, Nicaragua, Honduras, Mexico, the Dominican Republic, Uganda, Burundi, Zaire, Rawanda, Zambia and Egypt. In India, this disease was reported for the first time from Bombay by Uppal *et al.* (1946).

Diagnostic Symptoms

The disease affects stem, branches, leaves, pods and seed. The symptoms appear as characteristic water soaked spots on the under side of the leaves that enlarge and coalesce together to form large blighted areas. Infected areas often become flaccid and turn brown giving it a sunscald appearance. If infection occurs through the vascular tissues, reddish discolouration of veins and midribs is observed. In advanced stages, tissues become necrotic and defoliation may also take place. Lesions on stems appear as sunken water soaked areas, which gradually enlarge to form reddish streaks extending longitudinally in the stem. In early stages of pod formation, characteristic stem girdle or joint rot frequently develops. Infection on pods may occur at any place as small water soaked spots, which are surrounded by distinct zones of narrow reddish brown or brick red band of tissue. On seeds, discoloration appears especially near the hilum region. Plants arising from infected seeds exhibit characteristic wilting (Saettler, 1989).

Biology

The disease is caused by *Xanthomonas axanopodis* pv. *phaseoli* (Smith) Dye (XAP). The bacterium is a motile, non-capsulated, non-spore forming, gram-negative rod, measuring 1.9 x 0.8 µm in size with a single polar flagellum. The pathogen forms characteristic yellow coloured colonies due to a water insoluble eacotenoid and a mucoid growth on nutrient glucose agar medium (Abd Alla *et al.*, 2010). The characteristic yellow colour of the bacterial colonies is due to the production of brominated aryl polyenes, more frequently referred to as xanthomonadins, that have absorption peaks in between 443 and 446 nm (Starr, 1983). Some other useful properties of this bacterium include hydrolysis of starch, gelatin and casein, lipolysis of Tween 80, and lack of pectinase activity. Acid is produced as a metabolic by product when cells grow on media containing arabinose, glucose, mannose, galactose, trehlose or cellubiose. It also causes proteolysis of milk. The pathogen is a strict aerobe, and metabolism of glucose is through oxidative pathway.

Pazos *et al.* (1986) failed to differentiate in Cuban isolates of *X. c.* pv. *phaseoli* (XCP) isolated from bean and by using bacteriophages, 3 pathotypes were distinguished among 35 isolates. Highly significant differences among fifteen isolates of this pathogen were observed among varieties and isolates in relation to resistance and pathogenecity (Rava and Romeiro, 1990) and isolate Xp CNP 28, characterized by high degree of pathogenecity, was suggested as a new standard isolate to replace Xp CNF 15. Report regarding serological variability among *X. campestris* pv. *phaseoli* also exists in literature and 43 isolates from Sao Paulo and Perana States of Brazil were classified into 3 serological groups of which 22 did not produce melanin *in vitro* and reacted only with the homologous antiserum AS-Feij I (serogroup I), 17 reacted only with antiserum Feij 20 (serogroup II) and four did not react with the antisera tested (serogroup III) by using the agar double diffusion test (Maringoni *et al.*, 1994). The strains of XAP were also grouped in 3 groups through profiles of total proteins (PAGE-SDS) (Maringoni and Bach, 1999).

On *Phaseolus acutifolius*, 30 isolates of XAP interacted differently and 8 physiologic races were defined suggesting an underlying gene-for-gene relationship (Opio *et al.*, 1996). They further suggested that levels of resistance already available within *P. vulgaris* are adequate, so that breeder incorporating new complete resistance from *P. acutifolius* could risk destabilizing the host-bacterium relationship by introducing race-specific resistance that could be transient in agriculture. Genotypes useful for differentiation of XAP physiological races have been identified (Zapata, 1997). DNA probes were also used to differentiate *X. campestris* pv. *phaseoli* and *X. campestris* pv. *phaseoli* var. *fuscans* isolates originating from different geographical locations and RFLP analysis revealed DNA polymorphism within *X. campestris* pv. *phaseoli* isolates from the same geographical location (Otoya and Pastor-Corrales, 1992).

Disease Cycle

The bacterium overwinters in infected seed, crop debris and weed hosts (Irigoyen and Garbagnoli, 1998) of which contaminated seed has been observed to be the most efficient mode of survival and dispersal over long distances (Saettler, 1989). The pathogen survived for more than five months in the plant debris on the ground surface (Arnaud-Santana *et al.*, 1991) but the survival of the pathogen is influenced by tillage (Gilbertson *et al.*, 1990). Survival of the pathogen on the weed host *Parthenium hysterophorus* and subsequent transmission of the bacterium to *Phaseolus vulgaris* and *vice versa* has also been reported (Ovies and Larsinaga, 1988).

Upon germination, the bacterium from infected seed infects the developing cotyledons and can then enter the vascular system of the plant. Alternatively the bacterium can be splashed to the plants from the plant debris. Once inside the plant, the bacteria invade and cause breakdown of the middle lamella. Bacterial multiplication takes place, and masses of bacteria are formed that expand into what macroscopically appears as water-soaked spot. The bacteria also exude to the surface of the tissue and can be disseminated in the field by rain and wind to initiate new infections. Bacteria enter the plants via wounds caused by leaf eating insects or natural openings such as stomata or hydathodes (Saettler, 1989). The bacteria can infect stem via stomata or via the vascular system. Most vascular infections probably initiate from seedborne inoculum. In some of these cases, the bacteria can cause plugging of xylem elements, causing wilting. The bacteria can infect seed via the vascular system, moving through the pod pedicel into the funiculus and through the raphe into the seed coat, where they remain beneath the seed coat until seed germination.

Epidemiology

The bacterium is seed borne in nature. Infection of the pedicels of flower buds and small pods resulted in transmission of the bacteria through the vascular tissues of the pod to the seeds, causing internal infection without any external symptoms on either pods or seeds (Aggour *et al.*, 1989). The minimum population of *X.c. pv. phaseoli* required to initiate infection was 10^2 c.f.u./seed while seed infection levels of 0.2 per cent resulted in serious disease incidence (Opio *et al.*, 1993). The bacterium also enters plants through wounds and/or natural openings and infects stems via stomata, infected cotyledons or vascular system. The incidence and severity of the diseases increase with the increase in rainfall and temperature (Mayek-Perez *et al.*, 1995). Warm and humid conditions favour disease development and the disease can develop at temperatures ranging from 15-40°C and RH > 30 per cent with optimum 24°C and >50 per cent RH (Qadous and Khlaif, 1997). Wind splashed rains help to spread the pathogen from infected to healthy plants. Infected seed from resistant genotypes had no visible symptoms and such seeds may play an important role in the epidemiology of this disease because they are difficult to detect and may occur at low frequency in seed lots (Mabagala, 1997). The disease was more severe under short photoperiod and high temperature than longer photoperiod and lower temperature (Santana *et al.*, 1993). Akhavan *et al.* (2009) showed that the epiphytic population size and disease severity were higher on spray inoculated plants irrigated with overhead sprinkler system during pods filling.

Management

The common blight of beans is particularly a difficult disease to manage due to intricate association of the pathogen with the seed. An integrated approach is, therefore, useful to keep the disease under check.

Cultural Practices

Use of pathogen free seed is must because it eliminates an important source of primary inoculum for common blight (Irigoyen and Garbagnoli, 1997), which is an important component of effective disease management strategy for this disease. The disease free seed can be produced in arid zones with low rainfall but its use cannot guarantee a clean crop. Because blight bacteria can colonize symptomless plants and the bacteria can systemically invade seed via vascular elements, certified pathogen-free seed may still be contaminated by blight bacteria (Weller and Saettler, 1980). The reduction of inoculum achieved by the use of the pathogen free seed greatly reduces the probability of common blight epidemics developing in the field.

Deep ploughing, 2-3 years crop rotation and destruction of crop debris/ weed hosts, avoidance of irrigation with sprinklers help to reduce the probability of epidemics (Saettler, 1989; Gilbertson *et al.*, 1990). Bacterial population in plant debris decline over time (Gilbertson *et al.*, 1990) and atleast in temperate regions, the bacteria probably cannot survive for more than one winter. In such situations, crop rotation will be an effective management strategy. But this may not be an effective approach in tropical region particularly in case of the farmers with small land holdings. Moreover, the pathogen is having a vast host range; then again rotation will be less effective strategy.

Mixed intercropping of bean with maize was also found effective in reducing the incidence of this disease as compared to sole cropping and row intercropping, respectively (Fininsa, 1996). Delay in planting time to avoid frequent rains early in the season along with use of disease free seed has also been recommended to alleviate considerable yield loss and quality deterioration (Tu, 1999).

Host Resistance

Since there is no satisfactory chemical control available to combat the ravages of this disease, the use of resistant varieties/cvs. is an important control measure. Various varieties/lines *viz.*, Fullgreen, PLS-63, 101, MS 9471, DL 2434; P 597 (*P. acutifolius*) Feijao 60 Diaz, PI 207262, Coll. 10 B, Mexico, 168, Mexico 240, Desconhecido Amarelo, BR-40, Contista, Saginan, WW 55-0, Kuovskaya, Yushchoyasya; OAC 88-1; XR 235; Great Northern Jules, GN Nebr1, Sel. 27, Pinto Nebr, EP1, Walley, Xr1, Xr2; BAC-6, XAN-159; IIHR 220; T-589, T-769; T-352, T-647; Cut 54, XAN 280, 9356-26 which possess resistance to this disease has been reported from different parts of the world. Accession CF 800375 has also been reported to be resistant to this disease (Gupta and Thind, 2006).

Randomly amplified polymorphic DNA (RAPD) molecular markers have been effectively utilized in the identification of regions on the linkage map associated with resistance to common bacterial blight. The chromosomal region (a BCM interval in linkage group 5 was significantly associated with resistance to XAP strains DR-7 and EK-11 in leaves, pods and seeds. Moreover, regions in linkage group 1 were also significantly associated with resistance to XAP stains in more than one plant organ. The V locus of the flower colour gene (v^{lae}) was linked to a QTL (quantitative trait locus) with major effect on CBB resistance (Jung *et al.*, 1997). Five interspecific hybrids of *P. vulgaris* (Nim 1) x *P. coccineus* were resistant to this disease (Kumar *et al.*, 2005). Cultivars Bayo Madero, Negro 8025 and Pinto Villa were tolerant against common blight pathogen (Prudencio-Sains *et al.*, 2008). Plants of 17 bean genotypes were evaluated under greenhouse conditions for their reaction to *X. campestris* pv. *phaseoli* following leaf-spray inoculation with bacterial suspension (10^8 cfu/ml) and the cultivar Oreol was the most resistant (Todorovic *et al.*, 2008).

Biological Control

Bacterial isolates DFs093 offered 100 per cent control, DFs041 and DFs1297 offered 90 per cent and DFs490, DFs769, DFs831, DFs842 and DFs843 offered 80 per cent control of common blight pathogen in the cotyledonary leaves when bean seeds immersed in suspension of each one of these isolates were agitated for 5 hours at 10°C and sowed in non-sterilized soil (Zanatta *et al.*, 2007). *Paenibacillus polymyxa* strain HKA-15 showed strong antagonism against *Xanthomonas campestris* pv. *phaseoli* strains M-5 and CP-1-1. In agar diffusion assay, the antibacterial metabolite from *P. polymyxa* HKA-15 showed a clear zone of inhibition against M-5 and CP-1-1. Under phytotron conditions, the biocontrol activity of *P. polymyxa* HKA-15 against bacterial common blight pathogen *X. campestris* pv. *phaseoli* M-5 was assessed. At four days after inoculation of M-5, lowest mean disease rate (MDR) (1.13) and per cent disease incidence (PDI) (28.25) were recorded in streptomycin sulphate at 100 ppm as positive chemical treatment. The application of crude metabolite from *P. polymyxa* HKA-15 at 100 ppm was at par with positive chemical control in suppression of bacterial common blight disease in French bean plants (Mageshwaran *et al.*, 2012).

Chemical Method

Different chemicals have been used both as seed treatment or foliage application in attempts to control this disease. Although some success has been achieved but complete control is still intractable. Some chemicals used by various workers both seed and/or foliage application are mentioned as below:

Seed Treatment

Hot water seed treatment at 50°C for ten minutes followed by diping in streptocycline (10 µg/ml) solution has been found effective in reducing the seed infection (Jindal, 1991).

Foliar Sprays

Several fungicides/antibiotics have been applied to foliage to control this disease but complete control is still lacking. Application of copper fungicides to some extent checks this disease (Saettler, 1989). However, heavy use of copper based fungicides to control diseases caused by Xanthomonads can result in selection of copper-resistant strains of the bacteria (Stall *et al.*, 1986) hence need based use of copper fungicides is warranted. Moderate control of the disease coupled with higher yields has been achieved by three sprays of copper oxychloride + zineb or copper oxychloride + mancozeb (Maringoni, 1990).

The chemicals can form the effective part of integrated approach for this dreaded disease. Sometimes when the environment becomes abruptly favourable for the development/multiplication of bacteria, spot application of chemicals can keep the bacterial populations below damage threshold level.

Halo Blight

Halo blight has been recognized as a serious disease of beans the world over, causing considerable crop losses. The disease has been often referred to as the greasy spot disease due to the appearance of greasy water soaked lesions on affected organs. Halo blight is slightly less important as compared to common blight because it is more prevalent in those areas where temperature is low. Yield losses in the range of 23-43 per cent in Michigan fields have been reported (Saettler and Potter, 1970).

Host Range

The pathogen has a wide host range including *Neonotonia wightii, Parthenium hysterophorus, Phaseolus acutifolious, P. angularis, P. coccineus, P. lunatus, P. polyanthus, P. polystachyus, P. radiatus, Pueuaria hirsuta, P. thunbergia* and *Glycine max.* (Mabagala and Saettler, 1992).

Disease Occurrence and Distribution

Halo blight was first reported from New York, USA during the year 1926 (Burkholder, 1926). Subsequently it was reported from the Netherlands, Switzerland, New Zealand, France, Tasmania, Australia, Denmark, Poland, Brazil, Germany, Russia, South Africa, Belgium, England, French Morocco, Spain, Canada, Uruguay, Kenya, Mauritius, Chile and Italy (Bazzi and Calzolari, 1988).

Diagnostic Symptoms

Symptoms appear as small isolated water- soaked spots on the under surface of the leaves which gradually enlarge and coalesce. Later, these spots are surrounded by a characteristic greenish yellow halo measuring 2-3 cm in diameter, thereby giving the disease its name. Systemic plant chlorosis with leaf yellowing and malformation may also develop without much external infection. Later, the pathogen has been isolated from such leaves. At 20°C, the systemic chlorosis is more pronounced and uniform (Coyne and Schuster, 1974). During infection process, the pathogen produces a non-specific toxin "phaseolotoxin' which in turn produces systemic chlorosis and typical halo in the plant. Infections are initiated on the lateral veins which progress to the main vein often accompanied by a marked water-soaking and reddish discolouration of the interveinal regions. On young stems, sunken water- soaked spots gradually enlarge and appear as reddish streaks, extending longitudinally along the stem. In advanced stages of infection, the stem epidermis split open and are often accompanied by light cream to silver coloured bacterial exudations. Brown to red zonate lesions surround water- soaked spots on affected pods. The bacterium soon invades the tissue and progress to the developing seed often leading to shrivelling, discoloration and even rotting of seed.

Biology

The disease is caused by *Pseudomonas savastanoi* pv. *phaseolicola* (Burkholder) Garden *et al.* (Young *et al.*, 1996). It is a strictly aerobic, gram-ve, rod shaped, motile bacterium occurring singly or in pairs. The organism exhibits a smooth to rough colony variation in culture. The presence of D- gluconate, L (+) arabinose, sucrose, succinate, DL-β-OH butyrate, transaconitase, L-serine, L-alanine and P-hydroxybenzoate favour growth of the bacterium in culture while it does not utilize glutarate, mesotartarate, DL-glycerate, isoascorbate, erythritol, sorbitol and meso-inositol. White to cream coloured colonies with a bluish hue and accompanied by a green fluorescent pigment appear on agar medium. Sensitivity of bacterial strains to bacteriophages 11P, 12P, and 48P and serological tests also been suggested for the confirmation of identity of *P.s.* pv. *phaseolicola* (Taylor, 1970) which were later successfully used in the identification of this pathogen (Mabagala and Saettler, 1992; Fourie, 1998).

Variation among the populations of this pathogen is well known (Mabagala and Saettler, 1992). Jenson and Livingstone (1944) were the first to demonstrate host specificity among 13 strains of this pathogen. Later, the occurrence of races 1 and 2 in the United States on the basis of the reaction of bean cultivar Red Mexican U.I.-3, which is resistant to race 1 but susceptible to race 2 (Patel and Walker, 1965). The occurrence of these two races has been reported from Tanzania, Colombia, New Zealand and Kenya (Taylor *et al.*, 1987). The existence of a third race in Nabraska was first suggested by Schuster *et al.* (1979) and Coyne *et al.* (1979), who recovered strains of halo blight bacteria that were virulent to the usually resistant Great Northen bean cultivars such as Great Northern U.I.-59 and California Pink. Taylor *et al.* (1987) confirmed the existence of race 3 of *P.s.* pv. *phaseolicola* in South Africa. Race 3 causes a hypersensitive reaction in cv. Tendergreen, which is susceptible to race 1 and 2. Mabgala and Saettler (1992) also recorded the existence of races 1, 2 and 3 in northern Tanzania. Besides these, some strains of race 2, designated as 2P, produced a brown, diffusible pigment on agar media and it can grow at 34°C, whereas the non-pigment producing strains of race 2 and of races 1 failed to grow at this temperature. Taylor and Teverson (1996) extended the range of differentials to eight cultivars/lines, and accordingly identified nine races of this pathogen. Recently, in South Africa, seven races (1, 2, 4, 6, 7, 8 and 9) were recorded among 1128 isolates, with race 8 the most prevalent (Fourie, 1998). Races 1, 2, 6 and 8 were widely distributed through the whole production area, while races 4, 7 and 9 were restricted to 1 or 2 localities.

Disease Cycle

The bacterium enters directly through wounds or through stomata during periods of high relative humidity and free moisture and start multiplying inside the host tissue. Symptoms of blight appear on the leaf surface. The bacterium is an important seed-borne pathogen of this crop (Calzolari, 1997). Seed borne inoculum is carried on the surface of the seed or beneath the testa and in between the cotyledons. Pathogen also survives in bean debris and volunteer bean plants (Mabagala and Saettler, 1992). Within the same genotype, *P.s.*pv. *phaseolicola* tended to survive longer in the stems than in leaf debris.

Epidemiology

The bacterium is seed borne and also overwinters in infected plant debris. In the periods of high humidity and free moisture, the bacterium enters the host through stomata or wounds. Symptoms appear in 6-10 days at 24-28°C, but may be delayed for 2-3 days more at higher temperature (Zaumeyer and Thomas, 1957). Under high soil moisture conditions and low pH, survival of this bacterium did not exceed 2-3 months. The bacterium apparently cannot survive in colder regions in soil or plant debris.

Management

Cultural Practices

Cultural practices include use of disease free seed, a 2-3 year crop rotation, ploughing in plant debris immediately after harvest and avoiding irrigation with sprinklers (Calzolari, 1997).

Host Resistance

Breeding for halo blight resistance is generally considered the best method of disease control (Mabagala and Saettler, 1992) and has been used successfully in several areas of the world (Schwartz, 1989). The development and deployment of resistant cultivars requires knowledge of the distribution of races of the pathogen in a particular area. Resistance against this disease has been reported in *Phaseolus vulgaris* hybrid/cvs. Ph-64-23-144, Ph-65-9-9, Ph-64-9-61 and Ph-64-9-56 (Radkov *et al.*, 1978), *P. radiatus*, *P. calcaratus*, *Vigna angularis*, PI 150414, cv. Edmund (Kiessling and Skiebe, 1987), V4141had higher tolerance to race1 and 2 of the pathogen. (Hanan and Khan, 1999). Accession CF 800375 has also been reported to be resistant to this disease (Rava *et al.*, 2003). 12 accessions had resistance to halo blight from the common bean collections from the Iberian Peninsula (Monteagudo *et al.*, 2006). Cultivar 'Chase' has been found resistant to halo blight pathogen (Singh *et al.*, 2007).

Chemical Control

Seed Treatment

Hot air treatments (70°C for 2 h for moistened seed or 50°C for 3 h for dried seed) reduced 72-74 per cent seed infection. Hot water soaks at >48°C reduced the number of bacteria present in the seeds but tended to decrease germination and prolonged soaking rendered the seed unfit for sowing. Antibiotics, kasugamycin and streptomycin eradicated seed infection when used as a hot water soak (60 min) and at relatively higher concentration while a successive tap water rinse and a further 30 min soaking in sodium hypochlorite (0.5 per cent) improved the effectiveness of the streptomycin used at low dosage and reduced its phytotoxic effects (Tamietti and Garibaldi, 1984).

Foliage Treatment

Sprays of cupric hydroxide significantly reduced or limited *P. s.* pv. *phaseolicola* on bean foliage. Copper oxychloride (0.2-0.25 per cent) sprays starting from seedling stage to flowering are also found effective in reducing pod infection. Application (3-4 times) of copper products after seedling emergence was also recommended (Calzalari, 1997). Though, streptomycin is effective in reducing foliage infection but continuous use can induce resistant mutants in the bacterial population.

VIRAL DISEASES

Bean Common Mosaic

Bean common mosaic is one of the most commonly prevalent viral diseases of French bean, causing considerable losses in yield and quality. Yield losses up to 26 to 84 per cent have been reported to be caused by the disease (Sastry *et al.*, 1981; Sharma and Dhar, 1994).

Host Range

The virus is known to infect various cultivated species as well as weed hosts *viz.*; *Cajanus cajan; Cassia tora; Chenopodium amaranticolor; C. quinoa; Cicer arietinum; Crotolaria spectabilis; C. striata; Cyamopsis tetragonoloba; Glycine max; Gomphrena globosa; Lens esculenta; Lupinus albus; L. angustifolius; Macroptilium lathyroides; Melilotus alba; Nicotiana benthamiana; N. clevelandii; Phaseolus lathyroides; P. lunatus; P. vulgaris;*

Rhynchosia minima; Sesbania exaltata; Tetragonia expansa; Trifolium incarnatum; T. subterraneum; Vicia faba; V. sativa; V. villosa; Vigna angularis; V. radiata and *V. unguiculata* (Zaumeyer and Goth, 1964; Drijfhout and Bos, 1977).

Disease Occurrence and Distribution

Ever since its first record from Russia (Iwanowski, 1894) the disease has been reported from various countries *viz.*, U.S.A., Canada, Bermuda, Agrentina, Belgian Congo, Chile, Columbia, Czechoslovakia, Ecuador, France, Guatemala, Honduras, Italy, Japan, Java, The Netherlands, Peru and Uruguay, China, England, New Zealand, Australia, Brazil, Germany, Iran, Venezuella. Yaraguntaiah and Nariani (1963) were the first to report the occurence of the disease in India in 1963 from New Delhi and subsequently it was reported from Shimla, Himachal Pradesh, Bangalore, Karnataka, Pantnagar, Uttar Pradesh and Pune, Maharashtra.

Diagnostic Symptoms

Typical symptoms consist of well-defined dark green areas of foliar tissue against a lighter green background of affected leaf lamina. During early stages, affected leaves appear crinkled, chlorotic, stiff and accompanied by downward curling. In advance stages of infection much reduced leaves with sinuous or filiform leaflets are produced whereby plants become stunted and appear dwarfed and bushy. Seedlings from infected seed show characteristic mottling, blistering, downward cupping and chlorosis on the fruit and succeeding compound leaves. Such plants usually fail to set pods or develop undersized pods. The pods on severly-infected plants are usually undersized and contain fewer ovules than those produced on normal plants. They are occassionally covered with small, dark green spots and often mature very late. The seed may be aborted, smaller or malformed. The symptoms are temperature dependent and infection of this disease causes delayed seed maturity. The roots of affected plants turn dark and during periods of severe infection the tap root system turns almost black due to vascular necrosis giving rise to what are typically called Black root' symptoms. Plants affected by black root, initially exhibit slight wilting of the leaflets near the top of the plants, which subsequently loose their bright green colour and appear greyish, green. A permanent general wilting of all the leaves follows often resulting in death of the plants.

Biology

At least 20 viruses belonging to 12 taxonomic groups are shown to infect French bean under natural conditions. Dhanju *et al.* (1995d) reported that the viruses associated with mosaics of French bean in Himachal Pradesh are mostly strains of two viruses (*Bean common mosaic virus* and *Bean yellow mosaic virus* of which the former is most predominant.

The *Bean common mosaic virus* (BCMV) is an ubiquitous pathogen of worldwide occurence. Taxonomically, the virus belongs to the potyvirus group. Earlier two serotypes of the virus namely A and B had been described to cause mosaic diseases. However, those once grouped as serotype A are now considered to be isolates of a seperate potyvirus species and several strains (Drijfhout *et al.*, 1978) once considered to be distinct have now been shown to be strains of this virus (McKern *et al.*, 1992). BCMV consists of filamentous flexuous particles measuring 730-750 X 12-15 nm in size. The physico-chemical properties relating to this virus has been summarized in Table 7.1.

Table 7.1. Physico-chemical Properties of Bean Common Mosaic Virus

Genome	ssRNA, Monopartite
Transmission	Sap, Seed, Insects
Properties of particles in sap	
TIP	60°C
LIV	1-4 days
DEP	10-3 - 10-4
Physical Properties	
Sedimentation Coefficient	154-158 S
Density	1.31-1.32 g cm^{-3} in Cscl

TIP: Thermal inactivation point; LIV: Longevity *in vitro*; DEP: Dilution end point.

Source: Brunt *et al.*, 1996.

Molecular Characterization

Four strains (NL-1, NL-1n, NL-7 and NL-7n) of Bean common mosaic virus (BCMV) prevalent on common bean in Himachal Pradesh were compared at the 3′ terminal region of the viral genome to elucidate variation and relationship among these strains. 3′ terminal region comprising of partial nuclear inclusion b (NIb) (171–233 bp), complete coat protein (CP) (860 bp) and 3′untranslated region (UTR)

(208–244 bp) shared 96–98 per cent nucleotide and 96–99 per cent amino acids identity among various strains of BCMV. Multiple alignment as well as cluster dendrograms of the 3′ terminal region placed the test isolates in BCMV species of genus Potyvirus. Phylogenetic analysis of complete CP as well as 3′ UTR also showed Indian strains to be distinct strains (Sharma *et al.*, 2011).

Disease Cycle

Transmission by Insects and other Invertibrates

The efficient transmissibility of the virus through bean seed is responsible for the worldwide distribution of the disease (Schmidt, 1992). Seed transmission rates ranging from 0 to 93 per cent have been reported in cultivars of different hosts (Edwardson and Christie, 1991). In nature, aphids like *Acyrthosiphon pisum, Aphis fabae* and *Myzus persicae* also transmit the virusin a non-persistant manner.

Epidemiology

Infected seed, plants of susceptible French bean cultivars and weed hosts serve as natural sources of primary inoculum (Robertson, 1962). Aphid vectors are responsible for secondary field spread of BCMV. Gupta and Chowfla (1988) observed that nymphs and apterous aphids are more efficient as compared to winged adults. Air temperature and relative humidity significantly affected vector populations (Berg, 1984).

Management

No single method has proved effective to manage the disease. However, an integrated approach involving alteration of planting dates, manipulation of plant population density, use of insecticides to control insect vectors and breeding for disease resistance have been found to be effective (Jeyarajan and Doraisamy, 1986).

Cultural Practices

Use of virus free seed is essential because of efficient transmissibility of the virus through seed (Schmidt, 1992). The production of BCMV free seed is taken up in isolated regions away from other bean plantings and aphid vectors or under screenhouse/greenhouse conditions (Morales, 1989). Alteration of planting dates coinciding with periods of reduced vector activity, rouging of infected bean plants from the field or weed hosts from in and around the field help to eradicate natural reservoirs of the virus and planting of non host barrier crops especially maize, and closer spacing help in minimizing virus incidence by interfering with flight of the vector and thereby inactivating the virus (Dhanju *et al.*, 1995b).

Host Resistance

Phaseolus vulgaris cvs./lines *viz.*, Bean 1 49-3, Bean 2 119-3, Bean Br II 42-1, Contender, HPK 69, Iran 431, Kentucky Wonder, Master, Pusa Parvati, Turkey 326, Turkey 434 (Gupta and Chowfla, 1990); ESAL 547, ESAL 550, Jalo (Pellicano *et al.*, 1993); Topcrop, Masterpiece (Devanathan *et al.*, 1994); Premier (Sharma and Dhar, 1994); PDR 14 (Bhagawati and Bhagabati, 1994); EC94469, EC 94929 (Dhanju *et al.*, 1995a); IC 41654, IC 41661, P414, EC 19505, EC 27518, EC 24939, EC 43636, EC 271562, EC 271523, EC 271491 and EC 241107 (Dhanju *et al.*, 1995c) have also been reported to possess resistance to BCMV. Of late, fourteen cvs./lines like Contender, EC-3, EC-29, EC-41, EC-43, IC-7, IC-11900, IC-16453, IC-17851, IC-18546, IIHR-220, IIHR-909, Pant Anupama and Premier were found resistant to this disease (Ramanjula *et al.*, 2004).

Sharma *et al.* (2006) evaluated ninety four bean accessions comprising of land races, exotic collections and recommended cultivars against three strain groups (I, IV, VI) of Bean Common Mosaic Virus (BCMV). Two exotic varieties Top Crop and Amanda were found highly resistant to all the three strain groups. The existence of two pathogroups PG-I and PG-II and four strains (NL-1, NL-1n, NL-7 and NL-7n) in Himachal Pradesh, a North-Western Himalayan state of India was reported by Kapil *et al.* (2010). Two strains, NL-1 and NL-7 were identical to the previously described NL-1 and NL-7 strains from Europe and USA, whereas the other two designated as NL-1n and NL-7n differed from earlier identified strains with respect to their necrotic reaction on cultivar Jubila at high temperature (>30°C). Reverse phase HPLC peptide profiling of tryptic digests of coat protein of these strains further confirmed that NL-1, NL-1n, NL-7 and NL-7n are distinct from each other. 21 bean accessions *viz.*, KR 7, KR 225, KR 295, KRC 4, KRC 7, KRC 11, KRC 12, KRC 13, KRC 16, KRC 22, Amanda, Black Turtle Soup, Contender, Hans, Great Northern UI 123, Improved Tender Green 40031, Jubila, Kentucky wonder, Monroe, Premier and Sanilac, were found resistant to NL-1n and NL-7n strains of bean common mosaic potyvirus strains (Sharma *et al.*, 2008).

Chemical Control

Soil application of granular insecticides (Thimet @ 1.5 kg a.i./ha) at sowing followed by 2-3 foliar sprays of either monocrotophos (0.05 per cent) or dimethoate (0.05 per cent) (Singh *et al.*, 1981) has been seen effective to manage aphid population resulting in reduced spread of the virus.

Bean Yellow Mosaic

This disease can cause significant losses, if the favourable environment exists for its spread. The losses caused by BYMV are frequently very large. Since the symptoms of this virus and its strains and Common bean mosaic often overlap, it is almost impossible to assign any particular amount of loss caused by an individual virus.

Host Range

BYMV has a wide host range and can infect 233 species in 71 genera of 21 families (Edwardson and Christie, 1991). Besides French bean, its host range includes mung bean, lima bean, Pigeonpea, chickpea, lentil, peas, broadbean, soybean, fenugreek, *Cucurbitapepo*, *Gladiolus* spp. and *Robinia pseudoacacia* (Raizada *et al.*, 1991).

Disease Occurrence and Distribution

The occurrence of bean yellow mosaic virus (BYMV) was first reported in 1934 from USA (Bos, 1970). Susbsequently it was reported from East Africa, Germany, Canada, Holland, Italy and Switzerland.

Diagnostic Symptoms

Mosaic symptoms consist of contrasting dark and yellowish green areas, often accompanied by bright yellow spots, which intensify with the age of the plants. Some isolates cause only a mild and diffuse chlorotic mottle and limited plant stunting, while others cause a prominently coarse mosaic, rugosity, malformation and severe stunting. Some varieties may show necrotic spots, veinal and apical necrosis, wilting and premature death. Late infection generally causes less prominent foliar symptoms and if the infection coincides with pod formation, the pods eventually will exhibit a light green mottle and slight malformation. BYMV infected plants are usually stunted and bushy due to reduction of internode length and proliferation of lateral branches. Plant maturity is delayed and quality and quantity of the seed is affected. Certain strains of BYMV induce systemic necrosis that appear as a purplish discolouration at the base of the lower leaves, accompanied by veinal, stem and petiole necrosis. The necrosis at the terminal growing point consequently result in to death of the plant. Some strains incite local lesions and others produce malformed pods bearing reddish brown nrecrotic spots. However, type and severity of symptoms vary according to strain of virus and cultivar of French bean involved.

Biology

Bean Yellow Mosaic Virus (BYMV) belongs to the potyvirus group. The virus is characterized by elongate, flexuous, filamentous particles about 750-810 x 11-15 nm. The virus has TIP of 65°C, LIV of 2 to 4 days and DEP of 10^{3} to 10^{-5} (Brunt *et al.*, 1996).

Disease Cycle

Transmission by Insects and other Invertebrates

BYMV is readily transmitted through sap inoculation and by many species of aphids in non-persistent manner but not through the seeds of French bean though it has been reported in broadbean and other leguminous hosts (Bos, 1970). BYMV has been reported to be experimentally transmitted by 47 aphid species in 25 different genera (Edwardson and Christie, 1991) but *Acyrthosiphum pisum, Macrosiphum euphorbiae, Myzus persicae, Aphis craccivora* and *A. fabae* are the important vectors of this virus under field conditions.

Epidemiology

BYMV is not a seed borne in French bean though its seed borne nature has been reported in other leguminous hosts. The virus has a wide host range and aphid vectors transmit it. Field spread by aphid vectors from infected French bean and other collateral hosts is primarily responsible for the epidemics of this disease.

Management

Cultural Practices

Destruction of collateral hosts of this virus from in and around the field of French bean, plantation of barrier crops like maize and crop rotation reduce the infection of this disease.

Host Resistance

This is the most reliable method for the control of this disease. Sources of resistance to a BYMV strain including pod malformation have been identified in various Great Northern lines like GNUI No. 31, 59, 123 and 1140.

Chemical Control

Application of various systemic insecticides like metasystox, dimethoate *etc.* reduces aphid population in French bean fields or other host crops and reduces the spread of virus.

REFERENCES

Abd Alla, M. H., Bashandy, S. R. and Schnell, S. (2010). Occurrence of *Xanthomonas axonopodis* pv. *phaseoli*, the causal agent of common bacterial blight disease, on seeds of common bean (*Phaseolus vulgaris* L.) in Upper Egypt. *Folia Microbiol.* 55(1): 47-52.

Aggour, A.R., Coyne, D.P., Vidaver, A.K. and Eskridge, K.M. (1989). Transmission of the common blight pathogen in bean seed. *J. Am. Soc. Hortic. Sci.* 114(6): 1002-1008.

Aghora, T.S., Mohan, N., Somkuwar, R.G. and Girija, Ganeshan. (2007). Breeding French bean (*Phaseolus vulgaris* L.) for resistance to rust (*Uromyces phaseoli* Reben Wint.). *J. Hort. Sci.* 2(2): 104-107.

Ahuja, S.C. (1976). Banded leaf and sheath blight of maize. Ph.D. Thesis, Post-graduate School, IARI, New Delhi. 153p.

Akhavan, A., Bahar, M., Saeidi, G. and Lak, M. (2009). Factors affecting epiphytic population of *Xanthomonas axonopodis* pv. *phaseoli* in epidemiology of bean common blight. *J. Sci. Technol. Agric. Nat. Res.* 13(47): 265-277.

Alberini, J.L., Mohan, S.K., Menezes, J.R. de, Silva W.R. da, Oliari, L. and Meyer, R.C. (1987). IAPAR 8 - Rio Negro, a new French bean cultivar. *Pesquisa-Agropecuaria Brasileira* 22 (9-10): 995-998.

Allen, D.J. (1982). *Verticillium lecanii* on the bean rust fungus, *Uromyces appendiculatus. Trans. British Mycol. Soc.* 79(2): 362-364.

Anderson, A.L., Adams, M.W. and Whitford, G. (1963). Charlevoix anthracnose resistant dark red kidney bean. *Mich. State Univ. Agric. Stn. Res. Rep.* 6.

Anonymous. (1966). Division of Science Services, Biol. Branch, Plant Pathol. Section. Rep. Dep. Agric. NSW, 1964-65: 75-80.

Anonymous. (1989). French bean (*Phaseolus vulgaris*). *Pl. Varities J.* 2(2): 15-16.

Anoymous. (1953). Research work and workers. *Seed World* 72(11): 80-81.

Araiyo, G.A.de A., Viera, C. and Chagas, J.M. (1989). Performance of bean (*Phaseolus vulgaris* L.) cultivars in the Zona da Mata area of Minas Gerais V. *Revista Ceres* 36: 382-390.

Arnaud-Santana, E., Pena- Matos, E., Coyne, D.P. and Vidaver, A.K. (1991). Longevity of *Xanthomonas compestris* pv. *phaseoli* in naturally infested dry bean (*Phaseolus vulgaris*) debris. *Plant Dis.* 75(9): 952-953.

Azevedo, L.A.S.de. and Kushalppa, A.C. (1986). Quantification of the resistance of *Phaseolus vulgaris* cultivars to *U. phaseoli typical. Fitopatologia Brasileira* 11(3): 599-608.

Baker, C.J., Stavley, J.R., and Mock, N. (1985). Biocontrol of bean rust by *Bacillus subtilis* under field conditions. *Plant Dis.* 69(9): 770-772.

Baker, C.J., Stavley, J.R., Thomas, C.A. Sasser, M. and Macfall, J.S. (1983). Inhibitory effect of *Bacillus subtilis* on *Uromyces phaseoli var. typica* and on development of rust pustules on bean leaves. *Phytopathology* 73(8): 1148-1152.

Balardin, R.S. (1994). Effect of different fungicides on rust and powdery mildew control in common bean "Carioca" and "Goiano Precoce" cultivars. *Ciencia Rural* 24(2): 261-264.

Baldacci, E. and Betto, E. (1958). Tests of the fungicidal effect of some antibiotics applied by coverage and systemically. *Agricotura ital.* 58(NS 13): 289-301.

Bardas, G.A., Lagopodi, A.L., Kadoglidou, K. and Tzavella Klonari, K. (2009). Biological control of three *Colletotrichum lindemuthianum* races using *Pseudomonas chlororaphis* PCL1391 and *Pseudomonas fluorescens* WCS365. *Biol. Cont.* 49(2): 139-145.

Barrus, M.F. (1911). Variation of varieties of bean in their susceptibility to anthracnose. *Phytopathology* 1: 190-195.

Bashir, S., Ahmad, M., Sagar, V., Mohd, N. (2006). Identification of sources of resistance in French bean to *Phaeoisariopsis griseola* causing angular leaf spot disease. *Indian J. Pl. Protec.* 34 (2): 221-224.

Bazzi, C. and Calzolari, A. (1998). Bacterial diseases of bean: two years of observations in the Po valley. *Informatore Fitopatol.* 38(9): 49-51.

Beach, S.A. (1892). Some bean diseases. *New York State Agric. Expt. Stn. Bull.* 48: 305-333.

Becerra Leor, E.N., Lopezsalinas, E. and Acosta, G. J. A. (1994). Genetic resistance and chemical control of bean rust in the humid tropics of Mexico. *Revista Maxicana de Fitopatologia* 12(1): 35-42.

Berg, G.M. (1984). Nonpersistent viruses: epidemiology and control. *Adv. Virus Res.* 31: 387-429.

Bhagawati, R and Bhagabati, K.N. (1994). Incidence of bean common mosaic virus of French bean and its effect on grain yield. *Indian J. Virol.* 10: 141-143.

Bhardwaj, C.L., Kalia, N.R. and Singh, L.N. (1992). Reaction of French bean germplasm against angular leaf spot. *Indian J. Mycol. Pl. Pathol.* 22 (3): 270-271.

Bhardwaj, C.L., Nayital, S.C., Verma, S. and Kalia, N.R. (1994). Effect of sowing date, variety and management of angular leaf spot (*Phaeoisariopsis griseola*) on yield of French bean (*Phaseolus vulgaris*). *Indian J. Agric. Sci.* 64(5): 336-338.

Bhat, M.N. (2002). Chemical control of rust and angular leaf spot of French bean in Sikkim. *Pl. Dis. Res.* 17(1): 90-92.

Bhat, M.N., Singh, A.K. and Medhi, R.P. (1999). Reaction of French bean (*Phaseolus vulgaris*) cultivars/accessions to rust pathogen *Uromyces appendiculatus*. *Indian J. Agric. Sci.* 69(12): 849-850.

Borucka, K. and Marcinkowska, J.Z. (2001). Occurrence of *Colletotrichum lindemuthianum* on *Phaseolus* beans in Poland. *Veg. Crops Res. Bull.* 54(1): 147-151.

Bos, L. (1970). Bean yellow mosaic virus. In: *Description of Plant Viruses*. No. 73. CMI/AAB, Kew, Surrey, England.

Boudreau, M.A. (1993). Effect of intercropping beans with maize on the severity of angular leaf spot of beans in Kenya. *Plant Pathol.* 42(1): 16-25.

Bruggen, Van, A.H.O., Whalen, C.H. and Arneson, P.A. (1986). Emergence, growth and development of dry beans seedlings in respect to temperature, soil moisture and *Rhizoctonia solani*. *Phytopathology* 76: 568-572.

Brunt, A.A., Crabtree, K., Dallwitz, M.J., Gibbs, A.J. and Watson, L. (1996). In: *Viruses of Plants, Descriptions and lists from the VIDE Database,* CAB International, University Press, Cambridge, 1484p.

Burkholder, W.H. (1926). A new bacterial disease of the bean. *Phytopathology* 16: 915-927.

Buruchara, R.A., Gathuru, E.M. and Mukunya, D.M. (1988). Disease progress of angular leaf spot caused by *Isariopsis griseola* Sacc. and its implications on resistance of some bean (*Phaseolus vulgaris* L.) cultivars. *Acta Hortic.* 218: 321-328.

Butler, E.E. and Bracker, C.E. (1970). Morphology and cytology of *Rhizoctonia solani*. In: *Rhizoctonia solani, Biology and Pathology*. (Ed. Parmeter Jr. J.R.), University of California Press, Berkley, pp. 32-51.

Butler, E.J. (1918). *Fungi and Diseases in Plants.* Thacker Spink and Co., Calcutta, 547p.

Calzolari, A. (1997). Halo and common spot of beans. *Informatore Agrario.* 53(25): 66-67.

Campos, A.J. and Fucikovsky, Z.L. (1981). Study of some aspects of angular spot caused by *Isariopsis griseola* Sacc. on bean. *Fitopatologia* 16(1): 16-19.

Cardona-Alvarez, C. and Walker, J.C. (1956). Angular leaf spot of bean. *Phytopathology* 46: 610-615.

Carrion, G., Romero, A. and Rico-Gray, V. (1999). Use of *Verticillium lecanii* as a biocontrol agent against bean rust. *Association Latinoamericana de Fitopatologia (ALF) Fitopatologia* 34(4): 214-219.

Castro, J.L. de, Ito, M.F., Dudienas, C., Bulisani, E.A., D'Artegnan-de-Almeida, L. and De-Castro, J.L. (1991). Fungicide activity on two bean cultivars in Capao Bonito, Sau Paulo State, Brazil. *Bragantia* 50(2): 309-321.

Centurion MAP, D.A.C., Kimati, H. and Pereira, G.T. (1994). Machanism of action of antagonisits selected for the biological control of bean rust (*Uromyces phaseoli* (Reben.) Wint.). *Cientifica (Jaboticobal)* 22(2): 163-175.

Chakrabarty, P.K. and Shyam, K.R. (1988). Evaluation of systemic fungitoxicants against *Colletorichum lindemuthianum*, the incitant of bean anthracnose. *Indian J. Plant Pathol.* 6(1): 67-70.

Chakrabarty, P.K. and Shyam, K.R. (1990). Reaction of some promising lines of French bean to different isolates of *Colletotrichum lindemuthianum* in Himachal Pradesh. *Indian J. Plant Pathol.* 8(1 and 2): 86-88.

Cheo, P.C. (1955). The control of bean rust. *Tijdschr. Plziekt* 6I 1: 26.

Chhetry, G.K.N., Mishra, R.R. and Sharma, G.D. (1994). Effect of microclimatic factors on epidemiology of leaf spot disease caused by *Colletotrichum lindemuthianum* and *Phaeoisariopsis griseola* of French bean (*Phaseolus vulgaris*). *Indian J. Agric. Sci.* 64(9): 667-670.

Christen, R.and Echandi, E. (1967). The more common races of rust *Uromyces appendiculatus* in Costa Rica and evaluation of the rust resistance of some French bean cultivars. *Turrialba* 17(1): 7-10.

Cook, M.T. (1924). Common diseases of beans and peas. *Agric. Exp. Stn. Circ.* 142: 8.

Corrales, M.A. and Jara, C.E. (1995). Evolution of *Phaeoisariopsis griseola* with common bean in Latin America. *Fitopatol. Colombiana* 19(1): 15-24.

Correa-Victoria, F.J. (1987). Pathogenic variation, production of toxic metabolites and isozyme analysis in *Phaeoisariopsis griseola* (Sacc) Ferr. Michingan, USA, Michigan State University, Ph.D. Thesis.

Correa-Victoria, F.J. and Saettler, A.W. (1987). Angular leaf spot of red kidney beans in Michigan. *Plant Dis.* 71(10): 915-918.

Coyne, D.P. and Schuster, M.L. (1974). Breeding and genetic studies of tolerance to several bean (*Phaseolus vulgaris* L.) bacterial pathogens. *Euphytica* 23: 651-656.

Coyne, D.P., Schuster, M.L. and Erwin, D.C. (1979). Reaction of *Phaseolus vulgaris* germplasm to a new virulent strain of *Pseudomonas phaseolicola*. *Ann. Rep. Bean Improv. Coop.* 22: 20-21.

Davison, A.D. Vaughan, E.K. (1963). Longevity of uredospores of race 33 of *Uromyces phaseoli* var. *typica* in storage. *Phytopathology* 53(6): 736-737.

De Menezes, O.B. (1952). Identification of four physiologic races of bean rust. *Dusenia* 3(4): 309-312.

Devanathan, M., Muniyappa, V. and Peter, G.B. (1994). Screening of French bean (*Phaseolus vulgaris* L.) varieties for bean common mosaic virus (BCMV) resistance. *South Indian Hortic.* 42: 53-54.

Dhanju, K.S., Chowfla, S.C. and Garg, I.D. (1995d). Characterization of viruses causing French bean mosaic in Himachal Pradesh. *Int. J. Trop. Pl. Dis.* 13: 33-41.

Dhanju, K.S., Chowfla, S.C. and Handa, A.K. (1995a). Screening of French bean germplasm under field and glasshouse conditions against bean common mosaic virus. *Legume Res.* 18: 53-55.

Dhanju, K.S., Chowfla, S.C. and Handa, A.K. (1995b). Effect of barrier crops and spacings on the incidence of mosaic disease and yield of French bean. *Legume Res.* 18: 113-116.

Dhanju, K.S., Chowfla, S.C. and Handa, A.K. (1995c). Relative response of French bean germplasm against common mosaic under glass house conditions. *Legume Res.* 18: 135-136.

Dias, F.I.R. and Da Costa, J.C. (1968). Identification of physiologic races of *Uromyces appendiculatus* on bean (*Phaseolus vulgaris*) in two regions of Rio Grande do Sul, Brazil (RAM 1970: 3049)

Diaz-Plaza, R., Teliz, O.D. and Munoz, D.A. (1991). Effect of diseases of bean grown under rainfed conditions at Mixteca peblana. *Revista Mexicanade Fitopatol.* 9: 21-30.

Downing, K.J. and Thomson, J.A. (2000). Introduction of the *Serratia marcescens* Chi A gene into an endophytic *Pseudomonas flourescens* for the biocontrol of phytopathogenic fungi. *Can. J. Microbiol.* 46(4): 363-369.

Drijfhout E. and Davis J.H.C. (1989). Selection of a new set of homogeneously reacting bean (*Phaseolus vulgaris*) differentials to differentiate races of *Colletotrichum lindemuthianum*. *Plant Pathol.* 38: 391-396.

Drijfhout, E. and Bos, L. (1977). The identification of two new strains of bean common mosaic virus. *Neth. J. Pl. Pathol.* 83: 13-25.

Drijfhout, E., Silbernagel, M.J. and Burke, D.W. (1978). Differentiation of strains of bean common mosaic virus. *Neth. J. Pl. Pathol.* 84: 13-26.

Dumitras, L. and Sesan, T. (1990). Feature of antagonism of *Trichoderma viride* Pers ex Fr. to *Rhizoctonia solani* Kuhn. *Biologie Vegelata.* 32: 87-92.

Dundas, B. (1942). Breeding beans for resistance to powdery mildew and rust. *Phytopathology* 39(9): 828.

Dundas, B. and Scott, G.W. (1939). Physiologic strains of bean rust. *Phytopathology* 29(9): 820-821.

Eastman, J.W. (1943). Report of Provincial Plant Pathologist. *British Columbia Dept. Agri. Ann. Rept. 1943*: 59-63.

Edwardson, J.R. and Christie, R.G. (1991). *The potyvirus Group.* Vol. I-IV, *Fl. Agric. Exp. Stn. Monograph* 16.

Elad, Y., Chet, I. and Katen, I. (1980). *Trichoderma harzianum* biocontrol agent effective against *Sclerotium rolfsii* and *Rhizoctonia solani*. *Phytopathology* 70: 119-121.

Estevez, deJensen, C., Percich, J.A., Graham, P.H. and deJensen, C. Estevez. (2002). Integrated management strategies of bean root rot with *Bacillus subtilis* and *Rhizobium* in Minnesota. *Field Crop Res.* 74 (2-3): 107-115.

Faleiro, F.G., Paula Jr. T. J. De, Barros, E. G. De, Frietos, M. A. S. and Moreira, M. A. (1996). Common bean resistance to *Uromyces appendiculatus* from zona da Mata region of the state of Minas Gerais, Brazil. *Fitopatologia Brasileira* 21(1): 123-125.

Faleiro, F.G., Vinhadelli, W.S., Ragagnin, V.A., Paula, T.J.Jr., Moreira, M.A. and Barros, *E.G.* (1998). Resistance of the common bean to four races of *Uromyces appendiculatus*. *Revista Ceres*. 263: 11-18.

Faleiro, F.G., Vinhadelli, W.S., Ragagnin, V.A., Zambolin, L., Paula, T.J.Jr., Moreira, M.A. and Barros, *E.G.* (1999). Identification of physiological races of *Uromyces appendiculatus* in the state of Minas Gerais, Brazil. *Fitopatologia Brasileira* 24(2): 166-169.

Faure, B., Ponce, M. and Garcia, Y.E. (1989). Evaluation of common bean (*Phaseolus vulgaris*) vars. with black and white seeds. *Ciencia y Tecnica en la Agricultura* 8(1): 31-41.

Fenille, R.C. and Souza, N.L.de. (1999). The role of the organic material amended and the soil moisture on the pathogenecity of *Rhizoctonia solani* Kuhn AG-4 HGI in snap bean. *Pesquisa Agropecuaria Brasileira* 34(10): 1959-1967.

Fininsa, C. (1996). Effect of intercropping bean with maize on bean common bacterial blight and rust diseases. *Int. J. Pest Mgt*. 42(1): 51-54.

Fininsa, C. (1996). Effect of intercropping bean with maize on common bacterial blight and rust diseases. *Int. J. Pest Mgt*. 42(1): 51-54.

Flores Revilla, C. Neito Angel, D. and Torre Almaraz, R.de la. (1994). Effect of varieties, foliar nutrients and seedling rates on disease severity and yield in French bean (*Phaseolus vulgaris*). *Revista Mexicana de Fitopatologia* 12(1): 69-74.

Fortugno, C. (1977). Angular leaf spot (*Isariopsis griseola* Sacc.) of bean. *IDIA*317/320: 26-33.

Foster, H.H. (1947). Reaction to rust under Misssippi conditions, of certain lines and vars. of pole snap beans. *Pl. Dis. Reptr*. 31(10): 378-383.

Fouilloux, G. (1979). New races of bean anthracnose and consequences on our breeding programmes. Pp. 221-235 in: *Int. Symp. Dis. of Trop. Food Crops* (Eds. H. Maraite and J.A. Meyer). Universite Catolique de LouvainlaNeuve, Belgium.

Fourie, D. (1998). Characterization of halo blight races on dry beans in South Africa. *Plant Dis*. 82(3): 307-310.

Frazier, W.A. and Hendrix, J.W. (1949). Hawaiian Wonder, new rust resistant pole green bean. *Circ. Hawaii Agric. Exp. Sta*. 28: 7.

Garcia, Veloz R., Marin, M.R., Veloz, R.R., RodriguezGuerra, R., Torres, P.I., Gonzalez, C.M.M., Anaya Lopez, J.L., Guevara Olvera, L., Feregrino Perez, A.A., Loarca Pina, G. and Guevara Gonzalez, R.G. (2010). Antimicrobial activities of cascalote (*Caesalpinia cacalaco*) phenolics-containing extract against fungus *Colletotrichum lindemuthianum*. *Indust. Crops Prod.* 31(1): 134-138.

Ghimire, S.R., Pradhanang, P.M. and Lohar, D.P. (1995). Use of fungicides and identification of resistant varieties for the management of bean rust and anthracnose disease in common beans. *Working Paper Lumle Regional Agric. Res. Centre No*. 95/76: 42p.

Gilbertson, R.L., Rand, R.E. and Hagedorn, D.J. (1990). Survival of *Xanthomonas campestris* pv. *phaseoli* and pectolytic strains of *Xanthomonas campestris* in bean debris. *Plant Dis.* 74(4): 322-327.

Gonalves Vidigal, M.C., Sakiyama, N.S., Vidigal Filho, P.S., Amaral Junior, A.T., Poletine, J.P.and Oliveira, V.R. (2001). Resistance of common bean cultivar AB 136 to races 31 and 69 of *Colletotrichum lindemuthianum*: the Co-6 locus. *Crop Breeding App. Biotech.* 1(2): 99-104.

Gonzalez, M. and Garcia, C. (1996a). Evaluation of fungicides for the control of bean rust (*Uromyces appendiculatus*). *Agronomia Mesoamericana* 7(1): 86-89.

Gonzalez, M. and Garcia, C. (1996b). Evaluation of losses due to the rust on bean (*Phaseolus vulgaris* L.) in four sowing times in Cuba. *Agronomia Mesoamericana* 7(1): 95-98.

Gupta, S.K. and Mathew, K.A. (1999). Fungal and bacterial diseases of beans. In: *Diseases of Horticultural crops- Vegetables, Ornamentals and Mushrooms* (Eds. L.R. Verma and R.C. Sharma). Indus Publishing Co., New Delhi pp. 301-335.

Gupta, S.K. and Shyam, K.R. (2003). Angular leaf spot of French bean: Epidemiology and Management. *Annu. Rev. Plant Pathol.* 2: 221-237.

Gupta, S.K. and Thind, T.S. (2006). Disease Problems in Vegetable Production. Scientific Publishers, Jodhpur, 576p.

Gupta, S.K., Dohroo, N.P. and Shyam, K.R. (1991). Antagonistic studies on seed-borne mycoflora of French bean (*Phaseolus vulgaris* L.). *Indian J. Pl. Pathol.* 9(1 and 2): 62-63.

Gupta S K, Jarial K and Kumar M. 2008. Occurrence of French bean rust in Himachal Pradesh. *Journal of Mycology and Plant Pathology* 38(2): 352-353.

Gupta, S.K., Mathew, K.A and Shyam, K.R. (2000a). Seed transmission of *Rhizoctonia solani* in French bean cultivars. *Pl. Dis. Res.* 15(1): 83-84.

Gupta, S.K., Mathew, K.A. and Shyam, K.R. (2005). Influence of inoculum density and environmental factors on angular leaf spot of French bean. *Pl. Dis. Res.* 20(1): 25-28.

Gupta, S.K., Mathew, K.A. and Shyam, K.R. (2000e). Management of angular leaf spot of French bean through cultural practices. In: Proceedings of Indian Phytopathological Society-Golden Jubilee International Conference on "Integrated Plant Disease Management for Sustainable Agriculture" Vol. II, Indian Phytopathological Society, IARI, New Delhi: 701-704.

Gupta, S.K., Mathew, K.A., Shyam, K.R. and Gupta, D.K. (2000c). Role of fungicides in management of anthracnose of French bean. In: Proceedings of Indian Phytopathological Society-Golden Jubilee International Conference on *"Integrated Plant Disease Management for sustainable Agriculture"* Vol. II, Indian Phytopathological Society, IARI, New Delhi : 705-707.

Gupta, S.K., Mathew, K.A., Shyam, K.R. and Sharma, A. (1999). Fungicidal management of root rot (*Rhizoctonia solani*) of French bean. *Pl. Dis. Res.* 14(1): 20-24.

Gupta, S.K., Mathew, K.A., Shyam, K.R. and Sharma, A.K. (2000d). Evaluation of French bean germplasm against angular leaf spot. *Indian Phytopath.* 53(4): 488-489.

Gupta, S.K.; Mathew, K.A., Shyam, K.R. and Joshi, A.K. (1997). Screening of French bean germplasm against root rot and web blight. *Legume Res.* 20(1): 57-60.

Gupta, Y. and Chowfla, S.C. (1988). Susceptibility of French bean to bean common mosaic virus in relation to plant age, population age and age of *Myzus persicae. Indian J. Mycol. Pl. Pathol.* 18: 93-94.

Gupta, Y. and Chowfla, S.C. (1990). Screening of French bean germplasm for resistance to bean common mosaic virus. *Indian Phytopath.* 43: 434-436.

Guu, J.W. (1995). Development of rust resistant and stringless bean lines. *Bull. Taichung Distt. Agric Deptt. Sln. No.* 46: 11-18.

Guzman, P., Gilbertson, R.L., Nodari, R.l, Johnson, W.C., Temple, S.R., Mendala, D., Mkandawire, A.B.C. and Gupta, P. (1995). Characterization of variability in the fungus *Phaeoisariopsis griseola* suggests coevolution with the common bean (*Phaseolus vulgaris*). *Phytopathology* 85(5): 600-607.

Hagedorn, D.J. and Rand, R.E. (1980). Wisconsin RRR 46 Snap bean breeding line. *Hortic. Sci.* 15: 529-530.

Hanan, F. and Khan, I. (1999). Resistance of common bean (*Phaseolus vulgaris*) cultivars to two races of halo blight (*Pseudomonas phaseolicola* Burkh.). *Sarhad J. Agri.* 15(4): 305-309.

Harter, I.L., Andrus, C.F. and Zaumeyer, W.J. (1935). Studies on bean rust caused by *Uromyces phaseoli* var. *typica. J.Agric. Res.* 1(9): 737-759.

Harter, L.L. and Zaumeyer, W.J. (1941). Differentiation of physiologic races of *Uromyces phaseoli* var. *typica* on bean. *J. Agric. Res.* 6(12): 717-731.

Hazarika, D.K. and Das, K.K. (1998). Biological management of root rot of French bean (*Phaseolus vulgaris* L.) caused by *Rhizoctonia solani. Pl. Dis. Res.* 13(2): 101-105.

Hidalgo, R. and Araya, C.M. (1993). Optimum growth stage of common bean for chemical control of anthracnose (*Colletotrichum lindemuthianum*) and angular leaf spot (*Isariopsis griseola*) in San Carlos, Costa Rica. *Agronomic Costarricense* 17(1): 75-80.

Hocking, D. (1967). A new virulent form of *Phaeoisariopsis griseola* causing circular leaf spot of French bean. *Pl. Dis. Reptr.* 51: 276-278.

Hutchinson, C.M. and Ram Ayyar, C.S. (1915). The Indian rice beer ferment. *Mem. Dept. Agric. Indian Bact. Ser.* 1: 137-168.

Irigoyen, E.D. and Garbagnoli, C. (1997). Common bacteriosis in bean (*Xanthomonas compestris* pv. *phaseoli* (E.F. Smith) Dowson): detection, infection and transmission through seeds. *Fitopatologia* 32: 166-172.

Issa, E. (1985a). Resistance to four diseases in cultivars of French bean, *Phaseolus vulgaris* L., bred in Sao Paulo. *Biologico* 51: 113-117.

Issa, E., Cruz, B.P.B, Wattanabe, K. and Arruda, H.V. (1980). Damage done by bean (*Phaseolus vulgaris*) diseases. *Biologica* 46(7): 135-140.

Issa, E., Sinigaglia, C. and Oliveira, D.A. (1982). Chemical control of angular leaf spot of bean (*Phaseolus vulgaris* L.) due to *Isariopsis griseola* Sacc. *Biologica* 48(12): 299-303.

Iwanowski, D. (1894). Ueber die Mosaikkrankheit der tabaksflanze. *Acad. Imp. Sci. St. Petersbourg Bul.* 35(3): 67-70.

Jaffer, A.A. (1971). Effect of fungicides on bean rust (*Uromyces appendiculatus*) (Pers.) *Lev. Misc. Rep. Pestic. Res. Inst.* 765: 8p.

Jensen, J.H. and Livingstone, J.E. (1944). Variation in symptoms produced by isolates of *Phytomonas medicaginis* var. *phaseolicola. Phytopathology* 34: 471-480.

Jeyarajan, R. and Doraisamy, S. (1986). Virus diseases of pulses and their management. In *Pest* and *Disease Management Centre for Pl. Protect. Studies*, TNAU, Coimbatore, 109-114.

Jindal, K. K. (1991). Physical and chemical agents for the eradication of *Xanthomonas campestris* pv. *phaseoli* from bean seeds. *Pl. Dis. Res.* 6: 68-71.

Joshi Deeksha., Hooda, K.S., Bhatt, J.C., Mina, B.L. and Gupta, H.S. (2009). Suppressive effects of composts on soil-borne and foliar diseases of French bean in the field in the western Indian Himalayas. *Crop Protec.* 28(7): 608-615.

Joshi, D., Hooda, K.S. and Bhatt, J.C. (2008). Biocontrol potential of fungal antagonists from western Himalayan region against sclerotial plant pathogens. *Pesticide Res. J.* 20(1): 1-5.

Jung, G.H., Scrotch, P.W., Coyne, D.P., Nienhuis, J., Arnaud-Santana, E., Ariyarathen, H.M., Kaeppler, S.M. and Bassett, M.J. (1997). Molecular marker based genetic analysis of tepary bean derived common bacterial blight resistance in different developmental stages of common bean. *J. Am. Soc. Hortic. Sci.* 122(3): 329-337.

Kapil, R., Sharma, P., Sharma, S.K., Sharma, O.P., Sharma, O.P., Dhar, J.B.and Sharma, P. N. (2010). Pathogenic and molecular variability in *bean common mosaic virus* infecting common bean in India. *Archives Phytopathol. Pl. Protec.* 44(11): 1081-1092.

Kapoor, K.S., Gill, H.S. and Sharma, S.R. (1988). Screening of French bean against floury leaf spots. *Indian Phytopath.* 41: 650-651.

Karanja, T.W., Mwang'ombe, A.W. and Mibey, K. (1994). The effect of media and light regimes on cultural and morphological characteristics and sporulation of *Phaeoisariopsis griseola* Deighton. *East African Agric. For. J.* 59(3): 241-251.

Khlaif, H.M. and Qadous, N.S. (1995). Common blight of beans in the Jordon Valley: sources of inoculum. *Dirasat. Series B, Pure App. Sci.* 22(3): 717-728.

Kiessling, D. and Skiebe, K. (1987). Studies on the causes of variability in incomplete resistance of bean, *Phaseolus vulgaris* L. to *Pseudomonas syringae* pv. *phaseolicola* (Burkholder) Young *et al., Arch. Zichtungs-forschung.* 17: 191-200.

Kikoka, L.P., Katunzi, A.L. and Teri, J.M. (1989). Evaluation of effect of cropping systems on bean diseases and yield. In *Proc. Int. Pest Mgt. in Trop. Sub-trop. Crop. Syst.* 89(3): 853-858.

Kimati, H. and Galli, F. (1970). *Glomerella cingulata* (Stonem.) Spauld. *et* V. Schrenk. f. sp. *phaseoli* n.f., faseascegena de agente causal de anthracnose do feijoeiro. *Annuals da Escola Superior* de *Agricultura "Luisde Queiroz"* 27: 411-437.

Kishan, R. (1988). Loss in yield of French beans affected with common blight. pp, 29-31. In: *Advances in Research on Plant Pathogenic Bacteria.* (Eds. S.S. Gnanamanickan and A. Mahadevan). Today and Tomorrow's Printers and Publishers, New Delhi.

Kumar, A., Kaushal, R.P. and Sharma, R. (2005). Transfer of resistance to floury leaf spot and bacterial blight in French bean from alien germplasm. *Indian Phytopath.* 58(2): 136-142.

Kumar, A., Sharma, P.N., Sharma, O.P. and Tyagi, P.D. (1997). Resistance to *Colletotrichum lindemuthianum* in kidney bean accessions of diverse origin in Himachal Pradesh. *Indian Phytopath.* 50: 59-64.

Kumar, A., Sharma, P.N., Sharma, O.P. and Tyagi, P.D. (1999). Epidemiology of bean anthracnose *Colletotrichum lindemuthianum* under sub-humid mid-hill zone of Himachal Pradesh. *Indian Phytopath.* 52(4): 393-397.

Lecerda, J.T., Coetho, R.S.B., Mariano, R.L.R. and Menezes, M. (1994). Pathogenic variability of *Isariopsis griseola* on bean in the state of Pernambuco. *Summa Phytopathol.* 20: 93-96.

Lenne, J.M. (1992). *Colletotrichum* diseases of legumes. pp.134-166 In: *Colletotrichum: Biology, Pathology* and *Control.* (Eds. J.A. Bailey, and M.A Jeger), C.A.B. International, Wallingford, U.K.

Lopez, S.E., Acosta, G.J.A., Becerra, L.E.N., Frayre, V.G., Orozco, S.H. and Beebe, S.E. (1997). Registration of Negro Tacana common bean. *Crop. Sci.* 37: 1022.

Mabgala, R.B. and Saettler, A.W. (1992). Races and survival of *Pseudomonas syringae* pv. *phaseolicola* in Northern Tanzania. *Plant Dis.* 76(7): 678-682.

Mabgala, R.B. (1977). The effect of populations of *Xanthomonas campestris* pv. *phaseoli* in bean reproductive tissues on seed infection of resistant and susceptible bean genotypes. *European J. Plant Pathol.* 103(2): 175-181.

Macraw, W.M., Nasser, S.H., Sidky, S.T. and Faris, F.S. (1973). Producing a var. of beans, resistant to the rust disease. 1. Screening local and imported vars. against the disease. *Agric. Res. Rev.* 51(3): 145-152.

Mageshwaran V., Mondal, K.K., Kumar U. and Annapurna K. (2012). Role of antibiosis on suppression of bacterial common blight disease in French bean by *Paenibacillus polymyxa* strain HKA-15. *African J. Biotech.* 11(60): 12389-12395.

Maringoni, A.C. and Bach, E.E. (1999). Characterization of *Xanthomonas axonopodis* pv. *phaseoli* strains through electrophoretic analysis of total proteins. *Summa Phytopathol.* 25(2): 173-177.

Maringoni, A.C. (1990). Chemical control of the common bacterial blight of bean and its effect in the transmission of *Xanthomonas campestris* pv. *phaseoli* (Smith) Dye by seeds. *Pesquisa Agropecuaria Brasileira* 25(8): 1151-1156.

Maringoni, A.C., Barros, E.M. de and de Barros, E.M. 2002). Occurrence of isolates of *Colletotrichum lindemuthianum* resistant to benzimidazole fungicides. *Summa Phytopathol.* 28(2): 197-200.

Maringoni, A.C., Tofoli, J.G. and Fregonese, L.H. (1992). Fungicides action *in vitro* on *Rhizoctonia solani* and on control of dry bean damping off. *Summa Phytopathol.* 18: 269-275.

Mathew, K.A. and Gupta, S.K. (1998). Biological control of root rot of French bean caused by *Rhizoctonia solani. J. Mycol. Pl. Pathol.* 28(2): 202-205.

Mathew, K.A. and Gupta, S.K. (1996). Studies on web blight of French bean caused by *Rhizoctonia solani* and its management. *Indian J. Mycol. Pl. Pathol.* 26: 171-176.

Mathew, K.A., Gupta, S.K. and Shyam, K.R. (1998). New strategies in fungicidal management of angular leaf spot (*Phaeoisariopsis griseola*) of French bean. *J. Mycol. Pl. Pathol.* 28(2): 123-133.

Matz, J. (1921). Una enfermendad damina de la haleichuela. *Porto Rico Insular Stn. Cir.* 57: 8.

MayekPerez, N., PedrozaFlores, J.A., Villneal Garcia, L.A. and Valdes Lozano, C.G.S. (1995). Genetic and environmental factors related to temporal dynamics and diseases effects on dry beans (*Phaseolus vulgaris* L.) at Marin, Nuero, Mexico. *Revista Mexicana de Fitopatol. (RPP 98/415)*

McKern, N.M., Mink, G.I., Barnet, O.W., Mishra, A., Whittaker, L.A., Silbernagel, M.J., Ward, C.W. and Shukla, D.D. (1992). Isolates of bean common mosaic virus comprising two distinct potyviruses. *Phytopathology* 82: 923-929.

Menezes, J.R. and Dianese, J.C. (1988). Race characterization of Brazilian isolates of *Colletotrichum lindemuthianum* and detection of resistance to anthracnose in *Phaseolus vulgaris. Phytopathology* 78: 650-655.

Meyer, S., Saccardy Adji, K., Rizza, F. and Genty, B. (2001). Inhibition of photosynthesis by *Colletotrichum lindemuthianum* in bean leaves determined by chlorophyll fluorescence imaging. *Plant Cell Environ.* 24(9): 947-955.

Micro, H.N., Mohapatra, A.K., Sahoo, S. and Mohapatra A.K.B. (1998). Performance of French bean under different dates of sowing in eastern ghat high land zone of Orrisa. *Annals Agric. Res.* 19(1): 83-84.

Mizubuti, C.S.G., Maffia, L.A., Muchovej, J.J., Romeiro, R.S. and Batista, U.G. (1995). Selection of isolates of *Bacillus subtilis* with potential for the control of dry bean rust. *Fitopatologia Brasileira* 20(4): 540-544.

Mohan, N., Aghora, T.S. and Somkuwar, R.G. (1995). Reaction of French bean germplasm to angular leaf spot (*Isariopsis griseola* Sacc.). *Legume Res.* 18: 215-216.

Mohan, N., Aghora, T.S., Somkwar, R.G. and Girija Ganeshan. (1997). Sources of resistance to rust in French bean (*Phaseolus vulgaris*). *Indian J. Agri. Sci.* 67(2): 85-86.

Mondal, K.K. (2004). Integrated management strategies for fuscous blight and floury leaf spot of rajmash. *Indian Phytopath.* 57(2): 135-139.

Monteagudo, A. B., Rodino, A.P., Lema, M., Fuente, M de la., Santalla, M., Ron, A. M. de. and Singh, S.P. (2006). Resistance to infection by fungal, bacterial, and viral pathogens in a common bean core collection from the Iberian Peninsula. *HortScience* 41(2): 319-322.

Mora N.O.A., Vieira, C. and Zambolin, L. (1992). Differential bean vars. for determination of physiological races of *Uromyces phaseoli* var. *typica* Arth. *Revista Ceres* 224(39): 391-404.

Moralles, F.J. (1989). Bean Common Mosaic-Screening for Disease Resistance. *Centro Internacional de Agricultura Tropical* (*CIAT*) *Publ.* No. 33. Cali, Colombia. 28p.

Moreno, R.A. and Mora, L.E. (1984). Cropping pattern and soil management influence on plant disease 2. Bean rust epidemiology. *Turrialba* 34(1): 41-45.

Moreno, R.A. (1977). Efecto de diferentes sistemas de cultivo sobre la severidad de la mancha angular del frijol (*Phaseolus vulgaris* L.) caused for *Isariopsis griseola* Sacc. *Acta Agron. Palmira* 7: 164-169.

Mukherjee, S., Tripathi, H.S. and Rathi, Y.P.S. (2001). Integrated management of wilt complex in French bean (*Phaseolus vulgaris* L.). *J. Mycol. Pl. Pathol.* 31(2): 213-215.

Mundel, H.H., Kiehn, F.A., Huang, H.C., Conner, R.L., Saindon, G. and Kemp, G. (2004). Morden 003 navy bean. *Can. J. Plant Sci.* 84: 231-233.

Muyolo, N.G., Lipps, P.E. and Schmitthenner, A.F. (1993). Reactions of dry bean, lima bean, and soyabean cultivars to Rhizotonia root and hypocotyl rot and web blight. *Plant Dis.* 77: 234-238.

Navarro, A.R., Puerta-E, O.D. and Isaza, L. (1981). Combined effect of population density and chemical control of leaf diseases and yield of bean. *Fitopatol. Colombiana* 10(1/2): 7-9.

Nitshe, M.J. and Cafati, K.C. (1985). Identification of internal fungi present in bean seeds. *Agricultura Tecnica* 45: 227-234.

Ogoshi, A. (1987). Ecology and pathogenecity of anastomosis and intraspecific groups of *Rhizoctonia solani* Kuhn. *Annu. Rev. Phytopath.* 25: 125-143.

Oliveira, S.H.F., Barros, B.C. and Castro, J.L. (1992). Effect of fungicides on the control of bean foliar diseases and seed quality. *Summa Phytopathol.* 18(2): 178-184.

Oliviera, S.H.F. de and de Oliveira, S.H.F. (2003). New fungicides and spraying schedules in the control of anthracnose and angular leaf spot. *Summa Phytopathol.* 29: 45-48.

Opio, A.F., Allen, D.J. and Teri, J.M. (1996). Pathogenic variation in *Xanthomonas campestris* pv. *phaseoli*, the causal agent of common bacterial blight in *Phaseolus* beans. *Plant Pathol.* 45(6): 1126-1133.

Opio, A.F., Teri, J.M. and Allen, D.J. (1993). Studies on seed transmission of *Xanthomonas campestris* pv. *phaseoli* in common beans in Uganda. *Afr. Crop Sci. J.* 1(1): 59-67.

Otoya, M.M. and Pastor-Corrales, M.A. (1992). DNA probes for studies of genetic diversity of pathogens causing common bacterial blight of beans. *Fitopatol. Columbiana* 16(1/2): 149-157.

Ovies, J. and Larrinaga, L. (1988). Transmission of *Xanthomonas campestris* pv. *phaseoli* by a wild host. *Ciencia y Tecnica en la Agricultura Proteccion dePlantas* 11(2): 23-30.

Padder, B. A., Sharma, P.N., Kapil, R., Pathania, A., Sharma, O.P. (2010). Evaluation of bioagents and biopesticides against *Colletotrichum lindemuthianum* and its integrated management in common bean. *Nat. Sci. Biol.* 2 (3): xx-xx.

Padder, B.A. and Sharma, P.N. (2010). Assessment of yield loss in common bean due to anthracnose (*Colletotrichum lindemuthianum*) under glass house conditions. *Res. J. Agric. Sci.* 1(3): 184-188.

Padder, B.A., Sharma, P.N. Sharma, O.P. and Kapoor, V. (2007). Genetic diversity and gene flow estimates among five populations of *Colletotrichum lindemuthianum* across Himachal Pradesh. *Physiol. Molec. Pl. Path.*70: 8–12.

Padder, B.A., Sharma, P.N., Sharma, O.P. and Kapoor, V. (2008). Ribosomal DNA analysis of *Colletotrichum lindemuthianum virulences* from Himachal Pradesh (India). *Appl. Biol. Res.* 10: 6-10.

Pastor-Corrales, M.A. and Tu, J.C. (1989). Anthracnose. pp. 77-104 in: *Bean Production Problems in Tropics*. 2nd edition, (Eds. H.F. Schwartz, and M.A. Pastor-Corrales) CIAT, Cali, Colombia.

Pastor-Corrales, M.A. (1991). Estandarizacion de variedades differenciales y de designacion de razas de *Colletotrichum lindemuthianum*. *Phytopathology* 81: 694. (Abstr.).

Pastor-Corrales, M.A., Erazo, O.A., Estrada, E.I. and Singh, S.P. (1994). Inheritance of anthracnose resistance in common bean accession G 2333. *Plant Dis.* 78: 959-962.

Patel, P.N. and Walker, J.C. (1965). Resistance in *Phaseolus vulgaris* to halo blight. *Phytopathology* 55: 889-894.

Pathania, A. Sharma, P.N., Sharma, O.P., Chahota, R.K., Ahmad, B. and Sharma, P. (2006). Evaluation of resistance sources and inheritance of resistance in kidney bean to Indian virulences of *Colletotrichum lindemuthianum*. *Euphytica* 149: 97–103.

Paul, Y.S. and Sharma, R.C. (1990). Additions to the host range of *Rhizoctonia solani*. *Pl. Dis. Res.* 5: 234.

Paula, T.J. Jr., Pinto, C.M. F., Silva, M. B. da, Nietsche, S., Carvalho, G. A. de and Faleir, F.G. (1998). Resistance of snap bean cultivars to anthracnose, angular leaf spot and rust. *Revista Ceres* 45(258): 171-181.

Pazos, V., Lagomasino, I. and Martinez, R. (1986). Sensitivity of *Xanthomonas campestris* pv. *phaseoli* isolates to bacteriophages. *Revista de Proteccion Vegetal* 1(1): 21-24.

Pellicano, I.J., Ramalho, M.A.P., Figueira, A. and Dos, R. (1993). Evaluation of bean (*Phaseolus vulgaris*) cultivars for resistance to bean common mosaic virus. *Giencia Pratica* 13: 112-115.

Perfect, S.E., Green, J.R. and O'-Connell, R.J. (2001). Surface characteristics of necrotrophic secondary hyphae produced by the bean anthracnose fungus, *Colletotrichum lindemuthianum*. *Euro. J. Pl. Pathol.* 107(8): 813-819.

Persoon, C.H. (1795). Observations Mycologicae. In: Usteri, P., *Annalen der Botanic* 3(15): 1-43. Leipzig, Germany.

Petrak, F. (1950). Beitrage zur Pilsflora von Ekuador. *Sydowia* 4: 450-587.

Pinto, J.M.A., Souza, E.A. and Oliveira, D.F. (2010). Use of plant extracts in the control of common bean anthracnose. *Crop Protec.* 29(8): 838-842.

Ponnappa, K.M. Hiremath, P.C. and Sulladmath, V.V. (1976). Occurrence of *Phaeoisariopsis griseola* (Sacc.) Ferr. on French bean. *Curr. Sci.* 45: 836-837.

Prakasam, V. and Thamburaj, S. (1991). Efficacy of fungicides in the control of rust disease of French bean caused by *Uromyces phaseoli* (Reben.) Wimb. *Indian J. Mycol. Pl. Pathol.* 21(2): 158-159.

Prudencio Sains, J.M., Navarrete Maya, R., Navarrete Maya, J. and Acosta Gallegos, J.A. (2008). Dynamics of the bean common and halo blights in the valley of Mexico. *Agricultura Tecnica en Mexico* 34(2): 201-212.

Pyndji, M.M. and Trutmann, P. (1992). Managing angular leaf spot of common bean in Africa by supplementing farmer mixtures with resistant varieties. *Plant Dis.* 76 (11): 1144-1147.

Qadous, N. and Khlaif, H. (1997). Common blight of beans in Jordon valley: disease development and response of different host cultivars to the disease. *Dirasat. Agric. Sci.* 24(2): 260-267.

Radkov P., Nedelchev NK and Dekov D. (1978). Study of the field resistance of some bean hybrids and cultivars to bacteriosis under non-irrigated and irrigated conditions. *Rastenievdni Nauki* 15(9-10): 149-153.

Raffat, F.M. (1992). Efficiency of certain fugicides in controlling *Rhizoctonia* damping off of bean. *Assuit J. Agric. Sci.* 23: 37-47.

Raizada, R.K., Mohd Aslam, QMR Haq and Singh, B.P. (1991). *Vicia birsuta*: a natural host of bean yellow mosaic virus. *Int. Conf. Virol. Trop.* Dec. 5, NBRI, Lucknow, pp.37.

Rajnauth, G. (1987). Web blight, an important disease of bean and pak-choi in Trinidad. *Trop. Agric.* 84: 356-358.

Ramanjulu, V., Rajaram Reddy, D., Babu, G.P.and Gopal, K. (2004). Identification of sources of resistance to bean common mosaic virus in French bean (*Phaseolus vulgaris* L.). *Pl. Dis. Res.* 19(2): 134-139.

Rava, C.A. and Romeiro, R.S. (1990). Variability among isolates of *Xanthomonas campestris* pv. *phaseoli* in relation to their pathogenicity on *Phaseolus vulgaris* cultivars. *Summa Phytopathol.* 16(3/4): 225-232.

Rava, C.A., Costa, J.G.C.-da, Fonseca, J.R., Salgado, A.L. and da-Costa, J.G.C. (2003). Identification of sources of resistance to anthracnose, common bacterial blight and bacterial wilt in common bean. *Revista Ceres* 50: 797-802.

Ravi, S., Doraiswamy, S., Valluvaparidasan, V. and Jeyalakshmi, C. (1999). Effect of biocontrol agents on seed- borne *Colletotrichum* in French bean. *Pl. Dis. Res.* 14(2): 146-151.

Roberti, R., Ghisellini, L., Pisi, A., Flori, P. and Filippini, G. (1993). Efficiency of two species of *Trichoderma* as a biological control against *Rhizoctonia solani* Kuhn. isolated from string bean root rot in Italy. *Adv. Hortic. Sci.* 7: 19-25.

Robertson, R.S. (1962). The role of seed transmission in epidemiology of bean mosaics in central Washington. *Pl. Dis. Reptr.* 56: 71-72.

Rodriguez, A.M., Ovies, J. and Cruz, R. (1991). Two weed hosts of *Xanthomonas campestris* pv. *phaseoli*. *Proteccion de Plantas* 1(3/4): 43-48.

Rodriques, C.H., Zambolim, L. and Martins, H.C.P. (1987). Effectiveness of fungicides in the control of angular leaf spot (*Isariopsis griseola*) of bean (*Phaseolus vulgaris*) *Fitopatol. Brasileira* 12(1): 40-45.

Rolim, P.R.R., Brignani Neto, F., Roston, A.J. and Oliveira, D.A. (1981). Chemical control of bean diseases.1. Rust (*Uromyces phaseoli* (Pers.) Wint. Var. *typica* Arth.). *Biologica* 47(7): 201-205.

Saccardo, P.A. (1878). Fungi veneti novi criticivel Mycologiae venetae addendi *Isariopsis griseola* Sacc. *Michelia* 1: 273 p.

Saettler, A.W. and Corea, F.J. (1988). Transmission of *Phaeoisariopsis griseola* by bean seed. *J. Seed Tech.* 12: 133-142.

Saettler, A.W. and Potter, H.S. (1970). Chemical control of halo bacterial blight in field beans. pp. 1-8, *Michigan Agri. Exp. Stn. Res. Rep. 98.*

Saettler, A.W. (1989). Common bacterial blight. In: *Bean Production Problems in the Tropics*, 2nd ed. (Eds. H.F. Schwatz and M.A. Pasto-corrales), CIAT, Cali, Colombia, 726p.

Sanchez, A., Echavez, B.R. and Schroder, E.C. (1994). *Psedudomonas cepacia*, a potential biofungicides for root rot pathogens of beans. *J. Agric. Univ. Puerto Rico* 78: 55-57.

Sanchez, A.J.H. and Cardenas, A.M. (1988). Etiology and damage of root rot disease in beans *Phaseolus vulgaris* L. in the state of Durango. *Revista Chapingo* 12: 43-49.

Santana, E.A., Coyne, D.P., Beaver, J.S. and Zaiter, H.Z. (1993). Effect of photoperiod and temperature on common bacterial blight disease of common beans (*Phaseolus vulgaris* L.). *Euphytica* 66(3): 211-216.

Santos Filho, H.P., Ferraz S. and Sediyama, C.S. (1978). Effect of the time of inoculation of *Isariopsis griseola* Sacc. on three bean cultivars. *Fitopatol. Brasileira* 3(2): 175-180.

Sarotorato, A. (1988a). Angular leaf spot. Piracicaba Brazil, Asasociacao Brasileira para Pesquisa de Potassa e do Fosfato, pp. 491-501.

Sartarato, A., Zimmerman, M.J., Rocha, M. and Yamada, T. (1988). Web blight. In: *Cultura do Feijoero Factores que afetam a Productividade*, pp. 503-520.

Sastry, K.S.M., Singh, S.J. and Rajendran, R. (1981). Screening of different bean cultivars against bean common mosaic virus. *Pulse Crop Newsl.* 1: 30.

Schmidt, H.E. (1992). Bean Mosaics. In: *Plant Diseases* of *International Importance Vol II. Diseases of Vegetables and Oilseed Crops*. (Eds. H.S. Chaube, J. Kumar, A.N. Mukhopadhyay and U.S. Singh). Prentice Hall Inc., pp 40-73.

Schuster, M.L., Coyne, D.P. and Smith, C. (1979). New strains of halo blight bacterium in Nebraska. *Annu. Rep. Bean Improv. Coop.* 22: 19-20.

Schwartz, H.F. (1989). Halo blight. In: *Bean Production Problems in the Tropics*. 2nd ed. (Eds. H.F. Schwatz and M.A. Pastor-Corralles), CIAT, Cali, Colombia, pp. 285-301.

Shama, Saad and Shama, S. (1989). Transmission of *Rhizoctonia solani* Kuhn. in seeds of bean (*Phaseolus vulgaris* L.). *Curr. Sci.* 58: 972-974.

Sharma P.N., Sharma O.P. and Tyagi P.D. (1994). Status and distribution of bean anthracnose in Himachal Pradesh. *Himachal J. Agric. Res.* 20 (1 and 2): 91-96.

Sharma, A. K. (1998). Epidemiology and management of rust disease of French bean. *Veg. Sci.* 25(1): 85-88.

Sharma, K.D. and Kaushal, R.P. (1990). Studies on spore viability and perpetuation of *Mycovellosiella phaseoli* (Drummond) Deighton field. *Indian J. Plant Pathol.* 8(1 and 2): 28-32.

Sharma, K.D., Kaushal, R.P. and Singh, B.M. (1998). Effect of bioregulators on spore germination of *Mycovellosiella phaseoli* (Drummond) Deighton, a floury leaf spot pathogen of French bean. *Indian Phytopath.* 51(1): 43-46.

Sharma, M. and Gupta, S.K. (2003). Ecofriendly methods for the management of root rot and web blight (*Rhizoctonia solani*) of French bean. *J. Mycol. Pl. Pathol.* 33: 345-361.

Sharma, M., Gupta, S.K. and Sharma, T.R. (2005). Characterization of variability in *Rhizoctonia solani* by using morphological and molecular markers. *J. Phytopathol.* 153: 449-456.

Sharma, N., Gupta, S.K., Bharat, N.K. and Sharma, M. 2014. Evaluation of French bean germplasm against bean rust (*Uromyces appendiculatus*). *Pl. Dis. Res.* 29(1): 115-117.

Sharma, O.P. and Dhar, J. (1994). Studies on mosaic disease of French bean (*Phaseolus vulgaris* L.) in Himachal Pradesh. In: *Virology in the Tropics*. (Eds. N. Rishi; K.L. Ahuja and B.P. Singh), Malhotra Publishing House. New Delhi, pp. 319-325.

Sharma, P., Sharma, P.N., Kapil, R., Sharma, S.K. and Sharma, O.P. (2011). Analysis of 3′-terminal region of bean common mosaic virus strains infecting common bean in India. *Indian J. Virol.* 22(1): 37–43.

Sharma, P.N. and Sugha, S.K. (1995). Management of bean anthracnose through chemicals. *Indian Phytopath.* 48(3): 304-307.

Sharma, P.N., Kumar, A., Sharma, O.P., Sud, D. and Tyagi, P.D. (1999). Pathogenic variability in *Colletotrichum lindemuthianum* and evaluation of resistance in

Phaseolus vulgaris in the North-Western Himalyan region of India. *J. Phytopath.* 147(1): 41-45.

Sharma, P.N., Padder, B.A., Sharma, O.P., Pathania, A. and Sharma, P. (2007). Pathological and molecular diversity in *Colletotrichum lindemuthianum* (bean anthracnose) across Himachal Pradesh, a north-western Himalayan state of India. *Austral. Pl. Path.* 36: 191–197.

Sharma, P.N., Pathania A., Kapil, R., Sharma, P., Sharma, O.P., Patial, M. and Kapoor V. (2008). Resistance to bean common mosaic potyvirus strains and its inheritance in some Indian land races of common bean. *Euphytica* 164: 173–180.

Sharma, R.C., Barthe J.P., Lafitte, C. and Touze, A. (1986). *In vitro* absorption of endopolygalacturonase secreted by *Colletotrichum lindemuthianum* on cell walls of monogenically resistant and susceptible cultivars to anthracnose. *Can. J. Bot.* 64: 2903-2908.

Sharma, R.D. and Sohi, H.S. (1978). Evidence of a new strain of *Phaeoisariopsis griseola* Sacc. from India. *Curr. Sci.* 47: 348-349.

Sharma, S.R. (1977). Annual Report, IIHR, Bengalore, 1976.

Sharma, S.R. (1989). Yield loss caudsed by rust in French bean. *Pl. Dis. Res.* 4(1): 92- 94.

Sharma, S.R. and Sohi, H.S. (1989). Diseases of French bean in India and their control. *Indian Rev. Life Sci.* 9: 153-168.

Sharma, S.R. and Sohi, H.S. (1980). Assessment of losses in French bean caused by *Rhizoctonia solani* Kuhn. *Indian Phytopath.* 33: 366-369.

Sharma, S.R. and Sohi, H.S. (1981). Effect of different fungicides against Rhizoctonia root rot of French beans (*Phaseolus vulgaris* L.). *Indian J. Mycol. Pl. Pathol.* 11: 216-220.

Sharma, S.R. and Sohi, H.S. (1989). Diseases of French bean in India and their control. *Indian Rev. Life Sci.* 9: 153-168.

Shear, C.L. and Wood, A.K. (1913). Studies of fungus parasites belonging to the genus *Glomerella.U.S.D.A., Bureau Pl. Ind. Bull.* 252.

Sindhan G.S. and Bose S.K. (1981). Evaluation of fungicides against anthracnose of French bean caused by *Colletotrichum lindemuthianum. Indian Phytopath.* 34(3): 325-329.

Sindhan G.S. (1983). Effect of temperature and relative humidity on the development of anthracnose of French bean. *Prog. Hortic.* 15: 132-135.

Sindhan, G.S. and Bose, S.K. (1980). Effect of temperature and relative humidity on the development of angular leaf spot of French bean. *Prog. Hort.* 12: 18-24.

Sindhan, G.S. (1984). Evaluation of fungicides against occurrence of angular leaf spot of French bean caused by *Phaeoisariopsis griseola. Indian J. Plant Pathol.* 2(1): 39-42.

Singh, A.K. and Saini, S.S. (1980). Inheritance of resistance to angular leaf spot (*Isariopsis griseola* Sacc.) in French bean (*Phaseolus vulgaris* L.). *Euphytica* 29: 175-176.

Singh, B. M. and Sharma, Y.R. (1976). Screening of fungicides to control angular and floury leaf spots of beans. *Indian J. Mycol. Pl. Pathol.* 6: 148-151.

Singh, B.M. and Sharma, Y.R. (1975). Screening of bean lines for resistance to angular leaf spot caused by *Isariopsis griseola*. *Indian Phytopath.* 28: 435-436.

Singh, B.M., Agarwal, V.D. and Sharma, Y.R. (1974). Screening of bean cultivars to floury leaf spot caused by *Ramularia phaseoli*. *Indian Phytopath.* 27: 260-261.

Singh, J.P. and Musiyimi, A.B.K. (1984). Chemical control of bean rust in Kenya. *East African Agric. Forestry. J.* 45(3): 207-209.

Singh, J.P. and Musyimi, A.B.K. (1981). Effect of rust on bean yield. *Indian Phytopath.* 34(3): 378- 379.

Singh, J.P. and Musymi A.K.B. (1979). Control of rust and powdery mildews by a new systemic fungicide, Bayleton. *Pesticides* 13(4): 51-53.

Singh, R.Y., Tripathi, S.P. and Yadav, G.R. (1995). Evaluation of different fungicides against various leaf spot diseases of beans. *Recent Hortic.* 2: 111-112.

Singh, S.J.; Sastry, K.S. and Sastry, K.S.M. (1981). Field tests with insecticides and mineral oil for the protection of French beans from yellow mosaic virus disease. *Gartenbauwissenchaft* 46: 88-91.

Singh, S.P., Teran, H., Lema, M., Webster, D.M., Strausbaugh, C.A., Miklas, P.N., Schwartz, H.F. and Brick, M.A. (2007). Seventy-five years of breeding dry bean of the Western USA. *Crop Sci.* 47(3): 981-989.

Smolinska, U. and Kowalska, B. (2006). The effectivity of plant extracts and antagonistic microorganisms on the growth inhibition of French bean pathogenic fungi. *Veg. Crops Res. Bull.* 64: 67-76.

Sohi, H.S. and Sharma, R.D. (1974). Mode of survival of *Isariopsis griseola* Sacc. The causal agent of angular leaf spot of beans. *Indian J. Hortic.* 31: 110-113.

Sohi, H.S., Sharma, S.L., Verma, B.R. and Sachdeva, K.B. (1965). Floury leaf spot of bean caused by *Ramularia phaseolina* Pertak. *Indian Phytopath.* 18: 384.

Sokol, P.F., Gracia, R.M., Zimina, T.A. and Pivovarov, V.F. (1977). Some results of studies on the improvement of horticultural plants under Cuban conditions. *Inforem Cientifico- Tecnico Instituto de Investigaciones Fundamentales en Agricultura Tropical No.* 39: 15-19.

Srinivasan, K.V. (1953). *Isariopsis griseola* Sacc. on *Phaseolus vulgaris* L. *Curr. Sci.* 22: 20.

Srivastava, L.S. and Gupta, D.K. (1989). Aerial blight of French bean, groundnut, soyabean, black, green and horse gram - a new record from India. *Pl. Dis. Res.* 4: 163-164.

Srivastava, L.S. and Gupta, D.K. (1994). Management of angular leaf spot of 'rajma' through fungicides in Sikkim. *Indian Phytopath.* 47(2): 211-213.

Srivastava, L.S., Gupta, D.K., Dhiman, K.R. and Singh, G. (1995). Source of resistance in French bean (*Phaseolus vulgaris*) to angular leaf spot (*Phaeoisariopsis griseola*) in Sikkim. *Indian J. Agric. Sci.* 65(4): 305-307.

Srivsastava, L.S. and Gupta, D.K. (1989a). Farinose spot of French bean- a new disease. *Curr. Sci.* 58(14): 802.

Stall, R.E., Loschke, D.C. and Jones, J.B. (1986). Copper resistance and avirulence loci are linked to a self-transmissible plasmid in *Xanthomonas campestris. Phytopathology* 76: 240-243.

Starr, M.P. (1983). The genus *Xanthomonas*, In: *The Prokaryotes: A Handbook on Habitats, Isolation, and Identification of Bacteria*, (Ed. Starr, M.P.), Springer-Verlag, New York, 742 p.

Stavley, J.R., Steadman, J.R. and McMillan, R.T.Jr. (1989). New pathogenic variability in *Uromyces appendiculatus* in North America. *Plant Dis.* 73(5): 428-432.

Striab, W. (1943). Investigation on the biology and control of the bean rust (*Uromyces phaseoli* var. *typica* (Pers.) Wint.). *Gartenbavwirs* 17: 397-445.

Sud, A.K. and Sood, V.K. (2002). Screening of indigenous and exotic collections of kidney bean against bean anthracnose under dry temperate conditions of Himachal Pradesh. *Crop Res. Hisar* 24(1): 202-203.

Sugha, S.K., Sharma, O.P. and Sharma, P.N. (1998). Performance of 'vegetable type' French bean genotypes against floury leaf spot disease. *Pl. Dis. Res.* 13(1): 85-86.

Sumartini. (1998). Rust disease in French bean and its control. *J. Penelitian Pengembangan Pertanian* 17(4): 149-153.

Sutton, D.C. (1979). Systemic cross protection in bean against *Colletotrichum lindemuthianum*. *Aust. Pl. Pathol.* 8: 4.

Sutton, M.D. and Wallen, V.R. (1970). Epidemiological and ecological relations of *Xanthomonas phaseoli* var. *fuscans* in Southwestern Ontario, 1961-1968. *Can. J. Bot.* 48: 1329-1334.

Tamietti G and Garibaldi A. (1984). Eradication of *Pseudomonas syringae* pv. *phaseolicola* in bean seeds in Italy. In: *Phytopathology and Plant Bacterial Diseases of Quarentine Significance*, pp. 69-76.

Tan, C.S. and Tu, J.C. (1995). Tillage effects on root rot severity, growth and yield of beans. *Can. J. Pl. Sci.* 75: 183-186.

Taylor, J.D. and Teverson, D.M. (1996). Interactions between pseudomonads and *Phaseolus* beans and exploitation of genotype mixtures in subsistence agriculture. In: *Brighton Crop Prot. Conf. Pests Dis.-1996* Vol. 2.

Taylor, J.D. (1970). Bacteriophage and serological methods for identification of *Pseudomonas phaseolicola* (Burkh.) Dowson. *Ann. App. Biol.* 66: 387-395.

Taylor, J.D., Teverson, D.M. and Davis, J.H.C. (1987). Diseases of *Phaseolus* beans-biology, resistance and control. In: *Annu. Rep. Inst. Hortic. Res., Wellsbourne, Warwick*, England, pp. 63-64.

Thakur, R.S., Sugha, S.K. and Singh, B.M. (1991). Evaluation of systemic fungicides for control of *Rhizoctonia solani* under glass house conditions. *Indian J. Agric. Sci.* 61: 230-232.

Todorovic, B., Milijasevic, S., Rekanovic, E., Potocnik, I. and Stepanovic, M. (2008). Susceptibility of bean genotypes to *Xanthomonas campestris* pv. *phaseoli* in greenhouse conditions. *Pesticidi i Fitomedicina* 23(3): 167-173.

Trutmann O, Paul K.B. and Cishabayo (1992). Seed treatments increase yield of farmer varietal field bean mixtures in the Central African highlands through multiple disease and beanfly control. *Crop Prot.* 11(5): 458-464.

Trutmann, P. and Kayitare, E. (1991). Disease control and small multiplication plots improve seed quality and small farm dry bean yields in Central Africa. *J. App. Seed Prod.* 9: 36- 40.

Trutmann, P. and Pyndji, M.M. (1994). Partial replacement of local common bean mixtures by high yielding angular leaf spot resistant varieties to conserve local genetic diversity while increasing yield. *Ann. Appl. Biol.* 125(1): 45-52.

Tu, J.C. (1992). Management of root rot diseases of peas, beans and tomatoes. *Can. J. Pl. Path.* 14: 92-99.

Tu, J.C. and McNaughton, M.E. (1980). Isolation and characterization of benomyl resistant biotypes of the delta race of *Colletotrichum lindemuthianum Can. J. Pl. Sci.* 60: 585-592.

Tu, J.C. and Park, S.J. (1993). Root rot resistance in common bean. *Can. J. Pl. Sci.* 73: 365-367.

Tu, J.C. and Varrtaja, O. (1981). The effect of the hyperparasite (*Gliocladium virens*) on *Rhizoctonia solani* and *Rhizoctonia* root rot of the white beans. *Can. J. Bot.* 59: 22-27.

Tu, J.C. and Zheng, J. (1993). Compatibility of biocontrol agents with DCT. *Mededelingen van de Faculteit Landbouwwetenschappen Univrsiteit Gent.* 58: 1359-1364.

Tu, J.C. (1981). Anthracnose (*Colletotrichum lindemuthianum*) on white bean (*Phaseolus vulgaris*) in Southern Ontario: spread of the disease from an infection focus. *Plant Dis.* 5: 477-480.

Tu, J.C. (1983). Epidemiology of anthracnose caused by *Colletotrichum lindemuthianum* on white bean (*Phaseolus vulgaris*) in Southern Ontario: survival of the pathogen. *Plant Dis.* 67: 402-404.

Tu, J.C. (1999). Relationship between disease severity and yield loss of navy bean caused by bacterial blight (*Xanthomonas campestris* pv. *phaseoli*). In: 51st International Symposium on Crop Protection, Gent, Belgium, 4th May, 1999, Part II. *Mededelingen-Faculteit Landbouwkundige en Toegepaste Biologische Wetenschappen, Universiteit Gent* 64(3b): 637-642.

Tu, J.C., Sheppard, J.W. and Laidlaw, D.M. (1984). Occurrence and characterization of the epsilon race of bean anthracnose in Ontario. *Plant Dis.* 68: 69-70.

Upmanyu S. and Gupta, S.K. (2005). Influence of environmental factors on the progress of root rot and web blight (*Rhizoctonia solani*) of French bean. *Indian Phytopath.* 58(1): 79-83.

Upmanyu S. and Gupta, S.K. (2009). Physiological variation among French bean isolates of *Rhizoctonia solani*. *J. Pl. Dis. Sci.* 4(2): 160-163.

Upmanyu, S, Gupta, S.K. and Shyam, K.R. and Kaur, R. (2003). Patterns of variation among isolates of *Rhizoctoria Solani, Indian Phytopath.* 56: 308 (Abstr.).

Upmanyu, S, Gupta, S.K. and Shyam, K.R. (2002). Innovative approaches for the management of root rot and web blight (*Rhizoctonia solani*) of French bean. *J. Mycol. Pl. Pathol.* 32(3): 317-331.

Upmanyu, S., Gupta, S.K. and Sharma, M.K. (2004). Evaluation of French bean (*Phaseolus vulgaris*) germplasm against web blight (*Rhizoctonia solani*). *Indian J. Agric. Sci.* 74: 448-450.

Uppal, B.N., Patel, M.K. and Nigam, B.G. (1946). Bacterial blight of French bean in Bombay province. *Proc. Natl. Inst. Sci.*12: 351-359.

Verma, B.R. (1979). Studies on angular leaf spot of beans (*Phaseolus vulgaris* L.) caused by *Isariopsis griseola* Sacc. Ph.D Thesis, HPKVV, College of Agriculture, Solan. 221 p.

Vierra, R.F., Paula, T.J. Jr. and Berger, P.G. (1998). Effect of fungicides and cartap insecticides on the control of bean foliar diseases during the winter-spring season. *Summa Phytopathol.* 24(1): 17-22.

Villar- Sanchez, B. and Lopez-Salinas, E. (1993). Negro Inifap, a new var. of French bean for Chiapas state and similar tropical regions. *Revista Fitotecnia Mexicana* 16(2): 208-209.

Villegas, J.M.(1959). Variabilidad del *Isariopsis griseola* Sacc. agente caused de la mancha angular del frijol (*Phaseolus vulgaris* L.). Thesis Ing. Agr., Univesidad de Cldar, Mcnizales, Colombia, 61p.

Voland, R.P. and Epstein, A.H. (1994). Development of suppressiveness to diseases caused by *Rhizoctonia solani* in soils amended with composted and non-composted manure. *Plant Dis.* 78: 461-466.

Weller, D.M. and Saettler, A.W. (1980). Evaluation of seedborne *Xanthomonas phaseoli* and *Xanthomonas phaseoli* var. *fuscans* as primary inocula in bean blights. *Phytopathology* 70: 148.

Wimalajeewa, D.L.S. and Thavam P. (1973). Fungicidal control of bean rust disease. *Trop. Agriculturist* 129(1/2): 61-66.

Wolk, M. and Sarkar, S. (1994). Antagonism *in vivo* of *Bacillus* spp. against *Rhizoctonia solani* and *Pythium* spp. *Anzeiger fur Schadlingskunde pflanzenschutz Umweltschutz* 67: 1-5.

Wolk, M. and Sarkar, S. (1995). *In vivo* antagonism of fluorescent pseudomonads against *Rhizoctonia solani* and *Pythium aphanidermatum* on cucumbers and beans. *Anzeiger fur Schadlingskunde pflanzenschutz Umweltschutz* 68: 3-8.

Xin, F.G., Gouren, Z., Baodong, L., Jing, H.H. and Jine, Z. (2000). Microscopic and ultramicroscopic investigation into the infection process of snap bean rust (*Uromyces appendiculatus*). *Acta Phytopathologica Sinica* 30(1): 53-59.

Yaraguntaiah, R.C. and Nariani, T.K. (1963). Bean mosaic virus in India. *Indian J. Microbiol.* 3: 147-150.

Yoshii, K., Galvez, G.E. and Alvarez, G. (1976). Estimation of yield losses in beans caused by common blight. *Proc. Am. Phytopathol.* Soc. 3: 298-299.

Zambolin, L. and Chaves, G.M. (1974). Effect of low temperatures and of different combinations of temp. and RH on the viability of urediospores of *Hemileia vastatrix* Berk. and Br. and *Uromyces phaseoli* var. *typica* Arth. *Experientiae* 17(7): 151-184.

Zanatta, Z.G.C.N., Moura, A.B., Maia, L.C. and Santos, ASdos. (2007). Bioassay for selection of biocontroller bacteria against bean common blight (*Xanthomonas axonopodis* pv. *phaseoli*). *Brazilian J. Microbiol.* 38(3): 511-515.

Zapota, M.(1997). Identification of *Xanthomonas campestris* pv. *phaseoli* races in *Phaseolus vulgaris* leaves. *Agronomia Mesoamericana* 8(1): 44-52.

Zaumeyer, W.J. and Goth, R.W. (1964). A new severe symptom inducing strain of common bean mosaic virus. *Phytopathology* 54: 1378-1385.

Zaumeyer, W.J. and Thomas, H.R. (1957). A monographic study of bean diseases and methods for their control. *U.S.D.A. Tech. Bull. No.* 868: 255p.

Ziedan, E.H. and Mahamoud, S.Y.M. (2002). Calcium and sulfur soil treatment to improve biological control with *Trichoderma harzianum* for root rot disease control. *Assiut J. Agril. Sci.* 33(3): 149-160.

Chapter 8

Sclerotinia Stem Rot: A Serious Concern for Cultivation of Common Bean

Vibha Pandey

Common bean (*Phaseolus vulgaris* L.) rank third among important food grain legume crop and is cultivated in almost every part of the world (Singh, 1999 and Broughton *et al.*, 2003). Although, it is cultivated throughout the India but Himalayas hold the richest diversity of common bean. Its production is influenced by both biotic and abiotic stresses but biotic factors are responsible for major losses. Seven major diseases known to hamper the common bean production are *Sclerotinia* stem rot (SSR), anthracnose, rust, angular leaf spot, common bacterial blight and bean mosaic diseases. White mold, also known as *Sclerotinia* stem rot and *Sclerotinia* wilt, is caused by *Sclerotinia sclerotiorum* (Lib.) de Bary that belongs to an important family (*Sclerotiniacae*) of the class *Ascomycotina*. This fungus has wide hosts range and has a worldwide distribution on numerous field crops, vegetables and medicinal and aromatic plants (Perveen *et al.*, 2010). This disease occurs in almost every country of the world from the cool, moist regions to the hot, dry areas (Purdy, 1979). The incidence of disease in different fields ranged from a trace to 100 per cent and, this difference in incidence among fields were attributable to variations in precipitation, soil drainage, cultural practices, and sclerotial density in different localities (Tu, 1989a). This pathogen frequently causes serious and unpredictable yield losses in beans (dry and green) including *Phaseolus vulgaris* L. Symptoms of disease can be seen on leaves as water-soaked lesions that expand rapidly and move down the petiole into the stem. Stems of infected plants show two types of lesion *viz.*, dark stem lesions and water-soaked stem lesions. Lesions usually develop into necrotic tissues followed by development of patches of fluffy white mycelia growth which are the most obvious sign of plants infected with *S. sclerotiorum*. Plants may not appear to be affected by the disease during early lesion development. However, as the fungus progresses into the main stems, typical wilting symptoms occur. The presence of fungus on host can be easily be recognized by white fluffy mycelia growth on the surface of stem lesion. Infected tissues appear bleached and are often shredded, leaving only vascular tissues at later stage of development. Its fruiting bodies (sclerotia) are generally found to be associated with stem and pods and measure 0.15-0.5 cm (in diameter) x 0.3-3.8cm (in length). Sclerotia are typically formed inside infected tissue, often in the stem pith, but may form on the surface of tissues during high humidity conditions. Sclerotia commonly form on or in flowering and seed-producing portions of the plant and, therefore, are often found in harvest samples. Once the disease has been initiated in the host, infection can spread to adjacent plants through plant-to-plant contact. Besides, it decreases nutrient uptake causing rapid drying of leaves and reduction in seed yield.

The pathogen survives in the form of soil-hardy sclerotia in the soil for several years (Coley-Smith and Cooke, 1971), which makes the disease difficult to manage. The sclerotium germinates and produces apothecia that release large amount of ascospores, which can spread over a long distance and period and alight on plant canopies (Abawi and Grogan, 1979). New sclerotia develop within infected plants, leaving plant debris full of sclerotia at the end of the season. The disease cycle results in accumulation of sclerotia in the soil from season to season. *S. sclerotiorum* forms sclerotia, which can survive for several years in soil and germinate to produce apothecia (carpogenic germination) or mycelia (myceliogenic germination) under appropriate soil temperature and moisture conditions (Purdy, 1979). Both types of germination can result in the initiation of diseases (Morrall, 1977 and Huang and Kozub, 1993). Presence of oxalic acids, enzymes and other organic factor decides the virulence of pathogen and its establishment at early and subsequent stages. Out of two hypotheses (old and new) proposed for its infection and disease development; the new hypothesis proposed by Kabbage and co-workers (2015) have clearly arranged and presented the mechanism of switching of pathogen from a biotrophic to necrotrophic life style for disease establishment. Moreover, the regulation of defense mechanism by host through two innate immune systems to respond to pathogen infection, known as pathogen-associated molecular pattern (PAMPs)–triggered innate immunity and Effectors'-triggered immunity (Jones and Dangl, 2006) has also been discussed in this chapter.

Owing to inadequate levels of host resistance, application of fungicides became inevitable tool for controlling Sclerotinia diseases on most crops. Unfortunately, the extensive uses of few fungicides, populations of *S. sclerotiorum* have developed resistance to benzimidazole (carbendazim, MBC) and dicarboximide fungicides (dimethachlon). Hence, crop rotation with non host plant, maintenance of bioagents population in soil throughout the season along with judicious use of fungicides has been suggested as an effective method for the management of Sclerotinia stem rot in *Phaseolus vulgaris*.

Pathogen

S. sclerotiorum was first described in 1837 as *Peziza sclerotiorum* (Libert, 1837) and established under family Sclerotiniaceae by Whetzel (1945) of order Helotiales in the phylum Ascomycota. The Sclerotiniaceae includes species producing inoperculate asci from brownish, stipitate apothecia that arise from a sclerotial stroma within or associated with a host plant (Holst-Jensen *et al.*, 1997b and Whetzel, 1945). Additional criteria has also been included like characteristics of sterile tissues of both apothecia and sclerotia (Kohn, 1979a,b and Korf and Dumont, 1972); sclerotial ontogeny (Willetts and Wong, 1980); histochemistry and ultrastructure of sclerotia (Backhouse and Willetts, 1984); biochemical characteristics (Carbone and Kohn, 1993); and rRNA gene sequences (Holst-Jensen *et al.*, 1997a,b) to separate it into different genera. Currently, 33 genera are recognized (Willetts, 1997). *Sclerotinia sclerotiorum* is homothallic and exhibits clonality has been reported by Steadman and Boland (2005). However, contrast result of considerable genetic and phenotypic variability among 40 isolates of *S. sclerotiorum* was reported by Attanayake *et al.* (2012). They put-forth the evidence of outcrossing (heterothallism) within interbreeding subpopulations of *S. sclerotiorum*. Sclerotia are long-term survival and dissemination structures that can withstand adverse environmental conditions (Willetts, 1971) and are also known to play a crucial role in the life cycle of *S. sclerotiorum*, serving as the primary inoculum source during seasonal crop infection cycles by this pathogen (Bolton *et al.*, 2006). The sclerotium is a multiple-hyphal aggregate structure and consists of three distinct layers: a pigmented rind, a thin-walled cortex, and a large central medulla (Jones, 1970 and Colotelo, 1974). The biochemical composition of sclerotia has been shown to partially consist of carbohydrates, enzymes, free amino acids, and fatty acids (Cooke, 1969; Jones, 1970 and Colotelo *et al.*, 1971). The induction and development of sclerotia require specific metabolic and physiological processes. The initiation of sclerotial development is affected by several endogenous and exogenous factors including nutrients, light, temperature, pH, and aeration (Chet and Henis, 1975). Three distinct stages have been recognized during sclerotial development: (i) initiation, the formation of small and discrete initials through the aggregation of hyphae; (ii) development, increasing size along with the surface exudation of liquid droplets; and (iii) maturation, involving surface delineation and melanin deposition in peripheral rind cells followed by internal consolidation (Townsend and Willetts, 1954). Development and maturation of the sclerotia involves complex morphological and biochemical processes. These include the branching, interweaving and fusion of hyphae, translocation of nutrients and water, and the exudation of liquid droplets on the sclerotial surface (Chet and Henis, 1975 and Willetts and Bullock, 1992). Additionally, melanin is the darkly pigmented polymer whose deposition accompanies sclerotial development and maturation, and serves a critical role in fungal persistence and longevity by affording protection against adverse environmental conditions such as temperature extremes, desiccation, antagonistic or competitive microbes, toxic metals, and UV and visible irradiation (Bell and Wheeler, 1986). Fungal melanin is also involved in pathogenicity or virulence in some fungi, and is an essential requirement for appressorial formation and penetration of host plants (Butler and Day, 1998).

The survival of sclerotia is influenced by various factors, including structural and environmental factors such as moisture, aeration, and temperature (Coley-Smith and Cooke, 1971). Environmental conditions and crop canopies can lead to myceliogenic or carpogenic germination of sclerotia (Bardin and Huang, 2001). Myceliogenic germination results in the production of hyphae or mycelia that directly infect plant tissues (Huang, 1985). Carpogenic germination of sclerotia results in the formation of apothecia that release ascospores. Germination of ascospores under high moisture conditions, combined with the presence of senescent tissues that can serve as nutrient sources, leads to the infection of aboveground plant tissues (Tu, 1989a and Turkington and Morrall, 1993). Hyphae resulting from either germination type are hyaline, septate, branched and multinucleate with mycelium appearing white to tan in colour. Although, no asexual conidia are produced but microconidia are produced on hyphae or on the apothecial hymenium (Kohn, 1979b) that do not germinate.

Etiology and Epidemiology

This disease and its causal organisms have been subjects of intensive investigation. The etiology, biology, and epidemiology of white mold in beans have been studied extensively and summarized (Phillips, 1987; Purdy, 1979 and Roberts *et al.*, 1982). Sclerotia, the overwintering bodies of *S. sclerotiorum*, can germinate carpogenically or myceliogenically depending on environmental conditions and crops canopies. These two types of germination cause two distinct types of diseases. Sclerotia that germinate carpogenically produce apothecia and ascospores that infect aboveground tissues, resulting in stem blight, stalk rot, head rot, pod rot, white mold, and blossom blight of plants. The sclerotia serve as supporting structure for germination, resulting in the production of apothecia that produces million of airborne ascospores. This mode of infection by ascospores of *S. sclerotiorum* results in diseases on numerous crops including head rot of sunflower (Huang 1983), leaf blight of canola (Gugel and Morrall, 1986), head rot of safflower (Mündel *et al.*, 1985), pod rot of dry bean (Huang and Kemp 1989 and Boland and Hall 1987), stem and pod rot of soybean (Boland and Hall 1988a), blossom blight of alfalfa (Huang *et al.*, 2000), and lettuce crop (Mercier and Reeleder, 1987). Sclerotia that germinate myceliogenically produce mycelia that infect root tissues, causing crown rot and root rot of carrots (Finlayson *et al.*, 1989) and basal stalk rot and wilt of sunflower (Huang and Dueck, 1980) and Jerusalem artichoke (Huang and Stauffer, 1979). A direct relationship between myceliogenic germination of sclerotia and wilt of sunflower has been demonstrated (Huang and Dueck, 1980; Huang and Kozub, 1991b). Briefly, primary infections are initiated chiefly by ascospores, which germinate with the help of free moisture and exogenous energy sources of fallen flower petals and necrotic tissues (Steadman, 1979). Other means of primary infection are associated with ascospore contaminated leaves in contact with moist soil and leaves coming in contact with sclerotia on the soil surface (Tu, 1989). Secondary infections are achieved by natural contact of healthy plant parts with diseased ones. Later, many black resting bodies (sclerotia) of the fungus are formed on or in the infected stem tissues. The sclerotia eventually fall from the diseased plants or are incorporated into the soil with the infected plant debris, where they can survive for several years. Each year, sclerotia that are shallowly buried at a depth of less than 3 cm will germinate to produce apothecia, each containing millions of ascospores. The combined effect of light, temperature, and moisture are considered the most important factors conditioning carpogenic germination of scerotia and subsequent apothecial development in *S. sclerotium* (San and Yang, 2000).

The epidemiology of white mold is dependent on several factors: (a) soil inoculum, (b) soil moisture, (c) rainfall, (d) cultivar susceptibility, (e) row width, and (f) plant density. Fields with high inoculum density and high soil moisture have high disease incidence because disease initiation was favored by cool and damp soil conditions (Tu, 1997). In addition to the physiological mechanism of disease spread, *S.sclerotiorum* can spread between diseased and healthy plants by direct contact (Huang and Hoes, 1980), by transportation of infected pollen grains (Stelfox *et al.*, 1978), and through infected seed lots(Mueller *et al.*, 1999).Seedlings from infected seeds subsequently died from white mold at an early stage. Seeds that failed to germinate were rotted by *S. sclerotiorum*, and three to six sclerotia were formed in place of each seed (Tu, 1988). These sclerotia could become a source of inoculum. Boland and Hall (1988a, 1988b) found a quantitative relationship between number of apothecia in the field and the incidence of *Sclerotinia* diseases in soybean and bean. Ascospores require high moisture (Tu 1989a) and senescent tissues serve as nutrient sources for spores to germinate and infect aboveground plant tissues. Senescent petals of canola and bean plants (Turkington and Morrall 1993), stamens and styles of pea flowers (Huang and Kokko 1992), and pollen grains (Huang *et al.*, 1998b) have been cited as substrates for ascospore germination. It can also spread from field to field by internally infected seed, sclerotia mixed with seed, contaminated soil on farm equipment, and irrigation runoff water in addition to wind-blown ascospores (Steadman and Boland, 2005). Under suitable conditions of temperature, light, and moisture, sclerotia within 5 cm of the soil surface germinate to produce stipes and apothecia (Matheron and Porchas, 2005; Steadman and Boland, 2005). Ascospores are released from turgid asci, often simultaneously by "puffing." Ascospores germinate and colonize flowers and other tissues that are senescing, and the mycelium from colonized tissues invades adjacent host organs. Senescing flowers and flower parts often fall onto pods, leaves, branches, and stems and provide nutrients required by the fungus to develop appressorium to penetrate these organs. Infected tissues are rapidly killed and become dry and bleached. Limited spread of the fungus from one plant to another may occur by mycelia growth between tissues in contact. Sclerotia form in or on infected tissues and may fall to the soil, remain in crop debris, or be removed with harvested seeds or pods. The fungus may continue to develop and cause disease in green beans in transit and in storage (nesting) (Schwartz and Steadman, 1989 and Steadman and Boland, 2005).Infection by *S. sclerotiorum* of pollen grains of peas (Huang and Kokko 1993), canol (Huang *et al.*, 1998a), and alfalfa (Huang *et al.*, 1997) suggested that transport of pollen by wind or insects may be another means of dispersal for *Sclerotinia* diseases. Dormant mycelia of *S. sclerotiorum* in infected seeds can also be involved in the spread of the disease. The survival of dormant mycelia in infected seeds averaged 85 to 90 per cent, which did not change significantly over a 3-year period (Tu, 1988). When planted creating additional source of primary inoculums for the spread of the disease as these seeds failed to germinate but produced sclerotia.

Plant-pathogen Interactions

A major factor governing the pathogenic success of this fungus is the synthesis and secretion of oxalic acid, and *S. sclerotiorum* mutants defective in oxalate production causes limited lesions (Cessna *et al.*, 2000; Liang *et al.*, 2014). Oxalic acids (Cessna *et al.*, 2000; Kolkman and Kelly, 2002), hydrolytic enzymes (Yajima and Kav, 2006) and several other factors such as oleic acid and sclerin (Rollins, 2003; Harel *et al.*, 2006), are secreted virulence factors essential for *S. sclerotiorum* pathogenesis. The secretion of oxalic acid modifies the host cellular environment by interfering with host signaling, recognition and ultimately suppression of the host defense. In early stages of infection, *S. sclerotiorum* grow within the apoplastic space without crossing the plant cell wall, while host cell remain functional. During this time, the fungus secretes oxalic acid and other pathogenicity factors including potential effectors that modulate host cell defense response (Kabbage *et al.*, 2015). Effectors (small proteins) secreted by various phytopathogens, decides role in virulence in several biotrophic and hemi-biotrophic interactions. Effectors can be directly or indirectly recognized by resistance (R) proteins in the host, and activate a strong defense response, the effector-triggered immunity that often culminates in programmed cell death (PCD) of plant cells. The Oxalic acid contributes to pathogenesis by augmenting fungal colonization of host plants, the acidification of host tissues and sequestration of calcium from host cell walls, to facilitate the action of cell wall degrading enzymes. Full virulence of *S. sclerotiorum* is thought to be due to the synergistic activities of oxalic and endo-polygalacturonases (Rollins, 2003). These polygalacturonases (PGs) are encoded multigene families that grant adaptive flexibility to pathogens when infecting different host or plant organs (DeLorenzo *et al.*, 2001). Studies with oxalate deficient mutants have shown that this (oxalic acid) molecule is a key suppressor of plant defenses; thus allowing the fungus to evade recognition (William *et al.*, 2011). In the past few years, several additional functions associated with oxalate have been uncovered, including the induction of a PCD (Programme Cell Death) response in plant tissue that is required for disease development, adding to the continuing multi-functional nature of this "simple" organic acid (Dutton and Evans,1996) and (Kim *et al.*, 2008). In the absence of oxalate, the host plant is able to recognize the invading pathogen, and mount a defense response that includes a pronounced oxidative burst and callose deposition. Though models describing recognition and response to pathogens have mostly focused on plant interactions with biotrophs and hemi-biotrophs, it is evident that *S. sclerotiorum* is able to alter the plant recognition process at least in part, via the secretion of oxalate. The regulation of redox homeostasis plays a key role in this process; oxalate initially dampens the host oxidative response (as part of a potential biotrophic interaction) before triggering reactive oxygen species (ROS) at the later stages of infection, culminating in PCD (the advanced necrotrophic phase of the interaction) and disease (Williams *et al.*, 2011). *S. sclerotiorum*, via oxalate, has also been shown to induce apoptotic-like cell death in the host (Kim *et al.*, 2008)]. However, in its absence, the coordinated defense response against this pathogen involves autophagic cell death [Kabbage *et al.*, 2013], which has been suggested to be involved in the regulation of the plant hypersensitive response (Liu *et al.*, 2005). Therefore, the occurrence of PCD during *S. sclerotiorum* interactions with the host plant can confer both resistant and susceptible outcomes. The multi-functionality of oxalate including the suppression of defenses and autophagy, can allow the fungus valuable time for establishment of infection. This has been traditionally regarded as a feature associated with hemi-biotrophic growth.

In a new model (given below) suggested by Kabbage and co-worker (2015) propose that during the early stages of infection *S. sclerotiorum* grows within the apoplastic space without crossing the plant cell wall, while host cells remain functional. During this time, the fungus secretes oxalic acid and other pathogenicity factors including potential effectors that modulate host cell defense responses. The fungus then quickly switches to necrotrophic growth, while biotrophy is maintained at the leading edge of fungal colonization.

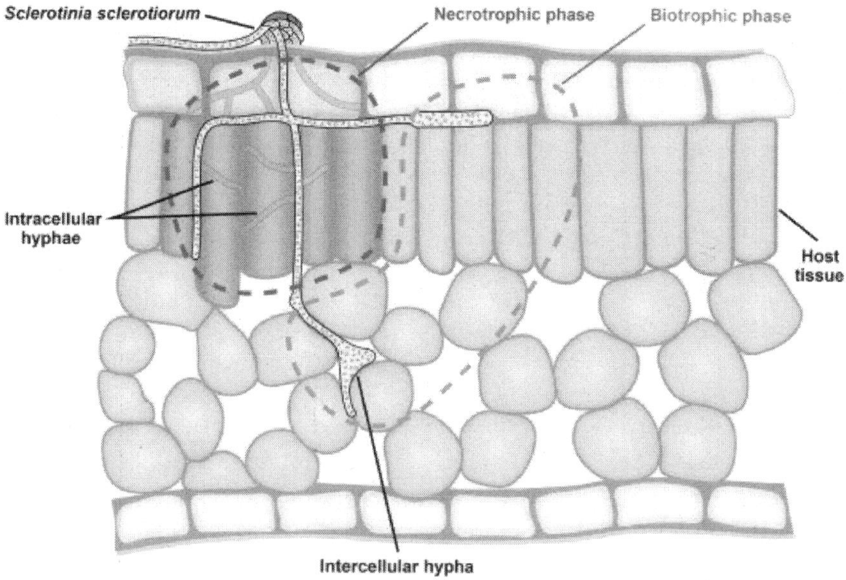

Figure 8.1. A New Model Depicting the Lifestyle Transition of *S. sclerotiorum*.

The *S. sclerotiorum* grows within the apoplastic space without crossing plant cell wall during the early stages of infection, while host cells remain alive. During this time, the fungus secretes oxalic acid and other pathogenicity factors that modulate host cell defense responses. The fungus then quickly switches to necrotrophic growth leaving a trail of dead cells, while biotrophy is maintained at the leading edge of fungal colonization (Kabbage *et al.*,2015). The two types of cell death observed are either host induced (autophagy) or are pathogen controlled (apoptosis-like), and occur during opposing pathogenic lifestyles. it is clear that not all cell deaths are equivalent, and though programmed cell death occurs in both situations, the outcome is predicated on the control of the cell death machinery and the type of cell death that is manifested. The types of cell death responsible for these opposing outcomes are clearly distinct (Kabbage *et al.*, 1913).

Plant Defense Systems

Abiotic and biotic stresses cause major losses in crop productivity worldwide. The stress responses of plants are regulated by multiple signaling pathways and there are significant overlaps between the patterns of gene expression that are induced in plants in response to different stresses. Plants are constantly exposed to microbes or many potential pathogens (*i.e.* bacteria, viruses, fungi, invertebrates, *etc.*). To be pathogenic, most microbes must access the plant interior, either by penetrating the leaf or root surface directly or by entering through wounds or natural openings such as stomata, pores in the underside of the leaf used for gas exchange. Therefore, plants need to continuously defend themselves against attack from these microbes and their ability to detect and mount a defense response to potential pathogenic microorganisms has been paramount to the evolution and developmental success of modern day plants (Chisholm *et al.*, 2006).

The plants have both innate and inducible/adaptive defence responses that are very different. These defence systems are generally effective in case the majority of fungi in the environment cannot cause disease. Plant cells have a thick, complex wall that acts as a barrier to invasion. Plants display inducible systemic acquired resistance, which occurs when previous exposure to a pathogen activates signaling pathways acting via molecules such as jasmonate, ethylene, and salicylic acid. These small and sometimes volatile molecules spread throughout the plant or even the plant population. This triggers responses such as the expression of "pathogenesis related proteins," including chitinases or glucanases, which can lead to the increased resistance of the whole plant against a subsequent pathogen attack. Apart from these, plants uses two innate immune systems to respond to pathogen infection, known as pathogen-associated molecular pattern (PAMPs)–triggered innate immunity and Effectors'-triggered immunity (Jones and Dangl, 2006). PAMP-triggered innate immunity is induced by recognition of PAMPs by trans-membrane receptors and gives early responses to pathogen infection. Effector-triggered immunity involves resistance (R) proteins as receptors that specifically recognize pathogen effectors, either directly or indirectly, and usually induces hypersensitive cell death (HR). The major (R) proteins are intracellular receptors and contain nucleotide binding site (NBS) and leucine-rich repeat (LRR) domains.

Pathogens use the many systems to deliver effector proteins that target multiple host proteins to suppress basal immune responses, allowing significant accumulation of pathogen in the plant apoplast. Plant resistance proteins (represented by CC-NB-LRR and TIR-NB-LRR) recognize effector activity and restore resistance through effector-triggered immune responses. Limited accumulation of pathogens occurs prior to effective initiation of effector-triggered immune responses (Chisholm *et al.*, 2006). Induction of cell death at the site of attempted attack by the pathogen is a typical feature of resistance in pant that is termed as hypersensitive response (HR). The HR is not due to the action of pathogen virulence factor that kill the plant cell, but rather appears to be earliest biochemical changes and resistance responses observed after pathogen recognition in the form of oxidative burst (*i.e.*, controlled release of O_2 and H_2O_2) at the site of pathogen attack. The oxidative brust is throught to be required for several subsequent defense responses and is expressed in most if not all plant species (Cessna *et al.*, 2000). Stress situation generate highly reactive oxygen species (ROS) such as hydroxyl radical, hydrogen peroxide, super oxide radicals and single oxygen that are known to cause several physiological disorder in plants (Scandalios, 1993). Several antioxidant enzymes like superoxide dismutase, catalase, glutathione, glutathione-S-transferase and different kind of

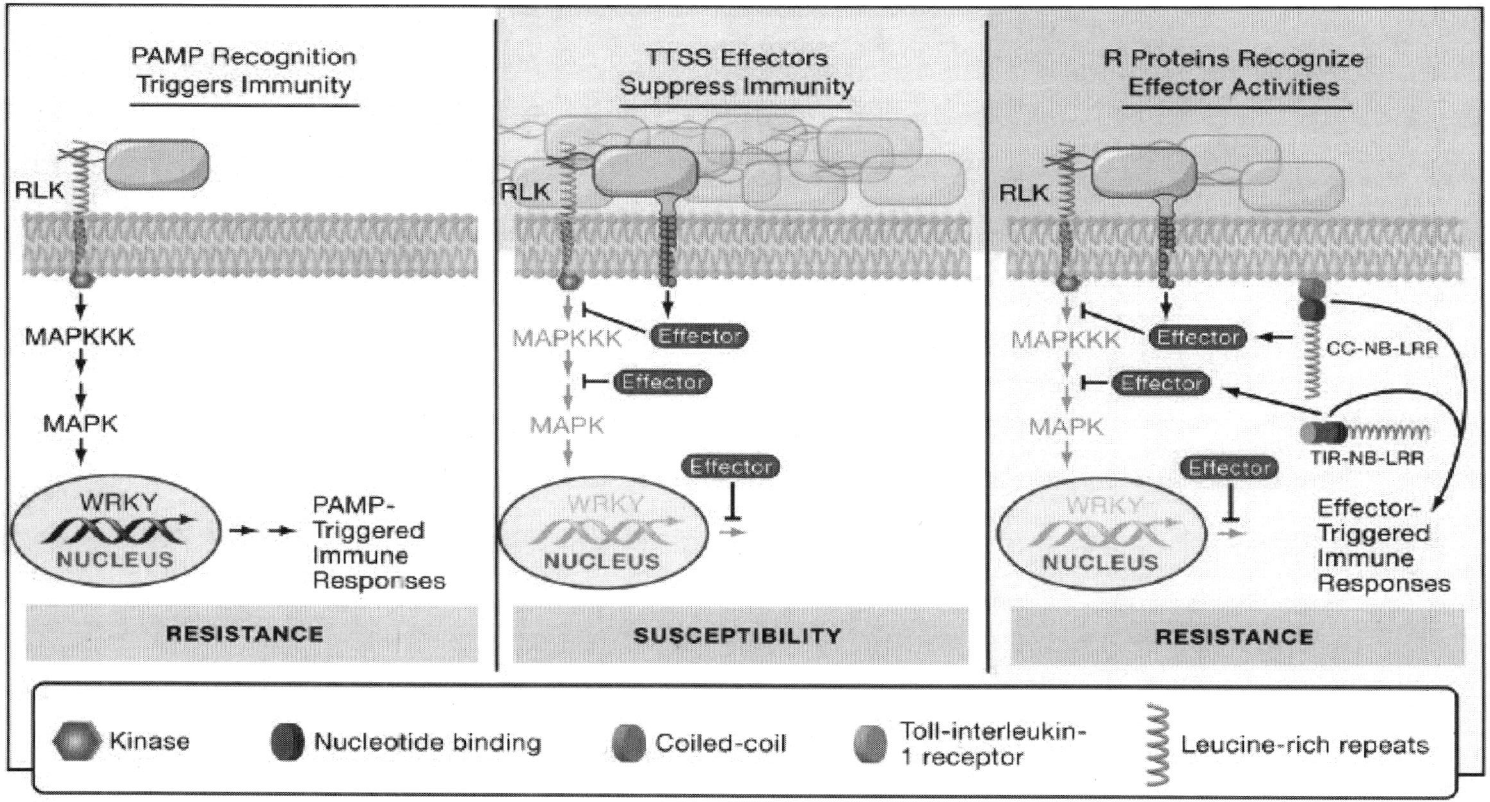

Figure 8.2: The Innate Immune System.

Left to right, Recognition of pathogen-associated molecular patterns by extracellular receptor-like kinases (RLKs) promptly triggers basal immunity, which requires signaling through MAP kinase cascades and transcriptional reprogramming mediated by plant WRKY transcription factors.

peroxidases (Px), as well as some non-enzymatic plant antioxidant systems such as glutathione, tocopherol, carotenoids, ascorbates, flavonoids and other phenolic compounds (Kang and Saltveit, 2002) have co-evolved with aerobic meta-bolisms to counteract oxidative damage from ROS. Susceptibility of plant to oxidative stress may be depend on the overall balance between factors that increase oxidant generation and those cellular components that exhibit an antioxidant capability (Sairam *et al.,* 2003).

Management

There are several disease-management strategies that focus on the elimination or reduction of the source of inoculum (the sclerotia) and; crop rotation using non-host crops to breaks the disease cycle to reduce the accumulation of sclerotia of *S. sclerotiorum* in the field can be effective option, as it will enhances temporal degradation of sclerotia by microorganisms in the soil. Wider plant spacing between and within rows can reduce disease spread by plant contact (Hoes and Huang 1985 and Tu and Zheng, 1997). Owing to long survival (up to eight years) of sclerotia of *S. sclerotiorum* and broad host range (Coley-Smith and Cooke, 1971), the one or two-year rotations are inefficient and unpractical method for this disease management. Therefore, scheduling the concomitant application of fungicide for a long period before planting makes the management more desirable. Direct disease-management strategies (through fungicide) by targeting either sclerotia or ascospores have been used to control Sclerotinia stem rot of *Phaseolus vulgaris*. Fungicides such as azoxystrobin, boscalid, thiophanate-methyl, iprodione and vinclozolin block various aspects of ascospore development and are moderately effective (Bradley *et al.,* 2006). Ascospores are released over a period encompassing up to about two weeks, Sclerotinia stem rot is difficult to manage using foliar application unless multiple fungicides applications are made (Bradley *et al.,* 2006; Mueller *et al.,* 2002). As Captan and thiophanate-methyl used in seed treatment were 100 per cent effective in eradicating the fungus from the infected seeds (Tu, 1989 a).Although soil fumigation can eliminate sclerotia in soil but it is not cost effective and may not meet the expectations of environmental stewardship (Budge and Whipps, 1991).

Indirect control of Sclerotinia stem rot (SSR) of *Phaseolus vulgaris* through the impacting the survival of sclerotia of *S. sclerotiorum* would be a good option, since the sclerotia are the source of apothecia and ascospores, are relatively static in the soil. These sclerotia can be easily targeted by BCA due to their large population size (especially in affected soils), position in the upper soil profile in minimally cultivated soils and relative abundance to deleterious pathogen in soils. Efficacy of BCAs is affected by disease pressure, soil temperature, moisture, resident microbial diversity and other soil conditions; that decide its (BCA) germination and colonization.

Biological control agents (BCAs) derived from antagonistic fungi and bacteria have effectively reduced sclerotia of *S. sclerotiorum* in many cropping systems such as lettuce (Budge and Whipps, 1991 and Chitrampalam *et al.,* 2008), soybean (del Rio *et al.,* 2002), dry bean (Huang *et al.,* 2000), and celery (Budge and Whipps, 1991) and it is therefore feasible to manage Sclerotinia diseases using BCA. *Coniothyrium minitans* Campbell is a mycoparasite of *S. sclerotiorum* (Campbell, 1947) and related species of *Sclerotinia* (Whipps and Gerlagh, 1992). Numerous reports indicated that *C. minitans* is a promising biocontrol agent of Sclerotinia diseases (Whipps and Gerlagh, 1992, de Vrije *et al.*, 2001and Whipps *et al.*, 2008). Use of *Coniothyrium minitans* (BCAs) for reduction of sclerotial density of *S. sclerotiorum* in soil has been studied under greenhouse (Zeng *et al.*, 2012) or field conditions on bean (Bremer *et al.*, 2000 and Gerlagh *et al.*, 2003), lettuce (Chitrampalam *et al.*, 2008) and celery (Budge and Whipps, 1991) and found to be resulted in reduction of disease incidence in some years. Although, *T. harzianum* has also been tested against the pathogen and found effective in reducing the disease in soybean (Menendez and Godeas, 1998) and celery (Budge and Whipps, 1991) but with less efficacy compared to *C. minitans*. Soil application of *C. minitans* is recognized as an effective and convenient approach for suppression or elimination of carpogenic and myceliogenic germination of *S. sclerotiorum* sclerotia (de Vrije *et al.*, 2001; Jones and Whipps, 2002; Huang and Erickson, 2008 and Yang *et al.*, 2009), thereby achieving biocontrol effect on *Sclerotinia* diseases. Commercial products of *C. minitans* including Contans® WG (Prophyta Biologischer Pflanzenschutz GmbH, Germany) and KONI® (Biovéd Biological Plant Protection Product Producing Co., Kemestaródfa, Hungry) have been developed and registered for control of *Sclerotinia* diseases in several European countries such as Germany, Switzerland, Norway and Hungary, and in the USA (de Vrije *et al.*, 2001). In comparison to the fungal biocontrol agents (*e.g., C. minitans* and *T. harzianum*.), the bacteria (*e.g., B. subtilis* and *S. lydicus*) have different modes of action against *S. sclerotiorum*. It has been reported that the population of *C. minitans* decreased gradually over the season, and it may continue to decrease through the winter (Gerlagh and Vos, 1991). Therefore, maintenance of high BCA populations is the key for effective control of sclerotia, because sclerotial reduction is highly related to the rates of BCA application (Zeng *et al.*, 2012). Applications of BCAs prior to tillage are preferred because tillage can enhance the even distribution of BCAs in soil and help direct contact between BCAs and pathogen. Early application at fall or at spring, benefits the establishment and growth of biocontrol strains in soil (Gerlagh and Vos, 1991). Despite local reduction in inoculums, crops remain at risk and some measure of integrated crop protection may be warranted; as external sources of inoculum (ascospores) can impact Sclerotinia stem rot development in any susceptible crop (Hammond *et al.*, 2008). Besides *bioagents*, some organic soil amendments and fertilizers have been reported to be able to suppress carpogenic germination of sclerotia of *S. sclerotiorum* when applied to soil. The reported organic soil amendments include seed meal of oilseed rape (*B. napus* cv. Dwarf Essex) (Dandurand *et al.*, 2000), S–H mixture (Huang and Sun, 1991) and aqueous extracts of barley, canola, lentil, oat, rye, sweet clover or wheat (Huang *et al.*, 2007). The reported suppressive fertilizers include urea (Huang and Janzen, 1991) and calcium cyanamide (Huang and Sun, 1991 and Huang *et al.*, 2006). Huang and Janzen (1991) reported that suppression of carpogenic germination of *S. sclerotiorum* sclerotia by urea is due to the release of volatile substances like ammonia. This mechanism may exist in suppression of the carpogenic germination of *S. sclerotiorum* sclerotia by the N–P–K compound fertilizer used in this study, as the compound fertilizer used in these studies contains ammonia hydrogen phosphate, which can be transformed into ammonia. Bennett *et al.* (2005) indicated that combinations of soil pasteurization and soil amendment with *C. minitans* enhanced infection of *S. sclerotiorum* sclerotia by *C. minitans*.

Conclusion

As discussed in this chapter, the fungus has wide hosts range and of worldwide distribution ranging from the cool, moist regions to the hot, dries areas. The pathogen survives in the form of soil-hardy sclerotia in the soil for several years. As disease initiation is favored by cool and damp soil conditions so fields with high inoculum density and high soil moisture make the environment conducive for high chances of disease incidence. A major factor governing the pathogenic success of this fungus is the synthesis and secretion of oxalic acid that modifies the host cellular environment by interfering with host signaling, recognition and ultimately suppression of the host defense. Owing to long survival of sclerotia of *S. sclerotiorum* and broad host range, crop rotation with non host plant, maintenance of a stable population of bioagents throughout the season along with scheduling the concomitant pre-planting application of fungicide with currently labeled rates and instructions for suppression of *Sclerotinia* stem rot in *Phaseolus vulgaris* appears to be critical for successful management of disease. Compared with major technological, environmental, and socioeconomic changes affecting agricultural production during next century, climate change may be more important; it will however, add another layer of complexity and uncertainty onto a system that is already exceeding difficulty to manage on a sustainable basis. Research on climate change and its interaction with pathogenically different isolates from different geographical regions of *S. sclerotiorum* could result in improved understanding and management of pathogen in face of current and future climate extremes.

REFERENCES

Abawi, G.S. and Grogan, R.G. (1979). Epidemiology of diseases caused by Sclerotinia species. *Phytopathology*, 69: 899–904.

Attanayake, R.N., L. Porter, D.A. Johnson, and W. Chen. (2012). Genetic and phenotypic diversity and random association of DNA markers of isolates of the fungal plant pathogen *Sclerotinia sclerotiorum* from soil on a fine geographical scale. *Soil Biology and Biochemistry* 55: 28–36.

Backhouse, D. and Willetts, H.J. (1984). A histochemical study of *Botrytis cinerea* and *Botrytis fabae*. *Canadian Journal of Microbiology*, 30: 171–178.

Bardin, S.D. and Huang, H.C. (2001). Research on biology and control of *Sclerotinia* diseases in Canada. *Canadian Journal of Plant Pathology* 23; 88–98.

Bell, A.A. and Wheeler, M.H. (1986). Biosynthesis and functions of fungal melanins. Annual Review of Phytopathology, 24: 411–451.

Bennett A.J., Leifert C., Whipps J.M. (2005). Effect of combined treatment of pasteurization and *Coniothyrium minitans* on sclerotia of *Sclerotinia sclerotiorum* in soil. *European Journal of Plant Pathology*, 113 : 197–209.

Boland, G.J., and Hall, R. (1987). Epidemiology of white mold of white bean in Ontario. *Canadian Journal of Plant Pathology*, 9: 218–224.

Boland, G.J., and Hall, R. (1988a). Relationships between the spatial pattern and number of apothecia of *Sclerotinia sclerotiorum* and stem rot of soybean. *Plant Pathology*, 37: 329–336.

Boland, G.J., and Hall, R. (1988b). Numbers and distribution of apothecia of *Sclerotinia sclerotiorum* in relation to white mold of white bean (*Phaseolus vulgaris*). *Canadian Journal of Botany* 66: 247–252.

Bolton, M.D., Thomma, B.P.H.J., Nelson, B.D. (2006). *Sclerotinia sclerotiorum* (Lib.) de Bary: biology and molecular traits of a cosmopolitan pathogen. *Molecular Plant Pathology*, 7: 1–16.

Bradley, C.A., Lamey, H.A., Endres, G.J., Henson, R.A., Hanson, B.K., Mckay, K.R., Halvorson, M., LeGare, D.G. and Porter, P.M. (2006). Efficacy of fungicides for control of Sclerotinia stem rot of canola. *Plant Disease*, 90 : 1129–1134.

Bremer, E., Huang, H.C., Selinger, L.J. and Davis J.S. (2000).Competence of *Coniothyrium minitans* in preventing infection of bean leaves by*Sclerotinia sclerotiorum*. *Plant Pathology Bulletin*, 9: 69–74.

Broughton, W.I., Hernandez, G., Blain, M., Beebe, S., Gepts, P., Vanderleyden, J. (2003). Beans (Phaseolus spp.) model food legumes. *Plant and Soil*, 252: 55-128.

Budge, S.P. and Whipps, J.M. (1991). Glasshouse trials of *Coniothyrium minitans* and *Trichoderma* species for the biological control of *Sclerotinia sclerotiorum* in celery and lettuce. *Plant Pathology*, 40: 59–66.

Budge, S.P. and Whipps, J.M. (2001). Potential for integrated control of *Sclerotinia sclerotiorum* in glasshouse lettuce using *Coniothyrium minitans* and reduced fungicide application. *Phytopathology*, 91: 221–227.

Butler, M.J. and Day, A.W. (1998). Fungal melanins: a review. *Canadian Journal of Microbiology*, 44: 1115–1136.

Campbell, W.A. (1947). A new species of *Coniothyrium* parasitic on sclerotia. *Mycologia*, 39: 190–195.

Carbone, I. and Kohn, L.M. (1993). Ribosomal DNA sequence divergence within internal transcribed spacer I of the Sclerotiniaceae. *Mycologia*, 85: 415–427.

Cessna, S.G., Sears, V.E., Dickman, M.B., Low, P.S. (2000). Oxalic acid, a pathogenicity factor for *Sclerotinia sclerotiorum*, suppresses the oxidative burst of the host plant. *Plant Cell*,12: 2191–2200.

Chet I., Henis, Y. (1975). Sclerotial morphogenesis in Fungi. *Annual Review of Phytopathology*, 13: 169–192.

Chisholm, S. T., Coaker,G., Day, B. and Staskawicz, B. J. (2006). Host-Microbe Interactions: Shaping the Evolution of the Plant Immune Response. *Cell:* 803–814.

Chitrampalam, P. Figuili, P.J. Matheron, M.E. Subbarao, K.V. and Pryor B.M. (2008). Biocontrol of lettuce drop caused by *Sclerotinia sclerotiorum* and *S. minor* in desert agroecosystems. *Plant Disease*, 92: 1625–1634.

Coley-Smith, J.R., Cooke, R.C. (1971). Survival and germination of fungal sclerotia. *Annual Review of Phytopathology*, 9 : 65–92

Colotelo, N. (1974). A scanning electron microscope study of developing sclerotia of *Sclerotinia sclerotiorum*. *Canadian Journal of Botany*, 52: 1127–1130.

Colotelo, N., Sumner, J.L., Voegelin, W.S. (1971). Chemical studies on the exudates and developing sclerotia of *Sclerotinia sclerotiorum* (Lib.) DeBary. *Canadian Journal of Microbiology*, 17: 1189–1194

Cooke, R.C. (1969). Changes in soluble carbohydrates during sclerotium formation by *Sclerotinia sclerotiorum* and *S. Trifoliorum*. *Transactions of the British Mycological Society*, 53: 77–86.

Dandurand, L.M, Mosher, R.D. and Knudsen, G.R. (2000). Combined effects of *Brassica napus* seed meal and *Trichoderma harzianum* on two soilborne plant pathogens. *Canadian Journal of Microbiology*, 46: 1051–1057.

De Lorenzo, G., D'Ovidio,R. and Cervone, F. (2001). The role of polygalacturonase-inhibiting proteins (PGIPs) in defense against pathogenic fungi. *Annual Review of Phytopathology*, 39: 313-335.

de Vrije, T., Antoine, N., Buitelaar, R.M., Bruckner, S., Dissevelt, M., Durand, A., Gerlagh, M., Jones, E.E., Lüth, P., Oostra, J.,Ravensberg, W.J., Renaud, R., Rinzema, A., Weber, F.J., Whipps J.M. (2001).The fungal biocontrol agent *Coniothyrium minitans*: production by solid-state fermentation, application and marketing. *Applied Microbiology and Biotechnology*, 56: 58–68.

del Rio L.E., Martinson C.A. and Yang X.B. (2002). Biological control of Sclerotinia stem rot of soybean with *Sporidesmium sclerotivorum*. *Plant Disease*, 86: 999–1004.

Dutton, M.V.and Evans, C.S. (1996). Oxalate production by fungi: its role in pathogenicity and ecology in the soil environment. *Canadian Journal of Microbiology*, 42: 881–895.

Finlayson, J.E., Rimmer, S.R., and Pritchard, M.K. (1989). Infection of carrots by *Sclerotinia sclerotiorum*. Canadian Journal of Plant Pathology,11: 242–246.

Gerlagh, M. and Vos, I. (1991). Enrichment of soil with Sclerotia to isolate antagonists of *Sclerotinia Sclerotiorum*. Developments in Agricultural and Managed Forest Ecology, 23: 165-171.

Gerlagh, M., Goossen-van de Geijn, H.M., Hoogland, A.E. and Vereijken P.F.G. (2003). Quantitative aspects of infection of *Sclerotinia sclerotiorum* sclerotia by *Coniothyrium minitans* – timing of application, concentration and quality of conidial suspension of the mycoparasite. *European Journal of Plant Pathology*, 109: 489–502.

Gugel, R.K., and Morrall, R.A.A. (1986). Inoculum-disease relationships in Sclerotinia stem rot of rapeseed in Saskatchewan. *Canadian Journal of Plant Pathology*, 8: 89–96.

Hammond, C.N., Cummings, T.F., Johnson, D.A. (2008). Deposition of ascospores of *Sclerotinia sclerotiorum* in and near potato fields and the potential to impact efficacy of a biocontrol agent in the Columbia Basin. *American Journal of Potato Research*, 85: 353–360.

Harel A., Bercovich S. and Yarden O. (2006). Calcineurin is required for the sclerotial development and pathogenicity of *Sclerotinia sclerotiorum* in an oxalic acid independent manner. *Molecular Plant-Microbe Interactions*, 16: 785-795.

Hoes JA and Hung HC. (1985). Effect of between row and within row spacing on development of Sclerotinia wilt and yield of sunflower. *Canadian Journal of Plant Pathology*, 7: 98-102.

Holley, J., Linowski, R., Gossen, B., and Harrison, L. (1995). *Sclerotinia sclerotiorum* causes blossom blight of alfalfa. *Canadian Journal of Plant Pathology*, 18: 489 (Abstr.).

Holst-Jensen, A., Kohn, L.M. and Schumacher, T. (1997b). Nuclear rDNA phylogeny of the Sclerotiniaceae. *Mycologia*, 89: 885–899.

Holst-Jensen, A., Kohn, L.M., Jakobsen, K.S. and Schumacher, T. (1997a). Molecular phylogeny and evolution of Monilinia (Sclerotiniaceae) based on coding and non-coding rDNA sequences. *American Journal of Botany*, 84: 686–701.

Huang H.C. and Kozub, G.C. (1993). Influence of inoculum production temperature on carpogenic germination of sclerotia of *Sclerotinia sclerotiorum. Canadian Journal of Microbiology*, 39; 548–550.

Huang H.C., Erickson R.S., Phillippe L.M., Mueller C.A., Sun S.K., Huang, J.W. (2006). Control of apothecia of *Sclerotinia sclerotiorum* by soil amendment with S-H mixture or Perlka® in bean, canola and wheat fields. *Soil Biology and Biochemistry*, 38: 1348–1352.

Huang, H.C. (1980). Control of *Sclerotinia* wilt of sunflower by hyperparasites. *Canadian Journal of Plant Pathology*, 2: 26–32.

Huang, H.C. (1983). Sclerotinia wilt and head rot of sunflower. Agriculture Canada, Ont., *Canadex* 632.145.

Huang, H.C. (1985). Factors affecting myceliogenic germination of sclerotia of *Sclerotinia sclerotiorum. Phytopathology*, 75: 433–437.

Huang, H.C. and Dueck, J. (1980). Wilt of sunflower from infection by mycelial-germinating sclerotia of *Sclerotinia sclerotiorum. Canadian Journal of Plant Pathology* 2: 47–52.

Huang, H.C. and Erickson, R.S. (2008). Factors affecting biological control of *Sclerotinia sclerotiorum* by fungal antagonists. *Journal of Phytopathology*, 156: 628–634.

Huang, H.C. and Hoes, J.A. (1980). Importance of plant spacing and sclerotial position to development of Sclerotinia wilt of sunflower. *Plant Dis.* 64: 81–84.

Huang, H.C. and Janzen, H.H. (1991). Control of carpogenic germination of sclerotia of *Sclerotinia sclerotiorum* by volatile substances from urea. *Plant Protection Bulletin* (Taipei), 33: 283–289.

Huang, H.C. and Kemp, G.A. (1989). Growth habit of dry beans (*Phaseolus vulgaris*) and incidence of white mold (*Sclerotinia sclerotiorum*). *Plant Prot. Bull.* 31: 304–309.

Huang, H.C. and Kokko, *E.G.* (1992). Pod rot of dry peas due to infection by ascospores of *Sclerotinia sclerotiorum*. *Plant Disease*, 76: 597–600.

Huang, H.C. and Kokko, *E.G.* (1993). Infection of pea pollen by *Sclerotinia sclerotiorum*. Plant Pathology (*Trends in Agriculture Sciences*), 1: 13–17.

Huang, H.C. and Stauffer, M.D. (1979). Sclerotinia wilt on Jerusalem artichoke. Agriculture Canada, Ont., Canadex 164.630.

Huang, H.C. and Sun S.K. (1991). Effect of S-H mixture or Perlka™ on carpogenic germination and survival of sclerotia of*Sclerotinia sclerotiorum*. *Soil Biology and Biochemistry*, 23: 809–813.

Huang, H.C. Erickson, R.S. and Moyer. J.R. (2007). Effect of crop extracts on carpogenic germination of sclerotia, germination of ascospores and lesion development of *Sclerotinia sclerotiorum*.*Allelopathy Journal*, 20: 269–278.

Huang, H.C., Acharya, S.N. and Erickson, R.S. (2000). Ethiology of alfalfa blossom blight caused by *Sclerotinia sclerotiorum* and *Botrytis cinerea*. *Plant Pathology Bulletin*, 9: 11–16.

Huang, H.C., and Kozub, G.C. (1991). Monocropping to sunflower and decline of Sclerotinia wilt. *Bot. Bull. Acad. Sin.* (Taipei), 32: 163–170.

Huang, H.C., Kokko, *E.G.*, and Erickson, R.S. (1997). Infection of alfalfa pollen by *Sclerotinia sclerotiorum*. *Phytoparasitica*, 25: 17–24.

Huang, H.C., Kokko, *E.G.*, Erickson, R.S., and Hynes, R.K. (1998a). Infection of canola pollen by *Sclerotinia sclerotiorum*. *Plant Pathology Bulletin*, 7: 71–77.

Huang, H.C., Kokko, *E.G.*, and Huang, J.W. (1998b). Epidemiological significance of pollen in fungal diseases. *Recent Research Development in Plant Pathology*, 2: 91–109.

Jones, D. (1970). Ultrastructure and composition of the cell walls of *Sclerotinia sclerotiorum*. *Transactions of the British Mycological Society*, 54: 351–360.

Jones, J. D. G. and Dangl, J. G. (2006). The plant immune system. *Nature*, 444: 323- 329.

Jones, E.E. and Whipps, J.M. (2002). Effect of inoculum rates and sources of *Coniothyrium minitans* on control of *Sclerotinia sclerotiorum* disease in glasshouse lettuce. *European Journal of Plant Pathology*, 108: 527–538.

Kabbage, M, Yarden, O. M., Dickman, B. (2015). Pathogenic attributes of *Sclerotinia sclerotiorum*: Switching from a biotrophic to necrotrophic lifestyle. *Plant Science*, 233: 53-60.

Kabbage, M., Williams, B., Dickman M.B. (2013). Cell death control: the interplay of apoptosis and autophagy in the pathogenicity of *Sclerotinia sclerotiorum*. PLoS Pathog.; p. e1003287.

Kang HM and Salveit M. (2002). Reduced chilling tolerance in elongating cucumber seedling radicles is related their reduced antioxidant enzymes and DPPH radical scavenging activity. *Physiol. Plant* 115: 244-250.

Kim, K.S., Min, J.Y., Dickman, M.B. (2008). Oxalic acid is an elicitor of plant programmed cell death during *Sclerotinia sclerotiorum*disease development. *Mol. Plant Microbe Interact.*, 21; 605–612.

Kohn, L.M. (1979a). Delimitation of the economically important plant pathogenic *Sclerotinia* species. *Phytopathology*, 69, 881–886.

Kohn, L.M. (1979b). A monographic revision of the genus Sclerotinia. Mycotaxon, 9, 365–444.

Kolkman, J.M. and Kelly, J.D. (2002). Agronomic traits affecting resistance to white mold in common bean. *Crop Science*, 42: 693–699.

Korf, R.P. and Dumont, K.P. (1972). Whetzelinia, a new generic name for Sclerotinia sclerotiorum and S. tuberosa. *Mycologia*, 64, 248–251.

Liang, X., Daniele, L., Li, M., Kim, Y.T., Hutchens, A., Wilson R. and Rollins. J. A. (2014). Oxaloacetate acetylhydrolase gene mutants of *Sclerotinia sclerotiorum* do not accumulate oxalic acid, but do produce limited lesions on host plants. *Mol. Plant Pathol.* http: / /dx.doi.org/10.1111/mpp.12211.

Libert, M.A. (1837). Plante crytogamicae arduennae (Exsiccati) no. 326. Published by the author.

Liu, Y., Schiff, M., Czymmek, K., Tallóczy, Z., Levine, B., Dinesh-Kumar, S.P. (2005). Autophagy regulates programmed cell death during the plant innate immune response. *Cell*, 121; 567–577.

Matheron, M.E., and M. Porchas. (2005). Influence of soil temperature and moisture on eruptive germination and viability of sclerotia of *Sclerotinia minor* and *S. sclerotiorum*. *Plant Dis.* 89: 50–54.

Menendez A.B. and Godeas, A. (1998). Biological control of *Sclerotinia sclerotiorum* attacking soybean plants. Degradation of the cell walls of this pathogen by *Trichoderma harzianum* (BAFC 742) – Biological control of Sclerotinia sclerotiorum by *Trichoderma harzianum*. *Mycopathologia*, 142 : 153–160.

Mercier, J., and Reeleder, R.D. (1987). Interactions between *Sclerotinia sclerotiorum* and other fungi on the phylloplane of lettuce. *Can. J. Bot.* 65: 1633–1637.

Morrall, R.A.A. (1977). A preliminary study of the influence of water potential on sclerotium germination in *Sclerotinia sclerotiorum*. *Canadian Journal of Botany*, 55; 8–11

Mueller D.S., Dorrance A.E., Derksen R.C., Ozkan E. Kurle J.E, Grau C.R., Gaska J.M., Hartman G.L., Bradley C.A. and Pedersen W.L. (2002). Efficacy of fungicides on *Sclerotinia sclerotiorum* and their potential for control of Sclerotinia stem rot on soybean. *Plant Disease*, 86 : 26–31.

Mueller D.S., Hartman, G.L and Pedersen W. L. (1999). Development of sclerotia and apothecia of *Sclerotinia sclerotiorum* from infected soybean seed and its control by fungicide seed treatment. *Plant Disease* 83: 1113-1115.

Mündel, H.H., Huang, H.C., and Kozub, G.C. (1985). *Sclerotinia* head rot in safflower: assessment of resistance and effects on yield and oil content. *Can. J. Plant Sci.* 65: 259–265.

Parveen K, Haseeb A. and Shukla P K. (2010). Effect of *Sclerotinia sclerotiorum* on the disease development, growth, oil yield and biochemical changes in plants of *Mentha arvensis*. *Saudi Journal of Biological Sciences* 17: 291-294

Phillips, A. J. L. (1987). Carpogenic germination of sclerotinia of *Sclerotinia sclerotiorum*: A review. *Phytophylactica* 19: 279-283.

Purdy, L.H. (1979). Sclerotinia sclerotiorum: history, diseases and symptomatology, host range, geographic distribution, and impact. *Phytopathology*, 69: 875–880.

Roberts, M. E., Dickson, M. H. and Hunter, J. E. (1982). Heritability of white mold resistance. *Annu. Rep. Bean Improv. Coop*. 25: 104.

Rollins J A. (2003). The *Sclerotinia sclerotiorum* pacl gene is required for sclerotial development and virulence. Molecular plant –Microbe interactions 16: 785-795.

Sairam, R.K., Singh D.V. and Srivastava G.C. (2003). Changes in activities of antioxidant enzymes in sunflower leaves of different ages. *Biol. Plant*, 47: 61-66.

Scandalios J G. (1993). Oxygen stress and superoxide dismutases. *Plant Physiology*, 101: 7-12.

Schwartz, H.F., and J.R. Steadman. (1989). White mold. In: H.F. Schwartz and M.A. Pastor Corrales, editors, Bean production problems in the tropics. 2nd ed. CIAT, Cali, Colombia. p. 211–230.

Singh, S.P. (1999). Integrated genetic improvement. In: Common bean improvement in the twenty-first century. (S.P. Singh ed.), Kluwer, Dordrecht, the Netherlands. pp. 133-165.

Steadman, J.R., and G. Boland. (2005). White mold. In: H.F. Schwartz, J.R. Steadman, R. Hall, and R.L. Forster, editors, Compendium of bean diseases. 2nd ed. *Am. Phytopath. Soc.*, St. Paul, MN. p. 44–46.

Steadman, J.R. (1979). Control of plant disease caused by *Sclerotinia* species. *Phytopathology* 69: 904-907.

Stelfox, D., Williams, J.R., Soehngen, U., Topping, R.C. (1978). Transport of *Sclerotinia sclerotiorum* ascospores by rapeseed pollen in Alberta. *Plant Disease Report* 62: 576–579.

Sun P. and Yang X. B. (2000). Light, temperature, and moisture effects on the apothecial production of *Sclerotinia sclerotiorum*. *Plant Disease* 84: 1287-1293.

Townsend, B.B. and Willets, H.J. (1954). The development of sclerotia in certain fung. *Transactions of the British Mycological Society* 37; 213–221

Tu, J.C. (1988). The role of white mold-infected white bean (*Phaseolus vulgaris* L.) seeds in the dissemination of *Sclerotinia sclerotiorum* (Lib.) de Bary. *Journal of Phytopathology* 121: 40–50.

Tu, J.C. (1989a). Modes of primary infection caused by *Sclerotinia sclerotiorum* in navy bean. *Microbios*, 57; 85–91

Tu, J.C. (1989b). Oxalic acid induced cytological alterations differ in beans tolerant or susceptible to white mould. *New Phytol.* 112: 519–525.

Tu, J.C. (1997). Biological control of white mould in white bean using *Trichoderma viride, Gliocladium roseum* and *Bacillus subtilis* as protective foliar spray. *In* Proceedings of the 49th International Symposium on Crop Protection. 6 May 1997, Gent. Part IV. Meded. Fac. Landbouwkd. *Toegep. Biol. Wet. Univ. Gent*, 62: 979–986.

Tu, J.C. and Zheng J. (1997). Relationship of row spacing and sowing rate to the development of white mold in Azuki and mungbean.

Turkington T.K. and Morrall, R.A.A. (1993). Use of petal infestation to forecast *Sclerotinia* stem rot of canola: the influence of inoculum variation over the flowering period and canopy density. *Phytopathology*, 83: 682–689.

Whetzel, H.H. (1945). Synopsis of the genera and species of the Sclerotiniaceae, a family of stromatic inoperculate discomycetes. *Mycologia*, 37: 648–714.

Whipps, J.M. and Gerlagh, M. (1992).Biology of *Coniothyrium minitans* and its potential for use in disease biocontrol. *Mycological Research*, 96: 897–907.

Whipps, J.M, Sreenivasaprasad, S., Muthumeenakshi, S., Rogers, C.W. and Challen M.P. (2008).Use of *Coniothyrium minitans* as a biocontrol agent and some molecular aspects of sclerotial mycoparasitism. *European Journal of Plant Pathology*, 121: 323–330.

Willetts, H.J. (1971).The survival of fungal sclerotia under adverse environmental conditions. *Biological Reviews*, 46: 387–407.

Willetts, H.J. (1997). Morphology, development and evolution of stromata/sclerotia and macroconidia of the Sclerotiniaceae. *Mycological Research*, 101: 939–952.

Willetts, H.J. and Bullock, S. (1992). Developmental biology of sclerotia. *Mycological Research*, 96;801–816.

Willetts, H.J. and Wong, J.A. (1980). The biology of *Sclerotinia sclerotiorum, S. trifoliorum* and *S. minor* with emphasis on specific nomenclature. *Botanical Review*, 46: 101–165.

Williams, B., Kabbage, M., Kim, H.J., Britt, R. and Dickman, M.B. (2011). Tipping the balance: *Sclerotinia sclerotiorum* secreted oxalic acid suppresses host defenses by manipulating the host redox environment. PLoS Pathog. 7.

Yang, L. Li, G.Q., Jiang, D.H. and Huang H.C. (2009). Water-assisted dissemination of conidia of the mycoparasite *Coniothyrium minitans* in soil. *Biocontrol Science and Technology*, 19: 779–796.

Yazima W. and Kav, N.N.V. (2006). the proteorne of the phytopathogenic fungus *Sclerotinia sclerotiorum*. Proteomics 6: 5995-6007.

Zeng, W., Wang, D., Kirk, W., Hao, J., (2012). Use of *Coniothyrium minitans* and other microorganisms for reducing *Sclerotinia sclerotiorum*. *Biological Control* 60, 225–232.

Chapter 9

Diseases of Fieldpea (*Pisum sativum* L.) and their Management

R.K. Mishra, Monika Mishra, Sonika Pandey & Naimuddin

Introduction

Fieldpea (*Pisum sativum* L.) is one of most important cool season food legume crops in the world and occupies a very prominent place because of its production and multiple uses. In India, the crop is cultivated in northern and central parts of the country. It is an important *rabi* pulse crop grown in about 0.97 m ha with annual production of 0.89 m tonnes. The major fieldpea growing states are Uttar Pradesh, Madhya Pradesh, Jharkhand, Bihar, Assam and Maharashtra. The average productivity of this crop has increased considerably over the years which is now to the tune of 1t/ha that is very much lower as compare to other countries likely France (>4300 kg/ha), Germany (>3000kg/ha), USA (>2100 ka/ha) and Canada (>2100 kg/ha). The reason behind low productivity of pea is its susceptibility to several diseases and insect pests. The fungal, bacterial, viral and nematode pathogen causing various diseases and insect-pests affecting the crop at different stage of its growth is given in Table 9.1. Diseases cause heavy reduction to both quality and quantity of pea seed and grain production. Some environmental conditions like high humidity and moderate temperature enhance the possibility of development of diseases. In this book chapter we elaborated about major diseases and their suitable management strategies.

Table 9.1. Major Diseases of Fieldpea in India

Sl.No.	Name of the Diseases	Causative Agents
1.	Rust	*Uromyces fabae*
2.	Powdery Mildew	*Erysiphe polygoni*
3.	Downy Mildew	*Perenospora pisi*
4.	Seedling rot	*Rhizoctonia solani*
5.	Aschochyta blight	*Aschochyta pinodes*
6.	Stem rot	*Sclerotinia scleriotiorum*
7.	Fusarium wilt	*Fusarium oxysporum f. Sp. pisi*
8.	Pea enation mosaic	*Pea enation mosaic virus*
9.	Pea leaf roll	*Pea leaf roll virus*

Interaction of disease and crops is a never ending process. Early farmers were in practice to use superstitious and supernatural ideas for the crop damage. Modern plant pathological detection systems have made the detection of plant disease based on scientific approaches. Filippo Re and Carlo Berti Pichat first started to classify the plant disease according to the symptoms. For the rapid management of disease we should know the severity of disease. So, without estimating the disease severity we cannot control it. Thus, analysis of disease and symptoms severity is very important for understanding the plant stress biology. Visual recognition of symptoms is essential for the diagnosis of Plant disease. However, this method is not very specific. Newer technologies offer disease assessment with great accuracy and precession. Visible light photography and digital image analysis have been used over the last 30 years (West *et al.*, 2003). Hyperspectral imagery is a recent technique and it is not using widely in the plant pathology. Plant diseases are main cause of crop destruction in the world. For example rice blast disease every year causes a great loss in rice production (Dean *et al.*, 2005). Plant disease causes a great loss in all food crops including fruit crops also. In south Western Europe Flavescence Doree, grape fruit disease cause a great loss in grape production. (Martinellli *et al.*, 2014). Early detection and diagnosis of disease is of key importance to prevent disease spread (Yang *et al.*, 2014). Traditional identification of disease is not enough for the early disease diagnosis and its management. Introduction of PCR techniques has a profound impact on the disease diagnosis. Through nucleic acid based techniques we can detect the pathogens that have not been cultured. According to European and Mediterranean Plant Protection Organization (EPPO) the protocol for detecting plant pathogen are-integrating phenotypic, serological and molecular techniques. For newly discovered pathogens molecular assays can be developed easily. Nucleic acid based techniques are easy to develop and are very accurate and sensitive. Present methods for plant disease detection include proximate detection, immunological and DNA based approaches, analysis of volatile compounds and genes remote sensing, spectroscopic methods sensors, biophotonics *etc.* There are several reports which emphasis the importance of biochemical and serological methods. (Hampton *et al.*, 1990., Schaad and Frederick., 2002) or applied RS techniques (West *et al.*, 2003; Bock *et al.*, 2010; Sankaran *et al.*, 2010; Mahlein *et al.*, 2012a). But there are many other methods which can be used for the disease diagnosis. Here we will discuss about those techniques.

Important Diseases of Fieldpea

Rust

Rust is one of the most economically important diseases of fieldpea. The disease is caused by *Uromyces fabae* (Pers. de Bary) It is an autoecious fungus completing its entire life cycle on the same host. The pathotypes develop particularly in regions with warm and humid climate and the disease generally appears at flowering or podding stage. In the tropical and subtropical regions such as India and China, the principal causal agent of pea rust has been reported as fungus *U. viciae-fabae*, where warm humid weather is suitable for the appearance of both the uredial and the aecidial stage. This disease is serious in North India (Mishra *et al.*, 2009). The stem of plant become malformed and the affected plant dies out. The earliest symptoms are the yellow spot having aecia in round or elongated clusters. Then the uredopustules develop which are powdery and light brown in appearance. High humidity and cloudy weather with temperatures of 20-22°C favor disease development. The plants give dark brown or blackish appearance visible as patches in the field (Mishra *et al.*, 2010).

Management Strategies

1. The affected plant debris and trash should be burnt after harvesting
2. Integrated management of rust includes control of volunteer plants over the summer and removal of infected plant debris. It is advisable to use

Figure 9.1. Rust Infected Fieldpea Leaves.

clean seeds without rust contaminations, and to treat the seed with a suitable fungicide such as diclobutrazole. Preventive fungicide sprays of mancozeb (0.25 per cent), Propiconazole (0.1 per cent) and Hexaconazole (0.1 per cent) at early disease development stage have been recommended (Mishra and Pandey, 2009). The use of host plant resistance is the best means of rust management (Mishra *et al.*, 2009).

Rhizoctonia Seedling Blight

It is caused by *Rhizoctonia solani*, soilborne fungus. It is also known as Rhizoctonia tip blight or stem rot of pea. Characteristic symptoms of the disease are the death of the growing tip of very young pea seedlings.

Management

Rotate peas with crops know to be relatively poor hosts of the fungus, such as grain crops and corn.

Bacterial Blights

Causative agent of disease is *Pseudomonas syringae*. Characteristic symptoms of the disease are the appearance of leaf and pod lesions (approx 3mm in diameter and angular shape).Lesions are brown shiny and translucent.

Management

1. Use of healthy and disease free seeds for planting helps to minimize the disease spread.
2. At least 3 year rotation with non host crops.

Ascochyta Disease

Mycosphaerella Pinodes

Phorma medicaginis var. pinodella and *Ascochyta pisi* are the main casual agents of aschyomycota disease. There are three main Ascochyta diseases- Mycospharella blight, Ascochyta foot rot and ascochyta leaf and pod spot. Out of all three, Mycosphaerella blight is most important.

Mycosphaerella Blight

Occurrence of small purple spots on the stem, pods and foliage are the typical symptoms of the disease. Leaf spot may enlarge upto5-6nm.Small lesions have no definite shape but larger lesions are circular with distinct margins.

Ascochyta Foot Rot

Similar to Mycosphaerella blight. But this is most often apparent on lower stem and upper root.

Management

1. Use of clean seeds and destroys old Pea crop residues.
2. Rotate crops so there is at least 3 years between pea crops and often 4 or 5 years may be required. Follow the recommended sowing dates for your district.
3. Avoid early sowing at high seeding rates. Don't mix fungicide with the inoculants as this will reduce the number of rhizobia.
4. Application of foliar fungicides before rain. *Pantoea agglomerans* and *Bacillus* species have been found effective in the control of Ascochyta disease of fieldpea.

Leaf and Pod Spot

Lesions of this disease are different from the above mentioned disease they are about 5-8mm in diameter and sunken and tan with a dark and distinct border. Lesions are oval on foliage and pods and are elongated on stems.

Management

1. Use western-grown disease-free seed.
2. Use a 3- or 4- year rotation. Mouldboard-plough pea profuse deeply and completely.
3. Isolate new pea fields ¼ mile (0.4 km) from old ones.

Anthracnose

Causative agent of this disease is *Collectotrichum pisi*. Characteristic symptoms of the disease are oval lesions on leaf and stipule which are around 2-8mm in diameter with brown margins and gray tan centers.

Management

Practice Crop rotation, use Chlorothalonil for the management of disease.

Downy Mildew

Downy mildew is caused *Peronospora viciae*. On the upper side of the foliage, lesions appear as irregularly shaped yellow to brown areas with indistinct margins. On the underside of foliage these discoloured areas have patches of mouse colour. Sometimes the symptoms of disease may appear on pods and not on foliage. Young pods are more susceptible for disease as compared to old ones. Yellow to brown sunken blotches are the characteristic symptoms of the disease. Oospores of the fungus sometimes present on the pod wall.

Figure 9.2. Greyish and Yellowish Symptoms on Fieldpea Leaves and Morphology of Pathogen.

Management

1. Since the causal organism survives from season to season in the form of oospores in the plant debris, the destruction of previous year's plant debris and following crop rotation of two or three years are very effective control measures.
2. Spraying and dusting pea plants with fungicides are effective in limiting the spread of the disease. Deep tillage to bury crop residues.
3. Use tolerant cultivars.
4. Use metalaxyl for seed treatment and application of Chlorothalonil just appearance of the symptoms for the management of disease.

Powdery Mildew

Powdery mildew is the most important disease of fieldpea. This air borne disease is caused by *Erysiphe pisi,* which is a biotrophic ascomycete fungus. Usually it develops at pod formation stage or just before harvest of the crop. As the name implies, a white powdery coating covers the aerial plant parts including leaves and pods. Eventually, the speckles turn into spots and the plant turns purple, bluish-green and lastly brown. Warm-dry days and cool nights are critical for the development of diseases. Late-sown crop is typically at higher risk. Frequent heavy dews in the absence of rain accelerate the disease's progression. The number and weight of the pods are reduced. Yield reduction due to these diseases is very high within short period of time. It appears in severe form almost every year when the plants are in the pod stage towards the end of January and in February. The losses in yield in a 100 per cent infected crop were estimated by Munjal *et al.* (1963) to be 21-31 per cent in pod number and 26-47 per cent in pod weight. However, several powdery mildew resistant varieties have been developed by the NARS.

Figure 9.3. Powdery Mildew Infected Fieldpea Plants.

Management

1. Late planting should be avoided the disease severity
2. Collect and burn the plant debris/stubbles that left on the field after crop harvest
3. Grow early maturing short duration varieties in high disease prone areas
4. The diseases can be managed by alternate spray of any of the wettable sulphur (0.3 per cent), karathane (0.05 per cent) or carbendazim (0.05 per cent). Give the first spray after the appearance of the diseases in the crop. Second spray should be given 15 days after the first spray and the third spray only if there is a need for it.
5. Use of diseases resistant varieties and to avoid excess irrigation in the field during flowering stage of the crop.
6. Biocontrol agents like *Acremonium alternatum, Irpex lacteus, Paecilomyces fumosoroseus, Verticillum lecanii, sporothrix regulosa, Trichoderma harzianum, Stephanoascus* spp and *Tilletiopsis* spp. have been used for management of powdery mildew. Bcterisation by *P. fluorescens* and *P. aeruginosa* is also effective for management of powdery mildew disease (Mishra and Pandey, 2010). Other fungi such as *Acrodontium crateriforme, Dissoconium acicularae* and *Ramichloridium apiculatum* have also been reported to be the antagonist of powdery mildew pathogen but their efficacy as biocontrol agent have not been studied (Sara Fondevilla *et al.*, 2012)

Stem Rot/White Rot

Stem rot caused by *Sclerotinia sclerotiorum* (Lib.) de Bary, is a cosmopolitan necrotrophic fungal pathogen with a very broad host range (fieldpea, chickpea, pigeonpea, lentil, rajmash, bean, mustard, potato, sunflower, *etc.*) and it infects a variety of broadleaf crops and weeds. It often enters fields by means of contaminated seed lots or equipments. It survives as a mass of fungal hyphae covered by an environmentally resistant black rind called sclerotia. When conditions are favorable, such as at canopy closure when humidity levels in the canopy are high and the crop begins flowering, an inverted mushroom structure called an apothecium produces ascospores which then infect the flowers. Hyphae can also grow directly into plant tissues, especially when plant-soil contact is established. The fungus kills tissue and fills the stem with white hyphae and sclerotia. In this way it survives in the soil for many years.

Figure 9.4. Stem Rot Disease Infected Fieldpea Plants and Pods.

This fungus is a soil-borne fungus that attacks plant in all stages of growth including young seedlings, mature plants and their harvested products distributed worldwide (Agrios, 2005; Purdy, 1979). It is a major disease of several leguminous crops such as fieldpea, chickpea, pigeonpea, lentil, soybean, fababean and alfalafa (Nene *et al.*, 1996, Mishra *et al.*, 2015). It is a major disease of fieldpea under favourable conditions; it may cause serious losses to the crop. Symptoms appear as wet, soft and white rotting of the tissues on all arial part of the plants including pods. In India, regular occurrence of the disease has been reported in some districts of Himachal Pradesh, where it causes considerable yield losses (Grewal and Pal, 1986). The yield loss ranges from 21 to 47 per cent at pod bearing stage of the crop (Sharma, 1990). Limited variety resistance may be available in dry pea and lentil cultivars.

Management

1. Seed cleaning is essential to remove the sclerotial mixer during threshing and processing period.
2. Deep ploughing, low land with continuous stagnation of water reduces the sclerotial population.

3. Alternate spray of recommended fungicides at 10-15 days intervals is essential.
4. All infected plant parts must be collected and burnt before drying of the plants.
5. Removal of the weeds from the fields.

Fusarium Root Rot

This is the most important disease of pea. Causative agent of the disease is *Fusarium solani* f. sp. *pisi*. Primary symptoms of the disease appear along the taproot and side roots of plant as brown slender lesions. Red discolouration of vascular tissue is also visible up to the three nodes above soil line.

Management

1. Soil amended with decomposed *Sesbenia aculata*, mustard cake and Farm yard manure
2. Deep summer ploughing during the month of May-June followed by cultural practices like long crop rotation is recommended to reduce the diseases severity.
3. Seed treatment with fungicides (Carbendazim @2.5g/kg seeds) is feasible and protects the seedlings from infection and ensures better plant stand.
4. Treatment of fieldpea seeds with bioagents (*Trichoderma harzianum* @ 6 g/kg) and fungicide (vitavax 1.0g/kg).
5. Advanced molecular methods used for disease detection and diagnosis
6. Early sowing should be avoided in badly infested areas.

Other Molecular Approaches

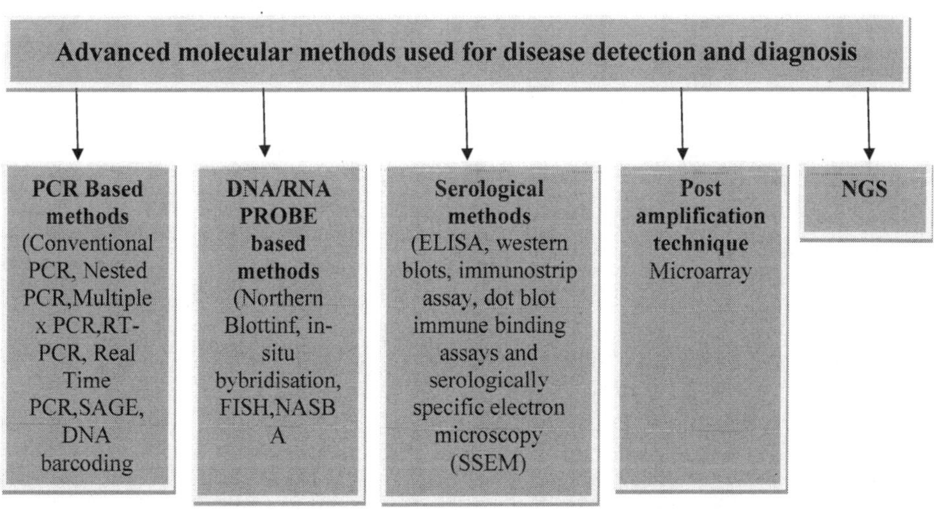

Serological Assays

Serological assays were developed for the detection of viruses. Using polyclonal and monoclonal antisera techniques such as ELISA, western blots, immunostrip assay, dot blot immune binding assays and serologically specific electron microscopy (SSEM) (VanVuurde *et al.*, 1987, Hampton *et al.*, 1990). In 1970 ELISA was first employed, it is the most widely used immunodiagnostic techniques. The sensitivity of ELISA depends upon the organism, sample and titer value. For viruses and fungi polyclonal antisera have been developed and they are been using for numerous protocols. Monoclonal antibodies recognize one epitome only and are expensive while polyclonal antibodies recognize multiple epitomes and are less expensive. In present time for phytopathogenic bacteria ELISA procedures using polyclonal and monoclonal antibodies are been employed.

Nucleic Acid-based Methods

DNA based detection methods are- fluorescence in situ hybridization (FISH) and PCR variants (PCR, NESTED PCR, CO-PCR MULTIPLEX PCR, REAL TIME PCR DNA Fingerprinting) RNA Based techniques are reverse transcriptase PCR, NUCLEIC ACID sequence-based amplification (NASBA) and Aplidet RNA. Molecular detection methods are based on the accurate design of oligonucleotides and probes. PCR offers several advantages over immunoassays. In PCR based diagnostic techniques primers are designed to pair with the target DNA. Specific amplification of Target DNA sequence is widely used to detect and identify the plant pathogens.(Mumford *et al.*, 2006). Amplification confirms the presence of organism in the tested sample. PCR reliability for pathogen specificity has been improved by the use of dye quenched probes (Morris *et al.*, 1996; Thelwell *et al.*, 2000). For plant virus detection RT-PCR is most reliable. By using n-PCR we can increase the specificity and sensitivity. Co- PCR techniques was especially developed for plant virus and bacteria detection. Co-PCR is carried out in single reaction, it reduces the rate of contamination and sensitivity is also similar to the n-PCR and RT-PCR. Colorimetric detection increases the sensitivity of virus detection upto 1000 times. In M-PCR, two or more than two target sequences are amplified simultaneously.

M-PCR is useful because several pathogens infect the same plant at a time (Davino *et al.*, 2012), so with M-PCR we can easily detect more than one pathogen. Unlike conventional PCR, RT-PCR allows pathogen quantification. In RT-PCR Mainly two chemistry works SYBR green and Taq Man. Taq man used most commonly. In DNA fingerprinting identification is done through the identification of unique patterns present in DNA. There are many fingerprinting techniques such as RFLP, SNP, STR, rep-PCR, amplified 16S ribosomal DNA restriction analysis (ARDRA) (Scortichini *et al.*, 2001), and amplified fragment length polymorphism (AFLP) (Clerc *et al.*, 1998; Manceau and Brin., 2003).Technique choice depends upon the aim.

NASBA is a technique used to amplify RNA sequences. This technique was developed in 1990s, it does not require PCR machine it works on water bath only. This technique commonly used to detect bacteria and viruses (Klerks *et al.*, 2001; Rodriguez-Làzaro *et al.*, 2006). Specificity and speed of this technique is very high. Loop-mediated isothermal amplification (LAMP) this technique is based on photometery (Mori *et al.*, 2001).This technique is a simple screening assay which eliminates the need of thermocycler. This technique is used for the plant virus detection (Varga and James., 2006). DNA microarray technique came into the light in 1990, and it creates a revolution in nucleic acid detection. Microarray techniques have been used extensively for Plant pathogen detection. There are many factors on which choice of molecular techniques is dependent such as budget, time and no. of species (Lopez *et al.*, 2009). Molecular methods are highly efficient and specific. But there are many limitations in molecular techniques also such as sometimes PCR give false results due to impairing and primer dimerization, dead pathogens can also give false positive results. Costing of these techniques is also a very important issue.

REFERENCES

Abhishek S; and Sobita S. Eco-friendly management of powdery mildew and rust of garden pea (*Pisum sativum* L.) *Journal of Pharmacognosy and Phytochemistry*. **2017,** 6,5,90-93.

Agrios, G. N. Plant Pathology (Fifth edition), Academic Press, London, **2005.**

Bock C.H.; Poole G.H; Parker P.;E, Gottwald T.R. Plant disease severity estimated visually, by digital photography and image analysis, and by hyperspectral imaging. *Crit Rev Plant Sci.* **2010,** 29,59–107, doi: 10.1080/07352681003617285

Carter D.J.; Cary R.B. Lateral flow microarrays: a novel platform for rapid nucleic acid detection based on miniaturized lateral flow chromatography. *Nucleic Acids Res.* **2007,**35,e74, doi: 10.1093/nar/gkm269

Clerc A; Manceau C; Nesme X. Comparison of randomly amplified polymorphic DNA with amplified fragment length polymorphism to assess genetic diversity and genetic relatedness within genospecies III of Pseudomonas syringae. *Appl Environ Microbiol.* **1998,** 64,1180–1187.

D'hondt L; Hofte M; Van Bockstaele E; and Leus L. Applications of flow cytometry in plant pathology for genome size determination, detection and physiological status. *Mol. Plant Pathol.* **2011,** 12,815-828.

Danks C; and Baker I. On-site detection of plantpathogens using lateral-flow devices. *EPPO Bull.* **2000,** 30,421- 426

Davino S; Miozzi L; Panno S; Rubio L; Davino M; Accotto G.P. Recombination profiles between Tomato yellow leaf curl virus and tomato yellow leaf curl Sardinia virus in laboratory and field condition: evolutionary and taxonomic implications. *J Gen Virol.* **2012,** 93,2712–2717, doi: 10.1099/vir.0.045773-0

Dean R.A; Talbot N.J; Ebbole D.J; Farman M.L; Mitchell T.K; Orbach M.J. *et al.*, The genome sequence of the rice blast fungus Magnaporthe grisea. *Nature.* **2005,** 434,980–986, doi: 10.1038/nature03449

Hampton R.; Ball E.; De Boer S.Serological methods for detection and identification of viral and bacterial plant pathogens. A laboratory manual. *APS Press St. Paul.* **1990.**

Klerks M.M; Leone G; Lindner J.L; Schoen C.D; van den Heuvel JFJM (2001). Rapid and sensitive detection of Apple stem pitting virus in apple trees through RNA amplification and probing with fluorescent molecular beacons. *Phytopathology*. **2001,** 91,1085–1091, doi: 10.1094/21 PHYTO. 2001.91.11.1085

López M.M; Llop P; Olmos A; Marco-Noales E; Cambra M; Bertolini E. Are molecular tools solving the challenges posed by detection of plant pathogenic bacteria and viruses? *Mol Biol.* **2009,** 11,13–46, swfrec.ifas.ufl.edu/hlb/database/pdf/00002423.pdf

Mahlein A.K; Oerke E.C; Steiner U.; Dehne H.W.Recent advances in sensing plant diseases for precision crop protection. *Eur J Plant Pathol.* **2012a,** 133,197–209, doi: 10.1007/s10658-011-9878-z

Manceau C; Brin C. Pathovars of Pseudomonas syringae are structured in genetic populations allowing the selection of specific markers for their detection in plant samples. In: Iacobellis NS *et al.* (eds) Pseudomonas syringae and related pathogens. *Kluwer*. **2003,** 503–512.

Martinelli F.; Uratsu S.L.; Albrecht U.; Reagan R.L.; Phu M.L.; *et al.,* Transcriptome profiling of citrus fruit response to Huanglongbing disease. *PLoS One.* **2012b,** 7,e38039, doi: 10.1371/journal.pone.0038039.

Mishra, R.K, K.K.Pandey and P.K.Pandey. Screening of different pea genotypes against Rust of Pea caused by *Uromyces fabae. Indian Journal of Agricultural Sciences*. **2009**. 79 (5): 403-405

Mishra, R.K. and K.K.Pandey. Effect of application of PGPR and Neemazal on management of pea rust (*Uromyces fabae*). J. Basic and Appl. Mycobiol. **2010,** Vol. 1 and 2: 102-106.

Mishra, R.K. and P.K.Pandey. Efficacy of fungicides against pea rust caused by *Uromyces fabae* Annals of Plant Protection Sciences. **2009**, (2): 54-55

Mishra, R.K., K.K.Mishra, K.K.Pandey and U.P.Singh). Biochemical changes in Pea infected by *Uromyces fabae. Indian Phytopathology.* **2010**, 65 (2): 98-101.

Mishra, R.K. and P.K.Pandey. Effect of rust on nodulation and nitrogenase activity of Pea (*Pisum sativum*). *Pest management in Hort. Ecosystem.* **2009**, 15 (1): 60-62

Mori Y; Nagamine K; Tomita N; Notomi T. Detection of loopmediated isothermal amplification reaction by turbidity derived from magnesium pyrophosphate formation. *Biochem Biophys Res Commun.* **2001,** 289,150–154, doi: 10.1006/bbrc.2001.5921

Morris T; Robertson B; Gallagher M. Rapid reverse transcription- PCR detection of hepatitis C virus RNA in serum by using the TaqMan fluorogenic detection system. *J Clin Microbiol.* **1996,** 34,2933– 2936, jcm.asm.org/content/34/12/2933.

Munjal, R. L., V. V Chenulu, and I. S. Hora. Assessment of losses due to powdery muldew *Erysiphe polygoni* on pea. Indian Phytopath., **1963**, 16: 260-267

Mumford R; Boonham N; Tomlinson J; Barker I. Advances in molecular phytodiagnostics—new solutions for old problems. *Eur J Plant Pathol* **2006,**

116,1–19, doi: 10.1007/s10658-006-9037-0

Na L; Shengchun Xu; Xiefeng Y; Guwen Z; Weihua M; Qizan IIu; Zhijuan F; and Yaming G. (2016). Studies on the Control of Ascochyta Blight in Fieldpeas (*Pisum sativum* L.) Caused by Ascochyta pinodes in Zhejiang Province, China. *Front Microbiol*. **2016,** 7.

Nene YL, Sheila YK and Sharma SB. A world list of chickpea and pigeonpea pathogens. 5th edn. Patancheru 502 324, **1996**, Andhra Pradesh, India: International Crops Research Institute for the Semi-Arid Tropics.

L.H.Purdy. *Sclerotinia sclerotiorum*: History, Diseases and Symptomatology, Host Range, Geographic Distribution, and Impact. *Phytopathology*, **1979**, 69 (8): 875-880

Rodriguez-Làzaro D; Hernàndez M; D'Agostino M; Cook N. Application of nucleic acid sequence-based amplification for the detection of viable foodborne pathogens: progress and challenges. *J Rapid Meth Aut Mic*. **2006,** 14,218–236, doi: 10.1111/j.1745-4581. 2006. 00048.x

Sankaran S.; Mishra A.; Ehsani R.; Davis C. A review of advanced techniques for detecting plant diseases. *Comput Electron Agric*. **2010,** 72, 1–13, doi: 10.1016/j. compag.2010.02.007

Sara F; Diego Rubiales A. Powdery mildew control in pea. A review. *Sustain. Dev*.**2012,** 32,401–409, DOI 10.1007/s13593-011-0033-1

Schaad N.W.; Frederick R.D. Real-time PCR and its application for rapid plant disease diagnostics. *Can J Plant Pathol*. **2002,** 24,250–258, doi: 10.1080/07060660209507006

Scortichini M; Marchesi U; Rossi M.P; Di P. P. Bacteria associated with hazelnut (Corylus avellana L.) decline are of two groups: Pseudomonas avellanae and strains resembling P. syringae pv. syringae. *Applied Appl Environ Microbiol* **2001,** 68,476–484, doi: 10. 1128/AEM.68.2.476-484. 2002

Thelwell N; Millington S; Solinas A; Booth J; Brown T. Mode of action and application of Scorpion primers to mutation detection. *Nucleic Acids Res*. **2000,** 28,3752–3761, doi: 10.1093/nar/28.19.3752

Van Vuurde J.W; Ruissen M.A; Vruggink H. Principles and prospects of new serological techniques including immunosorbent immunofluorescence, immunoaffinity isolation and immunosorbent enrichment for sensitive detection of phytopathogenic bacteria. In: Civerolo EL, Collmer A, Davis RE, Gillaspie AG (eds) Plant Pathogenic Bacteria. *Curr Plant Sci Biotech Agric*. **1987,** 4,835–842.

Varga A; James D. Use of reverse transcription loop-mediated isothermal amplification for the detection of Plum pox virus. J Virol Method, **2006,** 138-184–190, doi: 10.1016/j.jviromet. 2006.08.014

West J.S.; Bravo C.; Oberti R, Lemaire D.; Moshou D.; McCartney H.A. The potential of optical canopy measurement for targeted control of field crop diseases. *Ann*

Rev Phytopathol. **2003,** 41,593–614.

Yang W.; Chen J.; Chen G.; Wang S.; Fu F. The early diagnosis and fast detection of blast fungus, *Magnaporthe grisea,* in rice plant by using its chitinase as biochemical marker and a rice cDNA encoding mannose-binding lectin as recognition probe. *Biosens Bioelectron.* **2013,** 41,820–826, doi: 10.1016/j.bios.2012.10.032

Chapter 10

Symptomatology, Characterization and Management of *Rhizoctonia solani* Diseases of Pulse Crops

B.M. Bashyal, Bhupendra Singh Kharayat, S.C. Dubey & Rashmi Aggarwal

Biotic stresses are the major constraint for the low productivity of pulse crops in India. Major biotic stresses in pulse crops are insect pest and plant diseases caused by fungi, virus, bacteria, phytoplasma and nematodes. Amongst the diseases web/aerial blight and root rot caused by fungal pathogen *Rhizoctonia solani* Kühn [teleomorph-*Thanatephorus cucumeris* (Fr.) Donk] are a major limiting factor of higher pulse yield. *Rhizoctonia solani* is a ubiquitous basidiomycetous soilborne fungal pathogen causing damping-off of seedlings, aerial blights and postharvest diseases. *Rhizoctonia solani* (Teleomorph: *Thanatephorus cucumeris* [Frank] Donk) is an economically important soilborne fungus with a worldwide distribution. Pathogenic isolates of *R. solani* are known to attack 188 species of higher plants in 32 families, including various staple crops, ornamentals and turf grasses (Anderson, 1982). The genus *Rhizoctonia* was described by DeCandolle in 1815. He designated *R. crocorum* (Pers.) DC. as the type species, while *R. solani,* the most important species of *Rhizoctonia,* was described by Kuhn in 1858. Its wide host range, saprophytic and facultative parasitic nature and rapid adaptability to wide diversified ecosystem cause often failure of disease management strategies. It survives in soil by producing survival structure sclerotia and infected plant residue present in the field after crop harvest. *Rhizoctonia solani* causes severe diseases and considered major constraint in pulse production in India. Among the pulse crops, chickpea (*Cicer arietinum*

L), cowpea [*Vigna unguiculata* (L.) Walp.], mungbean [*Vigna radiata* (L) Wilczek], urdbean [*Vigna mungo* (L.) Hepper], lentil (*Lens culinaris* Medikus) and french bean (*Phaseolus vulgaris* L.) are severely affected by *R. solani*. In this chapter, we will cover the diseases caused by *Rhizoctonia sp.* in major pulse crop like mungbean, urdbean, chickpea, cowpea and soybean.

Host Range of *Rhizoctonia solani*

Rhizoctonia solani have very wide host range because of its necrotrophic nature and survival under unfavorable conditions. *Rhizoctonia solani* can also be classified according to infection they cause in host based on its anastomosis group (Table 10.1).

Table 10.1. Summary of the different Groups and Subgroups Recognized for Multinucleate *Rhizoctonia solani* Isolates (Table adapted from García *et al.*, 2006)

Group/Subgroup	*Diseases*	*Teleomorph*
AG 1 IA	Rice, corn, sorghum, bean, soybean, turfgrass, camphor seedlings	*Thanatephorus cucumeris* (=*Corticium sasakii; Hypochnum sasakii; Pellicularia sasakii*) *T. sasakii*?
AG 1 IB	Bean, rice, soybean, leguminous woody plants, lettuce, hortensia, cabbage	*T. cucumeris* *T. microsclerotius*?
AG 1 IC	Buckwheat, carrot, soybean, flax, pine	*T. cucumeris*
AG 2-1	Crucifers, strawberry, tulip	*T. cucumeris*
AG 2-2IIIB	Rice, mat rush, ginger, turfgrass, corn, sugarbeet, *Chrysanthemum*	*T. cucumeris*
AG 2-2IV	Sugarbeet, turfgrass	*T. cucumeris*
AG 3	Potatoes, tobacco, tomato, egg plant	*T. cucumeris*
AG 4 (HGI, HGII and HGIII)	Tomato, pea, potato, soybean, onion, cotton, snap bean, Loblolly pine seedlings	*T. cucumeris* (=*Pellicularia praticola*), *T. praticola*?
AG 5	Potato, turfgrass, bean, soybeans	*T. cucumeris*
AG 6 (HG-I and GV)	Non pathogenic	*T. cucumeris*
AG 7	Soybeans	*T. cucumeris*
AG 8	Poaceae	*T. cucumeris*
AG 9 (TP and TX)	Crucifers, potatoes	*T. cucumeris*
AG 10	Non pathogenic	*T. cucumeris*
AG 11	Wheat	*T. cucumeris*
AG 12	Cauliflower, radish	*T. cucumeris*
AG13	Non pathogenic	*T. cucumeris*
AGBI	Non pathogenic	T. cucumeris

Names between question marks are suggested to be biological species different to *Thanatephorus cucumeris*.

Disease Symptoms

Web Blight of Mungbean

Mungbean/green gram [*Vigna radiata* (L.) Wilczek] is one of the most important pulse crops. It is grown in almost all parts of the country and belongs to family leguminosae. Web blight/wet root rot disease of mungbean caused by *Rhizoctonia solani* Kühn (*Thanatephorus cucumeris* Donk) is a ubiquitous soil-borne, plant pathogenic fungus, causes severe loses in other agriculturally important food crops (Kharayat *et al.*, 2015) and also the major constraint to mungbean crop production in India. This causes infection at any growth pahse of mungbean plant and disease occurs in mainly three phase (a) seed rot (Figure 10.1), (b) root rot and (c) aerial or web blight. All of these phases of disease can cause severe losses. The seed rot and root phase generally occurs upto 21-25 days of seed sowing. Initially leaf symptoms appear as water-soaked, grayish green lesions that turn tan to brown at maturity (Figure 10.2). The web or aerial blight phase appears after 21-25 days of sowing at higher temperature (31.5 °C) and relative humidity (90.6 per cent). Web-like hyphae of *R. solani* can be seen spreading along the stems, branches and over the leaves of mungbean plants. Under favorable environmental conditions mycelium spread extensively and rapidly on the plant surface, forms localized mats of "webbed" foliage. *R. solani* form numerous sclerotia on leaves, stem, petiole and pods. The foliar part especially leaves and sometime branches merged with each other by the fungal mycelial mat. In case of aerial phase of necrotic blighted lesion appears on lamina, stem and pod surface. Only light mycelium may be seen on foliar parts but not web like structure. The pathogen may infect leaves, pods, and stems in the lower canopy. Reddish-brown lesions can form on infected petioles, stems, pods and petiole scars. Long strands of web-like hyphae can spread along affected tissue and small, dark brown sclerotia form on diseased tissue.

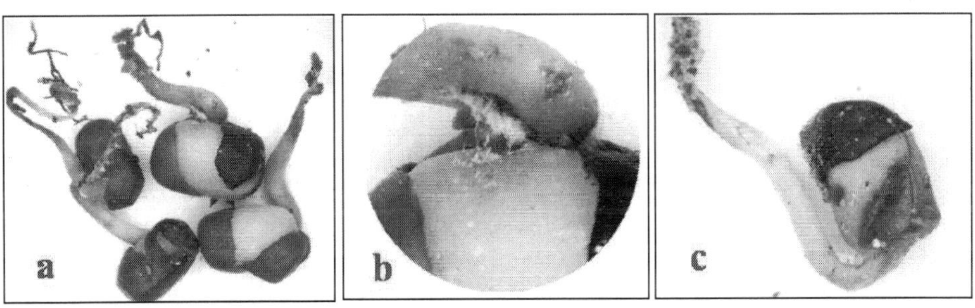

Figure 10.1. Mortality of Germinating Mungbean Seeds (cv. Pusa Vishal) Caused by *R. solani*.

(a) Brown necrotic rot lesion on radical, (b) White mycelial growth of *R. solani* on radical, (c) Dark brown lesion of on seed.

Web Blight/Aerial Blight of Soybean

Web blight/aerial blight of soybean is caused by a fungus, *Rhizoctonia solani* AG1-IA, AG1-IB, AG1-IC, AG4 and AG7, which overwinters as sclerotia in soil or

Figure 10.2. Typical Aerial/Web Blight Symptoms on Mungbean Cultivar (Pusa Vishal). White mycelial growth on stems and leaves. White to brown colored, scattered, micro- and macro-sclerotia on leaf surface. *R. solani* (AG-3) also causes defoliation of the mungbean leaves.

plant debris from the preceding crop (Faske *et al.*, 2014). This fungus can cause seed decay, pre-emergence damping-off of soybean seedlings, root rot on plants of all ages and aerial blight. The symptoms of web blight/aerial blight of soybean are similar to mungbean. Foliar symptoms often occur during late vegetative growth stages on the lower portion of the plant following canopy closure. Initially leaf symptoms appear as water-soaked, grayish green lesions that turn tan to brown at maturity. Symptoms on leaves are greenish lesions to gray brown necrotic with irregular shape and later turned reddish brown margins (Figure 10.3). Severely infected leaves collapse and fall with the petioles. Fluffy white mycelial growth typically develops on the necrotic parts (Rahayu, 2014). White to dark brown sclerotial structure with irregular shape was visible on infected plant parts including leaves, stems and pods. Severe infected pods showed empty and dry. Fungal structures have been found spread on soil surface around plants. Plant firstly infected on lower portions, then during plant growth stages the fungus spread along stem and infect the upper plants part (Rahayu, 2014). On the basis of disease symptoms and signs, this disease was named as web blight disease caused by *R. solani.* Localized reddish-brown lesions develop near the soil line can be observed on seedlings and older soybean plants. The development of red color of the lesions is considered a good diagnostic characteristic, but it is best to observe immediately after plants are uprooted from the soil because the color of this lesion fades as plants are exposed to the air. Lesions are usually confined to the cortical layer (outer layer) of the hypocotyl or stem. If lesions girdle the entire stem, plants may be extremely stunted or less vigorous than healthy plants, resulting in uneven stands. The foliage of infected plants may have an off-color or yellow cast, and the root systems may be poorly developed with decayed lateral roots. During warm, wet weather mycelium spreads extensively on the surface of plants, forming localized mats of "webbed" foliage. Spread from these localized areas can be rapid when conditions favor disease (High RH and 30 °C). Because this pathogen also causes sheath blight of rice, soybean fields that follow rice with a history of sheath blight are likely to have high incidence of aerial blight. Aerial blight may be present at low levels without being noticed, but when the plants are mature with a closed canopy of leaves across the field and when frequent rains occur, aerial blight will spread fast. Within one week or less an entire field will appear scorched. Small fields bordered by trees or poorly drained fields are more apt to have severe aerial blight (Kucharek, 2001).

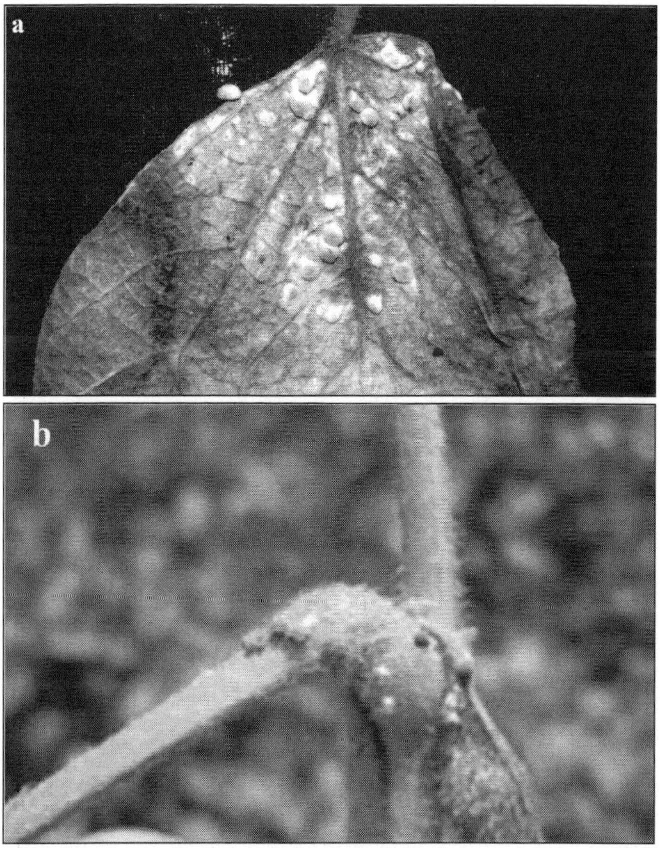

Figure 10.3. Sclerotia of *Rhizoctonia solani* on Soybean (a) Leaf and (b) Petiole. (Photograph: (a) Kharayat, B.S. (b) adapted from soybean diseases and courtesy of M. Emerson).

Wet Root Rot of Chickpea

Chickpea (*Cicer arietinum* L.), also known as the garbanzo bean, bengal gram, chana is an annual grain legume crop and leading winter legume crop in India. It is a good source of protein about 16-19 per cent protein. Wet root rot is a minor disease and is caused by *Rhizoctonia solani* Kuhn. Characteristic symptoms include root rotting, containing often originating at distal tip of the young root and gradual yellowing and wilting of foliage. The rotted, discolored tissues are soft and wet. In the field, scattered drying plants can be seen throughout the field. Affected seedlings gradually turn yellow and petioles and leaflets droop (Harveson, 2011). Seedlings do not usually collapse. A distinct dark brown lesion appears above the collar region on the main stem and can extend to lower branches in older plants. This disease is most often seen at the seedling stage (upto 6 weeks after sowing) in soils with relatively high moisture content. However, in irrigated chickpea, the disease may occur at later stages in the crop growth. The stem and root below the lesion shows rotting, frequently with pinkish mycelial growth. Sclerotia are not usually seen. The pathogen can infect at a wide range of soil temperatures, but cool (11-18 °C) wet soil conditions are optimum (Nene *et al.*, 2012).

Infection Process of *R. solani* on Mungbean

Infection strategies of *R. solani* were studied on mungbean cultivar (Pusa Vishal) by inoculating leaves with mycelial disk of RDLM-1 isolate. The hyphal tip growths radiating from the inculcated point were observed after 3 hrs of inoculation. Only the growing young mycelium proceeded towards stomata after 6 h of inoculation. Further branching of growing hyphae (90° angle) was observed. The growing hyphae became thick in width and the tip of hyphae swell in lobate or globose to form appresorium. Aggregations of hyphae forming infection cushion were also observed after 11 h of inoculation. The hyphae of *R. solani* were observed penetrating through two ways (i) direct penetration-by swollen hyphae, lobate appressorium and infection cushion (Figure 10.4 a and b) (ii) indirect-stomatal penetration.

Characterization of *Rhizoctonia solani*

Morphological Characterization

Rhizoctonia solani Kuhn is a plant pathogen that attacks a large number of plant species around the world with diverse symptoms (Ploetz *et al.*, 1985). *Thanatophorus cucumeris* (Frank) Donk is the sexual stage of this pathogen. So far, anastomosis of mycelia is the criterion most widely accepted and used to group the isolates of this fungus (Anguiz, 1989). In PDA, the color of colonies was light-brown in young colonies and darker in old colonies. It does not produce any kinds of conidia or spores. The microscopic features of the fungus are septated hyphae, branching near septa of parent hypha, formation of septa in the point of branching, constriction near branching, no conidia, no clamp connections, no rhizomorph, no differentiation into rind and medulla of sclerotia. But grouping by anastomosis does not always correspond to grouping by colony morphology, pathogenicity or other physiological features (Carling, 1996).

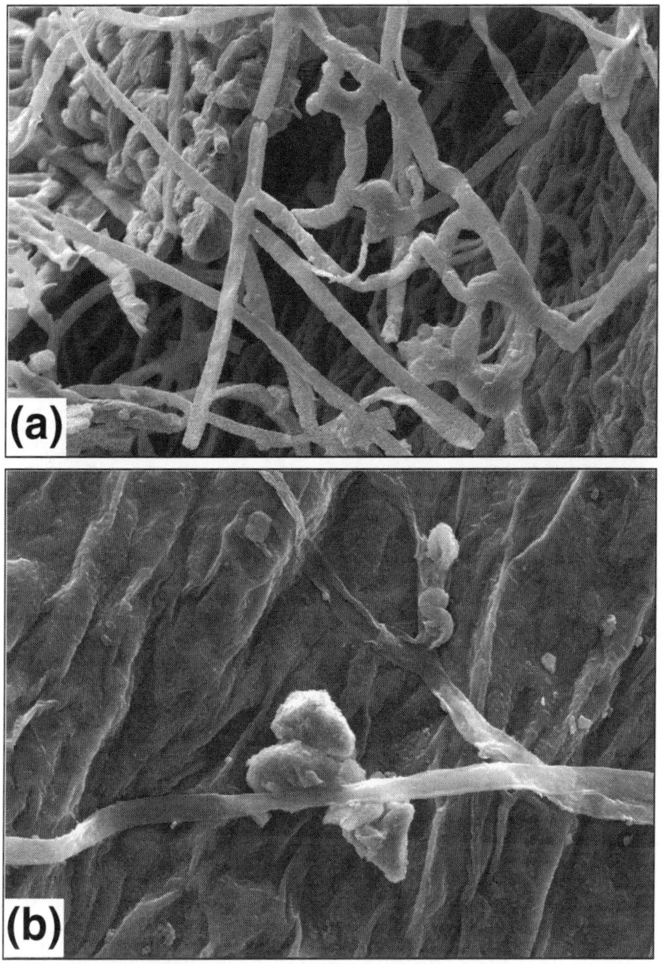

Figure 10.4. Scanning Micrographs Showing Infection Cushions and Appressorium of *R. solani* on Mungbean Leaves (a) Growing hyphae starting of infection cushion and lobate appressorium formation; (b) Infection cushion.

Anastomosis Grouping

Anastomosis grouping is based on the reaction of fusion of *Rhizcotonia* sp. Affinity for hyphal fusion (anastomosis) has been used to characterize isolates among *R. solani, R. zeae, R. oryzae, R. repens* and binucleate *Rhizoctonia* spp. with *Ceratobasidium* teleomorphs. To date, isolates of *R. solani* have been assigned to 13 anastomosis groups (AG) and those of *R. zeae* and *R. oryzae* have each been assigned to their own one group (Sneh *et al.*, 1998; Carling *et al.*, 1999, 2002). Anastomosis reactions between hyphae of paired isolates of *R. solani* consist of several types; such as perfect fusion, imperfect fusion, contact fusion and no reaction (Matsumoto *et al.*, 1932). At present, four categories of anastomosis (C3 to C0) defined by Carling *et al.* (1996) have been accepted by many researchers. These are useful for a better understanding of the genetic diversity of *R. solani* populations, because of the background genetically supported by vegetative or somatic compatibility (VC or SC) of confronted isolates (MacNish *et al.*, 1997). Each of categories is as follows: C3: walls fuse; membranes fuse, accompanied with protoplasm connection; anastomosis point frequently is not obvious; diameter of anastomosis point is equal or nearly equal hyphal diameter; anastomosing cells and adjacent cells may die, but generally do not. This category occurs for the same anastomosis group, same vegetative compatibility population (VCP) and the same isolate. C2: wall connection is obvious, but membrane contact is uncertain; anastomosing and adjacent cells always die. This category occurs in same AG, but not between different VCPs. C1: wall contact between hyphae is apparent, but both wall penetration and membrane-membrane contact do not occur; occasionally one or both anastomosing cells and adjacent cells die. This category occurs between different AGs or in the same AG. C0: no reaction. This category occurs between different AGs.

Molecular Characterization

Molecular Detection of *R. solani*

Molecular approaches have been very crucial and effective for answering basic questions about genetic variation and systematic relations in many groups of phytopathogenic fungi. Most of these molecular techniques are based on the detection and typing of genomic polymorphisms at several levels. Some of them include DNA/DNA hybridization, G+C content analysis or PCR based methods, including those that detect polymorphisms in the genome, like RAPDs and AFLPs, or in certain genomic regions or genes, like RFLPs (for ribosomal RNA) or direct sequencing of these and other genomic regions (ribosomal DNA, actin gene, transcription elongation factor (TEF), β-tubulin gene, Ras protein, chitin synthase, *etc.*) (Bruns, 1991; Mehmann, 1994; Thon and Royse, 1999; Helgason, 2003). Molecular approaches have been proved to be one vital tool for assessing genetic variations among isolates of *R. solani*. One direct approach for assessing genetic variation in fungi is by comparative study of molecular markers (Michelmore and Hulbert, 1987). A large number of genetic loci from the genome should be assayed so that genetic differences detected will reflect the variability of the genome as a whole. Molecular markers that are presumably selectively neutral satisfy such criteria. Many problems associated with studying differed levels of diversity in *Rhizoctonia* are best addressed through the use of molecular genetic markers.

Whole Genome Sequencing

Significant genomic resources for other AGs of *R. solani* have also recently become publicly available, formerly being limited to EST libraries of AG1-IA (Rioux *et al.*, 2011) and AG4 (Lakshman *et al.*, 2012). The recent generation of whole genome sequences of *R. solani* AGs presents new opportunities for comparative genomics between *R. solani* anastomosis groups (Table 10.2). The most comprehensive whole-genome study to date has been that of the rice pathogen AG1-IA (Zheng *et al.*, 2012) [GenBank: AFRT00000000]. The genome assembly of the closely related AG1-IB was published recently (Wibberg *et al.*, 2012) [GenBank: CAOJ00000000],

Table 10.2. Assembly Statistics of Four Draft Genome Assemblies of *R. solani*

Anastomosis Group	*Sequencing Method*	*Isolate*	*Total Assembly Length*	*Total Number of Scaffolds*	*Average Scaffold Length*	*SNPs*
AG8	Illumina: 75 bp paired-end, 200 bp insert size,	WAC10335	39.8 Mbp	861 2	46,260 bp	326180
AG1-IA	Illumina: 100 bp paired-end, 100 bp insert-size, 75 bp paired-end, 347 bp insert size	N/A	36.9 Mbp	2,648	13,949 bp	442164
AG1-IB	26 GS-FLX whole-genome shotgun libraries	7-3-14 (AJ868459)	47.65 Mb	1,600	N/A	191835
AG3	Sanger ABI 3730 XL: 4 and 10 kbp plasmid, 40 kbp fosmid	Rhs 1AP	51.0 Mbp	9,484	5,377 bp	866515

however full scaffold sequences were not in the public domain at the time of writing and thus AG1-IB data has not been used for synteny comparisons in this study. The mitochondrial genome sequence of the potato pathogen AG3 strain Rhs1AP and its comparison to that of AG1-IB has been published.

Variability Analysis

Upadhyay *et al.* (2013) analyzed genetic diversity of the *R. solani* infecting mungbean isolates using random amplified polymorphic DNA (RAPD), inter-simple sequence repeats (ISSR) and simple sequence repeat (SSR) markers. RAPD markers showed 98.5 per cent and ISSR showed 94.1 per cent polymorphism with the band size ranging from 0.1 to 5 kb whereas SSR markers showed 93.6 per cent polymorphism with the bands sizes ranging from 0.1 to 1.5 kb. Combined analysis of all the three markers by unweighted pair group method with an arithmetic average (UPGMA) grouped the isolates into 8 major clusters at 52 per cent genetic similarity and AGs specific sub-clustering was observed. The isolates showed high genetic variability and the molecular groups were partially corresponding to AGs and area of the origin but did not correspond to morphological characters. Recently, Tripathi and Dubey (2015) used SRAP markers for genetic diversity analysis of 89 isolates of *R. solani* belonging to 7 Anastomosis Groups (AGs) isolated from different pulse crops representing 21 states from 16 agro-ecological regions of India. Out of 30 SRAP primer combinations evaluated, 16 combinations provided amplification with 100 per cent polymorphism and the primer combinations Me1/Em1and Me1/Em4 provided the highest number of bands (14). The isolates of *R. solani* showed high level of genetic variability and grouped into 7 major clusters at 35 per cent genetic similarity by using unweighted pair group method with an arithmetic average analysis. Bootstrap analysis grouped the isolates into five major clusters at 28 per cent genetic similarity and about 95 per cent isolates shared common sub-grouping patterns in both the analysis.

Survival and Spread

R. solani survives by sclerotia in soil and by mycelium in infected crop residue in fields. As it has wide host range it may also survive in weeds. Yang *et al.* (1990) has reported that rain-splashed plant debris containing mycelia and sclerotia from the soil serves as the primary source of infection for *Rhizoctonia* aerial blight of soybean. There is also the possibility that *R. solani* has been survived a long time in soil. The results suggested that initiation growth and spreading of *R. solani* from soil and infected plant debris. Nelson *et al.* (1996) stated that the epidemics of *R. solani* in field crops were related to a function of environment condition, pathogen density in the soil, and as a result of rotation with host crops. According to Russin and Stetina (1999) there are two stages of *R. solani* disease development. The first is the soilborne phase, when disease loci became established and the second is sclerotia structure or mycelia fragments that are splashed onto the plants canopy (aerial plants part) by rain.

One interesting observation we have found during the investigation, close planting or higher plant density can lead to rapid spread or epidemic of web blight/ aerial blight of mungbean. If we assume this pot experiment mimics the natural or agroecosystems, we can halt the rapid spread of *R. solani* from infected mungbean leaves to touching other healthy leaves (Figure 10.6).

Figure 10.6. Mimicking Pot Experiment Showing Rapid Spread of *R. solani*.

Management of Diseases Caused by *Rhizoctonia* sp.

Host Resistance

Plant disease resistance is considered one of the safest, eco-friendly and economically feasible method of disease management. Disease management through resistant cultivars is still thrust research area in pulse crops. No effective resistant cultivars are available which can give control of *Rhizoctonia* caused web/aerial blight/root rot disease of mungbean, chickpea under field conditions. However, some workers have screened mungbean/chickpea against *Rhizoctonia*. Due to its necrotrophic nature, there does not appear to be qualitative resistance against *R. solani* (Govrin and Levine 2000; Glazebrook, 2005). Quantitative trait loci for resistance have been mapped in both potato (Leach and Webb, 2002) and rice (Liu *et al.*, 2009). Even with the availability of marker-assisted selection in rice, however, progress towards the development of *Rhizoctonia*-resistant cultivars has been limited (Liu *et al.*, 2009). Transgenic plants over-expressing defense-related genes have been shown to exhibit varying levels of resistance to *R. solani* but are not widely available due to lower efficacy in comparison with other methods, particularly chemical control (Datta *et al.*, 1999; Jha *et al.*, 2009).

Biological Control

Use of bio-agents such as *Trichoderma* spp. is helpful in managing fungal diseases in chickpea, lentil and pigeonpea and also enhances the growth of plants. Several isolates of *Trichoderma* spp. were characterized and evaluated against different fungal pathogens of pulse crops (Dubey *et al.*, 2007). Further, *Trichoderma harzianum* (wheat bran powder granules) has been used for management of pigeonpea and chickpea diseases (Prasad and Rangeshwaran, 1999). Tewari, and Mukhopadhyay (2003) showed that vitavax (carboxin @ 1g/kg seed) did not significantly affect the spore viability of *Gliocladium virens* (Gv). Application of carboxymethyl cellulose (CMC) with *G. virens* powder (10^9 spores/g) in combination with vitavax provided maximum protection (81.9 per cent) to the crop against chickpea root rot and collar rot pathogens in glasshouse. Chickpea seeds treated with Gv powder + CMC + vitavax significantly increased seedling emergence (47.9 per cent); final plant stand (85.8 per cent) and grain yield (79.7 per cent) which was statistically at par with the treatment Gv powder + vitavax and Gv suspension + vitavax in a sick plot. Dubey *et al.* (2011 and 2012) reported that *Trichoderma harzianum* and *Trichoderma virens* based seed dressing formulation Pusa 5 SD and soil application formulations Pusa Bio pellet 16G and Pusa Bio granule 6 were used for management of dry root rot and wet root rot of chickpea respectively. Three bio-control agents were evaluated in laboratory condition *Trichoderma viride, Trichoderma harzinum* and *Pseudomonas fleuresecense, Trichoderma viride* showed best performance against the pathogen, *Rhizoctonia bataticola*, followed by *Pseudomonas fluorescence* (Khan *et al.*, 2012). Parsada *et al.* (2002) evaluated two antagonistic fungi *viz., Trichoderma harzianum* (PDBCTH 10) and *T. viride* (PDBCTV) against natural incidence of wilt and wet root rot of chickpea in a field plot. Soil application of these bioagents, one week before sowing was found to be more effective than seed treatment in reducing wilt and wet root rot of chickpea. The two bioagents had positive effect on plant growth. *T. harzianum* soil treated plots recorded only 4.9 and 1.2 per cent root rot at 30 and 60 days respectively and was found to be superior to all other treatments. In pathogen check, the root rot incidence was 16.4 and 17.0 per cent at 30 and 60 days respectively. The seed yield recorded was significantly high in all treatments when compared to pathogen check and the highest yield of 965.4 kg/ha was obtained with soil treatment of *T. harzianum*.

Cultural Practices

Crop sanitation is one most important and best asset in reducing primary inoculum ($X_0 = 0$). The development of disease primarily depends on the amount of initial inoculum at the site of selection. As *Rhizoctonia* survive as sclerotia and dormant mycelium in crop residue. Therefore, burning and destroying of the crop residue may reduce inoculum pressure in field. Alteration in dates of sowing and manipulation in plant spacing are such techniques to reduce the crop losses from plant diseases. This technique works because there is no host pathogen interaction under favorable environmental conditions for the development pathogen and disease. The effect of dates of sowing on web blight have been studied and higher disease incidence is reported in early sown plants (July 1st and July 16th) whereas least disease severity was recorded on late sown crop (16th Aug.). However, maximum grain yield and comparatively lower disease severity was recorded in plants sown on August 1st as compared to those which were sown on July 1st and July 16 th, respectively (Singh, 2006). Spacing can also play a significant role in management of web blight disease as it spreads from disease to nearby healthy plants. Singh (2006) reported that wider row spacing of 50 cm or more resulted in decreased web blight infection and increased yield of mungbean.

Chemical Control

Undeniably chemical application is one of the most effective and economic means to manage crop losses incurred by plant diseases. Use of chemicals in plant disease control is aimed to (i) Creat a toxic barrier between the host surface or tissue and the pathogen, and (ii) eradicate the pathogen present at a particular site on the host such as seed, foliage, roots, *etc.* The chemicals used in disease control oppose the germination, growth and multiplication of the pathogen or directly destroy it by toxicity (Malavath *et al.*, 2013). Fentin acetate (0.16 kg/ha) or fentin hydroxide (0.20 kg/ha) applied after benomyl, gives good control. Thiophanatemethyl (0.5 kg/ha), carbendazim (0.5-1.0 kg/ha), and captafol (1.0-3.5 kg/ha) are also useful (Gálvez *et al.*, 1989). Dubey and Dwivedi (1989) found bavistin to be the most effective chemical against *Thanatephorus cucumeris* on *Vigna mungo*, followed by dithane Z-78 (Zineb) at 0.25 per cent and benelate (Benomyl) at 0.2 per cent in a pot experiments. Benomyl (0.25-0.5 kg/ha) helps manage the pathogen when it is applied at first-symptoms appearance and then every 15 days. Shailbala and Tripathi (2004) showed that tilt (0.1 per cent) gave lowest disease severity, highest grain yield and maximum thousand grain weight followed by bavistin (0.1 per cent), dithane M-45 (0.25 per cent), bavistin (0.05 per cent) sprayed plots, respectively. Antracol was found comparatively less effective in comparison to other treatments. Raxil was found least effective for controlling the web blight disease. Bavistin was found highly remunerative giving maximum returns.

Integrated Management

Dubey (2003) showed that the integration of soil application of *Karanj* cake (2 q ha'), seed treatment with G. *virens* (10^6spores/ml/10g seed) + carboxin (1 g/kg) + *Rhizobium* sp. (25 g/kg seed) and foliar spray of *Karanj* leaf extract (10 per cent) increased 35.7-36 per cent seed germination, reduced 92.4-93.7 per cent disease intensity and considerably enhanced plant vigour and grain yield of urd and mungbean. However, soil application of *Karanj* cake or *G. virens* multiplied on pulse bran-saw dust-tap water (3:1:4 *w/w/v*), seed treatment with *G. virens* + carboxin-*Rhizobium*, and foliar spray of *Karanj* leaf extract separately also proved effective for the management of disease, but integration of any two was more effective than anyone alone.

Conclusions

Rhizoctonia solani is an economically important soilborne fungus with a worldwide distribution. It survives in nature by infecting wild host and weed host, sclerotia in soil and as mycelium in infected host residues. Due to saprophytic nature, wide host ranges, high level of genetic diversity or variability, rapid genetic changes by anastomosis process, rapid adaptation on host, unavailability of resistant cultivars, the management strategy of *R. solani* often leads to failure and frustrating to growers.

REFERENCES

Anderson, N.A. (1982). The genetics and pathology of *Rhizoctonia solani*. *Annu. Rev. Phytopathol.* 20: 329–347. doi: 10.1146/annurev.py.20.090182.001553

Anguiz, R. (1989). Anatomosis groups, pathogenicity and other characteristics of *Rhizoctonia solani* isolated from potatoes in Peru. *Pl. Dis.* 73 (3): 199-201.

Bruns T.D., White T.J. and Taylor J.W. (1991). Fungal molecular systematics. *Ann. Rev. Ecol. Syst.* 22: 525-564.

Carling, D. E. (1996). Grouping in *R. solani* by hyphal anastomosis reaction. In *Rhizotonia* species. Taxonomy, Molecular biology, Ecology, Pathogenicity and Disease control. 37-47. Khiewer Academic Publishers, Netherlands.

Cubeta, M.A., Dean, R.A., Thomas, E., Bayman P, Jabaji S, *et al.* (2009). *Rhizoctonia solani* genome project: providing insight into a link between beneficial and plant pathogenic fungi. *Phytopathol.* 99: S166.

Datta, K., Velazhahan, R., Oliva, N., Ona, I., Mew, T., Khush, G.S., Muthukrishnan, S. and Datta, S.H. (1999). Over-expression of the cloned rice thaumatin-like protein (PR-5) gene in transgenic rice plants enhances environmental friendly resistance to *Rhizoctonia solani* causing sheath blight disease. *Theor. Appl. Genet.* 98: 1138–1145.

Dubey S.C., Tripathi A. and Singh B. (2012). Combination of soil application and seed treatment formulations of *Trichoderma* spp. for integrated management of wet root rot caused by *Rhizoctonia solani* in chickpea (*Cicer arietinum*). *Indian Journal of Agricultural Sciences* 82: 356-362.

Dubey S.C., Tripathi A., Bhavani R. and Singh R. (2011). Evaluation of seed dressing and soil application formulations of *Trichoderma* spp. for integrated management of dry root rot of chickpea. *Biocontrol Science* and *Technology,* 21: 93-100.

Dubey, S.C. (2003). Integrated management of web blight of urd and mung bean. *Indian Phytopath.* 56 (4): 413-417.

Dubey, S.C. and Dwivedi, R.P. (1989). Mode of penetration by *Thanatephorus cucumeris* (*Rhizoctonia solani*) in some leguminous host. *Indian Phytopath.,* 42 : 317.

Faske, T., Kirkpatrick, T. Zhou, J. and Tzanetakis, I. 2014. Soybean Diseases. Arkansas Soybean Production, University of Arkansas System. Handbook. 1-18p.

Galvez, G.E., Guzman, P. and Castano, M. (1979). *Web blight in bean production problems: Disease, insect, soil and climatic constraints of Phaseolus vulgaris* (eds. Schwariz, H.F. and Galvez, G.E.) Cali, Colombia, *Centro Inter. de Agric. Trop.,* pp. 101-110.

García, V. G., Portal Onco, M. A. and V. Rubio S. (2006). Biology and Systematics of the form genus *Rhizoctonia*. *Spanish J. Agric. Res.* 4(1), 55-79.

Genhua, Y. and Chengyun, Li (2012). General Description of Rhizoctonia Species Complex, Plant Pathology, Dr. Christian Joseph Cumagun (Ed.), ISBN: 978-

953-51-0489-6, InTech, Available from: http: / /www.intechopen.com/books/ plant-pathology/general-description-of-rhizocotonia-species-complex

Glazebrook J (2005). Contrasting mechanisms of defense against biotrophic and necrotrophic pathogens. *Annu. Rev. Phytopathol.* 43: 205–227.

Govrin E.M., Levine, A. (2000). The hypersensitive response facilitates plant infection by the necrotrophic pathogen *Botrytis cinerea. Curr Biol.* 10: 751–757.

Harveson, R. M. (2011). Soilborne root diseases of chickpeas in Nebraska. http: / / extension.unl.edu/pubications.

Helgason, T., Watson, I.J., Young, J.P.W. (2003). Phylogeny of the Glomerales and Diversisporales (Fungi: Glomeromycota) from actin and elongation factor 1-alpha sequences. *FEMS Microbiol. Lett.* 229, 127-132.

Jha, S., Tank, H.G., Prasad, B.D., Chattoo, B.B. (2009). Expression of Dm-AMP1 in rice confers resistance to *Magnaporthe oryzae* and *Rhizoctonia solani*. *Transgenic Res.* 18: 5969.

Kharayat, B.S., Kumar, J., Bashyal, B.M., Dubey, S.C., Sharma, S., Rawat, K. and Aggarwal, R. (2015). Evaluation of pathogenic potential and survival of *Rhizoctonia solani* isolates in rhizosphere of different wheat genotypes using conventional and real time PCR techniques. *Journal of Biotechnology and Biomaterial*. 5(6): 263. http: //dx.doi.org/10.4172/2155-952X.C1.044

Kucharek, T. (2001). Common soybean leaf and stem diseases. Plant Pathology Department, IFAS, University of Florida, Gainesville. *Plant Pathology Fact Sheet* PP-16.

Lakshman, D.K., Alkharouf, N., Roberts, D.P., Natarajan, S.S. and Mitra, A. (2012). Gene expression profiling of the plant pathogenic basidiomycetous fungus *Rhizoctonia solani* AG 4 reveals putative virulence factors. *Mycologia* 104: 1020–1035. doi: 10.3852/11-226

Lal, S., Baruah, P. and Butchaiah, K. (1980). Assessment of yield losses in maize cultivars due to banded sclerotial disease. *Indian Phytopath.*,33 (3) : 440-443.

Leach, S.S. and Webb, R.E. (2002). Evaluation of potato cultivars, clones and a true seed population for resistance to *Rhizoctonia solani*. *Am. J. Potato Res.* 70: 317–328.

Liu, G., Jia, Y., Correa-Victoria, F.J.C., Prado, G.A., Yeater, K.M., McClung, A., Correll, J.C. (2009). Mapping quantitative trait loci responsible for resistance to sheath blight disease. *Phytopathol.* 99: 1078–1084.

Losada L, Pakala SB, Fedorova ND, Joardar V, Shabalina SA, *et al.* (2014). Mobile elements and mitochondrial genome expansion in the soil fungus and potato pathogen*Rhizoctonia solani* AG-3. *FEMS Microbiol Lett* (In Press).

Malavath, Rajeshwar and Mani, S. (2013). Current status of web blight of mung bean. *Asian J. Soil Sci.*, 8(2): 495-504.

Mehmann, B., Brunner, I., Braus, G.H. (1994). Nucleotide sequence variation of chitin synthase genes among ectomycorrhizal fungi and its potential use in taxonomy. *Appl. Environ. Microbiol.* 60: 3105-3111.

Nelson, B., Helms, T., Christianson, T., Kural, I. (1996). Characterization and pathogenicity of *Rhizoctonia solani* from soybean. *Pl. Dis.* 80: 74-80.

Nene, Y.L., Reddy, M.V., Haware, M.P., Ghanekar, A.M., Amin, K.S., Pande, S, and Sharma, M. (2012). Field Diagnosis of Chickpea Diseases and their Control. Information Bulletin No. 28 (revised).

Ogosid, A. (1996). Introduction-the genus *Rhizoctonia, IN: Rhizoctonia* Species: Taxonomy, Molecular Biology, Ecology, Pathology and Disease Control. (edt. Sneh, B., Hare, S.J., Neate, S. and Dijst, G.). Springer Science and Business Media Dordrecht. 561p.

Ploetz, R.C, D. Mitchell and R.N. Gallaher. (1985). Characterization and pathogenicity of *Rhizoctonia* species from a reduced tillage experiment multicropped to Rye and soybean in Florida. *Phytopathol.* 75 (7): 833-837.

Prasad, R.D. and Rangeshwaran, R. (1999). Granular formation of *Trichoderma* and *Gliocladium* sp. In: Bio-control of *Rhizoctonia Solani* of chickpea. *J. Mycol. Pl. Pathol.* 29: 222-226.

Prasad, R.D., Rangeshwaran, R., Anuroop, C.P. and Rashmi, H.J. (2002). Biological control of wilt and root rot of chickpea under field conditions. *Ann. Pl. Protect. Sc.,* (10)1, 72 -75.

Rahayu, M. (2014). Identification and pathogenicity of pathogen responsible for aerial blight disease of soybean. *Journal of Experimental Biology and Agricultural Sciences.* 2(2S): 281-285.

Rioux, R., Manmathan, H., Singh, P., de los Reyes, B., Jia, Y.L. *et al.* (2011). Comparative analysis of putative pathogenesis-related gene expression in two *Rhizoctonia solani* pathosystems. *Curr. Genet.* 57: 391–408. doi: 10.1007/s00294-011-0353-3

Russin J.S., Stetina SR (1999). *Rhizoctonia* foliar blight. In: Hartman GL, Sinclair JB, Rupe JC (Eds). Compendium of soybean diseases. American Phytopathological Society, St. Paul, Minnesota. pp: 24-25.

Shailbala and H.S. Tripathi. (2004). Fungicidal management of web blight of urdbean. *Indian Phytopath.* 57 (1): 99-100.

Sharon, M., Kuninaga, S., Hyakumachi, M., Naito, S., and Sneh, B. (2008). Classification of *Rhizoctonia* spp. using rDNA-ITS sequence analysis supports the genetic basis of the classical anastomosis grouping. *Mycoscience* 49(2): 93-114.

Tewari, A. K. and Mukhopadhyay, A. N. (2003). Management of chickpea root rot and collar rot by integration of biological and chemical seed treatment. *Indian Phytopathol.* 56 (1): 39-42.

Thon, M. R. and Royse, D. J. (1999). Partial β-tubulin gene sequences for evolutionary studies in the Basidiomycotina. *Mycologia* 91: 468-474.

Tripathi, A. and Dubey, S.C. (2015). Sequence-related amplified polymorphism-PCR analysis for genetic diversity in *Rhizoctonia solani* populations infecting pulse crops in different agro-ecological regions of India. *Plant Pathology Journal* 14 (4): 234-241.

Upadhyay, B. K., Dubey, S.C., Singh, R. and Tripathi, A. (2013). Morpho-molecular characterization of Indian isolates of *Rhizoctonia solani* infecting mungbean. *Research Journal of Biotechnology*. 8 (11): 92-99.

Wibberg, D., Jelonek, L., Rupp, O., Hennig, M., Eikmeyer, F., *et al.* (2012). Establishment and interpretation of the genome sequence of the phytopathogenic fungus *Rhizoctonia solani* AG1-IB isolate 7/3/14. *J. Biotechnol*. 167: 142–155. doi: 10.1016/j.jbiotec.2012.12.010

Yang, X.B., Snow, J.P. and Breggen, G.T. (1990). Analysis of epidemics of *Rhizoctonia* aerial blight of soybean in Louisiana. *Phytopathol*. 80: 386-392.

Zheng, A., Lin, R., Zhang, D., Qin, P., Xu, L., *et al.* (2013). The evolution and pathogenic mechanisms of the rice sheath blight pathogen. *Nat. Commun.* 4: 1424. doi: 10.1038/ncomms2427

Chapter 11

Sclerotium rolfsii: A Threat to Legume Crops

M.K. Pandey & R.K. Mishra

Sclerotium rolfsii Sacc. (Teleomorph: *Athelia rolfsii* (Curzi) Tu and Kimbrough) is one of the devastating, cosmopolitan, ubiquitous, noxious, serious soil-borne and omnivorous pathogens with a diversified host range including both monocotyledonous and dicotyledonous plants encompassing more than 500 host species (Aycock, 1966; Punja, 1985). Bilici *et al.*, 1994 reported that *S. rolfsii* prevail at 95 per cent in the pea nut fields and produced on average 8.6-21.9 sclerotia/700 g soil samples, and 2.8-3.5 loci/30 metre (m) peanut rows in Adana province of Turky. Sclerotium is a hard, compact, dark coloured resting body resistant to unfavourable environmental conditions, which may remain dormant for long periods of time and germinate on the return of favourable conditions. Sclerotium is formed by many fungi like, *Rhizoctonia, Sclerotinia* and *Claviceps etc.*, whereas formation of sclerotium (pl. sclerotia) is a typical/basic character in *Sclerotium* for which the same word has been given to this genus. The major difference between sclerotia of *Rhizoctonia solani* and *S. rolfsii* is undifferentiation of *R. solani* into a rind and a medulla while, differentiation of sclerotia of *S. rolfsii* into a rind a cortex and a medulla (Singh, 1982). This fungus is mostly facultative parasite, growing saprophytically after killing their host. The pathogen has white thread like mycelium and tan to dark brown sclerotia produced on the colony. The pathogen does not produce inoculum for secondary infection in the neighboring plants in the same season (Fry, 1982). It becomes severe only when large inoculum is present in the soil. Root diseases caused by *S. rolfsii* generally become severe in moist soil at warm temperature (Cook and Papendick, 1972). Due to wide range, crop rotation is not an adaptable cultural practice that is usually recommended for other soil-borne diseases. Sclerotia formed on undecomposed tissues in the field are capable for initiating infection and serve as the primary inoculum of the disease in the field (Punja and Grogan, 1981).

Peter Henry Rolfs in 1892 first time reported this fungus on tomato in Florida causing more than 70 per cent loss (Aycock, 1966). While, the binomial of Rolfs fungus (*Sclerotium rolfsii*) was first time ascribed by Saccardo (1913) to include those fungi with no known sexual state but formed small, tan to dark-brown spherical sclerotia (0.5 to 2 mm diameter) comprised of rind, cortex and medulla. The large number of sclerotia produced by *S. rolfsii* and their ability to persist in the soil for several years associated with the prolific growth rate of the fungus (2-3 cm per day in culture) make it well-suited facultative and a pathogen of major importance throughout the world. Some workers (Harlton *et al.*, 1995; Nalim *et al.*, 1995; Okabe *et al.*, 1998) have studied the variability in this fungus taking mycelia as the experimental tool, but involvement of basidiospores in variability of this fungus has not yet been thoroughly explored. In addition to collar rot infection, report on this fungus causing spotted leaf rot disease in many taxonomically unrelated plant species is also available in literature (Singh and Pavgi, 1965). Several hypothesis exist on the possible functions and use of sclerotial exudate during sclerotial development without conclusive reasons. Polyphenoloxidase (PPO) production by the pathogen is reported earlier (Punja and Damiani, 1996). But its involvement in pathogenicity affecting the other plant phenolic is not clearly understood which unequivocally needs further study. Despite continuous research over the past one century, this pathogen continues to plague growers causing considerable economic loss. Among the methods employed to manage *S. rolfsii* are the following: fungicide applications, solarization, use of antagonistic microorganisms, deep ploughing, crop rotation, and incorporation of organic and inorganic residues (Punja, 1985). Management efforts have often met with limited success, due to the extensive host range. Furthermore, control measures effective for a particular crop or area may not be adaptable elsewhere due to regulatory or economic constraints. Several fungicides effectively controlled this pathogen on various crops in the field (Brown and Hendrix, 1980, Punja *et al.*, 1982), but the major limitations to their widespread use to control this pathogen are requirement of large amounts of chemicals, variable effects on different crops, and variation in results in different seasons. Biological control of phytopathogens by microorganisms has been considered to be more natural and an environmentally acceptable alternative to the existing chemical treatment methods. Several applications have been reported using plant-beneficial fungi and bacteria strains to protect plants against soil-borne pahtogens. The main mechanisms involved in the antagonism of biocontrol agents are mycoparasitism, competition for space and nutrients, stimulation of the plant's defensive capacity and secretion of bioactive compounds such as antibiotics and cell wall degrading enzymes. Chitinolytic enzymes have been considered for biocontrol of pathogenic fungi due to their ability to degrade chitin, which is a major structural component of most fungal cell wall. Several chitinolytic bacteria and fungi have been reported to be potential biological agents. In this chapter we discussed about Distribution, Host range, Symptomatology, Biology, Disease cycle and Management aspect of this fungus. Other aspects like ultrastructure and morphogenesis of sclerotia, soil biology, pathogenesis and disease development have been reviewed by Aycock, 1966, Punja, 1985, Sarma *et al.*, 2002 and Arunasri *et al.*, 2011.

Distribution

S. rolfsii continues to be a problem in a variety of crops when conditions are warm, humid and rainy. This disease occurs around the world in the equatorial zone between the 45° North and South latitudes. Diseases caused by *S. rolfsii* are rampant in the tropics, subtropics and other warm temperate regions, especially in areas of the southern and southeastern United States, Central and South America, the West Indies, Southern European countries bordering the Mediterranean sea, Africa, India, Japan, Australia, New Zealand, Philippines and Hawaii, where temperature are sufficiently high to permit the growth and survival of the fungus (Punja, 1985; Ferreira and Boley, 2004). There has been limited report from Siberia and North China. The pathogen rarely occurs where average winter temperatures fall bellow 0° C. In India, this pathogen has been reported from Uttar Pradesh, Madhya Pradesh, Bihar, West Bengal, Assam, Uttranchal, Jharkhand, Gujrat, Tamil Nadu, Rajsthan and almost all states.

Host Range

S. rolfsii has an extensive host range; at least 500 plant species in 100 families are susceptible. The most common hosts are the legumes, crucifers and cucurbits (Ferreira and Boley, 2004-internet; Punja, 1985). French bean (*Phaseolus vulgaris*), chickpea (*Cicer arietinum*), cowpea (*Vigna sinensis*), lentil (*Lens esculentum*), mustard (*Brassica compestris*) and pea (*Pisum sativum*) *etc.* are important host for *S. rolfsii.*

Diseases Caused by *S. rolfsii*

S. rolfsii causes different types of diseases in legume plants, i. e., collar rot, leaf spots, southern stem rot, southern wilt, peanut pod break down, root rot, seedling blight, damping off seedlings, white mould, crown rot, stem rot, foot rot and fruit rot *etc.* (Singh and Pavgi, 1965, Punja, 1985).

Symptomatology

Sclerotium rolfsii Sacc. causes various types of diseases in different plants. The symptoms appears on plants parts varies on types of infection. It infects seedlings, herbaceous plants, woody plants, roots, bulbs and fruits *etc.* It infects plant parts near or at the soil surface more frequently, but this fungus may also infect any part of susceptible plant as long as favourable environmental conditions exits. Following symptoms are produce by *S. rolfsii.*

Leaf Spot

Generally low growing plants, their leaves touching the soil may be infected with *S. rolfsii*. Usually the high humidity near the soil surface is an important factor in disease development. Main source for leaf spot development in the plants in nature may be landing of sclerotia with rain splash. The infection of *S. rolfsii* usually starts as a chlorotic translucent spot on the upper surface of the lower leaves closer to the ground (about 3-7 cm above the ground level) but it also infect the leaves which is 30 to 50 cm above the ground level. The spot gradually becomes necrotic measuring 5 to 10 mm in diameter. The spots are circular, medium brown to light brown or straw-coloured with narrow borders. On the basis of distance of leaves from the soil, number of infected leaves in a plants may vary. Humid weather is favourable for development of infection while dry weather check the development of infection. Normally only one leaf spot is formed on a leaf but some times two or more spots develop closely coalesce to each other and form bigger spots. Concentric rings like a target-board invariably appear in the necrotic tissues and small clumps of mycelium on them often develop into tiny pin head-like sclerotia. The sclerotia are pinkish to mustard brown in colour. Concentric rings are dark in colour with dull grey appearance in the center. Drooping do not take place in infected leaves but occasionally shot holes appear in the dead tissues. The disease was observed in many hosts while, in *Sphaeranthus indicus*, *Spathodia compoleta*, *Blepharis boerhaviaefolia*, *Ficus infectoria*, *Morus nigra*, *Artrica* sp., *Solanum melongena*, *Bombax malabaricum*, *Rauvolfia serpentina* and *Ficus religiosa* hosts are reported for the first time.

Collar, Crown, Root and Stem Rot

S. rolfsii infects the collar region of many plants. It causes total yield loss in the infected plants because whole plants dies before flowering due to infection. Disease begins with a small, water-soaked lesion on lower stem/collar region at or near soil surface. The brown lesion spreads rapidly to girdle the stem. The girdling lesion will quickly cause the herbaceous plant to wilt and fall over. After stem rot of herbaceous plant, a white mat of mycelium develops at the lesion site and later spread onto nearby soil surface. After some time white, round and fuzzy mycelial bodies begin to appear which later turn tan or brown sclerotia resembling mustard seeds. The fungus is sometimes referred to as the "mustard seed fungus". Sclerotia develop on infected stems, roots and the surrounding soil help in the identification of the disease. During rotting of lower part of the stem, plants usually remain erect and wilting of foliage start. Wilted leaves gradually become brown and remain hanging on the plant after infection in many host plants. Sclerotia present on diseased tissues, soil surfaces or in soil crevices serve as primary source of inoculum. Generally, white thread-like mycelia radiate from a mycelogenous germination of sclerotium, infect collar region of the plants under favourable conditions. The collar gets constricted and turns grey to brown in colour. The fungus grows upwards (2-5 cm) on the dried stem depending on the soil moisture. Occasionally it also moves downwards and kills the roots. In case of root rot the leaves eventually die, and branch dieback develops. The lower stem (trunk) above the girlding lesion usually appears normal for a long time and will be the last part of the plant to die. In case of woody plants infection of *S. rolfsii* begin as a crown rot. Characteristic mycelial mat and sclerotia appear at the infection site of crown. In monocots symptoms are totally different, wilting of foliage and dieback develop after rotting of the crown tissues. Brown lesions occur at the crown and lower parts of the culm. Lesions are often small, but they may extend into the hollow part of the culm. Strands of mycelium grow inside the lower internodes. Seed heads may appear normal, but they are devoid of grain. Premature ripening may also occur. The fungus secretes several enzymes, which dissolve various constituents of the host tissues before deep penetration of the hyphae. It takes about 3-5 days for infecting the plants. The affected tissues decay and the mycelium grows very rapidly on such rotten tissue.

Infected plants topple down as the constricted collar region becomes weak, hence, unable to bear the weight of the foliage. Sometimes, the plants do not fall on the ground even after complete drying. In case of severe infection patches of grayish brown wilted plants are seen in the field.

Fruit Rot

Fruits of many crops plant present at or near soil surface may be infected with *S. rolfsii*. Fruit rot symptoms differ with plant species. Fruit of tomato show soft, water-soaked, sunken yellow lesions. These lesions spread quickly to almost all fruits and mycelium and sclerotia develop on lesions.

Biology of the Pathogen

Systemic Position

Sclerotial stage of *S. rolfsii* belong to Kingdom-Mycota, Division-Eumycota, Sub-division-Deuteromycotina, Class-Hyphomycetes, Order-Agnomycetales (Mycelia sterilia), Family-Agnomycetaceae (Hawksworth *et al.*, 1983, Alexopolus, 1962). Sexual stage *Athelia rolfsii* belongs to higher Basidiomycotina (having true

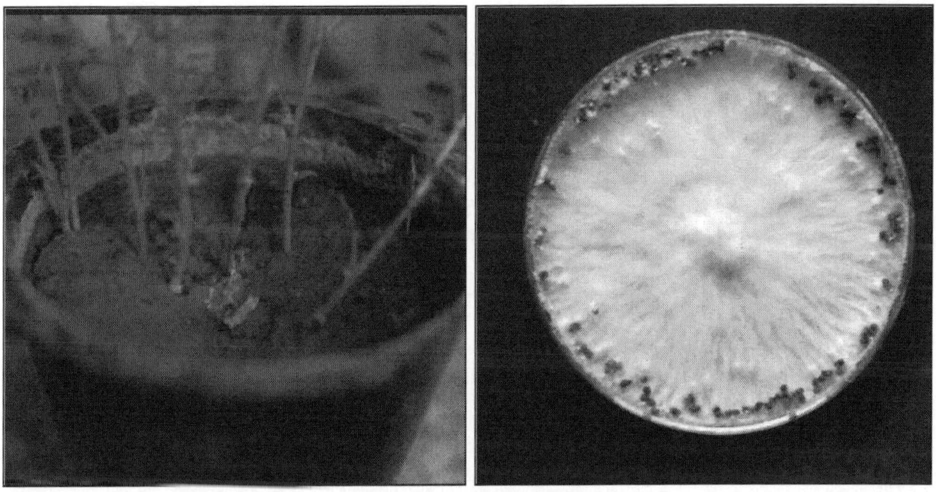

Figure 11.1. *In vitro* Study of *Sclerotium rolfsii.*

Figure 11.2. Infection Cause by *Sclerotium rolfsii.*

Figure 11.3. Seed Treatment and Soil Application of *Trichoderma* for Management of *Sclerotium rolfsii.*

basidia; eubasidia), Class-Hymenomycetes, Sub-class-Holobasidiomycetidae, Order-Aphyllophorales, Family-Corticiaceae (Talbot, 1968, Punja, 1985, Singh, 1982). Generic descriptions of both stages are given below.

Asexual Stage (Anamorph)

Mycelium

In early stage of growth in pure culture the fungus mycelium is at first silky-white but gradually loses its lustre and becomes somewhat dull in appearance. The mycelium radiating with abundant aerial hyphae appear as dense tufts dispersed all over the culture medium. Individual hyphae are hyaline, thin walled, branching at acute angles, sparsely septate when young, X-ray diffraction pattern indicate the presence of chitin and β-(1-3) glucan in both hyphal and sclerotial wall (Chet and Henis, 1968). The cells have been reported to be 60 to 350 micron (μ) long and 2 to 8 μ wide. The broader hyphae often may have clamp connections at the septa. The slender hyphae often lack clamp connections. Number of nuclei per cell is variable but in the secondary and tertiary branches it is mostly two although as many as 50-60 nuclei have been reported in all the cells.

Sclerotia

Sclerotial initials are formed from hyphal strands, which consist of 3-12 hyphae lying parallel. A spherical mycelial mass is soon assumed even if it is only a loose mass of hyphal network. After hardening sclerotium shows an outer layer of polyhedral cells. Rind, cortex and medulla are differentiated as the sclerotium darkens. The rind consists of thickened cells, 2-4 cells broad and the cortex consists of thin-walled cells with densely staining contents 6-8 cells broad. The medulla (central core) contains loosely arranged filamentous hyphae. Mature sclerotia are reddish to dark brown in colour and usually round in shape, surface smooth or pitted. In the beginning they are white but soon become dark and internal tissues of the sclerotium is white. Mature sclerotia germinate through mycelogenous germination. Melanin like pigment is reported to present in sclerotial wall (Chet and Henis, 1968).

Sexual Stage (teleomorph)

The mechanism of sexual stage formation in many plant pathogenic fungi, especially facultative parasites, are not well understood. Goto (1930) described the perfect stage of *S. rofsii* for the first time and named it provisionally as *Corticium centrifugus* (Lev.) Bres. Curzi (1931) erected a new species *Corticium rolfsii* Curzi. West (1947) transferred these fungi to *Pellicularia.* Venkatarayan (1950) followed another change by keeping them in the genus *Botryobasidium*. Talbot (1973) suggested that the perfect stage of *S. rolfsii* belong to the genus *Athelia,* and then Tu and Kimbrough (1978) described *Athelia rolfsii* as the perfect stage of *S. rolfsii* while studying *Rhizoctonia-Sclerotium* complex. The currently accepted binomial for the sexual stage of *S. rolfsii* is *Athelia rolfsii* (Curzi) Tu and Kimbrough (Punja *et al.,* 1982).

Various types of media were used to induce athelial stage by many workers. Goto (1930) used onion agar, Mundkur (1934) onion-asparagine-peptone and Chet *et al.* (1966) Joham's agar medium containing amino acids. Punja *et al.* (1982) used 8 different media including 2 per cent activated charcoal in potato dextrose agar (C-PDA) medium for inducing athelial stage. They found nutritional or environmental conditions that restricted optimal growth and sclerotial production by the fungus favoured hymenial formation in general and found no apparent correlation between host or area of isolation and ability of an isolate to form hymenia. They also found more extensive hymenial formation in cultures incubated at reduced light intensities (about 80-100 lux) than at higher light intensities (about 1200 lux). Singh *et al.* (1996) used *Cyperous rotundus* rhizome meal agar (CRMA) medium which is easiest and quickest for athelial stage induction.

Basidiocarp

Basidiocarp resupinate, effused, in patches usually 1-2 cm in diameter (sometimes upto 6 cm), pellicular, usually 50-100 um thick and loosely adherent to the substrate. Subicular hyphae small, hyaline, binucleate, loosely arranged, simple septate with rare clamp connections. Hymenophore cariaceous memberanous, meruloid, but becoming smooth on drying, whitish to creamy-white.

The hyphae from which basidia developed initially appear slender with pro basidial swellings. As the terminal portions enlarge, clavate basidia are discernable. Sterigmata developed as the basidia matured and basidiospores subsequently formed. The athelial stage grows as a spreading white, yellow, or buff-coloured granular or encrusted area with a slightly wavy surface of hymenium (unprotected layer of basidia) on the host surface while in culture it appears as dense white crustiform masses in small irregular area. The hymenium closely adhere to the substrate and always gymnocarpous. In appearance the hymenium may be coarsely aerolate with clusters of basidia arranged on a tenuous subiculum. The hymenial thickness may be upto 30-40 micron and almost white in culture. The hymenia may be grey, yellow or buff coloured on the host. The sub-hymenial layer is composed of loosely interwoven hyphae which branch monopodially. Branches are septate and later develop into basidia (Singh, 1982). Basidiocarp usually well developed, effused and mostly saprobe, some times parasite (facultative parasite). Cystidia and rhizomorphs are absent (Punja *et al.,* 1982).

Basidia

Basidia originate as the terminal cell of a dicaryotic hyphae. Basidia are typical holobasidia (aseptate basidia). Basidia are in cluster, relatively short, clavate, mostly 4-spored, without a basal clamp, 15-19X5-6 um in size. Basidium bears 4 parallel or divergent, slender, tapering sterigmata, which are 2.5-4 micron long (Punja *et al.*, 1982).

Basidiospores

Basidiospores are formed on sterigmata, and are ballistospores. Basidiospores are haiploid, unicellular, elliptical to obovate, sometimes rounded or pyriform, smooth, thin walled, 5-7X3-5 um in size, hyaline and apiculate at base. Germinating basidiospores produce 1-3 germ tubes and subsequently an extensive mycelium was produced (Punja *et al.*, 1982).

Variability

Prasad *et al.* (2012) studied variability in different isolates of *Sclerotium rolfsii* collected from infected samples of sunflower from various sunflower growing areas of India. The isolates were studied for morphological and cultural characters, vegetative compatibility and pathogenic variability. Variations were observed in colony morphology, mycelial growth, sclerotium formation, sclerotial size, colour and number of sclerotia. The number of sclerotia in different isolates varied from 57 (*Sr* 3) to 306 (*Sr* 11). Eighty four combinations showed compatible reaction out of 231 pairings in mycelial compatibility studies. Pathogenic variability of isolates with different sunflower cultivars was carried out by soil infestation method. Based on pathogenicity, *S. rolfsii* isolates were divided into four groups. DRSH 1 showed resistant to moderately resistant reaction to most of the isolates, while Morden variety was susceptible to all isolates.

Tortoe and Clerk (2012) studied the physiological and morphological characteristics of five strains of *Sclerotium rolfsii* identified by the phenomenon of aversion. The appearance of the five strains of *Sclerotium rolfsii* on PDA plates was similar and mycelial dry weights in PDB after 6 day incubation varied statistically and the dry weight declined due to autolysis after 6 days differed according to the strains. Autolysis was moderate in strain SrXLL and extreme in strains SrEL1 and SrEL2.

Prasad and Basha (2010) studied genetic variability among the virulent isolates of *Sclerotium rolfsii* using molecular techniques like RAPD, ITS-PCR and RFLP. The RAPD banding pattern reflected the genetic diversity among the isolates by formation of two clusters. A total of 221 reproducible and scorable polymorphic bands ranging approximately as low as 100 bp to as high as 2500 bp were generated with five RAPD primers. ITS region of rDNA amplification with specific ITS1 and ITS4 universal primers produced approximately 650 to 700 bp in all the isolates confirmed that all the isolates obtained are *Sclerotium rolfsii*. Studies by ITS-RFLP indicated that there was no polymorphism in restriction banding pattern among the isolates with the restriction endonucleases used.

Reddi Kumar *et al.* (2014) studied morphological and cultural variability of eight isolates of *Sclerotium rolfsii* based on their growth rate, colony colour, mycelial dispersion and appearance and sclerotium formation, colour, weight and number of sclerotia, arrangement and maturity days of sclerotia using two solid media *viz.;* PDA and CDA. Significant variability with reference to mycelial and sclerotial characters across isolates of *S.rolfsii,* isolated from different locations of Southern zone of Andhra Pradesh was observed. *In vitro* evaluation of new fungicide mixtures revealed that, carbendazim + Mancozeb and Hexaconazole + Zineb could control the test pathogen even at 250 ppm.

Moussa and Tharwat (2013) studied the ability to produce cellulose degrading enzymes by pathogen *Sclerotium rolfsii* Sacc. In liquid synthetic media with carboxymethy cellulose (CMC) as inducer was studied. Several cultural conditions were examined to assess their effect in optimizing enzymes production. Shaking cultures gave higher yields of cellulases compared with static ones. Aspargine supplement was the best nitrogen source, especially at 3.0 g/l concentration, in promoting enzyme production. Variation of cellulose/xylan ratio in the culture medium showed that cellulose and xylan both induced cellulases synthesis but cellulose being the most effective specific substrate. The influence of different inhibitors on enzymes production by *S. rolfsii* was also studied. Cyclohexmide and ethidium bromide inhibited protein synthesis by *S. rolfsii*. Moreover, glucose repressed cellulase synthesis in *S. rolfsii*.

Disease Cycle and Epidemiology

In the rainy season, hyphal growth resumes from infected tissues and germinating sclerotia. Hyphae and germinated sclerotia come into contact with susceptible crown, root, bulb, fruit tissues and penetrate directly but wounds facilitate infection. Infection of stem, roots, bulbs, fruits and leaves may occur in susceptible plant if temperature, humidity and other environmental conditions are favourable. Hyphae may be intracellular or intercellular. The fungus produces a variety of biochemicals like, polygalacturonase and oxalic acid, which help in pathogenicity. Secondary cycles occur and disease spreads as hyphae contact new healthy susceptible tissues. Also, basidiospore production may contribute to secondary cycles. The fungus overwinters as sclerotia and mycelia in infected plants and plant debris and some times as developing hymenial layers. It has been reported that sclerotia may pass through the digestive tract of cattle or sheep and still be viable.

The pathogen survives as mycelium on dead organic matter when living plants are not available. Fungal infection is enhanced when dead organic matters are available around susceptible plants. Optimum growth and sclerotia production of the fungus occurs between 27-35°C. In addition to temperature, hyphal growth and sclerotia germination require a water-saturated soil; high humidity also favoures fungal development. Formation of sclerotia and its survival is favoured by soil pH below 7, light, well-aerated and well-drained soil.

Sarker *et al.* (2013) conducted an experiment to evaluate the effect of different pH levels (4.5, 4.0, 5.5, 6.0, 6.5, 7.0, 7.5 and 8.0) on radial growth, and formation of sclerotia, weight of sclerotia of *Sclerotium rolfsii* on PDA (Potato Dextrose Agar) medium and fresh weight and dry weight on PDB (Potato Dextrose Broth) medium. Radial growth, fresh weight and dry weight, at different pH levels differed significantly at 1 per cent level of significance but number of sclerotia formation and their weight did not differ. Optimum pH for radial growth, fresh weight and dry weight were found 5.0 to 6.0, 4.5 to 6.5 and 4.5 to 6.5, respectively.

Zape *et al.* (2013) found that the most suitable medium for the better growth of *Sclerotium rolfsii* was Potato Dextrose Agar (PDA) (90.00 mm) and Malt Extract Agar (MEA) medium (82.00 mm). They also found that Potato dextrose agar (PDA) and Peptone Sucrose Agar (PSA) medium were suitable for the sclerotial production. The *S. rolfsii* showed rapid mycelia growth at 30°C and maximum sclerotial production was recorded at 25°C. The pathogen was unable to grow and produce sclerotia at minimum and maximum temperatures of 10 and 40°C. The optimum temperature for the better radial growth of *S. rolfsii* ranged between 25 to 35°C, whereas, for sclerotial production, it ranged from 20 to 30°C. Cultural studies of *S. rolfsii* on different pH levels revealed that if had a wide range of pH. The maximum radial growth of *S. rolfsii* was observed at pH 6.5 followed by pH 6.0 and 7.0 and whereas, for formation of sclerotia, it was at pH 7.0. Optimum pH for the mycelia growth and formation of sclerotial by *S. rolfsii* was pH 5.5 to 7.5.

Management

Management efforts have often met with limited success, due to the extensive host range, prolific growth and ability to produce large number of sclerotia that may persist in the soil for several years. Furthermore, control measures effective for a particular crop or area may not be adaptable elsewhere due to regulatory or economic constraints. Several fungicides effectively controlled this pathogen on various crops in the field (Blackman and Rodridguez-Kabana, 1975; Brown and Hendrix, 1980; Haas, 1976; Punja *et al.*, 1982), but the major limitations to their widespread use to control this pathogen are requirement of large amounts of chemicals, variable effects on different crops, and variation in results in different seasons.

The fungicide quintozene is registered for the control of S. *rolfsii* (Bot *et al.*, 1986) efficiently on various crops. Phillips (1986) found that *Gliocladium virens* parasitizes sclerotia of *Sclerotinia sclerotiorum* and sclerotia of S. *rolfsii* in the laboratory. Preliminary studies at the VORl indicated that *Bacillus subtilis,* isolated from vineyard soil, had an inhibitory effect on the mycelium growth of S. *rolfsii in vitro.* Wong and Hughes (1986) isolated a *Bacillus* species from soil antagonistic to S. *cepivorum* on onion. *B. subtilis* is a well-known fungal antagonist and has been proven by several workers to be a successful biological control agent against different fungal diseases (Baker and Stavely, 1985).

The management of this fungus classified into i. Cultural management, ii. Biological management, iii. Chemical management, iv. Integrated management.

Cultural Management

Fertilizer and Manure

In case of fertilizer, mechanism of nitrogen (N) action, both in form and level, was studied *in vitro* and it was shown that lower levels of NO_3- or NH_4 (50 ppm) were inhibitory, while higher levels were directly toxic to sclerotial germination. Effect of NO_3- were not dependent on soil pH. A wide variety of nitrogenous substances including inorganics like NH_3, NH_4NO_3-, $Ca(NO_3)_2$ and organics like chitin, peptone and ammonium acetate affect germination. Such inhibition of germination was attributed to a) build up of antibiotic producing bacteria in the mycosphere, b) release of NH_3 that temporarily raised soil pH to above 8.5 and caused direct toxicity and c) direct toxicity of the ions (Sen, 1983). Use of farm yard manure reduces root rot of *Cyamopsis psoraloides* and chick pea. While Sahni *et al.*, 2008 found significant protection of chick pea from *S. rolfsii* after use of vermicompost. Aqueous extract *in vitro* and some crop residue in soil appeared to inhibit the growth of the pathogen in soil. Alfalfa hay distillate containing volatiles like methanol, acetaldehyde, isobutyraldehyde and isovalerldehyde stimulated germinationlysis in the absence of the host. Decomposition of some crop residue like alfalfa hay, legume residue and oilcakes like those of groundnut and *Sesamum indicum* released NH_3 that was directly toxic to sclertia.

Use of Micro-elements and Macro-elements

Management of diseases caused by *S. rolfsii* through application of micro element calcium and macro element, nitrogen in different forms has been described by many workers (Punja, 1989). The earliest observations on the influence of nitrogenous fertilizers on development of *S. rolfsii* were made by Leach and Davey (1942) who showed that heavy post-plant applications of anhydrous ammonia and ammonium sulphate reduced disease on sugar beet from 30 per cent to 7.6 and 9.2 per cent, respectively. Similarly, reduced susceptibility of hosts to the pathogen following application of calcium compounds is mediated through host as calcium forms crystals of insoluble calcium oxalate results from the sequestering of oxalic acid produced by fungus (Punja *et al.*, 1985).

Soil Solarisation

Soil solarisation show reduction in inoculums of several pathogens including *S. rolfsii*. Perhaps, Jones *et al.*, 1966 were first to demonstrate the covering soil with black plastic film of 1.5 mm thickness reduced southern blight of tomato. Mulching with polythene, grass cuttings, coconut fronds and stones appear to reduce root rot in beans, but for some treatments disease incidence increases at a later stage. It was also shown that alternate wetting and drying of soil led to the release of nutrients from sclerotia that were later colonized by bacteria inhibitory to their germination.

Biological Management

Biological management of *S. rolfsii* mainly includes fungi and bacteria. Among the fungal bioagents, *Trichoderma* and *Gliocladium* species are the major antagonists which are ubiquitous in natural soil Papavizas, 1985. Antibiosis and mycoparasitism by these fungi are the major characteristics through which they reduce the inoculums of *S. rolfsii* (Elad *et al.*, 1982, Mukherjee *et al.*, 1995, Upadhyay and Mukhopadhyay, 1986). During antibiosis, both volatile and nonvolatile secondary metabolites are implicated in restricting the vegetative growth of pathogenic fungi (Claydon *et al.*, 1987, Dennis and Webster, 1971, Upadhyay and Mukhopadhyay, 1986). *Trichoderma* spp. attack the pathogen by excreting lytic enzymes including â-1,3- glucanase(s), protinase(s) and chitinase(s), enabling them to degrade host cell walls and thus reduce inoculum potential (Chet *et al.*, 1993, Geremia *et al.*, 1993, Goldman *et al.*, 1994). Elad *et al.* (1982) found that *Trichoderma* isolates produces chitinases and glucanases when grown on live mycelium of *S. rolfsii.* Numerous reports reveals in literature citing effective biocontrol of *S. rolfsii* by *Gliocladium virence in vitro* as well as under field conditions (Pierto *et al.*, 1993, Lumsden *et al;.*, 1992, Mukherjee *et al.*, 1995, Wilhite and Straney, 1996). The main mode of action of the biocontrol agent is parasitism and antibiosis. The biocontrol agent colonizes sclerotia of *S. rolfsii* immediately when it comes in contact with sclerotia and penetrates to the inner layers of sclerotia where it sporulates (Mukherjee *et al.*, 1995). Similarly, the biocontrol agent produces chitinolytic enzymes that include chitin 1,4- â-chitobio and glucan N-acetyl1- â-D-glucosaminidase and glucan 1,3- â-glucosidase, known to be deleterious for *S. rolfsii* (Pierto *et al.*, 1993). Similarly, the biocontrol agent also produces an antibiotic, gliotoxin which is inhibitory to *S. rolfsii* at 50 µg/ ml. Several compounds like some mixtures of fatty acids, viridian, dimethyl gluiotoxin, viridiol, mixture of phenolic acids like ferulic acid are also detected in the culture filtrate of biocontrol agent. But inhibitory effect of all the compounds is not the same and depends on the interacting pathogen and the biocontrol agent isolate (Lumsden *et al.*, 1992). Anitha Chowdary (1997) isolated mycoflora from the rhizosphere of bell pepper antagonistic to *S. rolfsii* which included *Aspergillus flavus, A. niger, Cladosporium* sp., *Fusarium* sp., *Pencillium* sp., *Rhizopus* sp., and *Trichoderma viride*. The rhizosphere mycoflora of ginger *viz., Rhizopus* sp., *Aspergillus carneus, A. niger, A. fumigatus, A. flavus* (Sclerotial stage), *Eupenicillium javanicum, Eupenicillium javanicum* and four isolates of bacteria were found to be antagonistic to *Pythium aphanidermatum.* Pandey and Upadhyay (2000) isolated hizosphere fungi of Pigeonpea on Martin medium *viz., Aspergillus niger, A. fumigatus, Penicillium* sp., *Fusarium udum, Pythium* sp., *Rhizopus* sp., *Trichoderma harzianum, Gliocladium virens* and bacteria were isolated on soil extract agar medium which included 3 isolates B1, B2 and B3. Kishore *et al.*, 2005 reported *Pseudomonas aeruginosa* inhibits the plant cell wall degrading enzymes of *S. rolfsii* and reduces the severity of ground nut stem rot. He also observed 393 ground nut associated bacterial strains, applied both as seed treatment, soil amendment, for control of stem rot disease of ground nut in control environment. Twelve strains significantly reduced the incidence of stem rot, rot of which ground nut seed endophytes *P. aeruginosa* GSE 18 and GSE 19 reduced the seedling mortality by 54 per cent and 58 per cent, compared to control. In dual culture the 12 bio-control strains reduced the mycelial growth of *S. rolfsii* by 32 per cent - 74 per cent as compared to the control. Cell free culture filtrate of *Pseudomonas aeruginosa* GSE 18 and GSE 19 inhibited the activity *in vitro* of the cell wall degrading enzyme (CWDE) polygalacturonase and cellulose by *S. rolfsii* up to a maximum of 55 per cent when measured 6 days after inoculation. *Pseudomonas aeruginosa* GSE 18 and GSE 19 with a known tolerance to thiram, a commonly used seed dressing fungicide suppressed the growth of *S. rolfsii* inhibited the activity of CWDE and reduced the incidence of stem rot. Anahosur (2001) isolated antagonists like *Trichoderma* sp., *Gliocladium virens* and *Pseudomonas fluorescens* from rhizosphere soil of sunflower infected with *S. rolfsii*. The potential for the use of fungal antagonists as bio-control agents of plant diseases was suggested more than 70 years ago by Weindling (1932), who was the first to report the parasitic activity of *Trichoderma* spp. Against *Rhizoctonia solani* and *S. rolfsii*. Morton and Stroube (1955) screened the antagonistic fungi bacteria and actinomycetes against *S. rolfsii* of soybean by dual culture technique. Muthamilan and Jeyarajan (1992) tested antagonism of *T. viride, T. harzianum* isolates and *Laetisaria arvalis* against *S. rolfsii* by dual culture on Potato Dextrose Agar medium. They observed a reduction of 68.7 per cent in sclerotial production in presence of *T. viride* isolate 2. In dual culture, out of 11 isolates of *T. harzianum*, three isolates *viz.*, T8, T10 and T2 were found effective against groundnut isolate of *S. rolfsii* and they over grew the pathogen up to 79-92 per cent (Kumar and Sen, 2000). Dutta and Das (2002) studied the antagonistic potential of *T. harzianum, T. viride* and *T. koningii* against tomato isolate of *S. rolfsii* by dual culture technique. They observed that the three antagonists reduced the growth of the pathogen by 61.5, 59.1 and 57.2 per cent, respectively and sclerotial production by 94.2, 86.8 and 84.1 per cent, respectively. Rajalakshmi (2002) observed 37.5 to 41.3 per cent reduction in growth of mycelium of groundnut, tomato and crossandra isolates of *S. rolfsii* by *T. viride* and *T. harzianum*. Rajalakshmi (2002) also observed significant reduction in sclerotial production by the antagonists in dual culture. Pant and Mukhopadhyay (2001) studied the mechanisms of biocontrol *viz.*, antibiosis, competition, mycoparasitism or other form of direct exploitation of *G. virens* and *T. harzianum* on *S. rolfsii*. They observed infrequent coiling of hyphae of *T. harzianum* around *S. rolfsii* hyphae resulting in coagulation of protoplasm and shrunk hyphae leading to lysis. Dube (2001) described the mechanisms of disease suppression by rhizobacteria. He stated that rhizobacteria antagonize soil borne pathogens through production of antibiotics or lytic enzymes (chitinase) and through competition for nutrients, notably iron as well as by inducing systemic resistance in the plant against subsequent infection by pathogens. Rakh *et al.* (2011) reported effective biocontrol system for management of stem rot disease caused by *Sclerotium rolfsii* in groundnut, 11 *Pseudomonas spp.* isolated from rhizospheric soil, were evaluated for their antagonistic activity against *Sclerotium rolfsii*. A soil bacterium identified as, *Pseudomonas cf. monteilii 9*, showed highest antagonistic activity against *Sclerotium rolfsii*. In dual cultures, the *Pseudomonas cf. monteilii 9* inhibited the *Sclerotium rolfsii* up to 94 per cent in terms of dry weight. *Pseudomonas cf. monteilii 9* produced diffusible antibiotic, volatile metabolites, hydrogen cyanide and siderophore which affect *Sclerotium rolfsii* growth *in vitro*. This strain also produced a clear halo region on skim milk agar plates, indicating that it excretes protease which played vital role in inhibition of *S. rolfsii*. In pot assay for control of *Sclerotium rolfsii, Pseudomonas cf. monteilii 9* treated seeds showed decrease in incidence of disease up to 45.45 to 66.67 per cent in comparison to untreated seeds.

Plant growth promoting rhizobacteria (PGPR) are rhizosphere inhabiting bacteria that protect the plants from various pathogens through various mechanism, *viz.*, induced systemic resistance in the host (Maurhofer *et al.*, 1994, van Peer *et al.*, 1991), antibiotic production (Maurhofer *et al.*, 1995), growth promotion of the host (Schippers *et al.*, 1987), competition for nutrients with the pathogens (Duijff *et al.*, 1993, Leeman *et al.*, 1996), *etc. Pseudomonas fluorescens* is one that has been reported to protect hosts against pathogen attack through these mechanisms (Duijff *et al.*, 1993, van Peer *et al.*, 1991, Keel *et al.*, 1992). Antagonistic effect of *P. fluorescens* against *S. rolfsii* has been reported by Rangeshwaran and Prasad (2000) under dual culture. They have also reported that the biocontrol agent is also effective under field conditions as 60-63 per cent *S. rolfsii* infection free sunflower plants were obtained in comparison to control following treatment with biocontrol agent. Similarly, in other report, Thiribhuvanamala *et al.* (1999) also reported disease suppression in tomato upto 80 per cent following treatment with *P. fluorescens* under pot culture. Similarly, Patil *et al.* (1998) showed that *P. fluorescens* strain FDP-15 significantly improved groundnut seed germination, nodulation, dry matter production pod yield and protected seedlings from *S. rolfsii* infection. Singh *et al.* (2003) reported for the first time that *P. fluorescens* could be effectively used to control collar rot disease of betelvine (*Piper betle*).

The possibility for preventing the establishment of *S. rolfsii* infection through the induction of impermeable layer of phenolic compounds in the host had been by Punja (1985). Sarma and Singh (2003) studied the phenolic acid profile in different parts of chickpea plants following infection with *S. rolfsii* and compared with similar non-infected ones. Three major phenolic acids (gallic, vanillic and ferulic acids) were found to be present consistently in the plant parts. Gallic and ferulic acid concentrations were found in increased amount in leaves and stems of infected plants. Interestingly, the amount of ferulic acid just above the infected collar region was found to be significantly high as compared to the healthy ones suggesting that the non-migration of *S. rolfsii* infection to leaves of infected seedlings due to high accumulation of antifungal phenolic acids particularly ferulic acid. Similarly, reduced plant mortality (30-40 per cent) in chickpea following seed bacterization with different strains of *P. fluorescens* under field conditions was shown by Sarma *et al.* (2002). HPLC analysis of the chickpea plants revealed PGPR elicited alterations in the phenolic compounds of the host and induced synthesis of total as well as specific phenolic acids, *viz.*, gallic, ferulic, chlorogenic and cinnamic acids.

Babu and Paramageetham (2013) isolated *Pseudomonas aeruginosa* from forest litter from nearby forests. They were assessed antagonistic activities *in vitro*. All antagonistic isolates produced an inhibition zone varied from 2.1 cm to 4.2cm and radial growth inhibition percentage from 47.5 per cent to 73.7 per cent. However isolate PATPT6 was found to be potential antagonists against *Sclerotium* with almost 73.7 per cent of radial growth inhibition. The isolates also showed the production of volatile toxic substances and cell wall degrading enzymes which play a vital role in disease suppression.

Pattanapipitpaisal and Kamlandharn (2012) isolated two hundred and eighty three strains of chitinase-producing actinomycetes from rhizoshere-associated soils, using Enrichment Media for isolation of Chitinase-producing Actinomycetes agar (EMCA agar). All strains were screened for chitinolytic activity and sixty eight strains gave significant clear zone on EMCA agar plates. The selected chitinolytic strains were assayed for *in vitro* antagonism against *Sclerotium rolfsii* using cornmeal agar (CMA agar) assay procedure and the result showed that thirteen isolates have remarkable inhibiting the growth of the fungus and the top five antagonistic actinomycetes were PACCH 277, PACCH129, PACCH225, PACCH24 and PACCH246, respectively. The result indicated that these actinomycetes produce chitinase which catalyze the degradation of chitin, resulting in inhibition of *S. rolfsii* growth. Their abilities to control the disease development were tested for *in vivo* bio-control assay on chilli seedlings. Two out of thirteen, PACCH24 and PACCH225, antagonists reduced the disease development at 90 per cent. They suggested that the ability to inhibit the growth of pathogen *in vitro* was not related to the disease reduction *in vivo*. The strain PACCH24 was further identified as *Streptomyces hygroscopicus* according to morphological characteristic, cell wall and cellular sugar analysis and 16S rDNA sequencing. The study implies a novel chitinolytic actinomycete which could be developed to be a biological agent which would be included as a complement with organic fertilizers in order to control stem rot disease and promote growth of chilli.

Rakh *et al.* (2011) in an attempt to develop effective bio-control system for management of stem rot disease caused by *Sclerotium rolfsii* in groundnut, 11 *Pseudomonas* spp. isolated from rhizospheric soil, were evaluated for their antagonistic activity against *Sclerotium rolfsii*. A soil bacterium identified as, *Pseudomonas cf. monteilii 9*, showed highest antagonistic activity against *Sclerotium rolfsii*. In dual cultures, the *Pseudomonas cf. monteilii 9* inhibited the *Sclerotium rolfsii* up to 94 per cent in terms of dry weight. *Pseudomonas cf. monteilii 9* produced diffusible antibiotic, volatile metabolites, hydrogen cyanide and siderophore which affect *Sclerotium rolfsii* growth *in vitro*. This strain also produced a clear halo region on skim milk agar plates, indicating that it excretes protease which played vital role in inhibition of *S. rolfsii*. In pot assay for control of *Sclerotium rolfsii, Pseudomonas cf. monteilii 9* treated seeds showed decrease in incidence of disease up to 45.45 to 66.67 per cent in comparison to untreated seeds.

Chanutsa *et al.* (2014) studied highest potential activities of bacteria against *Sclerotium rolfsii*. Thirteen of 329 isolates were screened as antagonistic bacteria to inhibit *S. rolfsii* by dual culture assay. High percentages of inhibition were found in three isolates of UD1EBa-2, KK1EBa-3 and KK11EBa-3 with 51.25 per cent, 56.25 per cent and 60.00 per cent, respectively. The culture filtrate of each bacterium was prepared to test the mycelial growth inhibition of *S.rolfsii*. The results showed that the culture filtrate could inhibit mycelial growth with 100 per cent. The most effective isolate KK11EBa-3 was characterized to be rod shape, gram negative and non spore forming. The colony on Nutrient agar was circular, smooth, translucent, flat elevation. Based on molecular sequencing data, it was identified as *Pseudomonas aeruginosa*.

Deepthi (2013) studied total 40 antagonists from groundnut rhizosphere and root endophytes. When tested *in vitro*, the potential biocontrol agents GRE-9 (100 per cent) and GRB-16 (85 per cent) were predominantly aggressive in inhibiting the mycelia growth of the pathogen in dual culture against *Sclerotium rolfsii*. Among the fungicides tested, Mancozeb was most compatible with all antagonists.

Rivera *et al.* (2013) evaluated the effects of fermented onion decoctions on the growth of *Rhizoctonia solani* and *Sclerotium rolfsii* and disease incidence. Broth (B) and sterilized broth (SB) were respectively obtained by boiling chopped yellow onions in water, and incubating for 14 days at room temperature, with or without subsequent sterilization. The pathogens were grown on potato dextrose agar supplemented with B and SB, diluted at 1.7, 3.3, 8.3, 16.7 and 25 per cent (v/v). Their growth was reduced by B at 8.3, 16.7 and 25 per cent. Production of sclerotia by *S. rolfsii* was diminished by B, but stimulated by SB. *Penicillium purpurogenum*, *P. simplicissimum* and *Aspergillus niger* obtained from B behaved as antagonistic against both pathogens, showing antibiosis, competition and hyperparasitism in dual confrontations with them. Broth and SB at 10 and 50 per cent dilutions were sprayed on chard (*Beta vulgaris*), tomato (*Solanum lycopersicum*), pepper (*Capsicum annuum*) and eggplant (*Solanum melongena*) seedlings cultivated in pathogen-colonized soil. Damping off incidence in the B treatment at 50 per cent dilution was markedly lower than that in SB. It is concluded that B antifungal activity depends on its dilution and mycota. Broth sprays should be considered as a tool to control damping off in low-environment-impact crop production.

Barakat *et al.* (2006) isolated forty-seven local *Trichoderma* isolates from one hundred sixty nine soil samples from irrigated fields by using dilute plate techniques on *Tricchoderma* selective media (TSM).The antagonistic potential of the local isolates against the phytopathogenic *Sclerotium rolfsii* was investigated in dual culture and bioassay on bean plants. Application of testing isolates as a conidial suspension (3*106) greatly reduced the disease index of bean plants in different rates by a percentage of 67 per cent.The results revealed that the variation of antagonistic potential between isolates was due to the variation in mycelium-coiling rate, sporulation, fungitoxic metabolites, induced growth response and temperature effect. The Effect of *Trichoderma* on bean seedlings growth was obvious; height was nearly doubled (160 per cent - 200 per cent), while fresh and dry weight were increased in plant by 133 per cent - 217 per cent respectively. Germination of bean seeds treated with *Trichoderma* isolates occurred about four days earlier than those in untreated soil.

Seed Treatment with Bio-control Agents

Muthamilan and Jeyarajan (1992) found that the seed pelleting with *T. harzianum* of 5 x 10^9 conidia/ml as the best treatment in controlling root rot of groundnut caused by *S. rolfsii* and seed treatment with *P. fluorescens* was found to be effective in controlling collar rot of groundnut incited by *S. rolfsii.* Hari Narayana (1999) evaluated the efficacy of seedling root dip in *T. viride* and *P. fluorescens* in controlling sclerotial wilt of bell pepper incited by *S. rolfsii.* Seedling root dip in *T. viride* was found to be superior. Seed treatment with *P. fluorescens* and *P. putida* effectively reduced sclerotium rot of sunflower caused by *S. rolfsii* (Rangeshwaran and Prasad, 2000).

Soil Application of Bio-control Agents

Soil application of inoculum of *Trichoderma* spp. at the time of transplanting reduced collar rot of tomato caused by *S. rolfsii* and increased dry mass of root and shoot and yield (Pranab Dutta and Das, 2002). Biocontrol by adding large amounts of *Trichoderma* preparations on solid media (ground annual ryegrass seed) for field control of *S. rolfsii* were reported. Mixing of wheat bran culture of *T. harzianum* (Th-2), *T. viride* (Tv-2) and *Gliocladium* sp. @ 10 g/kg soil reduced the seedling mortality by 12.71, 14.72 and 17.98 per cent, respectively, in brinjal due to collar rot caused by *S. rolfsii* (Singh and Singh, 1994). Application of *T. harzianum* inoculums to soil at the time of sowing was better than other treatments including seed treatment with antagonist in controlling root rot of groundnut caused by *S. rolfsii* (Muthamilan and Jeyarajan, 1996). Soil application of *P. fluorescens* was effective in controlling collar rot of groundnut incited by *S. rolfsii* (Patil *et al.*, 1998). Soil application of *T. harzianum* (20 g/plant) and *B. subtilis* (20 ml/plant) significantly reduced collar rot of elephant's foot yam caused by *S. rolfsii* (Gogoi *et al.*, 2002). Sclerotial wilt of groundnut caused by *S. rolfsii* was effectively reduced (92.58 per cent) when *T. harzianum* was applied @ 10 g/kg soil (Patibanda *et al.*, 2002).

Chemical Management

Anitha Chowdary (1997) evaluated *in vitro* sensitivity of bell pepper isolate of *S. rolfsii* to captan, thiram @ 25, 50, 100, 250, 500 and 1000 ppm and propiconazole @ 10, 20, 25, 50, 100, 250 and 500 ppm. She reported that propiconazole at a concentration of 250 ppm was effective in complete inhibition of *S. rolfsii*. Dutta and Das (2002) studied *in vitro* efficacy of thiram and mancozeb at 0.1 per cent concentration against tomato isolate of *S. rolfsii* and reported that thiram had inhibited 70.3 per cent of mycelial growth and 96.5 per cent of sclerotial production of *S. rolfsii*. Narayana Bhat and Srivastava (2003) evaluated *in vitro* efficacy of captan, thiophanate-methyl and propiconazole at 250, 500 and 1000 ppm concentrations against *S. rolfsii*. They found that propiconazole was effective even at 250 ppm concentration against *S. rolfsii*. Bhardwaj *et al.* (1983) evaluated *in vitro* the efficacy of thiophanate- methyl at 100 ìg/ml against two strains of *S. rolfsii* and found that potato isolate was more sensitive to thiophanatemethyl than rice isolate. Yaqub and Shahzad (2006) were tested six fungicides *viz.*, Benomyl, Sancozeb, Thiovit, Dithane M-45, Carbandazim and Topsin-M against *Sclerotium rolfsii* by food poison method. At low concentration, no fungicide inhibited the growth of *S. rolfsii*, however, at high concentration Dithane M-45 and Sencozeb significantly reduced the growth.

Deepthi (2014) studied *Sclerotium rolfsii* Sacc causing stem rot groundnut crops. They isolated 40 antagonists from soil rhizosphere and root endophytes of the groundnut crop. When these isolates are subjected to in vitro studies and hence GRE-9 and GRB-4 has shown 100 per cent inhibition of the mycelial growth against *Sclerotium rolfsii*. The efficacy of six fungicides *viz.*, Carbendazim, Hexaconazole, Propiconazole, Thiophenate Methyl and COC and Mancozeb were tested @ 0.1 per cent, 0.2 per cent and resulted that Mancozeb and propiconazole were found 100 per cent effective in checking the growth of the pathogen while carbendazim was not effective against pathogen.

Das *et al.* (2014) evaluated the potential of six systemic fungicides (*i.e.* Propiconazole 25 per cent EC, Hexaconazole 5 per cent EC, Mycobutanil 10 per cent WP, Thiophanate Methyl 70 per cent WP, Tebuconazole 25.9 m/m EC and Carbendazim 50 per cent WP); three non-systemic fungicides (*i.e.* Captaf, Mancozeb 75 per cent WP and Copper oxychloride) and three combo fungicides (*i.e.* Metalaxyl 8 per cent +Mancozeb 64 per cent, Carbendazim 12 per cent + Mancozeb 63 per cent and Carboxin 37.5 per cent + Thiram 37.5 per cent). These fungicides were evaluated at different concentrations (*i.e.* 1, 10, 50 and 100 ppm), except copper-oxychloride (contact) at 0.1 per cent, 0.25 per cent, 0.5 per cent and 1.00 per cent on the growth of *Sclerotium rolfsii* in Potato dextrose agar (PDA) medium using poisoned food technique, *in vitro*. The result shows that the effect of hexaconazole (systemic) has been highly effective in suppressing radial expansion as well as per cent inhibition of the fungus at all the concentrations used followed by Carboxin 37.5 per cent + Thiram 37.5 per cent (combo Fungicide) and tebuconazole (systemic). The results suggest that some of these fungicides may be tried against *S. rolfsii* in susceptible crop plants (*i.e.* brinjal) under field condition.

Seed Treatment

Charde *et al.* (2002) found that seed treatment with propiconazole and hexaconazole were superior in checking stem rot of groundnut caused by *S. rolfsii* and increasing the shoot and root length. Corm treatment with captan @ 0.2 per cent significantly reduced collar rot of elephant's foot yam incited by *S. rolfsii* (Gogoi *et al.*, 2002). Seedling root dip in mancozeb (0.1 per cent) and thiram (0.1 per cent) effectively reduced collar rot of tomato caused by *S. rolfsii* (Pranab Dutta and Das, 2002). Seedtreatment of soybean with hexaconazole and propiconazole inhibited *S. rolfsii*. These fungicides were found to be absorbed by roots and translocated to shoot and leaf (Tajane *et al.*, 2002). Muthamilan and Jeyarajan (1992) found that seed treatment with captan @ 4 g/kg seed significantly decreased root rot of groundnut caused by *S. rolfsii*. Dip treatment (3 g/litre water) or seed dressing (3 g/kg seeds) with thiophanate-methyl, carboxin and thiram were evaluated for their effectiveness in controlling soil borne pathogens of stevia *viz.*, *S. rolfsii, Macrophomina phaseolina, Fusarium oxysporum, F. semitectum, F. solani* and *R. solani*. Thiophanate-methyl was found to be effective in controlling these pathogens (Hilal and Baiuomy, 2000).

Soil Treatment

Charde *et al.* (2002) reported that soil treatment with propiconazole and hexaconazole effectively reduced stem rot of groundnut caused by *S. rolfsii*. Gogoi *et al.* (2002) reported that soil-drenching with captan (0.2 per cent) resulted the lowest collar rot incidence of elephant's foot yam caused by *S. rolfsii*.

Integrated Management

The old adage that prevention is better than cure best applies to the control of *S. rolfsii*. One of the most effective cultural methods for controlling *S. rolfsii* under field condition is the soil solarisation whereby heating of moistered soil under polythene mulches has been reported to result in reduction both in sclerotial numbers as well as disease incidence. Another effective method of removing the inoculum (sclerotia, debris of infested crops) from the infectious one (upper 8-12 cm of soil) is deep ploughing/and or deep summer ploughing which buries the sclerotia. However, ploughing alone had no effect on collar rot of carrot but reduced the disease incidence in conjuction with fungicides or fertilizers. Late sowing (mid October) of lentil and staking of plants to keep the fruit from touching the ground also play role in reduction of disease incidence. With several approaches to control, integration among these was attempted to improve efficiency of management. In soil *T. viride* become more active with Nitrogen + Atrazine. A combination of *T. harzianum* in wheat-bran culture and Dinitamine more effectively reduced the southern blight of groundnut than applied alone. In field trials, a combination of sublethal levels of Quintozene + wheat-bran formulation of *T. harzianum* controlled seedling blight of bean considerably. Methyl bromide, *T. harzianum*, and solar heating all help reduce the inoculum density of this pathogen in soil. By combination solarization and *T. harzianum* both at sub-optimal levels, 90- 100 per cent control of bean root rot was achieved in glasshouse. A combination effect of Methyl bromide and quintozene was assessed in the control of crown rot of onions. Both PCNB + MB were almost equally effective. However, Br toxicity appeared even in the succeeding season rendering the approach ineffective. It was also suggested that biological control of this pathogen can be achieved by altering the permeability properties of sclerotia with the help of sub-lethal dosage of CS2 followed by drying. The fumigant at sublethal dose caused nutrient leakage from sclerotia, which stimulated other microorganisms that kill the sclerotia. Papavizas (1985) emphasized the importance of integration of *Trichoderma* or *Gliocladium* with compatible fungicides or other practices in controlling soil borne pathogens. Muthamilan and Jeyarajan (1996) reported that integration of *T. harzianum, Rhizobium* and carbendazim remarkably reduced the root rot of groundnut caused by *S. rolfsii*. Seed treatment with Vitavax 200 (carboxin 37.5 + thiram 37.5) 75 WP @ 0.2 per cent + *T. harzianum* or *G. virens* effectively controlled seed and seedling rot of soybean caused by *S. rolfsii* (Uma Singh and Thapliyal, 1998). Seed and soil treatment with *T. harzianum* plus neem cake significantly reduced stem rot of groundnut caused by *S. rolfsii* with increased dry shoot and root weight (Sakthi Kumaran, 2000). Deepak Kumar and Dubey (2001) found that seed treatment with captan or thiram @ 1 g/kg along with *T. harzianum* @ 106 spores/ml gave not only good germination but also effective control of collar rot of pea caused by *F. solani* f.sp. *pisi*. Integration of *G. virens* and *T. harzianum* with carboxin (0.1 per cent) effectively controlled seed and seedling rot of soybean incited by *S. rolfsii* (Pant and Mukhopadhyay, 2001). Gogoi *et al.* (2002) reported that corm treatment with captan (0.2 per cent) integrated with soil application of *T. harzianum* or *B. subtilis* reduced collar rot of elephant's foot yam caused by *S. rolfsii*. Patibanda *et al.* (2002) observed effective control of sclerotium wilt of groundnut caused by *S. rolfsii* when seed coating with thiram (0.1 per cent) integrated with soil application of *T. harzianum* @ 4.0 g/kg soil. Rama Yalla Reddy (2002) found that integrated use of *T. viride, P. fluorescens*, commercial neem product Niwaar and thiram for seed treatment of groundnut improved pod yield and controlled soil mycoflora *viz., A. niger, Alternaria* sp., *Curvularia* sp., *Fusarium* sp., *Drechslera* sp., *Penicillium* sp., *R. stolonifer, Rhizoctonia* sp., *S. rolfsii* and *Verticillium* sp. Duffy *et al.* (1996) reported that combined application of *T. koningii* and fluorescent *Pseudmonads* effectively controlled take-all on wheat. Rajendran and Ranganathan (1996) reported that combined seed treatment of *T. viride* and *P. fluorescens* reduced the onion basal rot incidence incited by *F. oxysporum* f.sp. *cepae* both in pot culture and in field conditions. Seed treatment with both *G. virens* and *B. subtilis* suppressed *Fusarium* colonisation of cotton root (Xhang *et al.*, 1996). Anith and Manomohandas (2001) reported that combined application of *T. harzianum* and *Alcaligenes* sp. strain AMB8 effectively controlled nursery rot disease of black pepper caused by *P. capsici*. Damping - off of brinjal and tomato was effectively controlled by combined seed treatment with *P. fluorescens* and *T. harzianum* (Vishwakarma *et al.*, 2002). Ganesan *et al.* (2007) reported that combined application of selected symbiotic N_2 fixing and antagonistic *Rhizobium* isolates and biocontrol agent *Trichoderma harzianum* controls the pathogen significantly and increases the plant growth. It may be due to the competition for space and nutrient, parasitism, production of enzymes, volatile and non-volatile metabolites or combined action of these mechanisms of bio-control organisms against the pathogen. The growth-promoting activity may be due to the solubilisation and sequestration of inorganic plant nutrient, production of unidentified growth promoting chemicals or by the N2 fixing activity of the *Rhizobium* isolates. Hence these *Rhizobium* and *T. harzianum* can be used as bio-control agents and bio-fertiliser for the management of groundnut crop. Further research has also been undertaken to find out the mechanism of action and efficacy of the combined application of these microbes at field level. Sahni *et al.* (2008) reported two non-conventional chemicals, $ZnSO_4$ (10^{-4} mM) and oxalic acid (4 mM) alone as well as in combination with seeds bacterized with *Pseudomonas syringae* strain PUR46 and vermicompost substitution in the potting soil, for their ability to suppress collar rot of chickpea (*Cicer arietinum*) caused by *Sclerotium rolfsii* under greenhouse conditions. $ZnSO_4$ and oxalic acid they applied as pre-inoculation foliar spray on chickpea and subsequently challenged with *S. rolfsii*. They observed both the chemicals provided significant protection to chickpea compared to control (100 per cent plant mortality) when used alone as well as in combination with PUR46 and vermicompost. However, $ZnSO_4$ was more effective than oxalic acid against *S. rolfsii*. Amongst the treatments tried, plant mortality was least when $ZnSO_4$ was used in combination with seed bacterization with PUR46 and 25 per cent vermicompost substitution. Their findings indicate the utility of integration of the above factors in managing collar rot efficiently. Control of *S. rolfsii* by transgenic plants has recently been shown by Punja and Raharjo (1996). They used three lines of cucumber cv. Endeavor, each transformed with a chitinase gene originating from *Petunia* (acidic), tobacco (basic) or *Phaseolus vulgaris* (basic) using *Agrobacterium tumefaciens* and compared with non-transgenic plants for response to inoculation with *S. rolfsii*. In growth chamber studies using whole plants and *in vitro* inoculations with detached leaves, no differences in disease development (rate and final levels) were detected between transgenic and non-transgenic plants. Carrot cultivars Nanco and Golden State transformed with 2 chitinase genes (from *Petunia* and tobacco) were evaluated for response to inoculation with *S. rolfsii*. A detached petiole inoculation method was used to compare transgenic and non-transgenic plants. The rate and final extent of lesion development after 7 days was significantly lower ($P = 0.01$) in transgenic plants expressing the tobacco (basic) chitinase gene upon inoculation with *S. rolfsii* but not in plants expressing *Petunia* (acidic) chitinase gene. It is concluded that the efficacy of chitinase gene transformations strategy for enhancing disease tolerance in plants can be influenced by the plant species used, the type of chitinase protein expressed and the characteristics of the fungal pathogen.

Fery and Dukes (2011) conducted field studies to characterize the genetic nature of resistance to southern blight (caused by *Sclerotium rolfsii* Sacc.) exhibited by the cowpea [*Vigna unguiculata* (L.) Walp.] cultivars Carolina Cream and Brown Crowder and to determine if a genetic relationship exists for this resistance between the two cultivars. Examination of the comparative frequency distributions of the parental and progeny populations of the "Carolina Cream" x "Magnolia Blackeye" and "Brown Crowder" x "Magnolia Blackeye" crosses and the corresponding segregation data indicates that the southern blight resistances exhibited by "Carolina Cream" and "Brown Crowder" are conditioned by single dominant genes. Examination of the segregation data from the parental and progeny populations of the "Carolina Cream" x "Brown Crowder" cross suggests that the two resistance genes are not allelic. The availability of each of the resistance genes in cultivar-type genetic backgrounds should allow for rapid incorporation of southern blight resistance genes into other cowpea cultivars by the application of conventional plant breeding methodologies.

Mahato and Mondal (2014) conducted survey of *Sclerotium rolfsii* Sacc causing collar rot prevalent in varying percentage (0.12 to 6.14), and has been recorded on groundnut, both cultivated and uncultivated wild elephant foot yam [*Amorphophallus paeoniifolius* (Dennst.) Nicolson], tomato [*Solanum lycopersicum* L.], potato [*Solanum tuberosum* L.], brinjal [*Solanum melongena* L.] and mesta [*Hibiscus subdariffa* L.]. Disease incidence recorded from uncultivated wild elephant foot yam was maximum (50.22 per cent). No much variation was recorded when characterization of four isolates of *S. rolfsii* from groundnut was done. Pathogenicity through cross inoculation technique did not show any host specificity. But the isolates showed different degrees of virulence. Highly virulent isolate from uncultivated wild elephant foot yam developed symptom rapidly on different crops. They also conducted a field trial against the pathogen causing collar rot of brinjal with some non-chemical agents. The per cent disease control (PDC) after 65 days of observation was found highest (84.66 per cent) in Plaster of Paris ($CaSO_4$.1/2 H_2O) treated plots followed by *Azotobacter chroococcum* Beijerinck (84.09 per cent), Vermicompost (83.28 per cent) and *Pseudomonas fluorescens* Migula (82.19 per cent). FYM (74.10 per cent) and *Trichoderma viride* Pers. (73.27 per cent) was recorded effective than Neem cake (64.19 per cent) as well. Lowest PDC (50.70 per cent) was recorded in VAM [*Glomus fasciculatum* (Thaxt.) Gerdem. and Trappe] treated plot that was less effective. They suggested highly efficacious treatments *i.e.* Plaster of Paris, *Azotobacter chroococcum*, Vermicompost and *Pseudomonas fluorescens* can be incorporated into Integrated Disease Management Programme.

Amin *et al.* (2013) evaluated the effect of different plants extracts and namely rhizome of turmeric, rhizome ginger, neem leaf, tobacco leaf, tobacco leaf extract in water, tobacco leaf extract in cow's urine, and cow's urine at different concentrations

(70 per cent, 60 per cent, 50 per cent, 40 per cent and 30 per cent) on the growth and sclerotia formation of *Sclerotium rolfsii*. In all the cases, the growth inhibition was found to be increased with the increase of concentration. No growth of the tested fungi was observed in all concentrations of tobacco leaf extract in cow's urine, 70 per cent, 60 per cent, 50 per cent, 40 per cent concentration of cow's urine alone, and 70 per cent and 60 per cent concentration of tobacco leaf extracts in water and consequently no sclerotia were formed. Considerable growth inhibitions were observed in all concentrations of turmeric rhizome which were 47.16 per cent, 50.04 per cent, 55.97 per cent, 56.43 per cent and 57.13 per cent at 30 per cent, 40 per cent, 50 per cent, 60 per cent and 70 per cent concentrations respectively. Few sclerotia were formed in the plates treated with tobacco leaf extract in water at 50 per cent and 40 per cent (14 and 53 respectively). Less than 50 per cent sclerotia were formed at all concentration of turmeric rhizome extract.

Mahato *et al.* (2014) conducted an *in-vitro* experiment to determine the effect of different fungicides, plant oils and plant extracts on radial colony growth of *Sclerotium rolfsii* Sacc. following poison food technique in PDA medium. One systemic (Carbendazim 50 per cent WP as Bavistin @ 0.1 per cent), three contact (Mancozeb 75 per cent WP as Indofil M-45 @ 0.25 per cent, Copper oxy chloride 50 per cent WP as Blitox @ 0.4 per cent, Chlorothalonil 75 per cent WP as Kavach @ 0.2 per cent) and three combination of systemic and contact fungicides (Carboxin 37.5 per cent + Thiram 37.5 per cent WP as Vitavax Power @ 0.2 per cent), Metalaxyl 8 per cent +Mancozeb 64 per cent WP @ 0.25 per cent as Krilaxyl Gold, Cymoxanil 8 per cent + Mancozeb 64 per cent WP as Curzet @ 0.25 per cent) were evaluated against *S. rolfsii* in laboratory. Besides, four plant oils (Karanja oil from *Pongamia pinnata* L., Mahua oil from *Madhuca indica*, Neem oil from *Azadirachta indica*, Chaulmoogra oil from *Hydnocarpus wightiana* Blume) @ 5 per cent (v/v) and two plant extract (Kamini – *Murraya exotica* L. and Nayantara – *Vinca rosea* L.) @ 10 per cent (v/v) were also evaluated in laboratory. Viavax Power (95 per cent) was the best fungicide to restrict the fungal growth effectively followed by Krilaxyl Gold (94.04 per cent), Kavach (93.55 per cent), Indofil M-45 (90.66 per cent) and Curzet (87.77 per cent). Blitox had negligible effect (20.17 per cent) on reduction of mycelial growth. Among the plant oils and plant extracts, Karanja oil (88.49 per cent) and *Murraya exotica* leaf extract (86.15 per cent) were found effective in reducing the growth of *S. rolfsii*.

Conclusion

Sclerotium rolfsii Sacc. which is a serious soil borne pathogen common in tropical and sub-tropical regions of the world where high temperature coupled with high humidity is prevalent during the rainy season causing severe damage to the crop with yield losses. Since beginning deep ploughing, use of nitrogenous fertilizers and use of Quintozene appeared to be important in reducing losses from this pathogen. These still remain the most promising method of cultural and chemical control. *T. harzianum* has emerged as the most promising antagonist of *S. rolfsii*. Despite continuous research, this pathogen continues to plague growers causing considerable economic loss. Management efforts have often met with limited success, due to the extensive host range, prolific growth and ability to produce large number of sclerotia that may persist in the soil for several years. Exploitation of rhizosphere microflora was found to be an effective tool for management of the disease under field condition. Among the non-chemical and biological agents, Plaster of Paris, *Azotobacter chroococcum*, Vermicompost and *Pseudomonas fluorescens* were highly efficacious. FYM and *Trichoderma viride* were also effective than neem cake and VAM. More investigation is required to find out modus operandi of Plaster of Paris and *Azotobacter*. Therefore, detailed investigation on induced systemic resistance, screening of resistant variety and development of transgenic plant for the management of *S. rolfsii* are needed in future.

REFERENCES

Amin, R., Sarker, B. C., Adhikary, S. K., Sultana, S. and Zubair, T. (2013). Effect of some botanical extracts and cow's urine on *Sclerotium rolfsii* causal agent of foot and root rot of betel vine. *The Int. J. Engineering and Sc.* 2 (9) : 77-82.

Anahosur, K. H. (2001). Integrated management of potato Sclerotium wilt caused by *Sclerotium rolfsii*. *Indian Phytopathol.* 54: 158-166.

Anith, K. N. and Manomohandas, T. P. (2001). Combined application of *Trichoderma harzianum* and *Alcaligenes* sp. strain AMB8 for controlling nursery rot disease of black pepper. *Indian Phytopathol.* 54: 335-339

Anitha Chowdary, K. (1997). Studies on Sclerotial wilt of bell pepper (*Capsicum annuum* L.). M.Sc.(Ag.) Thesis submitted to Acharya N.G.Ranga Agricultural University, Hyderabad (A.P).

Arunasri, P., Chalam, T.V., Eswara Reddy, N. P. and Tirumala Reddy, S. (2011). Collar rot disease of crossandra induced by *Sclerotium rolfsii* and its management: a critical review. *International J. Appl. Biol. Pharma. Tech.* 2(2): 307-314.

Aycock, R. (1966). Stem rots and other diseases caused by *Sclerotium rolfsii*. North Carolina Agricultural Experiment Station Technical Bulletin No. 174, p. 202.

Baker, C.J. and Stavely, J.R., 1985. Biocontrol of bean rust by *Bacillus subtilis* under field conditions. *Plant Disease 69*, 770-772.

Barakat, Radwan M., Fadel Al-Mahareeq and Mohammad I. AL- Masri (2006). Biological control of *Sclerotium rolfsii* by using indigenous *Trichoderma* spp. isolates from Palestine. *Hebron University Research Journal*, 2 (2). 27 – 47.

Bhagat, Indramani (2011). Factors influencing mycelial growth of *Sclerotium rolfsii. Nepalese J. Biosciences* 1: 26-31.

Bhardwaj, C. L., Shyam, K. R. and Singh, B. M. (1983). Evaluation of fungicide for the control of seedling blight of rice caused by *Sclerotium rolfsii*. *Indian J. Mycol. Plant Pathol.* 13: 256-261.

Bindu, Madhavi, G. and Bhattiprolu, S.L. (2011). Integrated disease management of dry root rot of chilli incited by *Sclerotium rolfsii* (sacc.). *Int. J. Plant, Animal and Environmental Sc.* 1 (2): 31-37.

Bilici, Mehmet, Cinar, Ozdem and Erkilic Ali. (1994). Cultural, chemical, physical and biological methods to control stem rot, *Sclerotium rolfsii* Sacc., in peanut. *Turkish J. Agric, Forestry*. 18: 423-435.

Bot, J., Sweet, S. and Hollings, N., 1986. A guide to the use of pesticides and fungicides in the Republic of South Africa 31st ed. Directorate Agriculture Information, Private Bag X144, 0001 Pretoria, Republic of South Africa.

Brown, E.A. and Hendrix, F.F. (1980). Distribution and control of *Sclerotium rolfsii* on apple. *Plant Dis*. 64: 205-206.

Chanutsa, Natedara, Nutchanat Phonkerd and Wandee Bunyatratchata (2014). Potential of Pseudomonas aeruginosa to control *Sclerotium solfsii* causing stem rot and collar rot disease of tomato. *J. Advanced Agril. Tech.* 1 (2): 132-135.

Charde, J. D., Waghale, C. S. and Dhote, V. L. (2002). Management of stem rot of groundnut caused by *Sclerotium rolfsii*. *Plant Disease Research* 11: 220-221.

Chet, I., Barak, Z. and Oppenheim, A. (1993). Genetic engineering of microorganism for improved biocontrol activity. In: Biotechnological prospects for plant disease control. I. Chet, ed. Wiley-Liss, New York, pp. 211-235.

Claydon, N., Allan, M., Hanson, J.R. and Avent, A.G. (1987). Antifungal alkyl pyrones of *Trichoderma harzianum*. *Trans. Brit. Mycol. Soc*. 88: 503-513.

Cook, R.J. and Papendick, R.I. (1972). Influence of water potential of soils and plants on root disease. *Annu. Rev. Phytopathol*. 10: 349-374.

Das, N. C., Dutta, B. K. and Ray, D. C. (2014). Potential of some fungicides on the growth and development of *Sclerotium rolfsii Sacc. in vitro*. *Int. J. Scientific and Res. Publications*, 4 (12): 1-5.

Deepak Kumar and Dubey, S. C. (2001). Management of collar rot of pea by the integration of biological and chemical methods. *Indian Phytopathol*. 54: 62-66.

Deepthi, K C (2013). Effect of potential biocontrol agents against *Sclerotium rolfsii* causing stem rot of groundnut. *Int. J. LifeSc. Bt and Pharm. Res.*, 2 (2): 58-65.

Deepthi, K.C. (2014). *In vitro* evaluation of fungicides against *Sclerotium Rolfsii* Sacc causing stem rot of groundnut. *Int. J. Scientific Research* 3 (12): 1-2.

Dennis, C. and Webster, J. (1971). Antagonistic properties of species groups of *Trichoderma*, 1. Production of non-volatile antibodies. *Trans. Brit. Mycol. Soc.* 57: 25-39.

Dube, H. C. (2001). Rhizobacteria in Biological control and plant growth promotion. *J. Mycol. Plant Path.* 31: 9-21.

Duffy, B. K., Simon, A. and Weller, D. M. (1996). Combination of *Trichoderma koningii* and fluorescent Pseudomonads for controlling take-all on wheat. *Phytopath.* 86: 188- 194.

Duijff, B.J., Meijer, J.W., Baker, Pahm and Schippers, B. (1993). Siderophore-mediated competition for iron and induced resistance in the suppression of Fusarium wilt of carnation by florescent *Pseudomonas* spp. *Netherlands J. Pl. Pathol.* 99: 277-289.

Dutta, Pranab and Das, B. C. (2002). Management of collar rot of tomato by *Trichoderma* spp. and chemicals. *Indian Phytopathol.* 55: 235-237.

Elad, Y., Chet, I. and Henis, Y. (1982). Degradation of plant pathogenic fungi by *Trichoderma harzianum*. *Can. J. Microbiol.* 28: 719-725.

Fery, Richard, L. and Dukes, Philip D. (2011). Southern Blight (*Sclerotium rolfsii* Sacc.) of cowpea: genetic characterization of two sources of resistance. *Int. J. Agronomy* Vol. 2011.

Fry, W.E. (1982). Principle of plant disease management. Academic Press, New York.

Ganesan, S., Ganesh Kuppusamy, R. and Sekar, R. (2007). Integrated Management of Stem Rot Disease (*Sclerotium rolfsii*) of Groundnut (*Arachis hypogaea* L.) using *Rhizobium* and *Trichoderma harzianum*. *Turk J Agric For.* 31: 103-108.

Geremia, R., Goldman, G., Jacobs, D., Ardiles, W., Vila, S., Van Montagu, M. and Herrera-Estrella, A. (1993). Molecular characterization of protienase-encoding gene *prb* 1, related to mycoparasitism by *Trichoderma harzianum*. *Mol. Microbiol.* 8: 603-613.

Gogoi, N. K., Phookan, A. K. and Narzary, B. D. (2002). Management of collar rot of elephant's foot yam. *Indian Phytopathol.* 55: 238-240.

Goldman, H.G., Hayes, C. and Harman, G.E. (1994). Molecular and cellular biology of biocontrol by *Trichoderma* spp. *Trends Biotechnol.* 12: 478-482.

Hari Narayana, A. B. (1999). An integrated approach to the management of Sclerotium wilt disease in bell pepper caused by *Sclerotium rolfsii* Sacc. M.Sc. (Ag.) thesis submitted to Acharya N.G. Ranga Agricultural University, Hyderabad (A.P).

Harlton, C.E., Levesque, C.A.and Punja, Z.K. (1995). Genetic diversity in *Sclerotium (Athelia) rolfsii* and related species. *Phytopath.* 85: 1269-1281.

Hilal, A. A.and Baiuomy, M. A. (2000). First record of fungal diseases of Stevia (*Stevia rebaudiana* Bertoni.) in Egypt. *Egyptian J. Agri. Res.* 78: 1435-1448.

Jones, J.P., Overman, A.J. and Garaldson, C.M. (1966). *Phytopathology* 56: 929

Keel, C., Schnider, U., Maurhofer, M., Voisard, C., Haas, D. and Defago, G. (1992). Suppression of root diseases by *Pseudomonas fluorescens* CHA0: importance of the bacterial secondary metabolite 2,4-diacetylphologlucinol. *Mol. Plant Microbe Interact.* 5: 4-13.

Keyser, Helena A. and l.H.S. Ferreira (1988). Chemical and Biological Control of *Sclerotium rolfsii* in Grapevine Nurseries. *S. Afr. J. Enol. Vitic.*, 9 (1): 43-44.

Kishore, G. K., Pande, S., Narayana Rao, J. and Podile, A. R. (2005). *Pseudomonas aeruginosa* inhibits the plant cell wall degrading enzymes of *Sclerotium rolfsii* and reduces the severity of groundnut stem rot. *European J. Plant Pathol.* 113(3): 315-320.

Kumar, B. and Chitreswar, Sen. (2000). Management of stem rot of groundnut caused by *Sclerotium rolfsii* through *Trichoderma harzianum*. *Indian Phytopathol.* 53: 290-295.

Leach, L.D. and Davey, A.E. (1942). Reducing southern Sclerotium rot of sugar beets with nitrogenous fertilizers. *J. Agric. Res.* 64: 1-18.

Leeman, M., Den Ouden, F.M., Van Pelt, J.A., Dirkx, F.P.M., Stejil, H., Bakker, Pahm and Schippers, B. (1996). Iron availability affects induction of systemic resistance to Fusarium wilt of radish by *Pseudomonas fluorescens. Phytopath.* 86: 149-155.

Lumsden, R.D., Ridout, C.J., Vendemia, M.E., Harrison, D.J., Waters, R.M. and Walter, J.F. (1992). Characterization of major secondary metabolites produced in soilless mix by a formulated strain of the bjocontrol fungus *Gliocladium virens. Can. J. Microbiol.* 38: 1274-1280.

Mahato, A., Mondal, B., Dhakre, D. S. and Khatua, D. C. (2014). *In vitro* sensitivity of *Sclerotium rolfsii* towards some fungicides and botanicals. *Sch. Acad. J. Biosci.*, 2(7): 467-471.

Mahato, A. and Mondal, B. (2014). *Sclerotium rolfsii:* its isolates variability, pathogenicity and an eco-friendly management option. *J. Chem., Biol. and Physical Sc.* 4 (4): 3334-3344.

Maurhofer, M., Hase, C., Meuwly, P., Metraux, J.P. and Defago, G. (1994). Induction of systemic resistance of tobacco in tobacco necrosis by the root-colonizing *Pseudomonas fluorescens* strain CHA0: Influence of the *gacA* gene and of pyoverdine production. *Phytopath.* 84: 139-146.

Maurhofer, M., Keel, C., Haas, D. and Defago, G. (1995). Influence of plant species on disease suppression by *Pseudomonas fluorescens* CHA0 with enhanced antibiotic production. *Plant Pathol.* 44: 44-50.

Morton, D. J. and Stroube, W. H. (1955). Antagonistic and stimulatory effects of soil microorganisms upon *Sclerotium rolfsii. Phytopath.* 45: 417-420.

Moussa, Tarek A. A. and Nagwa A. Tharwat (2013). Optimization of cellulase and -glucosidase induction by sugarbeet pathogen *Sclerotium rolfsii. African J. Botany* 1 (4): 44-49.

Mukherjee, P.K., Mukhopadhyay, A.N., Sarmah, D.K. and Srestha, S.M. (1995). Comparative antagonistic properties of *Gliocladium virens* and *Trichoderma harzianum* on *Sclerotium rolfsii* and *Rhizoctonia solani*-its relevance to understanding the mechanism of biocontrol. *J. Phytopathol.* 143: 275-279.

Muthamilan, M. and Jeyarajan, R. (1992). Effect of antagonistic fungi on *Sclerotium rolfsii* causing root rot of groundnut. *J. Biological Control* 6: 88-92.

Muthamilan, M. and Jeyarajan, R. (1996). Integrated management of Sclerotium root rot of groundnut involving *Trichoderma harzianum, Rhizobium* and Carbendazim. *Indian J. Mycol. Plant Pathol.* 26: 204-209.

Nalim, F.A., Starr, J.L., Woodard, K.E., Segner, S. and Keller, N.P. (1995). Mycelial compatibility groups in texas peanut field populations of *Sclerotium rolfsii. Phytopath.* 85: 1507-1512.

Narayana Bhat and Srivastava, L. S. (2003). Evaluation of some fungicides and neem formulations against six soil-borne pathogens and three *Trichoderma* spp. *in vitro*. *Plant Disease Res.* 18: 56-59.

Okabe, I., Morikawa, C., Matsumoto, N., and Yokoyama, K. (1998). Variation in *Sclerotium rolfsii* isolates in Japan. *Mycoscience*. 39: 399-407.

Pandey, K. K. and Upadhyay, J. P. (2000). Microbial population from Rhizosphere and Non-rhizosphere soil of pigeonpea : Screening for resident antagonist and mode of mycoparasitism. *J. Mycol. and Plant Pathol.* 30: 7-10.

Pant, Rajeev and Mukhopadhyay, A. N. (2001). Integrated management of seed and seedling rot complex of soybean. *Indian Phytopathol.* 54: 346-350.

Papavizas, G. C. (1985). *Trichoderma* and *Gliocladium* : Biology, Ecology and Potential for Biocontrol. *Annu. Rev. Phytopathol.* 23: 23-54.

Patibanda, A. K., Upadhyay, J. P. and Mukhopadhyay, A. N. (2002). Efficacy of *Trichoderma harzianum* Rifai alone or in combination with fungicides against *Sclerotium* wilt of groundnut. *J. Biological Control* 16: 57-63.

Patil, R. P., Jagadeesh, K. S. and Kulkarni, J. H. (1998). Isolation of fluorescent pseudomonads associated with roots of different plants and their *in vitro* antagonism against groundnut collar rot pathogen *Sclerotium rolfsii* Sacc. *Karnataka J. Agri. Sc.* 11: 45-49.

Pattanapipitpaisal, P. and Kamlandharn, R. (2012). Screening of chitinolytic actinomycetes for biological control of *Sclerotium rolfsii* stem rot disease of chilli. *Songklanakarin J. Sci. Technol.* 34 (4): 387-393.

Phillips, A.J.L., 1986. Factors affecting the parasitic activity of *Gliocladium virens* on sclerotia of *Sclerotinia sclerotiorum* and a note on its host range. *Phytopathologische Zeitschrift. 116*, 212-220.

Pierto, A., Lorito, M., Hayes, C.K., Broadway, R.M. and Harman, G.E. (1993). Endochitinase from *Gliocladium virens*: isolation, characterization and synergistic antifungal activity in combination with gliotoxin. *Phytopath.* 83: 308-313.

Prasad, M. Santha Lakshmi, Sujatha, K., Naresh, N. and Chander Rao S. (2012). Variability in *Sclerotium rolfsii* associated with collar rot of sunflower. *Indian Phytopath.* 65 (2): 161-165.

Prasad, Saddala Durga and Shaik Thahir Basha (2010). Molecular variability among the isolates of *Sclerotium rolfsii* causing stem rot of groundnut by RAPD, ITS-PCR and RFLP. *EurAsian Journal of BioSciences,* 4: 80-87.

Prasada Babu, G. and Paramageetham, Ch. (2013). Biocontrol of *Sclerotium rolfsii* – a polyphagous plant pathogen by *Pseudomonas aeruginosa* isolated from forest litter. *Int. J. Research in Plant Science* 3 (1): 1-4.

Punja, Z.K. (1985). Biology, ecology and control of *Sclerotium rolfsii*. *Ann. Rev. Phytopathol.* 23: 97-127.

Punja, Z.K. (1989). Influence of nitrogen and calcium compounds on development of disease due to *Sclerotium rolfsii*. Pp. 75-89. In: Soilborne plant pathogens: Management of diseases with macro- and micro-elements. Ed. A.W. Engelhard, Scientific Publishers, Jodhpur.

Punja, Z.K. and Damiani, A. (1996). Comparative growth, morphology and physiology of three *Sclerotium* species. *Mycologia* 88: 694-706.

Punja, Z.K. and Grogan, R.G. (1982). Effect of inorganic salts, carbonate-bicarbonate anions, and ammonia and the modifying influence of pH on sclerotial germination of *Sclerotium rolfsii. Phytopathology* 72: 635-639.

Punja, Z.K. and Raharjo, S.H.T. (1996). Response of transgenic cucumber and carrot plants expressing different chitinase enzymes to inoculation with fungal pathogens. *Plant Dis.* 80: 199-215.

Punja, Z.K., Carter, J.D., Campbell, G.M. and Rossell, E.L. (1986). Effect of calcium and nitrogen fertilizers, fungicides, and tillage practices on incidence of *Sclerotium rolfsii* on processing carrots. *Plant Dis.* 70: 819-824.

Punja, Z.K., Grogan, R.G. and Unruh, T. (1982). Comparative control of *Sclerotium rolfsii* on golf greens in North California with fungicides, inorganic salts, and *Trichoderma* spp. *Plant Dis.* 66: 1125-1128.

Punja, Z.K., Huang, J.S. and Jenkins, F.S. (1985). Relationship of mycelial growth and production of oxalic acid and cell wall degrading enzymes to virulence in *Sclerotium rolfsii. Can. J. Plant Pathol.* 7: 109-117.

Rajalakshmi. (2002). Studies on variability among the isolates of *Sclerotium rolfsii* Sacc. M.Sc. (Ag.) Thesis submitted to Acharya N.G.Ranga Agricultural University, Hyderabad (A.P.).

Rajendran, K. and Ranganathan, K. (1996). Biological control of onion basal rot (*Fusarium oxysporum* f.sp. *cepae*) by combined application of fungal and bacterial antagonists. *J. Biological Control* 10: 97-102.

Rakh, R.R., Raut, L.S., Dalvi, S.M. and Manwar, A.V. (2011). Biological control of *Sclerotium rolfsii*, causing stem rot of groundnut by *Pseudomonas cf. monteilii 9. Recent Res. in Science Tech.* 3(3): 26-34.

Rama Yalla Reddy, M. (2002). Evaluation of certain bio-fungicides against predominant seed borne fungal disease of groundnut (*Arachis hypogaea* L.). M.Sc.(Ag.) Thesis submitted to Acharya N.G.Ranga Agricultural University, Hyderabad (A.P.).

Rangeshwaran, R. and Prasad, R. D. (2000). Biological control of Sclerotium rot of sunflower. *Indian Phytopathol.* 53: 444-449.

Reddi Kumar, M., Madhavi Santhoshi, M.V., Giridhara Krishna, T. and Raja Reddy, K. (2014). Cultural and morphological variability *Sclerotium rolfsii* isolates infecting groundnut and its reaction to some fungicidal. *Int.J.Curr.Microbiol. App.Sci.* 3(10) : 553-561.

Rivera, M.C., Wright, E.R., Fabrizio, M.C., Freixá, G., Cabalini, R. and Lopez, S.E. (2013). Control of seedling damping off caused by *Rhizoctonia solani* and *Sclerotium rolfsii* using onion broths. *Int. J. Exp. Bot.*, 82: 227-234.

Saccardo, P.A. (1913). *Sclerotium rolfsii.* Sylloge Fungorum XXII. Pavia, Italy. 1500 pp.

Sahni, S., Sarma, B. K. and Singh, K. P. (2008). Management of *Sclerotium rolfsii* with integration of non-conventional chemicals, vermicompost and *Pseudomonas syringae*. *World J. Microbiol. Biotech*. 24(4): 517-522.

Sakthi Kumaran (2000). Effect of *Trichoderma* spp. on *Sclerotium rolfsii* causing stem rot disease in groundnut (*Arachis hypogaea* L.). M.Sc.(Ag.) Thesis submitted to Acharya N.G. Ranga Agricultural University, Hyderabad (A.P.).

Sarker, B. C., Adhikary, S. K., Sultana, S., Biswas, A. and Azad, S. F. D. (2013). Influence of pH on growth and sclerotia formation of *Sclerotium rolfsii* causal agent of foot rot disease of betel vine. *IOSR J. Agri. and Veterinary Sc.*, 4 (1): 67-70.

Sarma, B.K. and Singh, U.P. (2003). Ferulic acid may prevent infection of *Cicer arietinum* by *Sclerotium rolfsii World J. Microbiol. Biotech*. 19(2): 123-127.

Sarma, B.K., Singh, D.P., Singh, H.B., Singh, Anand and Singh, U.P. (2002). *Sclerotium rolfsii* – a threat to crop plants. *Indian J. Plant Pathol*. 20(1 and 2): 1-14.

Schippers, B., Bakker, A.W., Bakker, Pahm. (1987) Interactions of deleterious and beneficial rhizosphere microorganisms and the effect of cropping practices. *Annu. Rev. Phytopathol*. 25: 339-358.

Sen, C. (1983). Advances in the management of *Sclerotium rolfsii*, 1965-1981. In: Recent advances in plant pathology ed: Husein, Akhtar, Singh, Kishan, Singh, B.P., Agnihotri, V.P. pp. 445-464.

Singh, A., Mehta, S., Singh, H.B. and Nautiyal, C.S. (2003). Suppression of collar rot pathogen of betelvine (*Piper betle* L.) by rhizosphere competent strains of fluorescent pseudomonads. *Current Microbiol*. 46:

Singh, Amar and Singh, Dhanbir. (1994). Biocontrol of *Sclerotium rolfsii* Sacc. causing collar rot of Brinjal. *J. Biological Control* 8: 111-114.

Singh, R. S. (1982). Plant pathogen: The fungi. Oxford and IBH Publishing Co. Pvt. Ltd., New Delhi pp 324-325.

Singh, U.P. and Pavgi, M.S. (1965). Spotted leaf rot – a new sclerotial disease. *Plant Dis. Rept*. 49: 58-59.

Singh, U.P., Prithiviraj, B. Vagner, K.G. and Schumechar, K.P. (1996). Induction of sexual reproduction in *Sclerotium rolfsii, Ustilago cynodontis* and *Cintractia limitata* by *Cyprus rotundus* rhizomes. *Can.J. Bot.*, 74: 803-806.

Tajane, V. S., Behere, G. T., and Aage, V. E. (2002). Efficacy and translocation of fungicides against collar rot of soybean caused by *Sclerotium rolfsii*. *Plant Disease Res*. 17: 196-197.

Thapliyal, Uma Singh. (1998). Effect of inoculum density, host cultivars and seed treatment on the seed and seedling rot of soybean caused by *Sclerotium rolfsii*. *Indian Phytopathol*. 51: 244-246.

Thiribhuvanamala, G., Rajeswari, E. and Sabitha, Doraiswamy (1999). Inoculum levels of *Sclerotium rolfsii* on the incidence of stem rot in tomato. *Madras Agricultural Journal* 86: 334.

Tortoe, C. and Clerk, G. C. (2012). Isolation, Purification and Characteristics of Strains of *Sclerotium rolfsii* in Ghana. *Global Res. J. Microbiol.*2(1): 76 –84.

Upadhyay, J.P. and Mukhopadhyay, A.N. (1986). Biological control of *Sclerotium rolfsii* by *Trichoderma harzianum* in sugarbeet. *Trop. Pest Management* 32: 215-220.

van Peer, R., Niemann, G.J. and Schippers, B. (1991). Induced resistance and phytoalexin accumulation in biological control of Fusarium wilt of carnation by *Pseudomonas* sp. Strain WCS417r. *Phytopath.* 82: 728-734.

Vishwakarma, S. N., Lalit Bhatt and Chaube, H. S. (2002). Effect of delivery systems of biocontrol agents on the incidence of damping - off of brinjal and tomato raised in nurseries. *Plant Disease Research* 17: 208.

Weindling, R. (1932). *Trichoderma lignorum* as a parasite of other soil fungi. *Phytopath.* 22: 837-845.

Wilhite, S.E. and Straney, D.C. (1996). Timing of gliotoxin biosynthesis in the fungal biological control agent *Gliocladium virens* (*Trichoderma virens*). *Appl. Microbiol. Biotchnol.* 45: 513-518.

Wong, W.C. and Hughes, LK., 1986. *Sclerotium cepivorum* Berk in onion (*Allium cepa* L.) crops: isolation and characterization of bacteria antagonistic to the fungus in Queensland. *1. Appl Bacteriol.*, 60, 57-60.

Worley, R.E. and Morton, D.J. (1964). Ineffectiveness of calcium nitrate and other calcium source in reducing southern blight incidence on Rutgers tomato under epiphytotic conditions. *Plant Dis. Rept.* 48: 63-65.

Xhang, I., Howell, C. R. and Starr, J. I. (1996). Suppression of *Fusarium* colonization of cotton roots and *Fusarium* wilt by seed treatment with *Gliocladium virens* and *Bacillus subtilis*. *Biocontrol Science and Tech.* 6: 175-187.

Yaqub, Fouzia and Saleem Shahzad (2006). Effect of fungicides on *in vitro* growth of *Sclerotium rolfsii*. *Pak. J. Bot.*, 38(3): 881-883.

Zape, A.S., Gade, R. M. and Singh, Ravindra (2013). Physiological studies on different media, pH and temperature on *Sclerotium rolfsii* isolates of soybean. *Scholarly J. Agril. Sc.* 2 (6): 238-241.

Chapter 12

Stem Rot or White Mold Disease in Grain Legumes and its Management

Sangita Sahni, Bishun D. Prasad, R.K. Mishra & Sunita Kumari

Over the last few years, White mold (also called Sclerotinia stem rot) is a hot topic in pulse crops, becoming a detrimental disease to many pulse producing regions. This disease, caused by the fungus *Sclerotinia sclerotiorum*, is one of the most ubiquitous, soil borne destructive diseases in areas, where pulse crops are grown annually as a major winter-season (rabi) pulse crops. This disease varies in incidence and severity from year to year because of its sensitivity to weather conditions. White mold has not only impaired yield but also had a huge impact on quality if left untreated. The disease caused by this pathogen has been reported from different parts of India (Saharan and Mehta, 2008). More than 60 names have been used to refer to the diseases caused by this fungal pathogen (Purdy, 1979 and Le Tourneau, D. 1979) including stem rot, cottony rot, watery soft rot, drop, crown rot, blossom blight and, perhaps most common white mould (Chet and Henis, 1975; Adams and Ayers, 1979; Willets and Bullock, 1992). It is recognized as an omnivorous plant pathogen with broad host range and worldwide distribution. This necrotrophic fungal pathogen is known to infect about 408 species of plants in 34 families (Boland and Hall 1994; Purdy 1979; Tu 1997), and is now considered a serious threat to many economically important pulse crops, including soybean, lentil, edible dry bean, chickpea, pea, Rajmash, faba bean and pigeonpea *etc.* Its infection in pulses leads to stem and pod rot, crop lodging and reduced seed fill (Figures 12.1–12.3) and causes considerable yield losses. Yield losses in susceptible crops vary and may be as high as 100 per cent (Purdy, 1979). The disease can cause major seed yield reductions or even completely destroy a crop when these pulse crops are planted in infested soil and there is a dense plant canopy with prolonged periods of wet weather.

Figure 12.1. Field Symptoms of Chickpea Stem Sot, Distal Portions of the Plant Show Wilt and then become Necrotic.

Figure 12.2. Field Symptoms of Fieldpea Stem Rot.

Figure 12.3: Field Symptoms of Rajmash White Mold.

Besides seed yield reductions, the disease also results in reduced seed quality and seed contaminated with the black sclerotia of the fungus. Furthermore, sclerotia returned to the soil can affect other crops in the rotation. This disease is responsible for reducing overall production as well as quality of the pulse crops produce. Developing an integrated disease management plan based on knowledge of field history and best disease management practices can help to reduce losses from white mold. Understanding how different environmental variables and management practices influence infection by *S. sclerotiorum* is important to optimize disease management and reduce yield loss. Therefore, there is a need to correct identification, diagnosis and adaptation of suitable management strategies against white mold disease of pulse crops. A brief account of this disease in India, including the symptoms, distribution, pathogenecity, disease cycle and the key strategies to help limit the negative impact of white mold in pulse crops, are discussed here.

Economic Importance

White mold disease, although, occur most frequently in cool and moist areas (Purdy 1979; Saharan and Mehta, 2008), now it has been reported from both cold and dry regions of the world. In India, during the eightees and ninettes, white mold of pulse crops was of very minor importance, but over the last few years, it is becoming a serious disease to many pulse producing regions including West Bengal, Madhya Pradesh, Uttar Pradesh and Bihar due to monocropping and cultivation of pulse crops under irrigated conditions. Infection by *S. sclerotiorum*, results in damage of the plant tissue, followed by cell death and development of soft rot or white mold (Purdy, 1979). Yield losses vary with the percentage of plant infected and the growth stage of the crop at the time of infection. Plant infected at the early flowering stage produce little or no seeds, and those infected at the late flowering stage will set seed and may suffer little yield reduction. This disease causes yield loss to pulse crops by reducing seed number and weight and affect both yield and quality of crops; the range of crop losses can vary from 0 to100 per cent (Purdy, 1979) depending upon factors such as crop species/cultivars, locations and environmental conditions.

Yield Loss

During periods of wet or humid weather, yield loss from stem rot can be as high as 50 per cent. In chickpea, the extent of damage ranged from 21.25 to 46.70 per cent at pod bearing stage of the crop (Sharma, 1995).

Grade Loss

Sclerotia (resting bodies) found in seed can reduce grades. Seed contamination with sclerotia can be a serious problem for exported seed since it may result in rejection of the seed lot at foreign ports of entry.

Host Range

S. sclerotiorum, an ubiquitous ascomycete has a wide host range and can infect and damage over 400 species, encompass 34 families, 225 genera, and 361 species of plants, including field crops, legumes, vegetables, fruits, and ornamentals found worldwide (Purdy, 1979; Boland and Hall, 1994) . *S. sclerotiorum* poses a threat to dicotyledonous crops such as sunflower, soybean, oilseed rape, edible dry bean, chickpea, peanut, dry pea, lentils and various vegetables, besides monocotyledons such as onion and tulip (Boland and Hall, 1994). The broad host range itself makes control of disease in pulse crops very difficult, because it restricts the number of non-host crops that can be included in crop rotations.

Symptoms and Signs

Signs and symptoms caused by *S. sclerotiorum* vary depending on the host. The pathogen can affect its hosts at any stage of growth, including seedlings, mature plants, and harvested products. It can usually be found on tissues with high water content and in close proximity to the soil. Sclerotinia white mold causes mold on stems, leaves and pods. When humidity is high, the pathogen produces white fluffy hyphae on the host surface (Figure 12.4). The hyphae produce enzymes and oxalic acid, creating water soaked or dry lesions on stem that rapidly progress above and below infected nodes, and eventually encircle the stem. Over time, infected stems become bleached and stringy, and lesions also can occur on stems, pods, petioles and rarely, leaves (Figures 12.4 and 12.5). Severe infection weakens the plant and can result in wilting, lodging and plant death (Figures 12.1–12.3). Mature lesions are easily shredded, resulting in premature ripening and lodging. The affected stems tend to thread and lodge under strong winds. Individual plants, and occasionally an entire crop, can be destroyed by the disease, resulting in the appearance of flat, brown patches in the field later in the growing season. At later stages of the disease, the cottony hyphae of the pathogen aggregate into typically pea-sized clumps of mycelium (Figure 12.6). These clumps eventually mature into hard black sclerotia (Figures 12.7 and 12.8), a reliable diagnostic sign of Sclerotinia, which are most commonly found on the outer surface of the diseased tissue, but sometimes inside or outside of stems and pods. These signs of *S. sclerotiorum* and symptoms of white

Figure 12.4. Infected Rajmash Plant Shows White Cottony Mycelium of the Pathogen that Grows on the Surfaces of Infected Flowers and Pod.

Figure 12.5. Infected Fieldpea Plant.

mold allow it to be distinguished easily from most other diseases. In beans, pea and soybean, *S. sclerotiorum* cause stem rot and infection is seen in pods both before and after harvest. Root system and base of plant become brown and rotten and may

Figure 12.6. Pea-sized Clumps of Mycelium on Chickpea Diseased Plants.

Figure 12.7. Infected Lentil Plant with Mycelial Growth and Hard Black Sclerotia of the Fungus. Arrows indicate sclerotia.

Figure 12.8. Well Developed Sclerotia on Rajmash Seeds.

have white mould growing on them. Plants first turn yellow, later infected areas become bleached and shredded. Wilting symptoms are observed in lentil, chickpea and pigeonpea around the time of flowering. The Sclerotinia spores are able to infect all portions of the plant including the stem, leaves and pods. This means that this disease can infect the pulse crops at any time in its life cycle. In rajmash, initial lesions usually develop at stem nodes during or after flowering. The lesions expand and the tops of the plants initially become grayish green and then wilt and die. Sclerotinia infection may be observed as individual plants scattered throughout the field, or in patches in the field where moisture was greatest such as in hollows or nitrogen-rich areas. Lodged crops are more susceptible to Sclerotinia mould development.

Pathogen Biology

S. sclerotiorum (Lib.) De Bary, is a necrotrophic fungal pathogens belongs to kingdom 'fungi', phylum 'ascomycota', class 'discomycetes' order 'Heliotiales', and family 'sclerotiniaceae'. Hyphae are hyaline, septate, branched and multinucleate. Mycelium may appear white to tan in culture and in planta. No asexual conidia are produced. Long-term survival is mediated through the sclerotium; a pigmented, multi-hyphal structure that can remain viable over long periods of time under unfavourable conditions for growth. Sclerotia can germinate to produce mycelia or apothecia depending on environmental conditions. Apothecia produce ascospores, which are the primary means of infection in most host plants. As a homothallic organism, *S. sclerotiorum* reproduces sexually by self-fertilization, resulting in the formation of apothecia and airborne ascospores, and asexually by formation of sclerotia. The sclerotium of *S. sclerotiorum* is a pigmented, asexual, multicellular, and firm resting structure composed of condensed vegetative hyphal cells which become interwoven and aggregate together, and it is capable of surviving years in soil in the absence of a plant host (Adams and Ayers 1979; Chet and Henis 1975; Le Tourneau 1979; Willetts and Bullock 1992). Sclerotia have a white to light coloured interior called medulla, and an exterior black protective covering the rind. The rind contains melanin pigments which are highly resistant to degradation, while the medulla consists of fungal cells rich in β- glucans and proteins. In *S. sclerotiorum*, sclerotia development can be divided into three distinguishable stages (Townsend and Willetts 1954): i) initiation, the appearance of small distinct initial forms of interwoven hyphae, which develop terminally by repeated branching of long, aerial, primary hyphae; ii) development, increase in size; and iii) maturation, characterized by surface delimitation, internal consolidation, and melanization, and often associated with droplet secretion. These phases are accompanied by both morphological and biochemical differentiation. The initiation and maturation stages of sclerotia development are affected by numerous factors, such as photoperiod, temperature, oxygen concentration, mechanical factors, and nutrients (Chet and Henis 1975). Sclerotia can vary in size and shape depending on the host tissue they grow in. They are irregularly shaped, and typically measure 2 to 5 mm in diameter and up to 25 mm in length. Sclerotia are resilient to adverse conditions and can survive in soil or plant tissue for three or more years. Sclerotia germinate by one of two methods, it can germinate to produce saprophytic or infectious mycelia (myceliogenic germination) or apothecia (carpogenic germination), depending on environmental conditions (Steadman, 1983).

Myceliogenic (Hyphal) Germination of Sclerotia

Sclerotia germinate by the direct emergence of hyphae (termed 'myceliogenic' or 'eruptive' germination). White hyphal strands extend from sclerotia that have been stimulated to germinate by host plant exudates. These hyphae invade nonliving organic matter, forming mycelium which then infects roots, and crowns and other plant parts that are touching the ground. Infection then spreads to above ground plant parts (Steadman, 1983; Saharan and Mehta, 2008).

Carpogenic (Apothecial) Germination of Sclerotia

Carpogenic germination usually takes about two to three weeks and occurs under prolonged conditions of moisture at the soil surface (at least 10 days) and temperatures of 11 to 15° C. At soil depths of up to 2 cm, apothecia can extend from the sclerotia to reach the soil surface. One or several apothecia can emerge from a single sclerotium. Apothecia are fleshy-colored mushroom-like structures lined with asci, which are filled with eight hyaline ascospores (Kohn, 1979). Ascospores are forcibly discharged and may release up to two million ascospores over a period of several days, which are blown by wind to the aerial portions of plants. Ascospores are hyaline (clear or non-pigmented), unicellular and thin-walled, and consequently, they survive for only a few days after release. Prolonged periods of leaf wetness (16-48 h) with temperatures in the range of 12°C to 24°C promote ascospore germination and infection (Adams and Ayres, 1979; Steadman, 1983). Ascospores are a major source of inoculums in pulse crops like bean and soybean, where aerial parts are affected (Steadman, 1983; Boland and Hall, 1988).

Disease Cycle/Epidemiology

The basic disease cycle of *Sclerotinia* diseases begins with the germination of sclerotium, which primarily survive in the soil over long periods of time, under unfavorable condition. Sclerotia get mixed with the soil through infected plant debris after the crop is harvested, or when contaminated seeds are sown in the soil. Sclerotia are conditioned to germinate by the overwintering process. At certain times during the growing season, depending on the inherent nature of the fungus and the various environmental factors, the overwintered sclerotia can germinate in one of two methods. Probably the most common is carpogenic germination. Under moist field conditions, dormant sclerotia within the upper several inches of soil will germinate to produce fruiting bodies called apothecia, which are small, thin stalks ending with a cup-like structure about 5-15mm in diameter. The cup of the apothecium is lined with asci, in which the ascospores are contained. A single sclerotium can produce several apothecia. The apothecia will forcibly eject their ascospores into the air where they will be carried by air currents to crop plants. Most ascospores fall on susceptible plants in the immediate vicinity of the apothecia, but some can travel long distances by wind. One apothecium can produce enormous quantities of ascospores over several days. Ascospores survive for short periods on plant tissue but do not overwinter. Ascospores require a film of water and a food base such as dead or senescing flower tissues to germinate and grow before they infect the plant. Flower tissue is the most important food base for initiating infections. Often infections will start in stem axils where senescing flower tissues have fallen and lodged. The ascospores germinate and ramify the nonliving plant parts, and invasion of healthy plant parts. Ascospore may directly penetrate healthy host tissues and establish infection in nearly all tissue types including stems, foliage, flowers, fruits, and roots. Eventually white, fluffy mycelium begins to grow in and on the surface of the infected tissues. A film of water on the plant surface promotes development of lesions and increases the amount of tissue damage. Initial disease development generally requires greater than 40 hours of continuous plant surface wetness, but once disease is started, shorter periods of wetness will allow lesion development. This is the reason disease is associated with lengthy periods of cloudy, humid, rainy weather. As plant surfaces dry, the progress of disease is slowed. Cool temperatures also favor disease development. Optimal temperatures for growth range from 15 to 21° C. Under wet conditions, *S. sclerotiorum* will produce an abundance of mycelium and sclerotia. The requirement of moisture and relatively cool temperatures under the plant canopy for carpogenic germination and growth of the pathogen are reasons why rainy periods or irrigation are associated with outbreaks of disease on most crops. It is a monocyclic disease, meaning that it has only a primary cycle of infection, and once the host tissue is diseased, it is not contagious (*i.e.*, spore-producing) within the same season. Plant to plant spread of *S. sclerotiorum* occurs occasionally. Most severe disease situations in a field develop from sclerotia that are local within the field soil. In the case of carpogenic germination of sclerotia, ascospore releases into the air typically occur at ground level under closed canopies, and consequently, sharp foci of infection are often observed within the field. Nonetheless, infections initiated by wind-disseminated ascospores from another location have been observed occasionally. Airborne ascospores derived from the apothecium are the most important means of *S. sclerotiorum* dispersal (Abawi and Grogan 1979; Steadman 1979). The other method of germination is myceliogenic, where the scerotium produces mycelium. Infection of host plants by mycelium often occurs at or beneath the soil-line. Sclerotia germinate in the presence of exogenous nutrients and produce hyphae which invade nonliving organic matter, forming mycelium which then infects living host tissues (Saharan and Mehta, 2008). At the end of the growing season, *S. sclerotiorum* once again produce sclerotia, most are produced in the decayed stem pith and on the roots as the plant dies. The sclerotia then remain on the surface of the ground or in the soil, on either living or dead plant parts until the next season. *Sclerotinia* passes winter as sclerotia in the soil or in plant debris. The lifecycle will then continue respectively.

There are many factors affecting survival of the sclerotia including soil type, previous crops, and environmental conditions. The fungus can survive in the soil mainly on the previous year's plant debris. Introduction of the pathogen from outside sources also may occur by moving contaminated soil (on farm equipment, shoes, infected seedlings) and fertilizing with manure from animals fed infected plant debris, are two common ways of spreading sclerotia or mycelium from place to place. Irrigation also has been shown to be involved in the spread of *Sclerotia* spp. from field to field. Irrigation transported sclerotia remained viable for at least 10-21 days in flowing water. In some pulse crops, seed may be internally infected with mycelium of *Sclerotinia*, or inadequately cleaned seed may be contaminated with seed-sized sclerotia. Numerous broadleaf weeds also are hosts of *Sclerotinia*, and may contribute to a persistence of sclerotia within a field. Like most fungi, *S. sclerotiorum* prefers darker, shadier conditions as opposed to direct exposure to sunlight.

Pathogenicity

Sclerotinia sclerotiorum (Lib.) de Bary is responsible to secrete multiple pathogenicity factors. It secretes an array of cell wall-degrading enzymes and toxins. Enzymatic degradation of plant cell wall polymers and tissue maceration are important aspects of the pathogenesis of *S. sclerotiorum*. Effective pathogenesis by *S. sclerotiorum* requires the secretion of oxalic acid (Noyes and Hancock, 1981; Marciano *et al.*, 1983; Magro *et al.*, 1984; Cessna *et al.*, 2000; Billon-Grand *et al.*, 2012), and multiple extracellular lytic enzymes including pectinases, β-1,3-glucanases, glucosidases, cellulases, xylanases, proteases and endo polygalacturonases (Cotton *et al.*, 2002) - that facilitate penetration, colonization, and maceration of plant tissues to generate assimilable nutrients for fungal growth (Marciano *et al.*, 1983; Godoy *et al.*, 1990; Cessna *et al.*, 2000, Poussereau *et al.*, 2001a,b). A wide range of activities are attributed to these enzymes, from enzymatic cleavage of structural host cell wall polysaccharides, in which the integrity of the cell wall barrier to pathogen entry is compromised, to more complex roles within the sphere of host-pathogen interactions, such as the release of oligosaccharides that function as elicitors of plant defense responses (Cooper, 1983; Cote and Hahn, 1994; De Lorenzo *et al.*, 1994, 1997). An intricate exchange of molecular signals and responses between the pathogen and host determines the pathogenicity. *S. sclerotiorum* is renowned for the ability to acidify their environment through the secretion of organic acids. In particular, the production by *S. sclerotiorum* of oxalic acid (OA) and its association with white mold symptom development has been studied for over a century (Maxwell and Lumsden 1970; Marciano *et al.*, 1983; Godoy *et al.*, 1990). OA enhance *S. sclerotiorum* virulence in at least three ways (Dutton *et al.*, 1996). First, OA reduces pH. Several cell wall-degrading enzymes secreted during invasion of plant tissues (*e.g.*, polygalacturonase) have maximal activities at low pHs (Batemanet al., 1965). Second, OA may be directly toxic to host plants (Noyes *et al.*, 1981). Finally, chelation of Ca^{2+} ions present in the cell wall with oxalic ions, leading to loss of plant tissue integrity and rendering the stressed tissue susceptible to a battery of fungal degradative enzymes (Lumsden, 1979, Godoy *et al.*, 1990). OA directly limits host defense compounds by suppressing the oxidative burst. In conjuction, plant cell wall degrading enzymes, including cellulytic and pectinolytic, cause maceration of plant tissues, and necrosis followed by plant death (Collmer and Keen, 1986). In recent years, it is known that OA has several unique functions that contribute to the success of this phytopathogen. Oxalic acid enhances the activity of polygalacturonases to promote cell wall degradation (Bateman and Beer, 1965), inhibits plant–protection enzymes (Marciano *et al.*, 1983, Favaron *et al.*, 2004), suppresses the plant oxidative burst (Cessna, 2000), deregulates stomatal guard cell closure (Guimaraes and Stotz, 2004), mediates pH signalling (Rollins and Dickman, 2001), induces apoptosis-like cell death(Kim, 2008) and alters the cellular redox-status in the plant (Williams, 2011). Each of these activities separately facilitates infection (Bolton, 2006; Dickman, 2007). Oxaloacetate acetyl hydrolase (OAH) activity is associated with oxalic acid accumulation in *S. sclerotiorum (*Maxwell, 1973*)*. Genes encoding OAH activity have been identified in the *S. sclerotiorum* genomes (SS1G_08218.1). A deletion mutant in SS1G_08218.1 is deficient in oxalic acid biosynthesis (Han *et al.,* 2007). This observation, however, does not establish whether the primary source of oxaloacetate is the TCA cycle, the glyoxalate cycle or both. The dynamic role of oxalic acid in *S. sclerotiorum* physiology is further exemplified by the identification of genes encoding oxalate decarboxylase, which converts oxalate into formate and carbon dioxide. Such activity was previously reported from *S. sclerotiorum* (Magro *et al.,* 1988). The biological functions of oxalic acid rely on a dynamic balance of production and breakdown and not simply on a system of maximal synthesis and accumulation. The acidic environment created by oxalate secretion facilitates the optimal activity of certain sets of cell wall degrading enzymes and peptidases (following two sections). Degradation of OA produced by *Sclerotinia* may reduce the damage that this pathogen causes in the plant tissues (Lumsden, 1979; Noyes and Hancock, 1981; Godoy *et al.,* 1990), thereby slowing the advance of this necrotrophic fungus. Metabolism of OA may allow also a more rapid or powerful modification of plant defenses by counteracting the OA-mediated suppression of the oxidative burst (Cessna *et al.,* 2000). Cell wall reinforcement (Schweizer *et al.,* 1999) and direct antifungal activity of H_2O_2 (Lane, 1994; Zhang *et al.,* 1995) may also play a role. It is likely that constitutive expression of OXO may confer enhanced resistance to other OA-generating pathogens, such as *Cristulariella pyramidalis* (Kurian and Stelzig, 1979) and *Septoria musiva* (Liang *et al.,* 2001). Their results suggest that H_2O_2-generating enzymes such as OXO have potential utility for engineering resistance to a spectrum of pathogens in plants. These important roles of H_2O_2 have attracted molecular pathologists' interest in manipulating the H_2O_2 level by overexpressing an H_2O_2–generating enzyme, such as glucose oxidase (Wu *et al.,* 1995; Kazan *et al.,* 1998), to combat diseases in plants.

Oxalic acid production has also been found to be correlated with sclerotial development and has been shown to be an important factor for pathogenicity in *S. sclerotiorum* (Donaldson *et al.,* 2001; Godoy *et al.,* 1990; Kesarwani *et al.,* 2000; Rollins and Dickman 2001; Zhou and Boland 1999). Sclerotial development is a complex, multistage process that is thought to be regulated by signal transduction pathways such as MAPK and PKA (Chen and Dickman 2005; Chen *et al.,* 2004; Harel *et al.,* 2005; Rollins and Dickman 1998). Recently, Calcineurin-MAPK and Calcineurin-PKA –associated pathways have also been shown to play a central role in the regulation of cation homeostasis, morphogenesis, cell-wall integrity, and pathogenesis in various fungi (Fox and Heitman 2002). For example, in *S. cerevisiae,* PKA has been shown to phosphorylate and, consequently, negatively regulate the activity of the calcineurin-regulated Zn-finger transcription factor Crz1p by inhibiting its nuclear import (Kafadar and Cyert 2004). In human cells, transcriptional activity of NFATc2 (a Crz1p homolog) is upregulated by phosphorylation of the MAPK

JNK (Ortega-Perez *et al.*, 2005). If similar pathways exist in *S. sclerotiorum*, current analysis demonstrating that calcineurin plays a significant role in the regulation of morphogenesis and pathogenesis in this pathogen may serve as a basis for the further dissection of these pathways. Thus, the release of an array of lytic enzymes and the oxalic acid from the growing mycelium is the important pathogenicity factors that are required for the establishment of the host-parasite relationship. However, *S. sclerotiorum* is poorly characterized at the molecular level and only a few genes encoding hydrolytic enzymes (Reymond *et al.*, 1994; Fraissinet-Tachet *et al.*, 1995; Fraissinet-Tachet and Fevre, 1996; Poussereau *et al.*, 2001a; Cotton *et al.*, 2002; Li *et al.*, 2004) have been reported. Expressed sequence tag (EST) analysis has proved to be an efficient approach to identify genes expressed under a wide variety of conditions in other systems (Adams *et al.*, 1991). Indeed, cDNAs encoding four endo- (SSPG1063, SSPG544, SSPG427 and ZY210R), and two exo- polygalacturonases (SSPG851 and SSPG1033) were found; SSPG1063, and endo-polygalacturonase, denoted as SSPG1d, was nearly identical to SSPG1a-c (Reymond *et al.*, 1994; Cotton *et al.*, 2002) and BcPG1, which are responsible for full pathogenicity of *Botrytis cinerea* (Ten Have *et al.*, 1998). SSPG1 has also been implicated in the initiation and establishment of the infection as well as lesion progression by *S. sclerotiorum* (Li *et al.*, 2004). The importance of oxalic acid production and polygalacturonase synergy during colonization and symptom development in diseases caused by *Sclerotinia* spp. has been well established (Marciano *et al.*, 1983; Maxwell and Lumsden 1970). The UV-induced mutants deficient in oxalic acid accumulation have been demonstrated to be nonpathogenic (Godoy *et al.*, 1990). This finding further suggests its roles for oxalic acid during pathogenesis of *S. sclerotiorum*. Although the synergism of diverse physiological factors may be required for broad host-range pathogenicity, these factors undoubtedly rely upon common environmental signals and molecular response pathways to coordinate their production and activities. The identification and understanding of these pathways are prime importance in order to identify novel targets for blocking disease development. In addition to appropriately sensing the environment, *S. sclerotiorum* actively modifies its environment, making it acidic *via* oxalic acid secretion. An acidic host-pathogen environment *S. sclerotiorum* provides a suitable condition for endopolygalacturonase activity and may also serve as a signal to enhance the production of this virulence factor and coordinate the production of new factors. One member of this group, an endopolygalacturonase encoded by the *pg1* gene, is regulated by pH at the transcription level (Rollins and Dickman 2001). Accumulation of pg1 transcripts is maximal between pH 3.0 and 3.8 *in vitro* and decreases significantly above pH 4.2. This finding indicates a correlation between pH-regulated gene transcription and catalytic activity of the encoded enzyme. Another gene *pac1* appears to be necessary for the appropriate regulation of physiological processes important for pathogenesis and development of *S. sclerotiorum* (Rollins JA 2003). Understanding of the physiological and molecular mechanisms involved in sclerotial development and pathogenicity of *S. sclerotiorum* may well reflect the development and pathogenesis of additional sclerotium-producing fungi and may provide new avenues for intervention in these processes, leading to improved control of disease caused by sclerotium-borne fungi.

Disease Management

There are no simple ways of achieving total control of this disease. Management is difficult, inconsistent and uneconomical due to the presence of wide host range and long term survival of the resting sclerotia. It is important to understand how different environmental variables and management practices influence infection by *S. sclerotiorum* to optimize disease management and reduce yield loss. However, a well executed integrated control can reduce the disease significantly and keep the yield loss to a minimum. The integrated control measures include implementation the management tactics of *Sclerotinia* at several stages of crop development *viz.*, making the crop microclimate less conducive for infection, elimination of sources of inoculum, selection of resistant plants and utilization of effective fungicides and bioagents to protect susceptible plants. The integrated control measures include various ecofriendly measures.

Site Selection

The degree of field infestation by *S. sclerotiorum* varies greatly. Such variation contributes in part to the difference in disease incidence in the fields ranging from 0 to 85 per cent (Tu, 1986). Fields with a previous history of severe white mold should be planted with resistant cultivars.

Tillage

The impact of tillage on white mold development is inconsistent, although several studies have indicated lower levels of disease in no-till. Deep tillage may initially reduce white mold incidence by removing sclerotia from the upper soil profile which may reduce the number of apothecia produced. Although sclerotia can survive in soil for several years, only the sclerotia near the soil surface germinate to produce apothecia and ascospores. Therefore, burying infested residues with a moldboard plow can prevent the apothecial germination of sclerotia. Deeply cultivate with mouldboard plough to bury sclerotia laden soil 150 mm deep. However, sclerotia can remain viable for more than three years if buried 8 to 10 inches in the soil, and may be returned to the soil surface in subsequent tillage operations. Although more sclerotia are found near the soil surface in no-till systems, sclerotia may degrade faster in no-till soils compared to tilled soils. If pulse crops are planted in a field with a history of white mold, a combination of proper crop rotation and no-till operation can reduce the risk.

Soil Solarization

Soil mulching with transparent polyethylene film can raise soil temperatures, thus adversely affecting the survival of sclerotia. Several reports indicate that solarization was effective in reducing the inoculum of *S. sclerotiorum* (Porter and Merriman, 1985). The polyethylene mulching inhibited sclerotial germination, disturbed apothecial formation and trapped ascospores.

Soil Amendment

Organic or inorganic matter can be used in the management of soil borne sclerotinia diseases (Sun and Huang, 1985). Huang and Sun (1991) reported that amending soil with a formulated product, S-H mixture (Sun and Huang, 1985) was effective in suppressing apothecial production from sclerotia of *S. sclerotiorum*.

Crop Rotation

Fields that has a history of white mold should not be in two or three year rotations with broadleaf crops susceptible to *S. sclerotiorum*. These crops include edible beans, canola, cole crops, sunflowers and potatoes *etc*. Crop rotation with nonhost crops like corn, wheat, sorghum and small grain is generally considered as a good practice to reduce the number of sclerotia in the soil by loss of viability over time. In addition, sclerotia may germinate in the absence of a host crop, but without subsequent host infection, new sclerotia are not returned to the soil and numbers are gradually reduced. Crop rotation is most effective when initiated before white mold becomes a serious disease problem in a field. Even in the absence of a susceptible crop, sclerotia can remain viable in the plow layer of field soil for up to 5 years. If numbers of sclerotia in a field are low, rotations of 3 to 5 years with a nonhost crop may be sufficient. Once the pathogen is well established in a field, and soil is highly infested with sclerotia, crop rotation may be of less value because of the long survival time of these propagules, therefore, rotations of more than 5 years with a nonhost crop may help to reduce the inoculum.

Weed Control

The host range of *S. sclerotiorum* is very broad and includes many of the most important broadleaf weeds of cultivated crops: Canada thistle, Jerusalem artichoke, lambsquarters, mustard, cocklebur, Dandelion, Wild carrot, nightshade, pigweed, ragweed, shepherd's purse, sow thistle, velvet leaf, vetch *etc*. If these weeds occur in a field, they may provide inoculum for a host crop, favoring disease development.

Cover Crops

The use of small grain cover crops like oat, wheat or barley grown with pulse crops can stimulate earlier emergence of apothecia compared to pulse crops grown alone. This can potentially lower white mold incidence. Consider first how cover crops may affect soil moisture, availability of soil nutrients and shading before implementing.

Cultivars Selection, Planting Date and Relative Maturity

Cultivars that mature early and have a more upright, as opposed to a vining (prostrate) growth habit can provide 'avoidance' or 'escape' resistance, generally resulting in less disease. Early planting, late-maturing varieties and varieties with a bushy architecture or a tendency to lodge have higher disease incidence because it can contribute to more closed canopies that more readily create a microclimate within the crop canopy. A positive correlation between plant canopy and white mold incidence and severity has been established (Steadman, 1979). In general, upright types of plants have less severe white mold than bush (or viny) types (Saindon and Huang, 1992). Dry beans with an upright or open-bush growth habit are less susceptible to infection by ascospores of *S. sclerotiorum* than the dense bush or vine-like plants, because the former can create an environment that is not conducive to the production of apothecia and infection of ascospores (Coyne *et al.*, 1974). On the other hand, lines of dry bean with determinate growth habit and early maturity resulted in greater disease avoidance than lines with indeterminate growth habit or late maturity (Steadman *et al.*, 1973; Fuller *et al.*, 1984). To avoid white mold, selection of plant types with upright characteristics and thin canopy that allow adequate penetration of sunlight and under-the-canopy aeration is important. *Sclerotinia sclerotiorum* has a wide host range without known strain specificity in pathogenicity (Steadman, 1979). Many researchers formerly believed that resistance to *S. sclerotiorum* did not exist. More recently, however, field resistance to this fungus was observed in several crops (Steadman, 1979).

Plant Populations

High plant populations contribute to dense, closed canopies, have been associated with increased white mold incidence (Coyne *et al.*, 1974; Steadman, 1979). Consider decreasing plant populations in fields with a history of white mold; however, be sure populations maintain yield potential.

Plant Spacing and Row Direction

Sclerotinia disease incidence is lower in crops with wide row widths than those planted in narrow rows by providing adequate room to facilitate air movement through the crop canopy to reduce plant wetness. Narrow row and high plant density reduce air circulation and trap moisture in the canopy. Thus, they contribute to higher incidence and more severe white mold than conventional row width. Narrow row and high plant density not only increase early senescence but also increase contact of plant parts. Increase in senescence and in contact of plant parts promote infection and spread of white mold, respectively. Altered plant spacing is effective in reducing the disease caused by infection of airborne ascospores of Sclerotinia spp. For example, white mould of beans is less severe in wider rows (60-80 cm) than narrow rows (25-30 cm) (Steadman *et al.*, 1973). Changing row direction to reduce disease caused by airborne inoculum was reported for *S. sclerotiorum* on beans. Precision seeding helps to optimize plant spacing and avoids clumping of plants. Because infection by ascospores of *S. sclerotiorum* requires an extended period of free moisture, orienting rows parallel to the direction of the prevailing winds quickly drying the canopy after a rain or irrigation event.

Fertility and Plant Nutrition

High soil fertility, especially the use of nitrogen-rich manures and fertilizers, favors white mold development by promoting lush plant growth and early canopy closure. The application of manure should be avoided on fields with a history of white mold. Increased level P, K, Ca and sulphur fertilizer application reduced disease severity.

Irrigation Management

Avoid excessive irrigation that favors a dense canopy or high moisture following flowering, because that will create ideal conditions for disease development. High soil moisture is conducive to carpogenic germination of sclerotia of *Sclerotinia species* and to increased severity of white mould of beans (Weiss *et al.*, 1980) and sclerotinia stem rot of soyabeans (Grau and Radke, 1984). Carpogenic germination of sclerotia occurred more rapidly in the continuously irrigated soil (Kopmans, 1993). For irrigated crops such as dry beans, reducing the number of irrigations during the period with high risk of disease can reduce disease in the absence of rainfall (Steadman, 1979). When a crop is irrigated, the goal is to manage irrigation events to reduce the frequencies of 12 to 24 hour periods of leaf wetness, especially during the bloom period, when flower petals can become colonized by the ascospores of *S. sclerotiorum*. Water deprivation at bloom, however, may compromise the quality of the harvested product. Therefore, a strategy known as 'irrigation cut-off' can be used to manipulate the duration of time in a day that foliage remains wet after irrigation. This technique simply requires initiating daily irrigation events in the very early morning and stopping all irrigation near noon. The early afternoon 'cutoff' allows time for the sun to dry the foliage before nightfall, thereby helping to avoid prolonged periods of plant wetness. Each irrigation event must therefore provide sufficient water to allow for a prolonged dry period. Low moisture levels within the plant canopy are critical for reducing the potential for white mold development. Infrequent, heavy watering is better than frequent, light watering. Further, flooding of the field (once a week) prior to sowing resulted in least disease incidence and minimum lesion length.

Chemical Protection

Seed Treatment

Sclerotinia sclerotiorum is able to survive in infected seeds as dormant mycelia in testae and cotyledons for 3 years or longer (Tu, 1988). When infected seeds in a recent study were sown, 88-100 per cent failed to germinate. Seedlings from infected seeds subsequently died from white mold at an early stage. Seeds that failed to germinate were rotted by *S. sclerotiorum*, and three to six sclerotia were formed in place of each seed (Tu, 1988). These sclerotia could become a source of inoculum. Captan, and thiophanate-methyl used in seed treatment were 100 per cent effective in eradicating the fungus from the infected seeds (Tu, 1989a).

Foliar Application

Fungicide applications can be a component of an integrated management system for white mold. Some foliar-applied fungicides have efficacy against *S. sclerotiorum*, although none offers complete control. Proper timing of spray and method of application have a great impact on results. For example, aerial application by aircraft is relatively ineffective as compared with ground application by boom sprayer. The latter gives better penetration of the spray into the crop canopy. The number of fungicide applications required for disease control depends on the length of the crop season and the period of time that 'weak' tissues (flower petals) are available for colonization by ascospores. In order to be effective, it is necessary that fungicides penetrate deep into the canopy to adequately cover the flowers and the places on the plant where the senescing petals might adhere or become lodged.

Currently, several chemicals are used for control of *Sclerotinia* diseases, such as benomyl, chlorothalonil, thiophanate methyl, iprodione and dicloran are available.

Biological Control

Recently, scientists have researched various areas of biological control in an attempt to control this disease while reducing chemical contamination of the environment. Biocontrol measures using microorganisms against the white mold fungus have been researched extensively. Many of these results have yet to be put into practical application at present because bioagents are subjected to strict registration guidelines. The fungus *Coniothyrium minitans* and *Trichoderma* spp. are the most widely available and tested biological control fungus for managing *S. sclerotiorum* (Ayers and Adams, 1981; Whipps and Budge, 1993). The time to implement biological control is in the resting (or sclerotium) stage of *S. sclerotiorum,* during which the pathogen has little mobility, or at the germinating stage, during which the pathogen is most vulnerable to attack. Adams and Ayers (1979) stated that the most significant component of soil affecting the survival of sclerotia appears to be the biota. Any biological measures should target the dormant sclerotia in the field. Application of *C. minitans* should occur a minimum of three months before white mold is likely to develop. *Coniothyrium minitans,* secrete 1, 3 glucanase, which degrades and lyses sclerotial tissues and suppress white mold due to *S. sclerotiorum*. In practice, dried spores of *Coniothyrium minitans* are sprayed onto pathogen-infested crop debris either at the end of a season, or onto the soil surface before planting. Parasitization of sclerotia has been shown to reduce the number of apothecia formed by *S. sclerotiorum*. *Coniothyrium minitans* should be incorporated as thoroughly as possible to a depth of two inches. Avoid additional tillage that can bring uncolonized sclerotia to the soil surface. In several studies, soil treatment with *C. minitans* has lessened white mold severity in the crop that follows the treatment. This treatment is most useful when combined with other components of an integrated disease management program. Other fungi that have shown effectiveness against *Sclerotinia* in culture are: *Gliocladium roseum, Trichothecium roseum, Sporidesmium sclerotivorum, Fusarium* spp., *Mucor* spp., *Alternaria* spp., *Epicoccum* spp., and *Penicillium* spp. (Adams and Ayers, 1979, Makkonen and Pohjakallio, 1960).

Little information is available about the effect of bacteria on the survival of sclerotia. However, damaged sclerotia appear more susceptible to bacterial invasion. Bacteria are consistently found in rotted sclerotia as in many other saprophytic fungi. Thus, many bacteria, such as *P. fluorescence, P. putida,* and *B. subtilis,* may have a direct or indirect impact on the survival of the sclerotia, particularly those which produce antibiotics. More research in this area is needed. Huang *et al.* (1993b) reported that spraying pea plants with an antagonistic strain of *Bacillus cereus* was effective in reducing incidence of pod rot caused by infection with airborne ascospores of *S. sclerotiorum*. Control of ascospore infection may be feasible by application of biocontrol agents to senescent tissues such as flower petals (Zhou and Reeleder, 1991; Hutchins and Archer, 1994). Some reports indicate that seed treatment with a biocontrol agent (Illipronti and Machado, 1993) or a combination of a biocontrol agent and a fungicide (Zazzerini and Tosi, 1985b) is effective in controlling sclerotinia diseases.

Host Resistance

Cultivars able to resist infection by *S. sclerotiorum* would be a predictable, convenient, and inexpensive option for controlling sclerotinia diseases. Unfortunately, identification and selection of highly effective physiological resistance to this pathogen has proven to be a difficult challenge, especially under a variety of field conditions. As a consequence, commercial cultivars that show high levels of resistance to sclerotinia diseases are relatively uncommon. Some cultivars of common bean, soybean, and other crops have been found that show a low level of physiological resistance but at best of this generally has only a partial impact on the overall amount of disease. Breeding for resistance to white mold is difficult because resistance is believed to be controlled by multiple genes. Screening for resistance is complicated because infection and disease development in field plots can be inconsistent. Differences in plant maturity and architecture can also influence infection and disease development. Biotechnological techniques are being used to create novel types of resistance to diseases caused by *Sclerotinia* spp. For example, a gene from barley that encodes for oxalate oxidase, an enzyme that degrades oxalic acid, has been transferred into peanut, making it much more resistant to infection by *S. minor*. Such genetically modified cultivars may become a major means of *Sclerotinia* disease management in the future. Biotechnology opens a new avenue to control this pathogen. The transgenic strategies include detoxification of the pathogenicity factor OA, activation of endogenous defense pathways, and inhibition of *S. sclerotiorum* growth by anti-fungal proteins, have potential to engineer *Sclerotinia* resistance.

REFERENCES

Abawi, G.S., and R.G. Grogan. 1979. Epidemiology of diseases caused by *Sclerotinia* spp. *Phytopathology*, 69: 899-904.

Adams P.B., Wong J.A.L. 1991. The effect of chemical pesticides on the infection of sclerotia of *Sclerotinia minor* by the biocontrol agent *Sporidesminum sclerotivorum*. *Phytopathology*, 81: 1340-1343.

Adams, P.B. and Ayers, W.A. 1979. Ecology of *Sclerotinia species*. *Phytopathology*, 69: 896–899.

Ayers, W.A. and Adams, P.B. 1981. Mycoparasitism and its application to biological control of plant diseases. In: Biological Control in Crop Production (G. C. Papavizas, ed.). BARC Symposium Vol. 5, Allanheld Osmum, Totowa, NJ, pp. 91–103.

Bateman, D.F. and Beer, S.V. 1965. Simultaneous production and synergistic action of oxalic acid and polygalacturonase during pathogenesis by *Sclerotium rolfsii*. *Phytopathology*, 55: 204–211.

Billon-Grand, G., Rascle, C., Droux, M., and Rollins, J.A. 2012. pH modulation differs during sunflower cotyledon colonization by the two closely related necrotrophic fungi *Botrytis cinerea* and *Sclerotinia sclerotiorum*. *Mol. Plant Pathol.*, 13: 568-578.

Boland, G.J. and Hall, R. 1988. Epidemiology of sclerotinia stem rot of soybean in Ontario. *Phytopathology*, 78: 1241–1245.

Boland, G.J. and Hall, R. 1994. Index of plant hosts of *Sclerotinia*. *Canadian J Plant Pathol.*, 16: 93-108.

Bolton, M.D., Thomma, B.P. and Nelson, B.D. 2006. *Sclerotinia sclerotiorum* (Lib.) de Bary: Biology and molecular traits of a cosmopolitan pathogen. *Mol Plant Pathol.*, 7: 1–16.

Cessna, S.G., Sears, V.E., Dickman, M.B. and Low, P.S. 2000. Oxalic acid, a pathogenicity factor for *Sclerotinia sclerotiorum*, suppresses the oxidative burst of the host plant. *Plant Cell*,12: 2191-2200.

Chen, C., and Dickman, M.B. 2005. cAMP blocks MAPK activation and sclerotial development via Rap-1 in a PKA-independent manner in *Sclerotinia sclerotiorum*. *Mol. Microbiol.*, 55: 299-311.

Chen, C., Harel, A., Gorovoits, R., Yarden, O., and Dickman, M. B. 2004. MAPK regulation of sclerotial development in *Sclerotinia sclerotiorum* is linked with pH and cAMP sensing. *Mol. Plant-Microbe Interact.*, 17: 404-413.

Chet, I. and Henis, Y. 1975. Sclerotial morphogenesis in fungi. *Ann Rev Phytopathology*, 13: 169-192.

Collmer, A. and Keen, N.T. 1986. The role of pectic enzymes in plant pathogenesis. *Annu.Rev.Phytopathol.*, 24: 383-409.

Cooper, M.R. 1983. The Mechanisms and Significance of Enzymatic Degradation of Host Cell Walls by Parasites. In: Biochemical Plant Pathology (Callow JA, ed.). John Wiley and Sons, New York, 101-135.

Cote, F. and Hahn, M.G. 1994. Oligosaccharins: structures and signal transduction. *Plant Mol. Biol.* 26: 1379-1411.

Cotton, P., Rascle, C. and Fevre, M. 2002. Characterization of PG2, an early endo PG produced by *Sclerotinia sclerotiorum*, expressed in yeast. *FEMS Microbiol Lett*, 213: 239-244.

Coyne, D.P., Steadman, J. R. and Anderson, F.N. 1974. Effect of modified plant architecture of great northern dry bean varieties (Phaseolus vulgaris) on white mold severity and components of yield. *Plant Dis. Reptr.*, 58: 379–382.

De Lorenzo, G., Castoria, R., Bellincampi, D. and Cervone, F. 1997. Fungal Invasion Enzymes and their Inhibition. In: The Mycota: V. Plant Relationships (Carroll GC and Tudzynski P, eds.). Springer, Berlin, 61-83.

De Lorenzo, G., Cervone, F., Bellincampi, D. and Caprari, C. 1994. Polygalacturonase, PGIP and oligogalacturonides in cell-cell communication. *Biochem. Soc. Trans.*, 22: 394-397.

Dickman, M.B. 2007. Subversion or coersion? Pathogenic derminants in fungal phytopathogens. *Fungal Biology Rev.*, 21: 125–129.

Donaldson, P.A., Anderson, T., Lane, B.G., Davidson, A.L. and Simmonds, D.H. 2001. Soybean plants expressing an active oligomeric oxalate oxidase from the wheat gf-2.8 (germin) gene are resistant to the oxalate-secreting pathogen Sclerotinia sclerotiorum. *Physiol Mol Plant Pathol.*, 59: 297–307.

Dutton, M. V. and Evans, C.S. 1996. Oxalate production by fungi: its role in pathogenicity and ecology in the soil environment. *Can.J. Microbiol.*, **42:** 881–895.

Favaron, F., Sella, L. and D'Ovidio, R. 2004. Relationships among endo-polygalacturonase, oxalate, pH, and plant polygalacturonase-inhibiting protein (PGIP) in the interaction between *Sclerotinia sclerotiorum* and soybean. *Mol Plant Microbe Interact.*, 17: 1402–1409.

Fox, D.S., and Heitman, J. 2002. Good fungi gone bad: The corruption of calcineurin. *BioEssays* 24: 894-903.

Fraissinet-Tachet L., Reymond-Cotton P. and Fevre M. 1995. Characterization of a multigene family encoding an endopolygalacturonase in *Sclerotinia sclerotiorum. Current Genetics.* 29: 96-99.

Fraissinet-Tachet L., Reymond-Cotton P. and Fevre M. 1996. Molecular karyotype of the phytopathogenic fungus *Sclerotinia sclerotiorum. Current Genetics*.42: 881–895.

Fuller, P., Coyne, D. and Steadman, J. 1984. Inheritance of resistance to white mold disease in a diallel cross of dry beans. *Crop Sci.*, 24: 929–933.

Godoy, G., Steadman, J.R., Dickman, M.B. and Dam, R. 1990. Use of mutants to demonstrate the role of oxalic acid in pathogenicity of *Sclerotinia sclerotiorum* on *Phaseolus vulgaris. Physiol Mol Plant Pathol.*, 37: 179–191.

Grau, C.R. and Radke, V.L. 1984. Effects of cultivars and cultural practices on Sclerotinia stem rot of soybean. *Plant Dis.*, 68: 56–58.

Guimaraes, R.L. and Stotz, H.U. 2004. Oxalate production by *Sclerotinia sclerotiorum* deregulates guard cells during infection. *Plant Physiol.*, 136: 3703–3711.

Haas, J. H. and Bolwyn, B. 1974. Predicting and controlling white mold epidemics in white beans. *Can. Agric.*, 18: 28–29.

Han, Y., Joosten, H., Niu, W., Zhao, Z., and Mariano, P.S., *et al.*, 2007. Oxaloacetate hydrolase, the C–C bond lyase of oxalate secreting fungi. *J Biol Chem.*, 282: 9581–9590.

Harel, A., Gorovits, R., and Yarden, O. 2005. Changes in protein kinase A activity accompany sclerotial development in *Sclerotinia sclerotiorum. Phytopathology,* 95 (4): 397-404.

Huang, H. C. and Sun, S. K. 1991. Effects of S-H mixture or Perlka T M on carpogenic germination and survival of sclerotia of *Sclerotinia sclerotiorum. Soil Biol. Biochem.* 23: 809–813.

Huang, H.C., Kokko, E. G., Yanke, I. J. and Phillippe, R.C. 1993. Bacterial suppression of basal pod rot and end rot of dry peas caused by *Sclerotinia sclerotiorum. Can. J. Microbiol.* 39: 227–233.

Hutchins, J. D and Archer, S.A. 1994. Biocontrol of *Sclerotinia sclerotiorum* by manipulation of petal microflora. *Brighton Crop Protection Conference, Pests and diseases*, pp 935-936.

Illipronti, J.R. and Machado, J.C. 1993. Antagonismo de fungos a *Sclerotinia sclerotiorum* em soja e feijao. *Fitopatologia Brasileira* 18: 162-166.

Kafadar, K. A. and Cyert, M.S. 2004. Integration of stress responses: Modulation of calcineurin signaling in *Saccharomyces cerevisiae* by protein kinase A. *Eukaryot. Cell* 3: 1147-1153.

Kazan, K., Murry, F.R., Goulter, K.C., Llewellyn, D.J. and Manners, J.M. 1998. Induction of cell death in transgenic plants expression a fungal glucose oxidase. *Mol Plant-Microbe Interact* 11: 555–562.

Kesarwani, M., Azam, M., Natarajan, K., Mehta, A. and Datta, A. 2000. Oxalate decarboxylase from *Collybia velutipes*: molecular cloning and its overexpression to confer resistance to fungal infection in transgenic tobacco and tomato. *J Biol Chem.*, 275: 7230–7238.

Kim, K.S., Min, J.Y. and Dickman, M.B. 2008. Oxalic acid is an elicitor of plant programmed cell death during *Sclerotinia sclerotiorum* disease development. *Mol Plant Microbe Interact.*, 21: 605–612.

Kohn, L.M. 1979. A monographic revision of the genus *Sclerotinia*. *Mycotaxon.*, 9: 365-444.

Kopmans, E. 1993. Influence of irrigation on formation of apothecia from sclerotia of the stem rot pathogen, *Sclerotinia sclerotiorum*.*Vaxtsyddsnotiser.*, 57: 75-77.

Kurian, P. and Stelzig, D.A. 1979. The synergistic role of oxalic acid and endopoly-galacturonase in bean leaves infected by *Cristulariella pyramidalis*. *Phytopathology*, 69: 1301–1304.

Lane, B.G. 1994. Oxalate, germin, and the extracellular matrix of higher plants. *FASEB J.*, 5: 294–301.

Le Tourneau, D. 1979. Morphology, cytology and physiology of *Sclerotinia* species in culture. *Phytopathology*, 69: 887-890.

Li, R., Rimmer, R., Buchwaldt, L., Sharpe, A.G., *et al.*, 2004. Interaction of *Sclerotinia sclerotiorum* with a resistant *Brassica napus* cultivar: expressed sequence tag analysis identifies genes associated with fungal pathogenesis. *Fungal. Genet. Biol.*, 41: 735-753.

Liang, H., Maynard, C.A., Allen, R.D. and Powell, W.A. 2001. Increased *Septoria musiva* resistance in transgenic hybrid poplar leaves expressing a wheat oxalate oxidase gene. *Plant Mol Biol.*, 45: 619–629.

Lumsden, R.D. 1979. Histology and physiology of pathogenesis in plant diseases caused by *Sclerotinia species*. *Phytopathology.*, 69: 890–896.

Magro, P., Marciano, P. and Di Lenna P. 1988. Enzymatic oxalate decarboxylation in isolates of *Sclerotinia sclerotiorum*. *FEMS Microbiol Lett.*, 49: 49–52.

Magro, P., Marciano, P., and Di Lenna, P. 1984. Oxalic acid production and its role in pathogenesis of *Sclerotinia sclerotiorum. FEMS Microbiol. Lett. 24: 9-12.*

Makkonen, R. and Pohjakallio, O. 1960. On the parasites attacking the sclerotia of some fungi pathogenic to higher plants and on the resistance of those sclerotia to their parasites. *Acta Agril Scandinavica.,* 10: 105-126.

Marciano, P., Di Lenna, P. and Magro, P. 1983. Oxalic acid, cell wall-degrading enzymes and pH in pathogenesis and their significance in the virulence of two *Sclerotinia sclerotiorum* isolates on sunflower. *Physiol Plant Pathol.,* 22: 339–345.

Maxwell, D.P. 1973. Oxalate formation *in Whetzelinia sclerotiorum* by oxaloacetate acetylhydrolase. *Physiol Mol Plant Pathol.,* 3: 279–288.

Maxwell, D.P. and Lumsden, R.D. 1970. Oxalic acid production by *Sclerotinia sclerotiorum* in infected bean and in culture. *Phytopathology,* 60: 1395–1398.

Noyes, R.D. and Hancock, J.G. 1981. Role of oxalic acid in the sclerotinia wilt of sunflower. *Physiol. Plant Pathol.,* 18: 123–132.

Ortega-Perez, I., Cano, E., Were, F., Villar, M., Vazquez, J. and Redondo, J. M. 2005. c-Jun N-terminal kinase (JNK) positively regulates NFATc2 transactivation through phosphorylation within the N-terminal regulatory domain. *J. Biol. Chem.,* 280: 20867-20878.

Porter, I. J. and Merriman, P. R. 1985: Evaluation of soil solarization for control of root diseases of row crops in Victoria. *Plant Pathology,* 34: 108-118.

Poussereau, N., Creton, S., Billon-Grand, G., Rascle, C., *et al.,* 2001a. Regulation of acp1, encoding a non-aspartyl acid protease expressed during pathogenesis of *Sclerotinia sclerotiorum. Microbiology,* 147: 717-726.

Poussereau, N., Gente, S., Rascle, C., Billon-Grand, G., *et al.,* 2001b. aspS encoding an unusual aspartyl protease from *Sclerotinia sclerotiorum* is expressed during phytopathogenesis. *FEMS Microbiol. Lett.,* 194: 27-32.

Purdy, L.H. 1979. *Sclerotinia sclerotiorum,* History, diseases and symptomatology, host range, geographical distribution and impact. *Phytopathology,* 69: 875-880.

Reymond, P., Deleage, G., Rascle, C. and Fevre, M., 1994. Cloning and sequence analysis of a polygalacturonase-encoding gene from the phytopathogenic fungus *Sclerotinia sclerotiorum. Gene,* 146: 233-237.

Rollins, J. A. and Dickman, M. B. 2001. pH signaling in *Sclerotinia sclerotiorum*: identification of a pacC/RIM1 homolog. *Appl. Environ. Microbiol.,* 67: 75–81.

Rollins, J.A. 2003. The *Sclerotinia sclerotiorum* pac1 gene is required for sclerotial development and virulence. *Mol Plant-Microbe Interact.,* 16: 785–795.

Rollins, J.A., and Dickman, M.B. 1998. Increase in endogenous and exogenous cyclic AMP levels inhibits development in *Sclerotinia sclerotiorum. Appl. Environ. Microbiol.,* 64: 2539-2544.

Rollins, J.A., Dickman, M.B. 2001. pH signaling in *Sclerotinia sclerotiorum*: Identification of a pacC/RIM1 homolog. *Appl Environ Microbiol.,* 67: 75–81.

Saharan, G.S. and Mehta, N. 2008. Sclerotinia Diseases of crop plants: biology, eology and disease Management. LXII. Springer-Verlag GmbH, Heidelberg. *Sclerotium rolfsii. Phytopathology*, 55: 204–211.

Saindon, G. and Huang, H. C. 1992. Solid seeded upright beans avoid white mold. *Annu. Rep. Bean Improv. Coop.*, 35: 95–96.

Schweizer, P., Christoffel, A. and Dudler, R. 1999. Transient expression of members of the germin-like gene family in epidermal cells of wheat confers disease resistance. *Plant J.*, 20: 541–552.

Sharma, S.K. 1995. Management of the stem rot of gram caused by *Sclerotinia sclerotiorum* (Lib) de Bary. M.Sc. Thesis, Himachal Pradesh Krishi Vishvavidyalaya.: Palampur 87 pp.

Steadman, J.R. 1983. White mold - a serious yield-limiting disease of soybean. *Plant Dis.*, 67: 346-350.

Steadman, J. R., Coyne, D. P. and Cook, G. E. 1973. Reduction of severity of white mold disease on great Northern bean by wider row spacing and determinate plant growth habit. *Plant Dis. Reptr.*, 57: 1070–1071.

Steadman, J.R. 1979. Control of plant diseases caused by *Sclerotinia species. Phytopathology.*, 69: 904–907.

Sun, S.K. and Huang, J.W. 1985. Formulated soil amendment for controlling Fusarium wilt and other soil borne diseases. *Plant Dis.*, 69: 917-920.

Ten Have, A., Mulder, W., Visser, J. and van Kan, J. A. L. 1998. The endopolygalacturonase gene *Bcpg*1 is required for full virulence of *Botrytis cinerea. Mol Plant–Microbe Interact.*, 11: 1009–1016.

Townsend, B.B. and Willetts, H.J. 1954. The development of sclerotia of certain fungi. *Ann Bot.*, 21: 153-166

Tu, J. C. 1986. Integrated disease control of white mold (*Sclerotiniasclerotiorum*) in navy bean (*Phaseolus vulgaris*). Int. Symp. *Crop Prot.*, 39: 731-740.

Tu, J. C. 1988. The role of white mold infected white bean (*Phaseolus vulgaris* L) seeds in the dissemination of *Sclerotinia sclerotiorum* (Lib.) de Bary. *J. Phytopathol.*, 121: 40–50.

Tu, J.C. 1989. Oxalic acid induced cytological alterations differ in bean tolerance or susceptibility to white mould. *New Phytol.*, 112: 519–525.

Tu, J.C., 1997. Biological control of white mould in white bean using Trichoderma viride, Gliocladium roseum and Bacillus subtilis as protective foliar spray. In: Proceedings of 49th International Symposium on Crop Protection. 6 May 1997, Gent. Part IV. Meded. Fac. Landbouwkd. Toegep. Biol. Wet. Univ. Gent, 62 : 979–986

Weiss, A., Hipps, L.E., Blad, B.L., and J.R. Steadman. 1980. Conparison of within canopy microclimate and white mold disease (*Sclerotinia sclerotiorum*) development in dry edible beans as influenced by canopy structure and irrigation. *Agro. Meteorol.*, 22: 11-21.

Whipps, J.M. and Budge, S.P. 1993. Transmission of the mycoparasite *Coniothyrium minitans* by collembolan *Folsomia candida* (Collembola: Entomobryidae) and glasshouse sciarid *Bradsia sp.* (Diperta: Sciaridae). *Ann. Appl. Biol.*, 123: 165–171.

Willets, H.J. and Bullock, S. 1992. Developmental biology of sclerotia. *Mycol Res.*, 96: 801–816.

Williams, B., Kabbage, M., Kim, H., Britt, R. and Dickman, M.B. 2011. Tipping the balance: *Sclerotinia sclerotiorum* secreted oxalic acid suppresses host defenses by manipulating the host redox environment. *PLoS Path.*, 7: e1002107.

Wu, G., Shortt, B.J, Lawrence, E.B., Le´on, J., Fitzsimmons, K.C., Levine, E.B. and Raskin, I. 1995. Activation of host defense mechanisms by elevated production of H_2O_2 in transgenic plants. *Plant Physiol.*, 115: 427–435.

Zazzerini, A. and Tosi, L. 1985. Antagonistic activity of fungi isolated from sclerotia of *Sclerotinia sclerotiorum*. *Plant Pathology*, 34: 415–421.

Zhang, Z., Collinge, D.B. and Thordal-Christensen, H. 1995. Germin-like oxalate oxidase, a H_2O_2-producing enzyme, accumulates in barley attacked by the powdery mildew fungus. *Plant J.*, 8: 139–145.

Zhou, T. and Boland, G.J. 1999. Mycelial growth and production of oxalic acid by virulent and hypovirulent isolates of *Sclerotinia sclerotiorum*. *Can. J. Plant Pathol.*, 21: 93–99.

Zhou, T. and Reeleder, R.D. 1991. Colonization of bean flowers by *Epicoccum purpurascens*. *Phytopathology*, 81: 774–778.

Chapter 13

Viral Diseases of Pulse Crops

Mohd. Akram & Naimuddin

A number of viral diseases are known in pulse crops such yellow mosaic in mungbean, urdbean, cowpea, rajmash, pigeonpea, horsegram and mothbean, sterility mosaic in pigeonpea, leaf curl/necrosis in mungbean, urdbean, rajmash, cowpea and fieldpea, leaf crinkle in urdbean, mungbean and mothbean, common bean mosaic in rajmash, stunt in chickpea and lentil. In this chapter information on the diagnostics symptoms and the management of chickpea and lentil stunt, leaf curl/necrosis, leaf crinkle and pea mosaic are being described. Separate chapters on yellow mosaic and sterility mosaic diseases are given. In this chapter the typical symptom of the diseases and the cause of the diseases are being described. The major management options are also being discussed in this chapter.

Chickpea Stunt

Chickpea stunt disease was first reported from Iran, and it occurs in North Africa, the Middle East, South Africa, Australia, Indian subcontinent, Spain, Turkey, and the United States (Kumar *et al.*, 2008, Kanakala and Kuria 2019). Initially the causal agent of stunt disease was considered to be Bean leaf roll virus, (BLRV) (Reddy and Kumar, 2004). Subsequently, BLRV and Chickpea stunt luteovirus (CpLV) were regarded as causal agents. The stunt disease is more common in the kabuli genotypes than in the desi types. The yield loss was nearly 100 per cent if the infection occurred at early stage of crop growth. If the infection occurred at the flowering stage, the yield losses could be as high as 75 per cent (Horn *et al.*, 1995). A new member of the genus Polerovirus named Chickpea chlorotic stunt virus (CpCDV) has been reported to infect chickpea (Abraham *et al.*, 2006). It is persistently transmitted by *A. craccivora* and *Acyrthosiphon pisum*. CpCDV produces symptoms very similar to those of stunt that include leaf rolling, yellowing, necrosis, and stunting and was

Figure 13.1. Symptoms of Stunt Disease in Chickpea.

shown to be a mastrevirus. It was recorded in India (Horn *et al.*, 1993) and other part of the world. When the infection occurs before flowering, the yield loss was reported to be 100 per cent in chickpea (Horn *et al.*, 1993). Recently, five other mastrevirus species, Chickpea red leaf virus, Chickpea yellows virus, Chickpea chlorosis virus, Chickpea chlorosis Australia virus, and Tobacco yellow dwarf virus were found infecting chickpea in Australia (Hadfield *et al.*, 2012; Kanakala and Kuria 2019).

Diagnostic Symptoms and Cause

Plants are stunted due to shortening of internodes and show browning/ reddening of the leaflets in margins, branches proliferate making affected plant bushy, phloem showed browning in collar region, and plants bear only few pods (Kanakala *et al.*, 2012; Nahid *et al.*, 2008). This disease is caused by many viruses including Bean leaf roll virus (BLRV), Ckickpea chlorotic dwarf virus (CCDV) and Ckickpea chlorotic stunt virus (CCSV). This disease is transmitted by aphids (*Aphis craccivora, A. fabae, Acyrthosiphon pisum, Myzus persicae)* and leaf hopper Orasius albicinctus in a persistent manner) and not transmitted mechanically or through seeds (Abraham *et al.*, 2006, Naidu *et al.*, 1997, Kanakala and Kuria 2019).

Management

Among 10,000 germplasm lines screened for resistance to stunt disease, two lines (GG669 and ICCC10) were found to be field-resistant to CpCDV (Horn *et al.*, 1996). Kanakala *et al.* (2013) has developed an agroinoculation technique to screen chickpea genotypes against CpCDV. Using this technique three agroinoculated genotypes (L-550, GNG-1499 (Gauri), and IPC 09-07) showed virus resistance and these plants were found virus –free in PCR tests. The other genotypes such as Pusa 244, Girnar, Valai, and SAK19516 showed some level of resistance to stunt. Sowing dates are however known to influence incidence of stunt. Delay in the sowing reduces the diseases in northern India. Sowing chickpea in early November results in higher incidence of stunt as compared to crops sown in the second half of November in Madhya Pradesh (*Saxena et al., 1997*). Almost similar trend was observed in north eastern plains (*Kumar, 2016*). In Sudan also, early sowing (October) results in higher incidence of chickpea stunt as compared to late sowing (November and December) (*Hamed and Makkouk, 2002*). Practicing optimum sowing time may therefore help minimize incidence of stunt. This is in fact due to influence of weather on the activities of the vector of the stunt causing virus. Sowing when the aphid activity is low is also advocated (Ali and Habtewold, 1994, Bos *et al.*, 1988, Makkouk *et al.* 2003). Our experience is that the aphid infestation is more in chickpea fields with sparse plant population. Incidence of stunt is also high in such fields. It may, therefore, be stated that maintaining an optimum plant population in the field may result in containing the incidence of the stunt disease. Though there are no experimental data on vector control for managing the chickpea stunt, but in case of heavy infestation of aphids, neem oil can be used to reduce the aphid infestation.

Leaf Crinkle Disease

Leaf crinkle disease on urdbean is widely distributed in India and Pakistan and reported to cause crop losses up to 100 per cent (Biswas *et al.*, 2012; Reddy *et al.*, 2005, *Gautam et al.*, 2016). Based on the symptoms and transmission, the causal agent of this disease is considered to be a virus; however, there are no reports of

Figure 13.2. Symptoms of Leaf Crinkle Disease in Urdbean.

its isolation or characterization. However, recently Barnawal and coworker (2015) have reported the association of *Cowpea mild mottle virus* (CpMMV) with mungbean and urdbean samples showing leaf crinkle symptoms based on coat protein (CP) and nucleic acid binding protein (NB) gene sequences characteristics. Further its association was confirmed in many such samples of mungbean and urdbean using specific primer designed from the sequences generated through deep sequencing. This is the first report of CpMMV infecting mungbean and urdbean in India. Since most of the field samples analyzed for CpMMV showed symptoms of leaf crinkle, a disease of unknown etiology of mungbean and urdbean, the role of CpMMV if any, in causing leaf crinkle disease of these crops also needs to be further investigated.

Diagnostic Symptoms and Cause

Symptoms appear in three to four weeks after sowing. First symptom starts from third trifoliate as an increase in the size of the leaf and then the leaves become lighter green in colour. Typical leaf crinkle becomes more conspicuous and plant shows the bushy appearance and inflorescence malformed. It induces pollen sterility and pod setting is also highly curtailed. The identity of Urd leaf crinkle virus is still not confirmed. It is transmitted by sap, seeds, insects (aphids, white fly, beetle, hopper *etc.*).

Management

Use of certified seeds is advocated. Rouging was found to reduce the disease incidence. Plant extracts of *Mirabilis jalapa, Catharanthus roseus, D. metal, Bougainvillea spectabilis, Boerhavia diffusa* and *Azadirachta indica* were found effective in reducing the crinkle disease incidence in urdbean under field conditions (Chowdhury and Shah, 1985, Ravandar *et al.*, 2006, Karthikeyan *et al.*, 2009). There are reports that the application of plant oils may reduce the crinkle disease incidence (Hein 1975; Sharma and Varma 1982). Application of nutrients and fertilizers are also found effective in reducing the disease (Ahmad *et al.*, 2012, Huber and Graham 1999).

Leaf Curl/Necrosis Disease

Five tospovirus species have been reported from India from groundnut and many other plant species (Mandal *et al.*, 2012, Akram *et al.*, 2004; Jain *et al.*, 2007; Pappu *et al.*, 2009). Among them GBNV is emerging as an important pathogen of leguminous crops like mungbean (*Vigna radiata*), urdbean (*Vigna mungo*) and cowpea (*Vigna unguiculata*), particularly in peninsular India. This virus was recently reported on pea (*Pisum sativum*) at Kanpur (Akram and Naimuddin, 2010, 2012). Annual losses in Asia due to GBNV were estimated at over $89 million (Reddy *et al.*, 1995). In India, GBNV has been reported to cause 70 90 per cent losses in peanut (Singh and Srivastava, 1995) and up to 29 per cent in potato (Singh *et al.*, 1997) but there is no report about the estimated loss on mungbean and urdbean. All the tospoviruses are exclusively transmitted by thrips species (Whitefield *et al.*, 2005). Leaf curl caused by GBNV is widespread on mungbean and urdbean in the Indian subcontinent. Early infection results in crop losses of up to 90 per cent (Biswas *et al.*, 2012).

Figure 13.3. Symptoms of Leaf Curl/Necrosis Disease in Mungbean and Urdbean.

Diagnostic Symptoms and Cause

The symptoms start as veinal chlorosis either on the secondary or tertiary veins which, later become necrotic resulting in downward curling of the leaves. Young leaves show chlorotic and necrotic spots and finally killing the leaves and growing buds. Early infection causes death of plants (Mandal *et al.*, 2012). Based on ELISA test and nucleotide sequence of the S and M RNA segments, it was confirmed that the associated virus is Groundnut bud necrosis virus (GBNV), a tospovirus, family *Bunyaviridae*. It is transmitted by *Thrips palmi*.

Management

Cultural practices (Barrier crops, mixed cropping *etc.*) may reduce the disease incidence. No resistant cultivar is available against GBNV in mungbean and urdbean. Since the virus is transmitted by thrips, the chemicals used against them may be applied to reduce the disease incidence. Sowing of mungbean during second half of May to the first half of June in summer, late sowing in spring and intercropping with pearl millet (at 2:1 ratio) resulted in relatively low disease incidence. Among the various insecticides (imidacloprid, thiamethoxam, acetamiprid fipronil, dimethoate, fenvalerate, and azadirachtin and imidacloprid gave the most satisfactory control of *T. palmi* and low GBNV incidence in mungbean (Sreekanth *et al.*, 2004a, 2004b). Of 39 mungbean genotypes screened for resistance to GBNV under field conditions, accessions LGG 460, 480, 491, and 582 consistently showed lower disease incidence than the susceptible genotypes (Sreekanth *et al.*, 2002).

Pea Mosaic Disease

Pea is widely grown in the cool regions in the world, however, its cultivation in tropics and sub tropics including India is restricted to winter/cool season. Diseases caused by Pea seed borne mosaic virus (PSbMV), Pea enation mosaic virus (PEMV)-1, PEMV-2, and BLRV are economically important (Makkouk *et al.*, 2003, 2012). PSbMV is widely distributed and it is seed transmited, thus has the quarantine importance. PSbMV is also transmitted by aphids (*A. pisum, M. persicae,* and *A. gossypii*) in a non persistent manner (Kraft and Pfleger, 2001). ELISA, dot-immunobinding assay, and PCR-based tests have been developed for its detection and characterization (Makkouk *et al.,* 2012).

Figure 13.4. Symptoms of Rosseting and Mosaic in Pea.

Diagnostic Symptoms

A translucent clearing of pea leaf veins downward rolling leaflets curling of shoots and formation of rosette in apex are important symptoms. Early infections reduce flower and fruit formation.

Management

Infected seeds introduce primary inoculums in the crop and facilitate secondary spread of the virus in the crop through aphid vectors. Therefore, highest priority should be given to the use of healthy seed. The first infected plants in a crop must be rouged immediately to reduce potential secondary spread of the virus. The secondary spread of disease that resulted from planting of PSbMV infected seed could be reduced to some extent by applying the insecticides for controlling the aphid vectors. Planting resistant cultivars provides the best option for reducing the virus spread and its subsequent establishment especially in new areas where the virus was not previously reported (Kraft and Pfleger, 2001). In India, pea seed borne virus is not a considered a problem in fieldpea cultivation and there is no report on its management. However, its seed borne nature warrants constant watch.

Conclusion

It is well known that virus diseases have significant impact on the production and productivity of pulse crops throughout the world. Among the causal viruses, MYMIV, MYMV, GBNV, PSbMV, BCMV, BLRV, CpCDV and TSV are economically more important in pulse crops from Indian point of view. Most of them infect more than one legume crops. Those viruses which are seed borne in nature have quarantine importance as well as they also play an important role in disease epidemiology and serve as primary sources of inoculums. Intensive and mono cropping systems facilitate the vector multiplication, creation of new biotypes and spread. For the management of legume viruses, several control options are available. Among them selecting virus-free seed and planting materials, resistant cultivars, planting the crops at the suitable date, following appropriate agronomic practices, applying newer and need based insecticides and bio-pesticides for vector offer good results leading to minimize losses occurred by virus diseases. Sometimes integration of the available options provides good results in minimizing the yield loss.

REFERENCES

Abraham, A. D., Menzel, W., Lesemann, D.-E., Varrelmann, M., and Vetten, H. J. (2006). Chickpea chlorotic stunt virus: A new polerovirus infecting cool-season food legumes in Ethiopia. *Phytopathology* 96: 437- 446.

Ahmad ZM, Safdar A, Khan MA, Sahi ST (2012). Role of nutrients and naphthalene acitic acid in the management of urdbean leaf crinkle virus. *Pak J Phytopathology* 24(1): 79–81

Akram M and Naimuddin (2012). Biological characterization and variability of the nucleocapsid protein gene of *Groundnut bud necrosis virus* isolates infecting pea from India. *Phytopathologia Mediterrranean* 51 (2): 266-275

Akram M. and Naimuddin, (2010). First report of *Groundnut bud necrosis virus* infecting pea (*Pisum sativum*) in India. *New Disease Report* [doi: 10.5197/j.2044-0588.2010.021.010].

Akram M., R.K. Jain, V. Chaudhary, Y.S. Ahlawat and S.M.P. Khurana (2004). Comparison of Groundnut bud necrosisvirus isolates based on movement protein (NSm) gene sequences. *Annals of Applied Biology* 145, 285 289.

Ali, K., and Habtewold, T. (1994). Research on insect pests of cool-season food legumes. Pages 367-396 in: Cool-Season Food Legumes in Ethiopia. Proceedings of the First National Cool-Season Food Legumes Review Conference. A. Telaye, G. Bejiga, M. C. Saxena, and M. B. Solh, eds. Institute of Agricultural Research, Addis Ababa, Ethiopia/ICARDA, Aleppo, Syria.

Baranwal V.K., P. Jain, R.K. Saritha, R.K. Jain, N.K. Gautam (2015). Detection and partial characterization of Cowpea mild mottle virus in mungbean and urdbean by deep sequencing and RT-PCR. *Crop Protection* 75 : 77-79

Biswas, KK., Tarafdar, A., and Biswas, K. (2012). Viral diseases and its mixed infection in mungbean and urdbean: Major biotic constraints in production of food pulses in India. In Asha Sinha, B. K. Sharma, and Manisha Srivastava

(Eds.), Modern trends in microbial biodiversity of natural ecosystem (pp. 301–317). New Delhi: Biotech Books.

Bos, L., Hampton, R. O., and Makkouk, K. M. (1988). Viruses and virus diseases of pea, lentil, faba bean and chickpea. Pages 591-615 in: World Crops: Cool Season Food Legumes. R. J. Summerfield, ed. Kluwer Academic Publishers, Dordrecht/Boston/London.

Chowdhury AK, Shah NK (1985). Inhibition of urdbean leaf crinkle disease by different plant extracts. Indian Phytopathology 38: 566–568

Dodds JA, Morris TJ and Jordan RL (1984). Plant viral doublestranded RNA. Annual Review of Phytopathology 22: 151–168.

Gautam NK, Kumar K and Prasad M (2016). Leaf crinkle disease in urdbean (Vigna mungo L. Hepper): An overview on causal agent, vector and host. Protoplasma. DOI 10.1007/s00709-015-0933-z

Hadfield, J., Thomas, J. E., Schwinghamer, M. W., Kraberger, S., Stainton, D., Dayaram, A. (2012). Molecular characterization of dicot-infecting mastreviruses from Australia. Virus Research, 166, 13–22.

Hein A (1975). The effect of oil on virus transmission by aphids. *Phytopathology* Z 71: 42–48

Horn, N. M., Reddy, S. V., and Reddy, D. V. R. (1995). Assessment of yield losses caused by Chickpea chlorotic dwarf geminivirus in chickpea (*Cicer arietinum*) in India. *European Journal of Plant Pathology*, 101, 221–224.

Horn, N. M., Reddy, S. V., Roberts, I. M., and Reddy, D. V. R. (1993). Chickpea chlorotic dwarf virus, a new leafhopper transmitted geminivirus of chickpea in India. *Annals of Applied Biology*, 122, 467–479.

Horn, N.M.; Reddy, S.V.; van den Huevel, J.F.J.M.; Reddy, D.V.R. (1996). Survey of chickpea (*Cicer arietinum* L.) for chickpea stunt disease and associated viruses in India and Pakistan. Plant Dis. 80, 286–290

Huber DM, Graham RD (1999). The role of nutrition in crop resistance and tolerance to disease. In: Rengel Z (ed) Mineral nutrition of crops, fundamental mechanisms and implications. Food product press, New York, pp 205–226

Jain R.K., S. Bag, K. Umamaheshwaran and B. Mandal, (2007). Natural infection by Tospoviruses of cucurbitaceous and fabaceous vegetable crops in India. *Journal of Phytopathology* 155, 22 25.

Kanakala S and Kuria P (2019). Chickpea chlorotic dwarf virus: An Emerging Monopartite Dicot Infecting Mastrevirus. Viruses 11: 5; doi: 10.3390/v11010005

Kanakala, S.; Sakhare, A.; Verma, H.N.; Malathi, V.G. (2012). Infectivity and the phylogenetic relationship of a mastrevirus causing chickpea stunt disease in India. *Eur. J. Plant Pathol.*, 135, 429–438.

Kanakala, S.; Verma, H.N.; Vijay, P.; Saxena, D.R.; Malathi, V.G. (2013). Response of chickpea genotypes to Agrobacterium-mediated delivery of Chickpea chlorotic dwarf virus (CpCDV) genome and identification of resistance source. *Appl. Microbiol. Biotechnol.*, 97, 9491–9501

Karthikeyan G, Sabitha D, Rabindran R, Ganapathy T (2009). Evaluation of antiviral principles for the induction of systemic resistance in blackgram (Vigna mungo) against urdbean leaf crinkle diseases. *Archives Phytopatho Plant Protec* 42: 12

Kraft, J. M., and Pfleger, F. L. (2001). Compendium of pea diseases and pests (2nd ed.). Saint Paul, MN: APS Press.

Kumar, P. L., Jones, A. T. and Waliyar, F. (2008). Virus diseases of pigeonpea. In: *Characterization, Diagnosis and Management of Plant Viruses*. Vol. 3: Vegetable and Pulse Crops. Rao, G. P., Kumar, P. L., and Holguin-Pena, R. J. (Eds.) Studium Press, Texas, USA. Pp. 235-258.

Makkouk, K. M., Kumari, S. G., Huges, J. D. A., Muniyappa, V., and Kulkarni, N. K. (2003). Other legumes. In G. Loebenstein, and G. Thottappilly (Eds.), Virus and virus-like disease of major crops in developing countries (pp. 447–476). Dordercht: Kluwer Academic Publisher.

Makkouk, K., Pappu, H., and Kumari, S. G. (2012). Virus diseases of peas, beans, and faba bean in the Mediterranean region. *Advances in Virus Research*, 84, 367–402.

Mandal, B., Jain, R. K., Krishnareddy, M., Krishna Kumar, N. K., Ravi, K. S., and Pappu, H. R. (2012). Emerging problems of Tospoviruses (Bunyaviridae) and their management in the Indian Subcontinent. *Plant Disease*, 96, 468–479.

Nahid, N.; Amin, I.; Mansoor, S.; Rybicki, E.P.; van der Walt, E.; Briddon, R.W. (2008). Two dicot-infecting mastreviruses, family Geminiviridae occur in Pakistan. *Arch. Virol.*, 153, 1441–1451.

Naidu, R. A., Mayo, M. A., Reddy, S. V., Jolly, C. A., and Torrance, L. (1997). Diversity among coat proteins of luteoviruses associated withchickpea stunt disease in India. *Ann. Appl. Biol.* 130: 37-47.

Pappu H.R., R.A.C. Jones and R.K. Jain, (2009). Global status of tospovirus epidemics in diverse cropping systems: Successes achieved and challenges ahead. *Virus Research* 141, 219 236.

Ravandar R, Tonapi Vilas AC, Varanavasiappn S, Navi SS, Jayarajan R (2006). Management of urdbean leaf crinkle disease in urdbean (Vigna mungo L. Hepper). *Int J Agric Sci* 2: 8–10

Reddy D.V.R., A.A.M. Buiel T. Satyanarayana S.L. Dwivedi A.S. Reddy A.S. Ratna, K. Vijaylakshmi, G.V. Ranga Rao R.A. Naidu and J.A. Wightman, (1995). Peanut bud necrosis virus disease: an overview. In: *Recent Studies on Peanut Bud Necrosis Disease: Proceedings of a Meeting.* (A.A.M. Buiel, J.E. Parevliet, J.M. Lenne, ed.), March 20, 1995, International Crop Research Institute for the Semi-Arid Tropics ICRISAT, Asia Center, Patancheru, Andhra Pradesh, India, 3–7.

Reddy, C., Tonapi, V. A., Navi, S. S., and Jayaram, R. (2005). Influence of plant age on infection and symptomological studies on urdbean leaf crinkle virus in urdbean (*Vigna mungo*). *International Journal of Agricultural Science*, 1, 1–6.

Reddy, M.V., Raju, T.N. and Lenne, J.M. (1998). Diseases of pigeonpea. Pages 517-558, In: *The Pathology of Food and Pasture Legumes*. D. J. Allen and J. M. Lenne, (Eds.) CAB International, Wallingford, UK.

Reddy, S. V., and Kumar, P. L. (2004). Transmission and properties of a new luteovirus associated with chickpea stunt disease in India. *Current Science*, 86, 1157–1161.

Sharma SR, Varma A (1982). Control of virus diseases by oil sprays. Zentralblatt Microbiology 137(4): 329–347

Singh A.B. and S.K. Srivatava, (1995). Status and control strategy of peanut bud necrosis disease in Uttar Pradesh. In: *Recent Studies on Peanut Bud Necrosis Disease: Proceedings of a Meeting*.(A.A.M. Buiel, J.E. Parevliet, J.M. Lenne, ed.), March 20, 1995, International Crop Research Institute for the Semi- Arid Tropics ICRISAT, Asia Center, Patancheru, Andhra Pradesh, India, pp. 65–68.

Singh R.B., K.K. Srivastava, S.K. Pandey and K.S.M Paul, (1997). Assessment of yield losses due to potato stem necrosis disease. *Indian Journal of Virology* 13, 135 137.

Sreekanth, M., Sriramulu, M., Rao, R. D. V. J. P., Babu, B. S., and Babu, T. R. (2004a). Effect of intercropping on Thrips palmi (Karny) population and Peanut bud necrosis virus incidence in mungbean. *Indian Journal of Plant Protection*, 32, 45–48.

Sreekanth, M., Sriramulu, M., Rao, R. D. V. J. P., Babu, B. S., and Babu, T. R. (2004b). Evaluation of certain new insecticides against Thrips palmi (Karny), the vector of Peanut bud necrosis virus (PBNV) on mungbean (Vigna radiata L. Wilczek). International Pest Control,46, 315–317.

Sreekanth, M., Sriramulu, M., Rao, R. D. V. J. P., Babu, B. S., and Babu, T. R. (2002). Evaluationof greengram genotypes for resistance to Thrips palmi Karny and Peanut bud necrosis virus. *Indian Journal of Plant Protection*, 30, 109–114.

Whitefield A.E., D.E. Ullman and T.L. German, (2005). Tospovirus- Thrips interaction. *Annual Review of Phytopathology* 43, 459 489.

Chapter 14

Yellow Mosaic Disease in Legume Crops: Viruses and their Management

Naimuddin & Mohd Akram

Yellow mosaic disease (YMD) caused by a group of bipartite begomoviruses is an economically important disease of many leguminous crops like mungbean, urdbean, cowpea, horsegram, mothbean, soybean, rajmash and pigeonpea in India and some other parts of the world. Till 2007, only four virus species - *Mungbean yellow mosaic virus*, *Mungbean yellow mosaic India virus*, *Horsegram yellow mosaic virus* and *Dolichos yellow mosaic virus* were known to cause YMD in many legume crops. However, during the present decade, 6 more begomoviruses have been reported to cause YMD in other legume hosts. All these viruses have close phylogenetic relationship among them and are often collectively referred as Legumoviruses. All legumoviruses have their host range limited to plants of the family Fabaceae with few exceptions. They all have a bipartite genome consisting of single stranded circular DNA molecules (DNA-A and DNA-B) with similar genome organization and are transmitted by whiteflies in a circulative and propagative manner. Temporal and spatial variation in the disease reaction of known genotypes of the plant hosts points to the existence of diversity in YMD causing viruses. The presence of multiple legumovirus species or sometime multiple DNA-B molecules in YMD affected plant has also been observed and this is likely to have a bearing on the epidemiology of the YMD in pulse crops. This may also lead to the emergence of new virus species or recombinant strains.

YMD is known to cause substantial yield losses which depends on the stage at which crop is infected and may vary from 10-100 per cent. It adversely affects yield contributing parameters like pods/plant, seeds/pod and 1000 seed weight in both mungbean and urdbean. Early infected plants suffer more than those infected in later stages of growth (Singh *et al.,* 1982). Nair and Nene (1974) studied the relation of age of the plant at which it is infected (YMD) and yield of the affected urdbean. The yield loss was not significant if the plants are infected in the 8th week or later. Virus infection reduced number of pods/plant and it was responsible for the overall yield reduction rather than the seeds/pod. Shape, size and appearance of pods and seeds of diseased plants are considerably distorted although seed germination remains unaffected (Chand and Varma, 1983). Decrease in the grain yield and other yield attributing characters is not significantly correlated with the YMD incidence (Khattak *et al.,* 2000). It supports the view that it is the severity of the infection rather than the incidence that influences the level of adverse effect on the yield contributing parameters. An estimated annual monetary loss of over US$ 300 millions in mungbean, urdbean and soybean has been reported (Varma *et al.,* 1992). In fact the extent of yield loss depends on age of the plant at the time of infection and also on the extent of disease severity. Complete loss of the yield is not uncommon in YMD susceptible genotypes.

Of the different options available for the management of YMD, use of resistant cultivars is most suited one. Appropriate agronomic practices, applying newer and need based insecticides have also been reported to reduce losses due to YMD significantly in many legume crops. The various aspects of yellow mosaic disease are being discussed in this chapter.

History, Occurrence and Distribution

YMD was first reported in cowpea in India by Vasudeva (1942). Soon after, it was reported from Pune on *Phaseolus lunatus* and *Dolichos lablab* (Capoor and Varma, 1948). A similar yellow mosaic disease of mungbean was first reported by Nariani (1960) who named the putative causal agent as Mungbean yellow mosaic virus (MYMV). A similar disease of mungbean was later reported from Pakistan by Ahmad and Harwood (1973). YMD has been reported in urdbean (Williams *et al.,* 1968), mungbean (Nene, 1973), pigeonpea (Williams *et al.,* 1968, Nene, 1972), horsegram (Muniyappa *et al.,* 1975) and Frenchbean (Singh, 1979). It is reported to occur in all pulse growing areas of Indian subcontinent and is a major constraint to the production of most of the warm season legume crops (Qazi *et al.,* 2007). YMD in accessions of wild *Vigna* has also been described recently (Naimuddin *et al.,* 2011a, b, c, Gautam *et al.,* 2014). YMD of mungbean and urdbean is reported to occur only in Asian countries *viz.,* India, Sri Lanka, Bangladesh, Philippines, Pakistan, Myanmar, Thailand, Nepal, Indonesia, Malaysia, Taiwan (Karthikeyan *et al.,* 2014). In India, it is reported to occur in all the states of India where these crops are grown.

YMD Causing Viruses

The causal virus of the YMD in pulses has always been considered whitefly transmitted. It was however, Honda *et al.* (1983) who provided evidence that the MYMV had geminate particles. Infectivity of purified virus was also proved in mungbean seedlings. These finding unequivocally established that the causal agent of YMD in mungbean is a geminivirus. Presence of geminate particles was shown in plants inoculated with MYMV/MYMIV isolates from blackgram, cowpea and pigeonpea by immunosorbent electron microscopy using polyclonal antibody to *Squash leaf curl virus* (Srivastava, 1989; Varma *et al.,* 1992). Presence of geminate particles and their reaction to polyclonal antibodies to other whitefly transmitted geminiviruses (WTGs) further established that the aetiological agent of YMD not only in mungbean and urdbean but also in other legumes is a whitefly transmitted geminivirus, which is now recognized as the species of the genus *Begomovirus* under the family *Geminiviridae*. At present, four species of begomoviruses, *Mungbean yellow mosaic India virus* (MYMIV), *Mungbean yellow mosaic virus* (MYMV), *Horsegram yellow mosaic virus* (HgYMV), *Dolichos yellow mosaic virus* (DoYMV) are known to cause YMD in various leguminous crops (Qazi *et al.*, 2007, Malathi and John, 2008, Akram *et al.*, 2015).

Nariani (1960) named the putative virus causing YMD in mungbean as MYMV. For decades MYMV was considered as the causal agent of YMD in many pulse crops. The genome of the Thailand isolate of MYMV was characterized by Moringa *et al.* (1993). Soon after, genome of the virus causing YMD in blackgram in north India was also characterized (Verma *et al.*, 1991, Mandal *et al.*, 1997). Genome sequence data of north Indian isolate of MYMV was found to have <89 per cent similarity *i.e.* the begomovirus species demarcation limit (Fauquet *et al.*, 2008) with the Thailand isolate of MYMV and hence the former was considered as a distinct species and named as *Mungbean yellow mosaic India virus* (MYMIV). Subsequently, the virus causing yellow mosaic disease in horsegram was characterized and it also came out to be a distinct species differing from both MYMIV and MYMV and has been named as *Horsegram yellow mosaic virus* (Baranbas *et al.*, 2010). The virus causing yellow mosaic disease in dolichos has also been shown to be a distinct begomovirus species and is designated as *Dolichos yellow mosaic virus* (DoYMV) (Maruthi *et al.*, 2006). DoYMV is generally known to be confined to dolichos bean only. However, yellow mosaic disease of dolichos has been shown to be caused by MYMIV in Varanasi (Singh *et al.*, 2006) and a mixed infection of DoYMV and MYMIV has been detected in cowpea (Naimuddin and Akram, 2010).

Other viruses causing yellow mosaic in leguminous plants *viz.*, *Rhynchosia yellow mosaic virus* (RhYMV) and *Rhynchosia yellow mosaic India virus* (RhYMIV) infecting *Rhynchosia minima*, a weed belonging to the family *Fabaceae*, respectively, in Pakistan (Ilyas *et al.*, 2009) and India (Jyothsna *et al.*, 2011), *Velvet bean severe mosaic virus* (VbSMV) infecting velvet bean (*Mucuna pruriens*), a medicinal plant belonging to the family *Fabaceae* (Zaim *et al.*, 2011), *Kudzu mosaic virus* (KuMV) infecting kudzu (*Pueraria montana*) (Ha *et al.*, 2008) and *Soybean mild mottle virus* and *Soybean chlorotic blotch virus infecting soybean in Nigeria* (Alabi *et al.*, 2010) have also been reported. In phylogenetic studies, all these viruses clustered with LYMVs (Qazi *et al.*, 2007) or 'Legumoviruses' (Ilyas *et al.*, 2009, Briddon *et al.*, 2010).

Biology of Viruses Involved in YMD

Begomoviruses are categorized into three types based on their genome organization. The begomoviruses causing yellow mosaic disease in different leguminous plants including pulse crops fall in to type I that comprises virus species with bipartite genome, *i.e.* the genome of these species consists of two molecules of single stranded DNA referred as DNA-A and DNA-B. In begomoviruses with bipartite genome, DNA-A and DNA-B differ in sequence except for a common region (CR) of 200-250bp that is nearly identical in the genome components of any given virus, but differ between virus species. A unique feature of isolates of MYMIV and MYMV is the divergence in the nucleotide sequence of CR in their DNA-A and DNA-B. Though DNA-A contains all viral information necessary for replication and encapsidation (Rogers *et al.*, 1986, Sunter *et al.*, 1987, Townsend *et al.*, 1986), both components are required for infectivity (Hamilton *et al.*, 1983).

At present, genome of four begomoviruses commonly causing YMD in different pulse crops is known to be bipartite, *i.e.* consisting of two single stranded circular DNA molecules *viz.*, DNA-A and DNA-B each approximately of 2.7 kb in size (Figure 14.1). Bipartite nature of DoYMV was however, reported only in 2015. At IIPR, Kanpur DNA-B of the DoYMV (NCBI accession no. KJ481025) was characterized thus confirming genome of DoYMV also to be bipartite (Akram *et al.*, 2015).

In bipartite begomoviruses, the genes in DNA-A and DNA-B are separated by an intergenic region (IR) that includes a segment of ~200 nt called the common region (CR) region, a highly conserved region between both components. All the *cis* elements required for replication are located in the CR, which varies from virus to virus with the exception of a highly conserved ~30 nt long segment with the potential to form a stem loop structure, also called as hairpin structure. In all these viruses, there are generally six-seven open reading frames (ORFs) in DNA-A and two ORFs in DNA-B. Two of the ORFs located in DNA A, two (AV1 codes for coat protein) and AV2 (for pre coat protein) are in virion-sense orientation or positive sense and rest in complimentary or negative sense orientation (Figure 14.1). AC1 codes for replication initiator protein (*rep*, also known as AC1 or AL1), AC2 for transcription activator protein (*TrAP*, also known as AC2) and AC3 for replication enhancer protein (*REn*, also designated as AC3). Two other genes, AC4 and AC5 have been described but their role in replication of the virus or pathogenesis is not known. The ORF AV2 is said to be a signature gene of old world begomoviruses. Of the two ORFs in DNA-B, the one (BV1) is in virion sense and the other (BC1) in complimentary sense orientation. ORF BV1 codes for the nuclear shuttle protein which is required for the movement of viral DNA- both ss- and dsDNA- between nucleus and cytoplasm of host cells, whereas ORF BC1 codes for the movement protein which regulates cell to cell movement of the virus through plasmodesmata. It has been demonstrated that BV1 is also involved in long distance movement of the virus by allowing the spread of the virus through vascular system of the host (Lazarowitz, 1992, Hanley-Bowdoin *et al.*, 1999).

Transmission

YMD of mungbean is spread through white fly *Bemisia tabaci* Genn. Extensive work on host-virus-vector relationship was carried out at Pantnagar in north India by Nair (1971) and Rathi (1972). Whitefly may acquire MYMV/MYMIV within 15 minutes of acquisition feeding period and the incubation period is four hours. Whitefly requires a minimum of 15 minutes (inoculation feeding period) to transmit the virus to healthy host plant. Starvation (of whitefly) before acquisition and inoculation feeding increases efficiency of acquisition and inoculation, however, the starvation has more pronounced effect on the acquisition feeding than on inoculation feeding. Single whitefly can transmit the virus. Female whitefly is more efficient

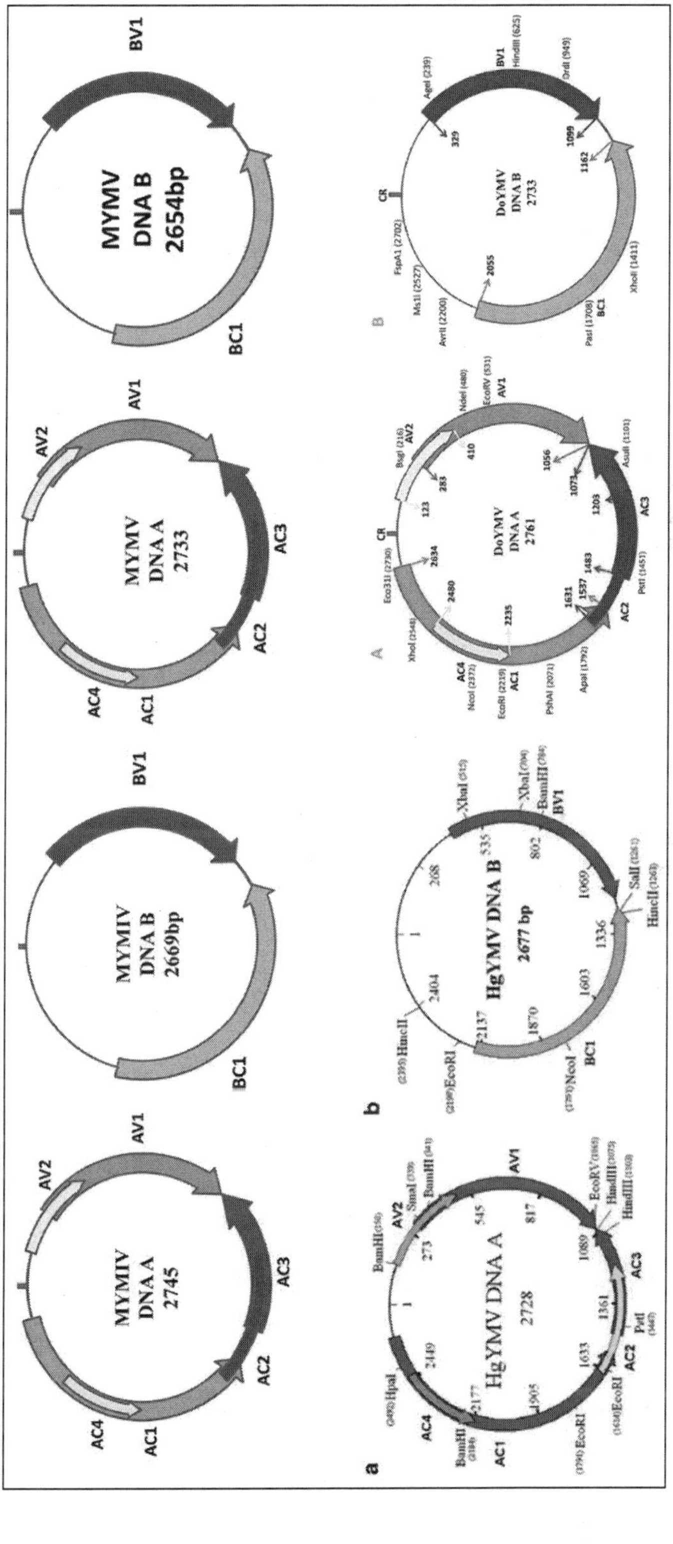

Figure 14.1. Genome Organization of LYMVs (MYMIV-DNA A and DNA B, MYMV-DNA A and DNA B, HgYMV-DNA A and DNA B and DoYMV-DNA A and DNA B).

in transmitting the virus and may retain it for up to 10 days. Males are however less efficient and generally retain the virus up to 1 day. MYMV/MYMIV is not transmitted to the offspring of the viruliferous whitefly indicating no transovarial transmission. *B. tabaci* is considered as a species complex containing numerous genetic and biological variants-bioptypes (Brown, 2007, Perring, 2001). Of the many biotypes recognized (Brown, 2007; Prasanna *et al.*, 2015), the 'B' biotype (silver leaf whitefly) due to its widest distribution, attributed to its ability to colonize many diverse plant hosts and remarkable ability to transmit a number of geminiviral diseases is considered a potential threat to many crops.

Symptoms of yellow mosaic may also be seen on pods and seeds of the YMD affected plants of mungbean and urdbean genotype. This led researchers to investigate seed transmission of YMD causal virus (MYMV/MYMIV). At ICAR-IIPR, Kanpur, seeds showing yellow patches on the seed coat collected from YMD affected mungbean genotype T44 and were subjected to PCR tests using MYMIV specific primers. Results indicated that the virus (MYMIV) was present in the seeds. However, the virus could not be detected in seedlings raised from MYMIV positive seeds indicating that the virus is lost during the process of germination and is not transmitted through seeds. In a recent publication Kothandaraman *et al.* (2016) concluded that MYMV is seed borne in urdbean. The seedlings from MYMV positive seeds did not show any symptoms of YMD though the virus (MYMV) was detected in 32 per cent of tested seedlings. They however did not explain the absence of symptoms in these seedlings. Also whitefly transmission from PCR positive symptomless seedlings was not demonstrated. Nevertheless they concluded MYMV to be seed borne in urdbean. It is well accepted that a virus is said to be seed borne only when it is carried to seedlings and serves as source of primary inoculum. This publication does not provide answer to it and hence this cannot be accepted as a case of seed borne virus simply due to the detection of a small fragment of the virus in the seed.

Diagnosis

Symptoms

Nariani (1960) and Williams *et al.* (1968) described symptoms of yellow mosaic disease on mungbean and urdbean, respectively. Symptoms first appear as small yellow flecks in young leaves. Subsequently emerging leaves exhibit more conspicuous and irregular yellow and green patches alternating each other. Leaf size is generally not much affected. In highly susceptible genotypes however affected plants bear few pods that are smaller in size and deformed. Yellow spots may also be observed on such pods. Seeds developing in such pods may also show yellow spots on the green seed coat. In urdbean two types of symptoms *viz.*, yellow mottle and necrotic mottle were considered to be caused either by different strains of the same virus or by different viruses. However, Nair *et al.* (1974a) clarified that these two types of symptoms were in fact due to the varied reactions to different genotypes to the same virus isolate. More or less similar types of symptoms of YMD are present in soybean, cowpea, dolichos, horsegram and other leguminous crops (Figure 14.2).

Figure 14.2. Yellow Mosaic Symptoms in Legume Crops.
A: Soybean, B: Rajmash, C: Mungbean, D: Horsegram, E: Urdbean, F: Dolichos, GvCowpea, H: Pigeonpea, I: Wild *Vigna.*

Yellow mosaic disease in mungbean, urdbean, rajmash, dolichos, horsegram, pigeonpea is caused by at least number different virus species but with similar symptoms (Figure 14.2) *viz.*, yellow mosaic. Identification of a virus based on the symptoms only may lead to an erroneous conclusion and hence is a challenge. This makes it imperative to have an accurate diagnosis of the viruses which is often the first step in disease management.

Molecular Diagnostics

Use of polyclonal and monoclonal antibodies to *African cassava mosaic* (ACMV) and Indian cassava mosaic viruses (ICMV) has been employed to ascertain the identity of the YMD causing agents as begomoviruses (Harrison *et al.*, 1991). Positive reaction with polyclonal and monoclonal antibodies to ACMV and ICMV established the etiological agents of YMD to be begomoviruses. Based on epitope profile, LYMVs isolates could be categorized broadly into two groups, one, comprising virus isolates causing yellow mosaic in dolichos and the second group consisting of LYMVs isolates infecting other legumes (Malathi, 2007). Species specific identification of the virus isolates based on serological reaction has not been possible due to conserved coat protein gene. Once this was realized, nucleic acid based identification of begomoviruses developed fast.

In fact the use of nucleic acid-based diagnostic methods for the detection of plant pathogens started in late 70s of the last century when the first double-stranded RNA and dot-blot hybridization protocols were developed for the detection of viruses and viroids. Development of polymerase chain reaction (PCR) protocol (Saiki *et al.*, 1985, 1988) in 90s of the last century was soon followed by its exploitation for the detection of plant pathogens (Puchta and Sanger, 1989). At present PCR is a widely accepted method for the detection of plant viruses.

Degenerate primers (Deng *et al.*, 1994, Rojas *et al.*, 1993) are commonly used to ascertain geminiviral nature of a virus. Use of degenerate primers however does not provide information about the species specific identity of the virus thereby limiting the utility of degenerate primers. PCR products from degenerate primers are required to be sequenced to identify the likely species of the *Geminiviridae*. Moreover, most of the geminivirus specific degenerate primers have been designed using genome sequences of New World Geminiviruses and their utility in detection of LYMVs has not been examined critically. There is however, limited information on specific primers for the detection and identification of pulse infecting begomoviruses wherein characterization of whole genome is targeted (Rouhibakhsh *et al.*, 2008, Malathi and Johns, 2008).

Differences in the reaction of urdbean and mungbean genotypes against YMD across the country necessitated the need to develop an easy and efficient PCR based protocols for differentiating begomovirus species involved in YMD. The option of sequencing the whole genome is a time consuming and costly affair for routine tests. Therefore, at IIPR, Kanpur attempts were made to develop PCR based diagnostics for species specific identification of viruses known to cause YMD in different pulse crops. Species specific primers were designed (Table 14.1) and PCR protocols developed for detection of viruses in YMD affected samples of different pulse crops. These primers are designed in such a way that complete gene can be extracted from the direct sequencing of amplified product without cloning. Using these protocols infection of MYMIV, MYMV, HgYMV and DoYMV has been confirmed in mungbean, urdbean, soybean, pigeonpea, dolichos, rajmash, horsegram, many wild accessions of *Vigna* and weeds (Naimuddin *et al.*, 2011a, b, c). At Indian Institute of Pulses Research, Kanpur, YMD affected samples of different leguminous plants collected from around the Kanpur in north India and Coimbatore, Vamban in South India were subjected to PCR tests using specific primers. Results indicated involvement of MYMIV in YMD affected samples of mungbean, urdbean, rajmash, cowpea, pigeonpea, soybean from north India and MYMV in samples of mungbean, urdbean, rajmash and HgYMV in horsegram from South India. Using primers (Table 14.1) MYMIV and DoYMV prevalent in and around Kanpur have been detected in different hosts. The protocols for the detection of these viruses are being described in following paragraphs.

Table 14.1. Details of the Primers Designed to get the Complete Coat Protein Gene and Movement Protein Gene of Four Bogemoviruses known to Infect different Leguminous Host in India and Expected Size of Amplicons

Primer ID	*Primer Sequence 5'_3'*	*AT (°C)*	*~ Size of DNA Fragment to be Amplified*	*Name of the Virus to be Detected/DNA Component*
NM 1/AV1PF NM 2/AV1PR	GTA TTT GCA KCA WGT TCA AGA AGG DGT CAT TAG CTT AGC	54	1000bp	MYMIV/DNA A
HYMV-CP-F HYMV-CP-R	ATG CTT GCA ATT AAG TAC TTG CA TAG GCG TCA TTA GCA TAG GCA	56	1050 bp	HgYMV/DNA A
DYMV-CP-F DYMV-CP-R	CTG TGA AAT TTG TGC AGG TAC GCG GTT GCG AAT ATG TAT	54	900bp	DoYMV/DNA A
MYMV-CP-F MYMV-CP-R	ATG GG (T/G) TCC GTT GTA TGC TTG GGC GTC ATT AGC ATA GGC AAT	54	1000bp	MYMV/DNA A
HYMV-MPF HYMV-MPR	ATG GAG CAT TAT TCC GGT GCA TTA CA(G/A) GGT TTT GTT TAC AGT	64	900bp	HgYMV/DNA B
MYMV-MPF MYMV-MPR	ATG GAG AAT TAT TCA GGC GCA TTA CAA CGC TTT GTT CAC ATT	58	900bp	MYMV/DNA B
MYMIV-MPF MYMIV-MPR	ATG GAA AAT TAT TCA GGT GCA CTA CAA CGC TTT GTT CAC ATT	53	900bp	MYMIV/DNA B
AC1PF AC1PR	AGT TGA TAT GGA TGT AATAGC3 ACA AAA ACG ACT TCA AATATG CCA A	49	1100bp	MYMIV/DNA A
AC2PF AC2PR	AGC TAA TGA CCC CTA AAT TAT GAG TAC TTG GAT GAA GAG AAC	49	480bp	MYMIV/DNA A
AC3PF AC3PR	TTA TGA TTC GAT ATT GAA TTA ATA CTG AAG TGTGGG TGT AGC TAT	48	450bp	MYMIV/DNA A
AC4PF AC4PR	CAA ATT ACAATT TAA GTT ATG ACT TCT AGCCTT GTC AAC ACC AG	48	390bp	MYMIV/DNA A

NM1/NM2=Used for CP gene amplification of *Mungbean yellow mosaic India virus*; MYMIV-MPF/MYMIV-MPR=used for movement protein gene amplification of *Mungbean yellow mosaic India virus* (DNA B component); HYMV-CPF/HYMV-CPR= Used for CP gene amplification of *Horse gram yellow mosaic virus*; HYMV-MPF/HYMV-MPR= Used for movement protein gene amplification of *Horse gram yellow mosaic virus*; MYMV-CP-F/MYMV-CP-R=Used for CP gene amplification of *Mungbean yellow mosaic virus*; MYMV-MPF/MYMV-MPR=Used for CP gene amplification of *Mungbean yellow mosaic virus*; DYMV-CP-F/DYMV-CP-R= Used for CP gene amplification of *Dolichos yellow mosaic virus;* AC1PF/AC1PR, AC2PF/AC2PR, AC3PF/AC3PR, AC4PF/AC4PR= Used for AC1, AC2, AC3 and AC4 gene amplification of MYMIV (Naimuddin *et al.,* 2011a, b, c).

Disease Cycle

Since a number of cultivated plant species and weeds are hosts of MYMV, availability of inoculums throughout the year can be explained. However, the differences in the incidence of the disease may be attributed not only to the

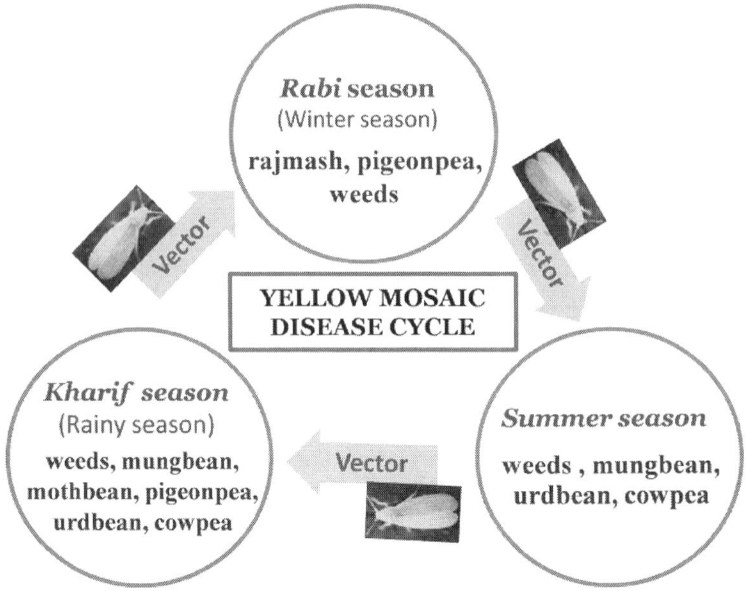

Figure 14.3. Yellow Mosaic Disease Cycle.

prevalence of alternate/collateral hosts (of MYMY) in the vicinity of the crop, but also to the availability of vector, whitefly also. The vector is polyphagous but its population is generally influenced by weather parameters. Under north Indian conditions, mungbean and urdbean are cultivated in spring/summer and *Kharif* seasons. Long duration pigeonpea is also a host of MYMV and remains in the field till April. The MYMV affected pigeonpea may provide primary inoculums of MYMV for spring/summer sown crop. Besides, weeds hosts may also act as source of primary inoculums of MYMV for spring/summer crops. Beans grown as winter season also get MYMV/MYMIV infection and may serve as a source of primary inoculum for spring/summer crop. Once the virus is introduced in to the spring/summer sown crops, it perpetuates and is carried over to *kharif* sown crops of not only mungbean and urdbean but also to *kharif* sown pigeonpea. Virus may also be introduced in to the *kharif* crop from weed hosts also through whitefly vector. Weeds reported to be the hosts of the MYMV are *Eclipta alba, Xanthium strumarium, Brachiara romosa, Cosmos bipinnatus* (Nene, 1972). Raychuadhari *et al.* (1977) reported 72 weeds to be the host of MYMV. This is, however, based on the yellow mosaic type symptoms in these weeds and cannot be accepted without unequivocal evidence for the identity of the virus in these weeds. At IIPR, MYMIV infection of the weed, *Ageratum conyzoides* has been proved unequivocally (Naimuddin *et al.*, 2014). This weed present throughout the year may be an important alternate host of MYMIV. Extensive studies are required for characterizing geminiviruses in weeds so as weeds harbouring the crop infecting viruses are identified. This will indeed help devising strategies for the management of YMD of pulse crops.

Management

Cultural Approaches

Nutrients applied to the soil affect YMD incidence in mungbean. Soil application of boron at 3 kg/ha and molybdenum at 2 kg/ha reduced incidence of YMD and increased grain yield in mungbean in Bangladesh (Jalaluddin *et al.*, 2006). Application of nitrogenous and phosphorous fertilizers was found to encourage the build-up of the whitefly population as well as the incidence of the disease in mungbean (Yein and Singh, 1982). On the other hand, in a pot experiment, Sekhar and Chand (2001) found highest reduction in yellow mosaic disease (53.34 per cent) when 20 Kg N+40 Kg P+50 Kg K/ha were applied. Disease was less in treatments involving K and P at 40 kg/ha. However, the increase in N rate with 40 kg P/ha increased the severity of YMD.

Certain intercrops like safflower, sunflower and maize are reported to be effective in delaying the onset and subsequent development of yellow mosaic disease in urdbean (Thakur and Agrawal, 1998). Borah *et al.* (1996) found that the use of cotton as a trap crop, sown one month ahead between the mungbean rows with a single spray of dimethoate 0.03 per cent at 15 days after germination of mungbean, effectively controlled both *B. tabaci* and yellow mosaic virus. Even the inter-varietal intercropping of mungbean and urdbean has been reported effective in reducing the incidence of YMD in both mungbean and urdbean (Dasgupta and Chowdhury, 1985). Marigold as a trap crop along with yellow sticky traps (8/ha) recorded reduced disease incidence (Salam, 2009).

Manipulation of sowing dates has also been exploited to reduce losses caused by YMD in mungbean and urdbean. Sowing crop early or late may, however, depend on the agro-ecological zone. In north India, early sown crops of mungbean and urdbean in *Kharif* season gets more YMD than the crops sown late. Dubey and Singh (2006) found that under Delhi conditions, incidence and severity of YMD was more in crops sown early (1 July) than crops sown late (15, 30 July, 15 Aug). 15 July was the best sowing date as the crop gave significantly higher grain yield and got moderate disease. Disease was reduced in crops sown on 30 July and later, but the yield was also declined. In Orissa also, the mungbean crop sown early (1 November) got less YMD than in crops sown late (16 November) (Nayak and Patra, 2000). In northern Telangana, sowing of rabi mungbean and urdbean early (25 September) results in higher incidence of YMD than in crops sown late (26 October) (Reddy *et al.*, 1996). In Assam, Nath (1994) reported that mungbean crop sown between 1-15 February got less population of whitefly and incidence of YMD than the crops sown later. Rouging of infected plants helped reduce incidence of yellow mosaic disease and increased yield of mungbean (Jalaluddin *et al.*, 2006, Islam *et al.*, 2002).

Chemical

Unlike fungal and bacterial pathogens, chemicals targeting virus are not available. However, chemicals have been used against the vector, whitefly thereby reducing the YMD. A number of insecticides are known as effective against whitefly and have been used as seed treatment, foliar spray and soil application for managing the YMD in mungbean and urdbean. Two foliar sprays (a fortnight apart) of dichlorvos, monocrotophos or phosphamidon effectively reduced the YMD in both mung and urd (Agrawal *et al.*, 1979). A combination of Aureofungin (0.003 per cent) and Dimecron (0.25 kg a.i./ha) decreased incidence of mung bean yellow mosaic virus in mungbean and increased yield (Ahmad and Gaur, 1982). In Himachal Pradesh, application of Octanol + demeton-S-methyl (as Metasystox) gave the best control of YMD and resulted in higher yields in blackgram (Singh and Sirohi, 1997). Ghosh (2008) found imidacloprid at 12.5 kg/ha, applied at 15- and 30-days stage of green gram crop grown as direct-sown under residual soil moisture conditions after the harvest of kharif rice (rain-fed lowland) effective in reducing YMD and increasing grain yield. In further studies, Ghosh *et al.* (2009) found Imidacloprid (Confidor @ 0.25 ml/l) to be effective in reducing the YMD incidence in both pre- and post- kharif season by more than 40 per cent. They advocated imidacloprid as an alternative to monocrotophos. No significant difference was found between Thiomethoxam and monocrotophos in reducing the disease and therefore, the former may also be a potential replacement for monocrotophos. Two sprays of imidacloprid @ 0.24 ml/l along with imidacloprid @ 5 g/kg seed treatment and 2 sprays of imidacloprid @ 0.24 ml/l alone recorded least mean disease incidence of 30.2 and 30.3 per cent with least vector populations of 1.87 and 1.73 whiteflies leaf (Salam *et al.*, 2009). Ganapathy and Karuppiah (2004) also found the application of thiamethoxam @ 0.2 g/litre at 15 DAS effective in decreasing the whitefly population and YMD incidence and increased yield. Panduranga *et al.* (2011) evaluated inecticides for managing the YMD and found foliar spray of thiomethoxam 25 WS @ 0.005 per cent, spirotetramat 150 OD @ 90 g a.i./ha and acetamprid 20 per cent SP @ 0.002 per cent to be the most effective treatments in reducing the whitefly population and YMD incidence.

Soil application of granular insecticides has also been used to manage YMD in mungbean and urdbean. Four granular insecticides, *viz.*, aldicarb 10G, carbofuran 3G, disulfoton 5G and phorate 10G applied by broadcasting before sowing, in furrows alongwith seed, in furrows below the seed and broadcasting at the time of hoeing applied at 1.5 kg a.i./ha significantly reduced not only YMD but also galerucid beetle (*Madurasia obscurella*), stemfly (*Melanagromyza phaseoli* [*Ophiomyia phaseoli*), pod borer (*Euchrysops cnejus*) in mungbean (Pandey *et al.*, 2007). Sharma and Varma (1982) found aldicarb, phorate or disulfoton granules at 2 kg a.i./ha or a combination of disulfoton granules and paraffin oil as most effective in controlling the spread of the YMD and consequently in increasing yield also. Application of aldicarb at a rate of 1.5 kg a.i./ha in the furrow before sowing gave the best control of *B. tabaci* populations (66-67 per cent at 5 weeks after sowing). Soil application of Solvirex at sowing followed by 3 weekly foliar sprays of Metasystox beginning after 20 days after sowing lowered the spread of YMD in mungbean crop and increased grain yield (Singh and Varma, 1977).

Based on two years' experimentation a treatment involving seed treatment with imidacloprid 17.8SL @ 5ml/kg seeds and two foliar sprays with the insecticide Nurelle D505 (50 per cent chlorpyriphos and 5 per cent cypermethrin) @0.1 per cent at 15 and 45 days after sowing was found best in significantly increasing the grain yield in mungbean cv. T44 at Kanpur. None of the treatments used significantly reduced the yellow mosaic disease (Akram and Naimuddin, 2016).

Botanicals

Of the many plant extracts tried for the management of yellow mosaic disease in mungbean and urdbean, neem products are the most commonly used plant products (Gupta and Pathak, 2009). Singh *et al.* (2004) found that six foliar sprays at weekly intervals of aqueous root extract of *Boerhaavia diffusa* starting from the seedling stage were most effective in delaying the symptom appearance, reducing the disease severity and decreasing the disease incidence by 80-90 per cent and also enhanced root nodulation, plant growth and yield parameters. Gupta and Pathak (2009) found an admixture of neem seed kernel extract (NSKE) (in cow urine) @ 3 per cent +dimethoate @ 0.03 per cent and of neem oil @ 0.5 per cent +dimethoate @ 0.03 per cent to be effective in reducing the incidence not only of whitefly and yellow mosaic but also of pod borer in urdbean in Bundelkhand zone of Madhya Pradesh. These treatments also increased grain yield and resulted in net profit of Rs 3934 and Rs 3320/ha with incremental cost benefit ratio of 11.2 and 10.9, respectively. Eight sprays of extracts of *Clerodendrum aculeatum*, root extract of *Boerhaavia diffusa* and neem leaf extract reduced the YMD incidence by 66.70, 60.27, 42.43 per cent, respectively in mungbean, and by 63.65, 58.20, 42.93 per cent, respectively in urdbean. All these treatments also increased plant growth and nodulation in both the crops (Singh and Awasthi, 2009). Foliar sprays of different formulations of neem oil @ 3 per cent, neem seed kernel extract @5 per cent at 35 and 55 days after sowing to be as effective as monocrotophos in reducing the YMD disease in urdbean (Sethuraman *et al.*, 2001). Hossain *et al.* (2010) found reduction in YMD incidence by 18.67 per cent and increase in grain yield by 17.65 per cent in mungbean by foliar sprays of neem leaf extract.

Host Plant Resistance

YMD is difficult to manage through chemicals. Though the insecticides kill the vector of the YMD and also the other insect pests on these crops; but whitefly, being polyphagous and highly active, re-emerges in crop fields as soon as effect of insecticides diminishes. Therefore, use of resistant varieties is a cheap and best mean of avoiding losses caused by this disease. A number of workers have identified sources of resistance against YMD in mungbean and urdbean using infector row technique. Infector row technique is commonly employed to screen pulse crops genotypes against YMD under field conditions using a highly susceptible genotype as infector row planted between two rows of test genotypes (Figure 14.4). However, in very few of them, identity of the causal virus of YMD has been unequivocally proved. Bag *et al.* (2014) screened 344 accessions of urdbean. Of the eight genotypes found resistant in field screening and tested using whitefly inoculation, only four accessions *viz.* (IC144901, IC001572, IC011613 and IC485638) showed resistant reaction to MYMIV. Resistance of these four genotypes was further confirmed through agroinoculation of the infectious cloned viral DNA under controlled conditions.

Figure 14.4. Infector Row Technique.

Many other workers have reported field resistance of urdbean genotypes from different parts of the country (Gupta and Mishra, 2014, Obaiah *et al.*, 2013, Sudha *et al.*, 2015, Suman *et al.*, 2015). Singh and Awasthi (2004) evaluated 98 germplasm accessions of urd bean against YMD and found thirteen genotypes (HPU-227, UH-86-40, IPU-94-2, IPU-981, KU-300, IPU-94-1, ND-6, Pant U-19, K-11686, PDU-2, PLU-800, PLU-603 and NDU-7) as resistant. Gupta (2003) reported 13 urdbean lines (KU 96, TU 98-14, KU 96-1, UG 774, UG 737, UPU 97-10, KU 315, NP 6, PLU 96-8, PLU 44, PLU 131, PLU 463 and KU 96-8) as resistant to YMD. Biswas and Verma (2001) found urdbean genotypes DPU 84-14', 'DPU 102', 'UG 218', 'UG 389', 'UL 257' and 'UL 310' as resistant to highly resistant to MYMIV under both the field and glasshouse conditions. Panigrahi and Baisakh (2013) identified urdbean lines PDU 1-15, PDU 1-9 and Sarala-13 as resistant to YMD in Odisha.

Sudha *et al.* (2013) screened 78 mungbean genotypes against MYMV and found the 28 resistant in the field screening. Agro-inoculation of these genotypes with the infectious cloned viral DNA of two strains (VA 221 and VA 239) revealed only three genotypes (ML 1108, KMG 189 and SP 84) to be resistant to the VA 221 strain while only one genotype, namely ML 818, showed resistance to the strain VA 239, but it was susceptible to strain VA 221. These findings highlight the importance of confirming resistance to YMD using infectious clones of the virus. Manivannan *et al.* (2001) screened 551 green gram genotypes in Vamban, Tamil Nadu, India and rated genotypes EC300072, K141, LGG424B and LM108B rated as resistant to the YMD. Singh and Awasthi (2009) screened 84 genotypes of mungbean and found ten (GM-13, GM-21, GM-125, GM-138, M-125, GM-164, GM-850, GM-899, M-126 and PLM-214) of them resistant to YMD. Kumar *et al.* (2006) screened 50 genotypes of mungbean and found NDM 2, ML 337, and PDM 90-1 to be resistant to YMD. Biswas *et al.* (2005) identified mungbean genotypes MGG 443, ML 5, ML 337, ML 459, ML

513, ML 610, P 9271 and P 9272 as resistant to five variants of MYMIV namely Pp1, Bg3, MbD, MoL and MbS through whitefly inoculation.

YMD is also prevalent in neighbouring Bangladesh, Nepal and Pakistan and sources of resistance have been reported from there also. Karim *et al.* (2010) found mungbean genotype, BMX-97008-8 as resistant to YMD in Bangladesh. Ilyas *et al.* (2007) screened 101 genotypes of mungbean and found mungbean genotypes CMG-523, CM-mung-97, 96002, 98010, C10/95-3-21 and C1/95-3-45 as resiatant to YMD in Pakistan. In Nepal, Neupane *et al.* (2004) found mungbean genotypes VC 6372 (45-8-1), VC 6370 (30-65), and VC 6368 (46-7-2) as resistant to YMD. A number of YMD resistant varieties of mungbean and urdbean have been developed and released for cultivation in different parts of the country (Anonymous, 2019).

Biotechnological Approach

Durability of the resistance in released varieties has always been a concern. Genetically engineered resistance mostly virus derived against YMD may therefore be exploited. Genetic engineering can be used to modify the genetic makeup of the existing varieties in such a way that the engineered plant does not allow multiplication of the target virus thereby rendering it resistant against the latter. The coat protein gene mediated resistance (CPMR) has been exploited in case of begomoviruses. It works in case of monopartite geminiviruses (Kunik *et al.*, 1994), but it is assumed that a CP-mediated strategy against bipartite geminiviruses will not produce a high level of resistance (Azzam *et al.*, 1996, Frischmuth and Stanley, 1998) because nuclear shuttle protein (NSP) (gene located on DNA B) can substitute for function of CP in transport. However, CPs may have the potential for transgenic interference as geminivirus CPs control specific interactions with the virus vector (Briddon *et al.*, 1990; Azzam *et al.*, 1996; Hofer *et al.*, 1997; Noris *et al.*, 1998; Haq *et al.*, 2011) and a mutated or non-functional CP may therefore impede the virus spread among its vectors in geminivirus infected fields.

The subject of engineering resistance to geminiviruses has been nicely reviewed by Vanderschuren *et al.* (2007). Pathogen derived resistance has been demonstrated in transgenic tobacco having MYMV-Vig genes for coat protein (*CP*), replication-associated protein (*Rep*)-sense, Rep-antisense, truncated Rep (*T-Rep*), nuclear shuttle protein (*NSP*) and movement protein (*MP*) (Shivaprasad *et al.*, 2006; Sunitha *et al.*, 2011, 2013). Virus (MYMV-Vig) accumulation was inhibited in transgenic tobacco plants harbouring the Rep-sense and T-Rep gene of MYMV-Vig agroinoculated with partial dimers of MYMV-Vig (Shivaprasad *et al.*, 2006). So far the results of genetic engineering or biotechnological approached have been encouraging in modifying the native immune system of the plant (transgenic) in a way that plants are protected against the geminivirus infection and it is very likely to lead researchers to use novel tools that can be exploited to stabilize the crop production in general and pulses in particular. However, it may be concluded that the various approaches available to manage YMD may be integrated to make the intervention farmers friendly which will help them avoid losses caused by YMD and result in obtaining good harvest thereby increasing farmers income.

REFERENCES

Agrawal HS, Gupta NK, Prasad VK and Vishwakarma S. 1979. Chemical control of yellow mosaic of moong. *Pesticides* 13: 44-47.

Ahmad M and Harwood RF. 1973. Studies on whitefly transmitted yellow mosaic of urdbean (*Phaseolus mungo*). *Plant Disease Reporter* 57: 800-802.

Ahmed SR and Gaur RB. 1982. Chemotherapeutic approach to check yellow mosaic virus of moong (*Phaseolus aureus* L.). *Hindustan Antibiotics Bulletin* 24: 18-20.

Akram M and Naimuddin. 2016. Management of mungbean yellow mosaic disease and effect on grain yield. *Indian Journal of Plant Protection* 44: 127-131

Akram M, Naimuddin, Agnihotri AK, Gupta S and Singh NP. 2015. Characterization of full genome of *Dolichos yellow mosaic virus* based on sequence comparison, genetic recombination and phylogenetic relationship. *Annals of Applied Biology* 67: 354-363.

Alabi, O.J., Lava Kumar, P., Mgbechi-Ezeri, J.U. *et al.*, Two new 'legumoviruses' (genus *Begomovirus*) naturally infecting soybean in Nigeria. *Arch Virol* 155, 643–656 (2010). https: / /doi.org/ 10.1007 / s00705-010-0630-3

Annonymous. 2019. Project Coordinator's Report, 2019.AICRP (MULLaRP). ICAR-IIPR, Kanpur.

Azzam O, Diaz O, Beaver JS, Gilbertson RL, Russell DR and Maxwell DP. 1996. Transgenic beans with the bean golden mosaic geminivirus coat protein gene are susceptible to virus infection. *Annual report of the Bean Improvement Cooperative* 39: 276–277.

Bag MK, Gautam NK, Prasad TV, Pandey S, Dutta M and Roy A. 2014. Evaluation of an Indian collection of black gram germplasm and identification of resistance sources to *Mungbean yellow mosaic virus*. *Crop Protection* 61: 92-101.

Barnabas AD, Radhakrishnan GK and Ramakrishnan U. 2010. Characterization of a begomovirus causing horsegram yellow mosaic disease in India. *European Journal of Plant Pathology* 127: 41-51.

Biswas KK and Varma A. 2001. Evaluation of resistance in blackgram (*Phaseolus mungo*) to variants of mungbean yellow mosaic geminivirus. *Indian Journal of Agricultural Sciences* 71: 215-218.

Biswas KK, Malathi VG and Varma A. 2005. Resistance in mungbean against five variants of Mungbean yellow mosaic virus. *Indian Journal of Virology* 16: 27-31.

Borah RK, Nath PD and Deka N. 1996. Effect of insecticides and crop trap on the incidence of white fly *Bemisia tabaci* (Genn.) and yellow mosaic virus in greengram *Vigna radiata* (L.) Wilczek. *Indian Journal of Virology* 12: 75-77.

Briddon RW, Patil BL, Basavaraj B, Nawaz-ul-Rehman MS and Fauquet CM. 2010. Distinct evolutionary histories of the DNA-A and DNA-B components of bipartite begomoviruses. *BMC Evolutionary Biology* 10: 97. DOI: 10.1186/1471-2148-10-97

Briddon RW, Pinner MS, Stanley J and Markham PG. 1990. Geminivirus coat protein gene replacement alters insect specificity. *Virology* 177: 85-94.

Brown JK. 2007. The *Bemisia tabaci* Complex: Genetic and Phenotypic Variability Drives Begomovirus Spread and Virus Diversification. Online. APS*net* Features. doi: 10.1094/APSnetFeature-2007-0107.

Capoor S.P. and P.M. Varma, 1948. A new virus disease of *Dolichos lablab*. *Current Science* 19, 248-249.

Chand P and Varma JP. 1983. Effect of yellow mosaic on growth components and yield of mungbean and urdbean. *Haryana Agricultural University Journal of Research* 13: 98-102.

Dasgupta B and Chowdhury AK. 1985. Use of intervarietal intercropping to minimize the yellow mosaic virus of urd and mung beans. *Indian Journal of Plant Pathology* 3: 100-101.

Deng D, Mcgrath PF, Robinson DJ and Harrison BD. 1994. Detection and differentiation of whitefly-transmitted geminiviruses in plants and vector insects by the polymerase chain reaction with degenerate primers. *Annals of Applied Biology* 125: 327–336.

Dubey SC and Singh B. 2006. Integrated management of cercospora leaf spots and yellow mosaic of urdbean (*Vigna mungo*). *Indian Journal of Agricultural Sciences* 76: 485-489.

Fauquet CM, Briddon RW, Brown JK, Moriones E, Stanley J, Zerbini FM and Zhou X. 2008. Geminivirus strain demarcation and nomenclature. *Archives of Virology* 153: 783–821.

Frischmuth S and Stanley J. 1998. Recombination between viral DNA and the transgenic coat protein gene of African cassava mosaic geminivirus. *Journal of General Virology* 79: 1265–1271.

Ganapathy T and Karuppiah R. 2004. Evaluation of new insecticides for the management of whitefly (*Bemisia tabaci* Genn.), mungbean yellow mosaic virus (MYMV) and urdbean leaf crinkle virus (ULCV) diseases in mungbean [*Vigna radiata* (L.) Wilezek]. *Indian Journal of Plant Protection* 32: 35-38.

Gautam, NK, Akram M, Akhtar J, Khan Z, Dwivedi NK, Latha M and Ram B. 2014. Responses of wild Vigna species/sub-species to yellow mosaic disease viruses, detected by a PCR-based method. *Phytopathologia Mediterrenea* 53: 428-437.

Ghosh A. 2008. Management of yellow mosaic virus by chemical control of its vector, Whitefly (*Bemisia tabaci*) and its impact on performance of green gram (*Phaseolus aureus*) under rainfed lowland rice fallow. *Archives of Phytopathology and Plant Protection* 41: 75-78.

Ghosh D, Laha SK and Biswas NK. 2009. Effect of different pesticides on incidence of mungbean yellow mosaic virus incidence. *International Journal of Plant Protection* 2: 67-70.

Gupta MP and Pathak RK. 2009. Bioefficacy of neem products and insecticides against the incidence of whitefly, yellow mosaic virus and pod borer in Black gram. *Natural Product Radiance* 8: 133-136.

Gupta O and Mishra M. 2014. Field resistance in mungbean and urdbean genotypes against yellow mosaic disease. *Journal of Food Legumes* 27: 80-81

Gupta O. 2003. Resistance to Mungbean yellow mosaic virus, phenotypic characters and yield components in urdbean. *Indian Phytopathology* 56: 110-111.

Gurha SN, Misra DP and Kamthan KP. 1982. Studies on some aspects of yellow mosaic disease of black gram (*Vigna mungo* (L.) Hepper). *Madras Agricultural Journal* 69: 435-438.

Ha, C., Coombs, S., Revill, P., Harding, R., Vu, M., Dale, J., 2008. Molecular characterization of begomoviruses and DNA satellites from Vietnam: additional evidence that the New World geminiviruses were present in the Old World prior to con-tinental separation. J. Gen. Virol. 89, 312–326.

Hamilton WDO, Bisaro DM, Coutts RHA and Buck KW. 1983. Demonstration of the bipartite nature of the genome of a single-stranded DNA plant virus by infection with the cloned DNA components. *Nucleic Acids Research* 11: 7387-7396.

Hanley-Bowdoin L, Settlage SB, Orozco BM, Nagar S and Robertson D. 1999. Geminiviruses: models for plant DNA replication, transcription, and cell cycle regulation. *Critical Reviews in Plant Sciences* 18: 71-106.

Haq QMI, Jyothsna P, Ali A and Malathi VG. 2011. Coat protein deletion mutation of *Mungbean yellow mosaic India virus* (MYMIV). *Journal of Plant Biochemistry and Biotechnology* 20: 182-189.

Harrison BD, Muniyappa V, Swanson MM, Roberts IM and Robinson DJ. 1991. Recognition and differentiation of seven whitefly-transmitted geminiviruses from India, and their relationship to African cassava mosaic and Thailand mungbean yellow mosaic viruses. *Annals of Applied Biology* 118: 299-308.

Hofer P, Bedford ID, Markham PG, Jesek H and Frischmuth T. 1997. Coat protein gene replacement results in whitefly transmission of an insect non-transmissible geminivirus isolate. *Virology* 236: 288-295.

Honda Y, Iwaki M, Saito Y, Thongmeearkom P, Kittisak K and Deema N. 1983. Mechanical Transmission, Purification, and Some Properties of Whitefly-Borne Mung Bean Yellow Mosaic Virus in Thailand. *Plant Disease* 67: 801-804.

Hossain ASMA, Akhter K, Akhtar N, Muqit A and Islam SMAS. 2010. Effect of some management practices on the incidence of mungbean yellow mosaic and on grain yield. *Bangladesh Journal of Plant Pathology* 26: 9-13.

Ilyas M, Qazi J, Mansoor S and Briddon RW. 2009. Molecular characterisation and infectivity of a "Legumovirus" (genus *Begomovirus*: family *Geminiviridae*) infecting the leguminous weed *Rhynchosia minima* in Pakistan. *Virus Research* 145: 279-284.

Ilyas MB, Chaudhry MA and Ghazanfar MU. 2007. Screening of mungbean germplasm for the sources of resistance against mung bean yellow mosaic virus infection. *Pakistan Journal of Phytopathology* 19: 101-104.

Islam MR, Ali MA, Islam MS and Hossain AFMGF. 2002. Effect of nutrients and weeding on the incidence of mungbean mosaic. *Pakistan Journal of Plant Pathology* 1: 48-50

Jalaluddin M, Khalil MI and Kashem MA. 2006. Management of yellow mosaic of summer mungbean with boron, molybdenum, rouging and insecticide. *Bangladesh Journal of Plant Pathology* 22: 17-20.

Jyothsna P., Rawat R., Malathi V.G. (2011) Molecular char-acterization of a new begomovirus infecting a leguminous weed *Rhynchosia minima* in India. *Virus Genes*, 42, 407–414.

Karim Z, Hossain MS and Islam MM. 2010. Screening of mungbean lines against mungbean yellow mosaic virus (MYMV) under field condition. *Bangladesh Journal of Plant Pathology* 26: 79-80.

Karthikeyan A, Shobhana VG, Sudha M, Raveendran M, Senthil N, Pandiyan M, Nagarajan P. 2014. Mungbean yellow mosaic virus (MYMV): a threat to green gram (*Vigna radiata*) production in Asia. *International Journal of Pest Management* 60: 314-324.

Khattak GSS, Haq MA, Rana SA, Abass G and Irfan M. 2000. Effect of mungbean yellow mosaic virus (MYMV) on yield and yield components of mungbean (*Vigna radiata* (L.) Wilczek). *Kasetsart Journal, Natural Sciences* 34: 12-16.

Kothandaraman SV, Devadason A and Ganesan MV. 2016. Seed-borne nature of a begomovirus, *Mung bean yellow mosaic virus* in black gram. *Applied Microbiology and Biotechnology* doi: 10.1007/s00253-015-7188-7.

Kumar R, Shamshad Ali and Rizvi SMA. 2006. Screening of mungbean genotypes for resistance against whitefly, *Bemisia tabaci* and mungbean yellow mosaic virus. *Indian Journal of Pulses Research* 19: 135-136.

Kunik T, Salomon R, Zamir D, Navot N, Zeidan M, Michelson I, Gafni Y and Czosnek H. 1994. Transgenic tomato plants expressing the tomato yellow leaf curl virus capsid protein are resistant to the virus. *Biotechnology* 12: 500–504.

Lazarowitz SG, Wu LC, Rogers SG and Elmer JS. 1992. Sequence-specific interaction with the viral AL1 protein identifies a geminivirus DNA replication origin. *Plant Cell* 4: 799-809.

Malathi VG and John P. 2008. Geminiviruses infecting legumes. In: Characterization, Diagnosis and Management of Plant Viruses, Volume 3: Vegetables and Pulse Crops (G.P. Rao, P. Lava Kumar and Ramon J. Holguin-Pena, Ed.), Stadium Press LLC, Texas, USA. Pp 97-123.

Malathi VG, Surendranath B, Naghma A and Roy A. 2005. Adaptation to new hosts shown by the cloned components of *Mungbean yellow mosaic India virus* causing cowpea golden mosaic in northern India. *Canadian Journal of Plant Pathology* 27: 439-447.

Malathi VG. 2007. Genetic identity of yellow mosaic viruses infecting legumes and their phylogenetic relationship. *Indian Phytopathology* 60: 143-155

Mandal B, Varma A and Malathi VG. 1997. Systemic infection of *Vigna mungo* using the cloned DNAs of the blackgram isolate of mungbean yellow mosaic geminivirus through agroinoculation and transmission of the progeny virus by whiteflies. *Journal of Phytopathology* 145: 505–510.

Manivannan N, Sethuraman K and Natarajan S. 2001. Screening of greengram (*Vigna radiata* (L.) Wilczek) germplasm for yellow mosaic resistance. *Legume Research* 24: 268-271.

Maruthi M.N., Manjunatha B., Rekha A.R., Govindappa M.R., Colvin J., Muniyappa V. (2006) *Dolichos yellow mosaic virus* belongs to a distinct lineage of Old World begomoviruses; its biological and molecular properties. *Annals of Applied Biology*, 149, 187–195.

Morinaga T, Ikegami M and Miura K. 1993. The nucleotide sequence and genome structure of mungbean yellow mosaic geminivirus. *Microbiology and Immunology* 37: 471-476.

Muniyappa V, Reddy HR and Shivashankar G. 1975. Yellow mosaic disease *Dolichos biflorus* Linn. (horsegram). *Current Research* 4: 176.

Naimuddin and Akram M. 2010. Detection of mixed infection of begomoviruses and their molecular characterization based on CP gene sequences. *Journal of Food Legumes* 23: 191-195.

Naimuddin, Akram M and Gupta S. 2011a. Identification of *Mungbean yellow mosaic India virus* infecting *Vigna mungo* var. *silvestris* L. *Phytopathologia Mediterranea* 50: 94–100.

Naimuddin, Akram M, Gupta S and Agnihotri AK. 2014. *Ageratum conyzoides* harbours *Mungbean yellow mosaic India virus*. *Plant Pathology Journal* 13: 59-64.

Naimuddin, Akram M, Pratap A, Chaubey BK and Joseph KJ. 2011c. PCR based identification of the virus causing yellow mosaic disease in wild *Vigna* accessions. *Journal of Food Legumes* 24: 14-17.

Naimuddin, Akram M and Pratap A. 2011b. First report of natural infection of *Mungbean yellow mosaic India virus* in two wild species of *Vigna*. *New Disease Reports* 23: 21.

Nair NG and Nene YL. 1974. Studies on yellow mosaic of urdbean (*Phaseolus mungo* L.) caused by mungbean yellow mosaic virus. Factors influencing transmission and symptom expression. *Indian Journal of Farm Sciences* 2: 42-47.

Nair NG. 1971. Studies on the yellow mosaic of urd bean (*Phaseolus mungo* L.) caused by mung bean yellow mosaic virus. Ph.D. Thesis, G.B. Pant Univ. Agriculture and Technology, Pantnagar, UP, India. Pp. 1-66.

Nariani TK. 1960. Yellow mosaic of mung (*Phaseolus aureus* L.). *Indian Phytopathology* 13: 24-29.

Nath PD. 1994. Effect of sowing time on the incidence of yellow mosaic virus disease and whitefly population on greengram. *Annals of Agricultural Research* 15: 174-177.

Nayak BC and Patra AK. 2000. Response of greengram (Phaseolus radiatus) cultivars to dates of sowing during winter. *Indian Journal of Agricultural Sciences* 70: 320-322.

Nene YL. 1972. A survey of viral diseases of pulse crops in Uttar Pradesh. G.B. Pant Univ. Agriculture and Technology, Pantnagar, UP, India. pp. 191.

Nene YL. 1973. Viral diseases of warm weather pulse crops in India. *Plant Disease Reporter* 57: 463–467.

Neupane RK, Darai R, Sah RP, Pokhrel DN and Shanmugasundaram S. 2006. Development of mungbean (*Vigna radiata* (L.) Wilczek) varieties for rice and maize-based systems in Nepal. In: AVRDC, Shanhua, Taiwan, Improving income and nutrition by incorporating mungbean in cereal fallows in the Indo-Gangetic Plains of South Asia DFID Mungbean Project for 2002-2004, proceedings of the final workshop and planning meeting, Punjab Agricultural University, Ludhiana, Punjab, India, 27-31 May 2004. 268-282pp.

Noris E, Accotto GP, Tavazza R, Brunetti A, Crespi S and Tavazza M. 1996. Resistance to tomato yellow leaf curl geminivirus in *Nicotiana benthamiana* plants transformed with a truncated viral C1 gene. *Virology* 224: 130–138.

Obaiah S, Reddy BVB, Reddy NPE, Prasad YS. 2013. Screening of some blackgram Vigna mungo (L.) Hepper) genotypes for resistance to yellow mosaic virus. *Current Biotica* 7: 96-100.

Pandey A, Pandy A and Srivastava SP. 2007. Management of pest complex of summer mung by granular insecticides. *Annals of Plant Protection Sciences* 15: 376-380.

Panduranga GS, Vijayalakshmi K and Reddy KL. 2011. Evaluation of insecticides for management of *Bemisia tabaci* and MYMV disease of mungbean [*Vigna radiata* (L.) Wilczek]. *Annals of Plant Protection Sciences* 19: 295-298.

Panigrahi KK and Baisakh B. 2013. Reaction of local and mutant lines of black gram (Vignamungo [L] Hepper) to yellow mosaic virus and powdery mildew. *Journal of Plant Protection and Environment* 10: 57-60.

Perring TM. 2001. The *Bemisia tabaci* species complex. *Crop Protection* 20: 725-737.

Prasanna HC, Kanakala S, Archana K, Jyothsna P, Varma RK and Malathi VG. 2015. Cryptic species composition and genetic diversity within *Bemisia tabaci* complex in soybean in India revealed by *mtCOI* DNA sequence. *Journal of Integrative Agriculture* 14: 1786–1795

Puchta H and Sanger HL (1989) Sequence analysis of minute amounts of viroid RNA using the polymerase chain reaction (PCR). Archives of Virology 106: 335–340.

Qazi J, Ilyas M, Mansoor S and Briddon RW. 2007. Legume yellow mosaic viruses: genetically isolated begomoviruses. *Molecular Plant Pathology* 8: 343–348.

Rathi YPS. 1972. Mungbean yellow mosaic virus: Host range and relationship with the vector, *Bemisia tabaci* Genn. Ph.D. Thesis, G.B. Pant Univ. Agriculture and Technology, Pantnagar, UP, India. Pp. 1-93.

Raychaudhari SP, Dhingra KL and Verma JP. 1977. Virus diseases of pulse crops in India. *Seed and Farms* 3: 7-17.

Reddy MD, Krishna A and Rao PS. 1996. Performance of promising greengram and blackgram varieties during *rabi* in red chalka soils of Northern Telangana. *Journal of Research ANGRAU* 24: 98-100.

Rogers SG, Bisaro DM, Horsch RB, Fraley RT, Hoffmann NL, Brand L, Elmer JS and Lloyd AM. 1986. Tomato golden mosaic virus A component DNA replicates autonomously in transgenic plants. *Cell* 45: 593-600.

Rojas MR, Gilbertson RL, Russell DR and Maxwell DP. 1993. Use of degenerate primers in the polymerase chain reaction to detect whitefly transmitted geminiviruses. *Plant Disease* 77: 340-347.

Rouhibakhsh A, Priya J and Perisamy A. 2008. An improved DNA isolation method and PCR protocol for efficient detection of multicomponents of begomovirus in legumes. *Journal of Virological Methods* 147: 37-42.

Saiki RK, Gelfand DH, Stoffel S, Scharf SJ, Higuchi R, Horn GT, Mullis KB and Ehrlich HA (1988) Primer directed enzymic amplification of DNA with a thermostable polymerase. Science 239: 487–491.

Saiki RK, Scharf S, Faloona F, Mullis KB, Horn GT, Erlich HA and Arnheim N (1985) Enzymatic amplification of Beta-globin genomic sequences and restriction site analysis for diagnosis of sickle cell anaemia. Science 230: 1350–1354.

Salam, S. A., Patil, M. S., and Byadgi, A. S. (2009). IDM of mungbean yellow mosaic disease. Annals of Plant Protection Sciences, 17, 157–160.

Sekhar NR and Chand H. 2001. Effect of host nutrition of mungbean yellow mosaic virus, anthracnose, root rot and leaf crinkle of mungbean [*Vigna radiata* (L). Wilczek]. *Indian Journal of Plant Protection,* 29: 88-91.

Sethuraman K, Manivannan N and Natarajan S. 2001. Management of yellow mosaic disease of urdbean using neem products. *Legume Research* 24: 197-199.

Sharma SR and Verma A. 1982. Control of yellow mosaic of mungbean through insecticide and oil. *Journal of Entomological Research* 6: 130-136.

Shivaprasad PV, Thillaichidambaram P, Balaji V and Veluthambi K. 2006. Expression of full-length and truncated Rep genes from Mungbean yellow mosaic virus-Vigna inhibits viral replication in transgenic tobacco. *Virus Genes* 33: 365-374.

Singh A and Sirohi A. 1997. Effect of various insecticides for the control of yellow mosaic of black gram. *Plant Disease Research* 12: 37-38.

Singh JP and Varma JP. 1977. Effect of some insecticides and oils on incidence of yellow mosaic and grain yield of mungbean cv. Varsha. *Proceedings of the National Academy of Sciences, India* 47: 219-225.

Singh RA, Sahambhi HS and Gurha SN. 1982. Investigations on yellow mosaic of urdbean (*Vigna mungo*) with particular reference to yield losses and sources of resistance. *Indian Journal of Mycology and Plant Pathology* 12: 115.

Singh RN. 1979. Natural infection of bean (*Phaseolus vulgaris*) by mungbean yellow mosaic virus. *Indian Journal of Mycology and Plant Pathology* 9: 124–126.

Singh S and Awasthi LP. 2009. Screening of mungbean [*Vigna radiata* (L.) Wilczek] germplasm against natural infection of plant viruses. *International Journal of Plant Protection* 2: 93-94.

Singh S, Awasthi LP and Verma HN. 2004. Prevention and control of yellow mosaic disease of mungbean by application of aqueous root extract of *Boerhaavia diffusa. Indian Phytopathology,* 57: 303-307.

Singh SK, Chakraborty S, Singh AK and Pandey PK. 2006. Cloning, restriction mapping and phylogenetic relationship of genomic components of MYMIV from *Lablab purpureus*. *Bioresource Technology* 97: 1807-1814.

Srivastava SK. 1989. Further studies on viruses of cowpea (*Vigna unguiculta* ssp. *unguiculta* (L.) Walp.) Ph.D. Thesis, IARI, New Delhi. 199p.

Stanley J. 1983. Infectivity of the cloned geminivirus genome requires sequences from both DNAs. *Nature* 305: 643-645.

Sudha M, Karthikeyan A, Nagarajan P, Raveendran M, Senthil N, Pandiyan M, Angappan K, Ramalingam J, Bharathi M, Rabindran R, Veluthambi K and Balasubramanian P. 2013b. Screening of mungbean (*Vigna radiata*) germplasm for resistance to Mungbean yellow mosaic virus using agroinoculation. *Canadian Journal of Plant Pathology* 35: 424-430.

Sudha M, Karthikeyan A, Shobhana VG, Nagarajan P, Raveendran M, Senthil N, Pandiyan M, Angappan K, Balasubramanian P, Rabindran R and Bharathi M. 2015. Search for Vigna species conferring resistance to Mungbean yellow mosaic virus in mungbean. *Plant Genetic Resources: Characterization and Utilization* 13: 162-167

Suman S, Sharma VK, Harsh Kumar and Shahi VK. 2015. Screening of mungbean [*Vigna radiata* (L.) Wilczek] genotypes for resistance to mungbean yellow mosaic virus (MYMV). *Environment and Ecology* 33: 855-859.

Sunitha S, Marian D, Hohn B and Veluthambi K. 2011. Antibegomoviral activity of the agrobacterial virulence protein VirE2. *Virus Genes* 43: 445-453.

Sunitha S, Shanmugapriya G, Balamani V and Veluthambi K. 2013. *Mungbean yellow mosaic virus* (MYMV) *AC4* suppresses post-transcriptional gene silencing and an *AC4* hairpin RNA gene reduces MYMV DNA accumulation in transgenic tobacco. *Virus Genes* 46: 496-504.

Sunter G, Gardiner WE, Rushing AE, Rogers SG and Bisaro DM. 1987. Independent encapsidation of tomato golden mosaic virus, a component DNA in transgenic plants. *Plant Molecular Biology* 8: 477-484.

Thakur MP and Agrawal KC. 1991. Appraisal of losses due to yellow mosaic of mungbean from Chhattisgarh region. International Conference on Virology in the Tropics, NBRI, Lucknow, India, 2-6 December, 1991, p. 31.

Thakur MP and Agrawal KC. 1998. Management of yellow mosaic disease of urdbean in Chhattisgarh region. *Indian Journal of Virology* 14: 85-87.

Townsend R, Watts J and Stanley J. 1986. Synthesis of viral DNA forms in *Nicotiana plumbaginafolia* protoplasts inoculated with cassava latent virus (CLV), evidence for independent replication of one component of the CLV genome. *Nucleic Acids Research* 14: 1253-1265.

Vanderschuren H, Stupak M, Fütterer J, Gruissem W and Zhang P. 2007. Engineering resistance to Geminiviruses-review and Perspectives. *Plant Biotechnology Journal* 5: 207-220

Varma A, Dhar AK and Malathi VG. 1991. Cloning and restriction analysis of mungbean yellow mosaic virus, p 114. In: *International Conference on Virology in the Tropics.* Lucknow, India.

Varma A, Dhar AK and Mandal B. 1992. MYMV transmission and control in India. In: *Mungbean Yellow Mosaic Disease* (S.K. Green, D. Kim, ed.), Asian Vegetable Research and Development Centre, Taipei, Taiwan, 8–27.

Vasudeva RS. 1942. A mosaic disease of cowpea in India. *Indian Journal of Science,* 12: 281-283.

Williams FJ, Grewal JS and Amin KS. 1968. Serious and new diseases of pulse crops in India in 1966. *Plant Disease Reporter* 52: 300-304.

Yein BR and Singh H. 1982. Effect of pesticides and fertilizers on the population of whitefly and incidence of yellow-mosaic virus in greengram. *Indian Journal of Agricultural Sciences* 52: 852-855.

Zaim M., Kumar Y., Hallan V., Zaidi A.A. (2011) Velvet bean severe mosaic virus: a distinct begomovirus species causing severe mosaic in *Mucuna pruriens* (L.) DC. *Virus Genes*, 43, 138–146.

Chapter 15

Bioecology and Management of Vectors Associated with Viral Diseases of Food Legumes

A.K. Panday, Rakesh Pandey & P.K. Dwivedi

Pulses, the food legumes, have been grown by farmers since millennia providing nutritionally balanced food to the people of India (Nene, 2006) and many other countries in the world. Legumes belong to the family Leguminosae (Fabaceae), consisting of four subfamilies; the Papilionoideae, Caesalpinoideae, Mimosoideae and Swartzioideae (Lewis *et al.*, 2005). The Papilionoideae includes the major food legumes, common bean (bean, French bean or kidney bean, *Phaseolus vulgaris*), cowpea (*Vigna unquiculata*), pigeonpea (red gram or arhar, *Cajanus cajan*), chickpea (Bengal gram, *Cicer arietinum*), pea (fieldpea, *Pisum sativum*), mungbean (green gram, *Vigna radiata*), urdbean (black gram, *Vigna mungo*), faba bean (broad bean, *Vicia faba*), and lentil (*Lens culinaris*). The Leguminosae are composed of some 690 genera and 18,000 species (Purseglove, 1968). It is the second largest family of seed plants followed by Gramineae (Aykroyd and Doughty, 1964). The seeds of legumes are second only to cereals as the most important source of food for humans and animals. Within the Leguminosae, there are 18-20 species that are cultivated widely for their edible seeds which are high in protein (17-25 per cent) (Aykroyd and Doughty, 1964). These are important crops in terms of the daily diet, contribution to human nutrition and also in terms of their contribution to farmer's income. They also contribute significantly through enriching the soils with nitrogen fixation and improve soil productivity.

Pulses are considered as the poor man's major source of protein in India. Pulses are grown on 25-30 million hectares of area with an annual production of

20-25 million tonnes. India accounts for 33 per cent of the world area and 22 per cent of the world production of pulses. About 90 per cent of the global pigeonpea, 65 per cent of chickpea and 37 per cent of lentil area falls in India, corresponding to 93 per cent, 68 per cent and 32 per cent of the global production, respectively (FAOSTAT 2011). The variation in production of pulses in India is common as pulses are cultivated mainly under rainfed conditions. The ever increasing population of the country and low productivity of pulses has resulted depletion of per capita availability to about 40 gm in recent year. Thus, it is necessitated to import pulses from neighboring countries. The gap between realizable and actual yields needs to be bridged up with appropriate technologies to meet the demand of pulses. The technological backstopping basically revolves around two aspects: production and protection. Although the former has been a subject of greater emphasis at all levels of strategy, yet the later continues to suffer from neglect and a sort of apathy under the cover of ignorance, economic status of the farmers, *etc.* In India, Acharya (1985) has pointed out that plant protection remains a most neglected aspect in pulse cultivation; further stating that only 5 to 6 per cent of the growers use plant protection measures in only 1.5 per cent of the total area under this crop.

The low productivity of the pulses is caused by the effect of several biotic stresses, of which diseases caused by viral pathogens are the most important especially in the tropics and subtropics (Loebenstein and Thottappilly, 2003; Rao, Kumar, and Holguin-Pena, 2008; Sastry and Zitter, 2014). Cultivated food legumes are susceptible to natural infection by at least 150 viruses, belonging to different genera (ICTV, 2012). These crops are attacked by many viruses, some individual viruses being important in several different crops (Table 15.1). For instance, peanut mottle potyvirus can spread from peanut crops into soybean, pea and French bean crops. Pasture legumes, such as clovers and medics, also often carry viruses such as alfalfa mosaic alfamovirus which can be carried by insect vectors into the legume crops. Further, some of the viruses are seed-borne in nature, making their control more difficult.

Plant Viruses

The word 'virus' means poison or the poisonous element by which infection is communicated. The virus can be defined as a transmissible parasite whose nucleic acid genome is less than 3×10^8 daltons in mass and that need ribosomes and other components of their host cells for multiplication (Gibbs and Harrison, 1976). Plant viruses are submicroscopic entities showing obligate relationship with living cells of the host and ability to cause specific diseases. Viruses are very small (submicroscopic) infectious particles (virions) composed of a protein coat and a nucleic acid core (Dickinson, 2005). The characteristic features have been given by Feakin, 1973. They contain one type of nucleic acid, either RNA, or DNA. They multiply only in the host cell (obligate pathogens). The RNA and protein subunits are formed separately in the host cell and combine to form intact virus particles. Viruses do not possess enzyme systems, required to perform metabolic processes. They are transmitted by graft inoculations, by sap, by insects and by nematodes.

Table 15.1: List of Major Viral Diseases and their Vectors

Crop	*Disease*	*Causal Organism*	*Virus Genus*	*Host Range*	*Geographical Distribution*	*Symptoms*	*Modes of Transmission*
Chickpea	Stunt	Bean leaf roll virus (BLRV), SbDV, Beet western yellows virus (BWYV), Legume yellows virus (LYV) and Chickpea luteovirus (CpLV)	Luteovirus	Chickpea	Asia and Africa		Aphids (C)
	Chlorotic stunt	Chickpea chlorotic stunt virus (CpCSV)	Polerovirus				Aphids (C)
	Chlorotic dwarf	Chickpea chlorotic dwarf virus (CpCDV),	Mastrevirus				Leaf hoppers (C)
		Chickpea red leaf virus (CpRLV), Chickpea yellows virus (CpYV), Chickpea chlorosis virus (CpCV), Chickpea chlorosis Australia virus (CpAV)					
		Bean yellow dwarf virus (BeYDV)					
	Mosaic	Alfalfa mosaic virus					
	Proliferation	Cucumber mosaic virus					
	Narrow leaf	Bean yellow mosaic virus					

Crop	Disease	Causal Organism	Virus Genus	Host Range	Geographical Distribution	Symptoms	Modes of Transmission
	Necrosis	Lettuce necrotic yellows virus					
Pea	Mosaic due to potyviruses	Pea seed-borne mosaic virus (PSbMV)	Potyvirus	*Lens esculenta, Pisum sativum, Vicia faba* and *V. villosa*	Asia (India, Japan, Taiwan), Australia, Europe, New Zealand, North Africa and North America.	Stunting, systemic vein clearing, leaf rolling, rosetting, flower distortion or abortion, small pods. Leafrolling	Sap, seed, aphids in the non-persistent manner
		Bean yellow mosaic virus (BYMV)					
	Enation mosaic	Pea enation mosaic virus-1 (PEMV-1)	Enamovirus				Sap, aphids (C)
		Pea enation mosaic virus-2 (PEMV-2)	Umbravirus				
	Top yellows	BLRV	Luteovirus				Aphids (C)
	Pea early-browning virus	Pea early-browning virus	Tobravirus group	Pea, common bean, faba bean (Bos and van der Want, 1963; Gibbs and Harrison, 1964; Lockhart and Fischer, '1976: Gerhardson and Ryden, 1979; Fiedorow, 1980, l983) and yellow lupin (Pospieszny and Frencel, 1985)	Europe (Harrison, 1973; Kowalska, 1979; Fiedorow, 1983) and Morocco (Lockhart and Fischer, 1976).	Pea irregular leaf, stem and pod necrosis; entire shoots may be killed; in some cultivars leaf mottling (Bos and van der Want, 1963). In common bean irregular leaf and stem necrosis with severe plant stunting (Gerhardson and Ryden, 1979; Bos and Huijberts, unpublished data). In faba bean infection is often symptomless	Trichodorid nematodes

Crop	Disease	Causal Organism	Virus Genus	Host Range	Geographical Distribution	Symptoms	Modes of Transmission
						(Fiedorow, 1980; Lockhart and Fischer, 1976), but plants may die prematurely if simultaneously infected by bean leafroll virus (Coskbain *et al.*, 1983)	
Mungbean and Urdbean	Yellow mosaic	MYMV, Mungbean yellow mosaic India virus (MYMIV) and Horsegram yellow mosaic virus (HgYMV)	Begomovirus				Whiteflies (C)
	Leaf curl	PBNV	Tospovirus				Sap, thrips (CP)
	Leaf crinkle	Urdbean leaf crinkle virus disease (ULCVD)	Tospovirus	Urdbean (*Phaseolus* (*Vigna*) *mungo*) cowpea, mungbean (*V. radiata*), pigeon pea (*Cajanus cajan*) and tepary bean (*Phaseolus* (*Vigna*) *aconitifolius*) (Kolte and Nene, 1975; Beniwal, 1983).	India.	Leaf rugosity, crinkling and distortion	Sap, seed, aphids, whitefly and leaf hoppers
	Blackgram mottle virus	Blackgram mottle virus	Carmovirus	Blackgram (urd) (*Vigna mungo*) Phatak, 1974).	Australasia, India, Thailand	Mottling and stunting	Beetles (*Cerotoma trifurcata* and *Epilachna varivestis*
Lentil	Yellows and stunt	BLRV	Luteovirus				Aphids (C)
		FBNYV	Nanovirus				

Crop	*Disease*	*Causal Organism*	*Virus Genus*	*Host Range*	*Geographical Distribution*	*Symptoms*	*Modes of Transmission*
	Mosaic and mottle	PSbMV	Potyvirus				Sap, seed, aphids (NP)
		BYMV					
		CMV	Cucumovirus				
		Broad bean stain virus (BBSV)	Comovirus				Sap, beetle (SP
Pigeonpea	Sterility mosaic (SMD)	Pigeonpea sterility mosaic virus (PPSMV)	Emaravirus		South Asia, Thailand and China		Sap, Eriophyid mites (SP)
	Yellow mosaic	MYMV, Rhycosia mosaic virus (RhMV) and Tomato leaf curl New Delhi virus (ToLCNDV)	Begomovirus		India		Whiteflies (C)
Cowpea	Mosaic	Cowpea aphid-borne mosaic virus (CABMV) and Bean common mosaic virus—blackeye cowpea mosaic strain (BCMV-BICM)	Potyvirus	cowpea (Anderson, 1955; Lima *et al.*, 1979), asparagus bean (*V. unguiculata* var. *sesquipedalis*) (Tsuchizaki *et al.*, 1984), common bean, mungbean (Green, 1985), soybean (deviant strain, Dijkstra *et al.*, 1987), and *Crotalaria spectabilis* (Anderson, 1955).	All the cowpea growing areas of the world	Prominent mosaic, mottle, green vein-banding and distortion	Sap, seed, aphids (NP) (*Aphis craccivora*, *Macrosiphum euphorbiae* and *Myzus persicae*) Lovisolo and Conti, 1966; Bock, 1973

Crop	*Disease*	*Causal Organism*	*Virus Genus*	*Host Range*	*Geographical Distribution*	*Symptoms*	*Modes of Transmission*
	Mosaic	Cowpea mosaic virus (CPMV) and CPSMV	Comovirus	Cowpea (Chant, 1959; Bock, 1971), groundnut and soybean in Japan, from *Crotalaria juncea* (Ladipo, 1988) and *Cajanus cajan* (Bock, 1971), sporadically found in soybean in Africa (Rossel and Thottappilly, unpublished).	Humid savanna and forest zones of West Africa. Also reported from some countries in East Africa: Kenya (Bock, 1971), Tanzania (Patel and Kuwite, 1982) and in Suriname, Cuba and the USA	Severe mosaic, mottle and distortion	Sap, seed, beetles (SP) chrysomelid beetles, *Ootheca mutabilis* and *Paraluperodes quaternus,* and by *Nematocerus acerbus* (Curculionidae Chant, 1959; Bock, 1971; Whitney and Gilmer, 1974).
	Mosaic	Southern bean mosaic virus (SBMV)	Sobemo virus	common bean, cowpea, black gram, mungbean, and, to a lesser extent, in soybean.	Warm, temperate and tropical regions of the Americas, India and Africa	Mosaic and mottle	leaf beetles (*Chrysomelidae*) in a circulative manner
	Stunt	CMV	Cucumo virus				Sap, seed, aphids (NP)
		BCMV-BICM	Poty virus				
	Golden mosaic and yellow mosiac	Cowpea golden mosaic virus (CGMV)	Begomo viurus				Whiteflies (C)
		Mungbean yellow mosaic India virus (MYMIV)					
		Dolichos yellow mosaic virus (DoYMV)					

Crop	Disease	Causal Organism	Virus Genus	Host Range	Geographical Distribution	Symptoms	Modes of Transmission
	Mild mottle	Cowpea mild mottle virus (CMMV)	Carla virus	Cowpea (Brunt and Menten, 1973), bambara groundnut (*Vigna subterranea*), soybean, winged bean (*Psophocarpus tetragonolobus*) (Fauquet *et al.*, 1979; Thouvenel *et al.*, 1982), groundnut (lizuka *et al.*, 1984), mungbean (Mink and Keswani, 1987), and some leguminous weed species (Anno-Nyako, 1984). Commonly found in common bean, and lima bean (*Phaseolus lunatus*) in Nigeria, in which it causes prominent disease symptoms (Rossel, unpublished).	Worldwide	Prominent chlorosis, stunt and rugose symptoms in common bean. bright yellow mosaic in soybean (Rossel and Thottappilly, unpublished).	Sap, seed, whiteflies (*Bemisia tabaci* (NP)
	Chlorotic mottle	Cowpea chlorotic mottle virus (CCMV)	Bromo virus				Sap, beetles (SP)
	Mottle	Cowpea mottle virus (CPMoV)	Carmo virus	Cowpea and bambara groundnut (*Vigna* (=*Voandzeia*) *subterranea*) (Robertson, 1966; Rossel, 1977; Shoyinka *et al.*, 1978).	Humid savanna and forest zones of West Africa	Severe mosaic, mottle and distortion	Sap, seed, chrysomelid beetle, *Ootheca mutabilis* (SP)

Crop	Disease	Causal Organism	Virus Genus	Host Range	Geographical Distribution	Symptoms	Modes of Transmission
	ringspot	Cowpea ringspot virus	Cucumovirus	Lima bean (*Phaseolus lunatus*) and winged bean (*Psophocarpus tetragonolobus*) (Rossel, unpublished).	All the cowpea growing areas of the world	very mild and consisting of characteristic patchy chlorosis or mottle	aphid species in the non-persistent manner
Common Bean	Common mosaic and black root	Bean common mosaic virus (BCMV) and Bean common mosaic necrosis virus (BCMNV)	Poty virus	*Phaseolus* species, mungbean (Kaiser *et al.*, 1968) and some wild legumes such as *Rhynchosia minima*. Also reported from *Lupinus luteus* (Frencel and Pospieszny, 1979).	Worldwide.	Vein-banding mosaic of dark green areas along main leaf veins, sometimes accompanied by leaf malformation (curling or blisters)	Sap, seed, non-persistent manner by several aphid species, mainly *Aphis fabae* and *Myzus persicae*
	Golden mosaic	BGMV	Begomo virus				Whiteflies (C)
	Golden yellow mosaic	Bean golden yellow mosaic virus (BGYMV)					
	Dwarf mosaic	Bean dwarf mosaic virus (BDMV)					
	Mosaic due to CMV	CMV					
	Beanpod mottle virus	Beanpod mottle virus	Comovirus group	Common bean and soybean		Plant stunting, severe leaf mosaic and pod mottle in common bean. Leaf mottle and puckering and pod and seed-coat mottling in soybean.	*Cerotoma trifurcata* and other leaf beetles

Crop	*Disease*	*Causal Organism*	*Virus Genus*	*Host Range*	*Geographical Distribution*	*Symptoms*	*Modes of Transmission*
Faba bean	Necrotic yellows	Faba bean necrotic yellows virus (FBNYV) and Faba bean necrotic stunt virus (FBNSV)	Nano virus				Aphids (C)
	Leaf roll	BLRV	Luteo virus				
	Mosaic and necrosis	BYMV	Potyvirus				Sap, seed, aphids (NP)
	Mottle	Broad bean mottle virus (BBMV)	Bromovirus	Chickpea, lentil and pea, which suffered severely, and soybean, *Phaseolus vulgaris, Trifolium* spp.	North Africa, Portugal, Sudan, Syria, UK.	Mottling, marbling or diffuse mosaic often associated with leaf malformation and sometimes with plant stunting and bushy growth	Sap, beetles (*Acalymma trivittata, Diabrotica undecimpunctata* and *Colaspis flavida*) and possibly weevils (*Sitona lineatus*) seed, beetles (SP)
	Broad bean stain virus	Broad bean stain virus	Comovirus group	Lentil, pea, vetch and hybrid clover (Makkouk *et al.*, 1986,1987; Musil *et al.*, 1983; Tapio, 1970)	Europe, North Africa, Sudan and West Asia (Makkouk *et al.*, 1987).	Mild mottling in faba bean and diffuse mottling in pea.	By weevils (*Apion vorax* and *Sitona* spp.
	Broad bean true mosaic virus	Broad bean true mosaic virus	Comovirus group	Faba bean (Gibbs and Paul, 1970) and pea	Europe and northwest Africa (Gibbs and Paul, 1970), and China (Ji, 1987).	Malforming leaf mottle and mosaic, often masked at high temperatur	

NP, nonpersistent; SP, semipersistent; C, circulative; CP, circulative and propagative.

Viruses cause many important plant diseases (Strange, 2003) and are responsible for huge losses in crop production in terms of quality and quantity in all parts of the world. In other words, plant viruses can cause severe yield losses to the food crops and substantially lessen the quality of crop products (Andret-Link and Fuchs, 2005; Agrios, 2014). Viruses are a significant production constraint in pulses. Virus diseases are often misdiagnosed as physiological, toxicological, or nutritional disorders because symptoms are similar and no pathogen is visible, except under an electron microscope.

Plant viruses rarely come out of the plant and it should therefore be noted that they cannot be disseminated as such by wind or water, but they are being transmitted from plant to plant in a number of ways such as; vegetative propagation, mechanically through cell sap, seed, pollen, insects, mites, nematodes, dodder and fungi, man, wind and ants (Singh, 2005; Mehrotra and Aggarwal, 2010; Pandey, 2012 and Agrios, 2014). However, the most common and economically most important means of transmission of viruses in the field is by insect vectors (Pandey, 2012). Since plant viruses cannot penetrate the intact plant cuticle and the cellulose cell wall, they are transmitted either vertically or horizontally by vectors (Hull, 2001). Most plant viruses are actively transmitted from infected to healthy plants by a living organism called a vector. Plant virus transmission by vectors can be divided into four components: transmissibility of the virus, transmitting ability of the vector, availability of virus in the diseased source plant, and susceptibility of healthy plants used as indicators of transmission.

Virus–Vector Relationships

Vector-virus transmission consists of several successive steps: acquisition of virions from an infected source, stable retention of acquired virions at specific sites through binding of virions to ligands, release of virions from the retention sites upon salivation or regurgitation, and delivery of virions to a site of infection in a viable plant cell. Each step of this sequence is needed for transmission to be successful (Andret-Link and Fuchs, 2005). According to Andret-Link and Fuchs (2005), a single mode of transmission is characteristic of most viruses, features of the different modes of virus transmission are important for transmission specificity and the different modes of virus transmission are characterized based on the acquisition and retention time, sites of retention, and internalization of virions by vectors. There are currently four described mechanisms of insect transmission of plant viruses. Originally, Watson and Roberts proposed a classification of plant viruses with regards to transmission into two groups, nonpersistent and persistent (Watson and Roberts, 1939). Later, it was recognized that an intermediate category of semipersistent viruses exist; these can be transmitted by the vector from a few hours to a few days post acquisition but are lost after molting (Kennedy *et al.*, 1962 and Sylvester, 1958) Based on observations that nonpersistent viruses are retained by the vector mainly in the stylet (food canal) (Ammar *et al.*, 1994 and Kennedy *et al.*, 1962), whereas semipersistent viruses are retained mainly in the foregut (Nault and Amir 1989), Nault, 1997 used the terms nonpersistent stylet-borne and semipersistent foregut-borne viruses for these two categories.

Non-Persistent Transmission

In this type of transmission, virus replicates in epidermal cells and/or parenchyma - accessed rapidly. Insect can inoculate a healthy plant in a few seconds. It takes less than one minute of feeding for an insect to acquire the virus. Retention lasts for only a few minutes to several hours. The feeding insect looses the virus rapidly when feeding on a non-infected plant (Ramusi and Flett, 2012; Agrios, 2014). Such a relationship is termed "non-persistent" (Strange, 2003). The best studied examples are of potyvirus transmission by aphids (Andret-Link and Fuchs, 2005). If aphid remains on a single plant for a few minutes, it loses the ability to transmit.

Helper Strategy (Indirect)

In most non-persistent viruses, helper components plays very important role in virus transmission. Potyviruses and caulimoviruses encodes HC component for virus transmission by aphids. The HC is a non-structural protein, to bind to both to virion and to the cuticular lining of aphid mouthparts (Bridge hypothesis), thus retaining the virion within the food canal of the aphid stylets *e.g.*, Potato Acuba Mosaic virus is transmitted by the aphid vector by the involvement of PVY which act as helper virus.

Capsid Strategy (Direct Strategy)

In some aphid transmitted viruses under the genera Cucumovirus, Alfamovirus, and Carlavirus, purified viral particles are readily acquired and transmitted by the vector. This indicated that coat protein (capsid protein) of these viruses must be capable of direct attachment to vector receptor and thus leading to their transmission. *e.g.* Cucumber mosaic virus.

Semi-Persistent Transmission

These aphids require a long time to acquire virus. The probability of transmission increases with feeding periods 12-24 hours. Once acquired, it retained for days, weeks, or even months. Virus appears to accumulate in the anterior gut - possibly binding at specific protein site. Virus presents deep in plant tissue – phloem. Aphids, whiteflies and leafhoppers of various genera and families are recognized as vectors transmitting in semi-persistent manner *e.g.*, beet pseudo yellows virus. Non-persistent viruses are retained by their vectors for a few hours, whereas semipersistent viruses are retained for days, weeks, or even months. Viruses in these two categories are acquired from infected plants and inoculated within seconds or minutes to recipient plants. In addition, they do not require a latent period, *e.g.* time interval between acquisition and transmission, and do not replicate in the vector. Nonpersistent and semipersistent viruses are specifically associated with the epicuticle that lines the stylets (mouthparts) or the foreguts of their arthropod vectors, respectively, or the cuticle lining of the feeding apparatus of their nematode vectors. Since the cuticle, including the lining of the mouthparts and foregut, is shed during molting, acquired viruses are lost at each molt (Strange, 2003; Andret-Link and Fuchs, 2005). The nonpersistent and semipersistent viruses are collectively, referred to as noncirculative because they are not internalized by vectors. In other words, they do not enter the hemocoel (vector body cavity) or cross any vector cell membrane (Gray and Banerjee, 1999).

Persistent Transmission

It takes several hours of feeding for an insect to acquire the virus. The virus is taken up into the vector, circulates within the vector body and is released through the salivary glands before transmission can occur. There is a latent period during which transmission cannot occur (while the virus particles travel through the insect's body) (Strange, 2003; Andret-Link and Fuchs, 2005; Dickinson, 2005). When the latent period is completed, the insect can then transmit the virus for many weeks or the rest of its life without needing to obtain more viruses from an infected plant such a relationship is termed "persistent" or "circulative" (Strange, 2003). The best studied examples are of luteovirus transmission by aphids. In some examples of this type (*e.g.* some hoppers and thrips), the virus multiplies within the vector and this is termed "propagative" (Strange, 2003). Successful transmission of persistent viruses requires an internalization of the ingested viruses that are actively transported across several cell membranes. Thus, they are found in the hemocoel of vectors and retained by vectors after molting (Andret-Link and Fuchs, 2005). Ultimately, they must associate with the vector salivary system to be transmitted into a new host. Persistent viruses are referred to as circulative. Gray and Banerjee (1999) further divided the persistent or circulative group of viruses in to two types:

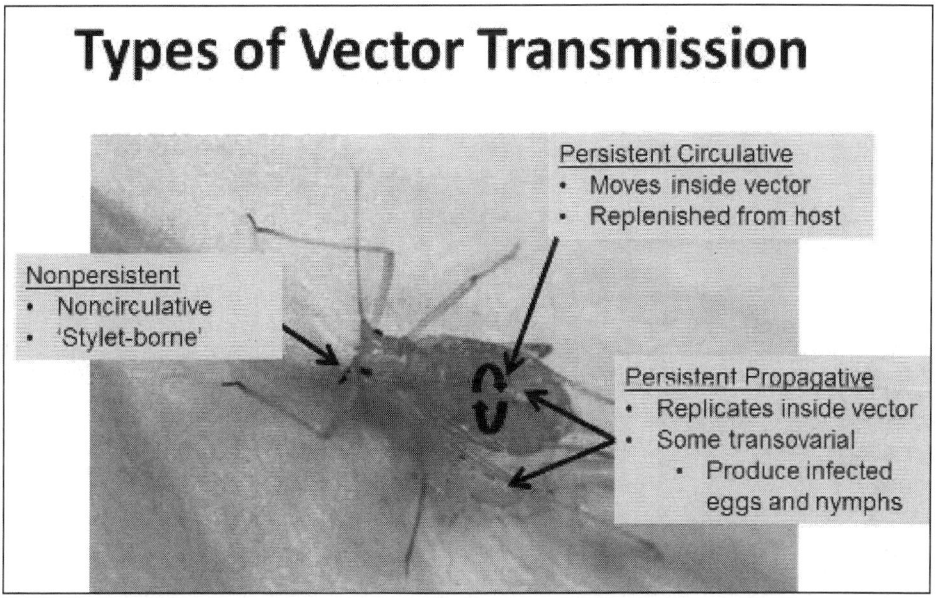

Figure 15.1. There are Multiple Types of Vector Transmission of Viruses. Tomato spotted wilt virus and impatiens necrotic spot virus are persistent propagative viruses. Pea aphid image courtesy of Joseph Berger, Bugwood.org.

Non-propagative

Virus circulates through the vectors body, but does not multiply *e.g.*, Potato leaf roll.

Propagative

Virus needs to multiply in the cells of the insect vector before transmission can occur. The transmission latent period is normally longer with these viruses *e.g.*, Tomato spotted wilt.

Many of the non- and semipersistently transmitted viruses *e.g.*, potyviruses and caulimoviruses, need one or more helper components for transmission. These helper components are proteins produced by the virus during plant infection and are required for attachment of virus particles to the inner cuticular lining of the vector's maxillary stylets, particularly near the tip where the food and salivary canals merge, via a "bridge structure" (Ng and Falk 2006). This attachment allows for temporary retention of virus particles inside the vector's food canal after which they are released into plants during feeding, probably through protease-mediated degradation of the attachment proteins.

Insect Vectors of Plant Viruses

Insects are the most important group of plant vectors both in terms of number of viruses transmitted and in the economic importance of the diseases concerned. Among arthropod vectors, approximately 99 per cent are insects. Insect vectors of plant viruses are found in seven orders of the class Insecta. The majority of vectors are found in the two orders of insects *i.e.*, Thysanoptera and Hemiptera. Some species of insects in orders: Orthoptera, Dermaptera, Coleoptera, Lepidoptera and Diptera, also act as vectors of diseases. Insect vectors that transmit most plant viruses are aphids, whiteflies, leafhoppers, thrips, beetles or weevil, mealy bugs, and mites. Sucking insects that feed on plants, vascular tissues (xylem and phloem) appear to be the most common vectors of plant viruses.

The order Hemiptera contains the greatest number of insect vectors of plant viruses. Hemipteran insects have distinct features that allow for efficient virus transmission. The predominant feature of these insects is that they have piercing-sucking mouthparts that include a needle-like stylet bundle consisting of two mandibular and two maxillary stylets (Cranston and Gullan 2003). The two maxillary stylets are interlocked and between them form two canals. The narrower salivary canal delivers saliva into the feeding puncture in plant tissues and the wider food canal takes up plant sap into the cibarium (the sucking pump), the esophagus, and the rest of the alimentary canal. Some plant-feeding hemipteran insects are specialized as phloem, xylem, or mesophyll feeders, whereas others can feed on a combination of these tissues (Chapman, 1998). Similarly, many plant viruses transmitted by hemipteran insects are phloem-limited, whereas others are not tissue specific and exploit almost all plant tissues. The plant-feeding hemipteran insects were formerly classified in the suborder Homoptera, which included aphids (Aphidoidea), whiteflies(Aleyrodoidea), jumping plant lice (Psylloidea), scale insects (Coccoidea) belonging to the Sternorrhyncha, and leafhoppers (Cicadelloidea), planthoppers (Fulgoroidea), froghoppers/spittlebugs (Cercopoidea), and cicadas (Cicadoidea) belonging to the Auchenorrhyncha (Gullan, 2001). The most common being aphids with more than 200 identified vector species (Ng and Perry, 2004). More than half of the nearly 550 vector transmitted viruses recorded so far are disseminated by aphids (55 per cent). Leafhoppers, beetles, whiteflies and thrips transmit about 11 per cent, 9 per cent, 7 per cent and 2 per cent viruses, respectively (Astier *et al.*, 2001).

Insect Vector–Virus Transmission

It consists of several successive steps:

1. Acquisition of virions from an infected source by vector.
2. Stable retention of acquired virions at specific sites through binding of virions to ligands.
3. Release of virions from the retention sites upon salivation or regurgitation.
4. Delivery of virions to a site of infection in a viable plant cell.
5. Each step of this sequence is needed for transmission to be successful.

Moreover, the relationship between viral accumulation and virulence may be obscured by potential heterogeneity of the hosts with respect to interactions with pathogens. Indeed, plants can be resistant, tolerant and/or susceptible to parasite infection. Resistance is the ability of the host to limit parasite burden, and tolerance is the ability to limit the harm caused by a given burden (Cooper and Jones 1983).

Plant Viral Diseases Transmission by Insects

After a successful transmission, infected plants may show a range of symptoms depending on the type of disease, but often there is leaf yellowing (either of the whole leaf or in a pattern of stripes or blotches), leaf distortion (*e.g.* curling) and/or other growth distortions (*e.g.* stunting of the whole plant, abnormalities in flower or fruit formation) (Strange, 2003; Dickinson, 2005; Pandey, 2012; Agrios, 2014).

Typical leaf symptoms of viral diseases include mosaic patterns, chlorotic or necrotic lesions, yellowing, stripes or streaks, vein clearing, vein banding, and leaf rolling and curling. Flower symptoms include deformation and changes in the color of the flowers including dramatic color mosaics called color breaking. Fruit and vegetable symptoms may include mosaic patterns, stunting, discoloration or malformation, and chlorotic ringspots. Stems of plants may develop stem pitting and grooving or tumors in response to virus infection. The symptoms induced by plant viruses lead to reduced crop quality and yield.

Major Plant Virus Vectors

Whitefly: *Bemisia tabaci* (Gennadius)

White fly born virus diseases of legumes are of considerable importance in all over the world and has large host range including cultivated crops, fruit trees, vegetables and ornamental plants. It occurs throughout the northern and western regions of India. It is a serious pest of mung bean, urad bean, arhar, cowpea, pea *etc.*

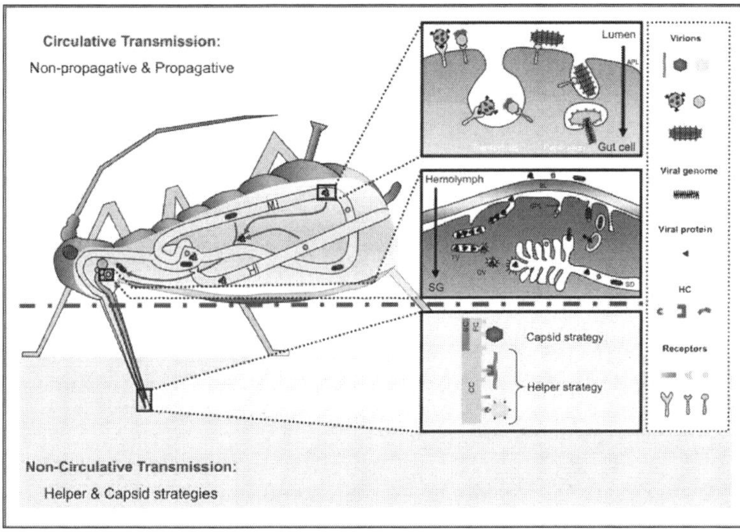

This figure is from the Review article: Aphids as Transport Devices for Plant Viruses from *Competes Rendus Biologie* (2010) Courtesy: Véronique Brault[a], Maryline Uzest[b], Baptiste Monsion[a], Emmanuel Jacquot[c], and Stéphane Blanc[b]

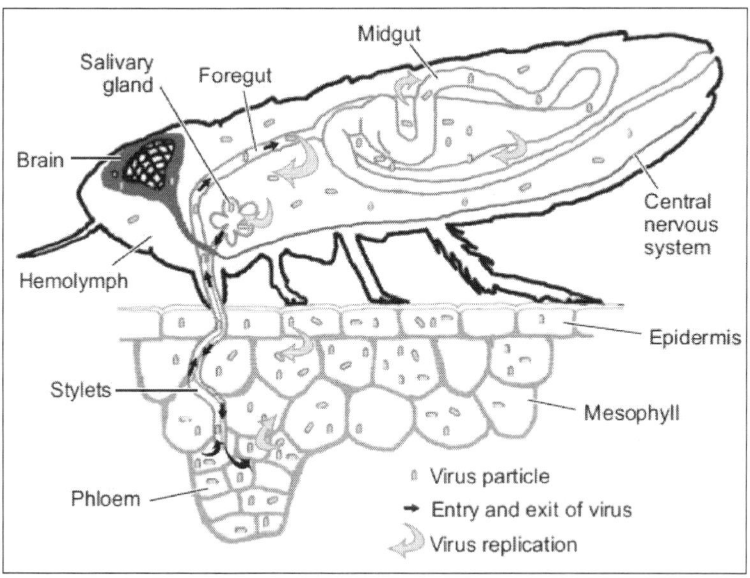

Figure 15.2. Schematic Representation of Persistent Virus Transmission by a Leafhopper; Viruses that are transmitted in a circulative persistent manner do not replicate in the insect and usually enter the salivary glands from the hemolymph. In contrast, most propagative viruses replicate in several plant tissues and in different organs of the insect vectors and may enter the salivary glands either from the hemolymph or from other connecting tissues. (Courtesy: Hogenhout *et al.,* 2008).

B. tabaci has been recorded feeding on at least 172 plant species of 34 families (Azab, *et al.,* 1970) and it transmits over 25 diseases throughout its pan – tropical range (Verma, 1963). White fly transmitted gemini viruses (WTGs, genus Begomovirus, family Geminiviridae) have emerged as serious threat to the cultivation of crop plants in various parts of the world (Varma and Malathi, 2003). The yellow disease in black gram (Gemini virus) is most serious disease (Nene, 1972).

Nature of Damage

It is a sucking pest and generally causes damage by sucking the sap from leaves and tender parts of the host plants. In case of food legumes, it damages the plants in three ways: (i) sucking of cell sap from leaves, thus lowering the vitality of the plants. Under high population pressure of the pest, the plant becomes very weak, giving a sickly look and may die. (ii) The insect excretes honey dew, on which growth of sooty mould takes place which result in blackening of leaves, resulting in decrease in photosynthesis. In severe attacks there is total blackening of the crop, drying of leaves and total failure of the crop. (iii) The whitefly is a vector for a number of viral diseases especially mung bean yellow mosaic virus (MYMV).

Mungbean yellow mosaic disease is transmitted by the vector, the whitefly (*Bemisia tabaci*). It is found to spread the begomoviruses, the major hazard to the flourishing production of mungbean in India, Sri Lanka, Pakistan, Bangladesh, Papu New Guinea, Philippines and Thailand (Honda *et al.,* 1983; Chenulu and Verma 1988; Varma *et al.,* 1992; Jones 2003; Haq *et al.,* 2011a). MYMV disease is transmitted principally by the polyphagous pest *Bemisia tabaci* in a persistent (circulative) manner and grafting but not by sap, seed or soil. The latent period of whitefly is less than four hours, so it is a tremendously efficient vector for virus transmission. Viruses are acquired from the phloem of plants during feeding. A single viruliferous adult is capable of transmitting the dreadful virus and it can transmit the virus within an acquisition and inoculation access period of 24 hours. Acquisition and inoculation by whitefly adults can be affected in a minimum of 15 minutes. The insect may attain the virus after a single bite and its transmission efficiency increases with time on the source plant of virus as well as on the healthy mungbean plant (Malathi and John, 2008a). The most efficient female and male adults in a population can retain infectivity for 10 and 3 days, respectively. Thus, the female adults are three times more efficient as vectors than males. Nymphs of *B. tabaci* can acquire the virus from diseased leaves and the virus does not transovarially transmitted (not pass through eggs of *B. tabaci*). Nevertheless, the cropping seasons highly influence the vector population. Whitefly is found to be high during summer season compared to spring and rainy seasons. They thrive best under hot and humid conditions and the population also towers with higher temperatures. Moreover, the spread of MYMV on a local (within and between fields) and a regional basis reflects its dispersal.

Its population is correlated with disease incidence. High populations appear on the crop that are 20-30 days old leading to higher disease incidence (on the 45th day). Spring and rainy seasons attribute to unfavourable conditions for the multiplication of the whitefly. Therefore, the disease incidence is not high during these seasons. Nath (1994) studied the effect of the weather parameters on the population of whitefly and the incidence of yellow mosaic virus on green gram. He reported a simple positive significant correlation between the disease incidence and the population of the fly, temperature, relative humidity, rainfall and the number of rainy days necessary for the infection. Honda *et al.* (1983) reported the mechanical transmission of the isolate of MYMV of Thailand and they obtained the highest transmission rates with 0.1 M potassium or sodium phosphate buffer of pH 7.8. Thereafter, none reported any infestation caused by mechanical transmission. Attempts were also made by Grewal (1988) to test the transmission of disease by sap inoculation by rubbing freshly extracted sap of mosaic affected leaves on the healthy young seedlings of mungbean. However, the disease could not be transmitted in this manner. The whitefly delivers the virus through proboscis to the phloem cells of the host plant where it gets multiplied. In leaf cells, the virus particles often form loose aggregates that sometimes fill the nuclei of infected phloem cells. In mungbean, hypertrophied nucleoli, aggregates of virus particles and fibrillar bodies appear in the nuclei of phloem cells as early as two days before the appearance of the symptom. Virus particles are often scattered in distribution but occasionally form aggregates having a paracrystalline or double cylindrical arrangement in the vacuoles of infected sieve elements (Thongmeearkom *et al.*, 1981). It causes yellow coloured spots scattered on young leaves followed by yellow mosaic pattern. Later, the spots gradually increase in size resulting in complete yellowing of leaves.

Biology

The insect breeds throughout the year and all the developmental stages are noticed. During winter season generally the adults are predominant but the many

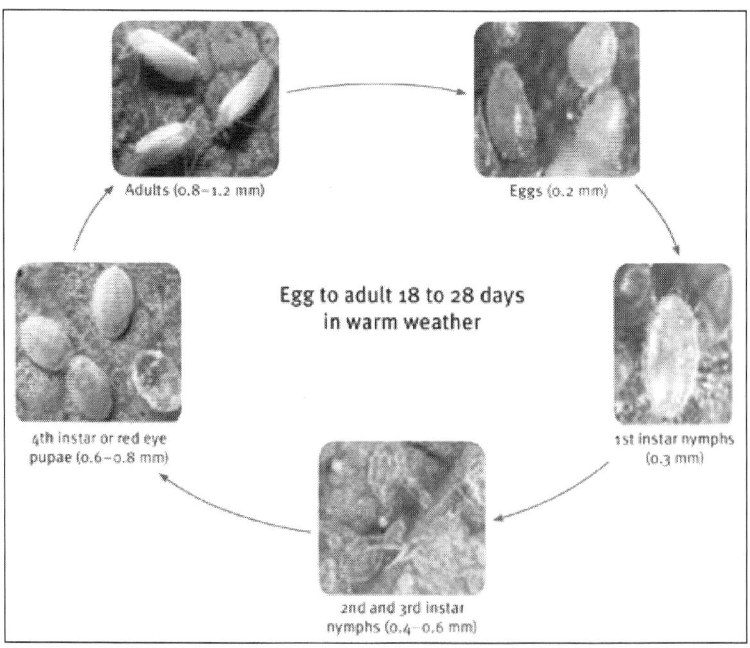

Figure 15.3. Life Cycle of White Fly (*Bemisia tabaci*)
www.daf.qld.gov.au.

workers have reported that female whiteflies are more efficient vectors than males. The female lays eggs singly on lower side of the leaves, on an average of 120 eggs per female. The eggs are light yellow first, and later turn brown. The eggs hatch in 3-5, 5-17 and 33 days in April-September, October- November and December-January, respectively. The nymphs soon after emergence fix their mouth parts into the plant tissues and feed on the cell sap. They grow into three stages to form the pupae within a period of 9-14 and 17-81 days in April –September and October-March, respectively. The pupal period generally lasts for 2-8 days. The total life cycle is completed in 14-122 days and there are 11-12 generations of the insect in a year. The effect of climate on biology and distribution of vector, (*Bemisia tabaci* Genn.) was earlier discussed by Singh and Gurha (1994) and Nath and Saikia (1995). It was noticed that the crop infected at early stages suffered more with severe symptoms with almost all the leaves exhibiting yellow mosaic and complete yellowing and puckering. Murugesan and Chelliah (1977) reported that higher temperatures prevailing during summer are favorable for whitefly vector (*Bemisia tabaci* Genn.) to develop and multiply.

Management for Whitefly

Host Plant Resistance

Many genotypes of mung bean and urd bean were screened for resistance to whitefly and jassids by Kooner (1998), Kooner and Cheema (2007a, b), where genotypes ML 508, ML 537, ML 1265, ML 1229, ML 1260, ML 1256 and ML 1191 of mungbean and were identified as resistant. Chhabra *et al.* (1981) while studying the mechanism of resistance in mungbean and blackgram against whitefly investigated that bio-chemicals like phenols, amino acids and nonreducing sugars were responsible for resistance to the insect. Kooner *et al.* (1994) studied the phytochemical basis of resistance in urd bean and concluded that high contents of phenols, free amino acids, total sugars and low content of non-reducing sugars governed the resistance.

Cultural Control

Ravindrababu (1987) reported that maize, sorghum and pearl millet when grown as barrier crops for mung bean, significantly reduced the population of *B. tabaci* and incidence of MYMV and resulted in higher yield. Intercropping rice bean with sorghum had a suppressing effect on the incidence of whitefly on rice bean (Satyanarayana *et al.*, 1997). Manipulation in date of sowing may help in minimizing the whitefly population and thus reducing the disease incidence but it may greatly differ from region to region. The incidence of MYMV is reduced in plots with inter-varietal intercropping of mungbean and urdbean (Das Gupta and Chowdhury, 1985).

Biological Control

Nene (1973) reported nuclear polyhedrosis virus (NPV) and a fungal pathogen, *Poecilomyces fairnosus*, as biocontrol agents of adult whitefly. Lal (1987) reported *Prospaltella flava* and *Geocaris tricolor* parasitising nymphal, pupal and immature stages. He further reported that *Brumus* sp. as a predator of immature stages of the pest.

Chemical Control

Chhabra *et al.* (1993) reported that whitefly can be controlled by spraying the mung bean, urd bean, soybean and rice bean with any of the following insecticides using 250 litres of water per hectare with manually operated sprayer: 950 ml of Malathion 50 EC (malathion), 625 ml of Rogor 30 EC (dimethoate) or Anthio 25 EC (formothion) or Metasystox 25 EC (oxydemeton -1 methyl), and 225 ml of Dimecron 85 WSC (phosphamidon). At the time of sowing, as basal application Disulfoton @ 1.5 kg a.i. per ha significantly reduced the population of whitefly (Chaudhary *et al.*, 1981). Combination of phorate 10 G and monocrotophos (Vadodaria and Vyas, 1987), endosulfan, monocrotophos, chlorpyriphos and dimethoate (Yein, 1983) were the most effective for the control of the pest. Cheema *et al.* (2009) reported that pest can be controlled by spraying the crop with 1.25 litre triazophos 40 EC or 100 g thiamethoxam 25 WG/ha by using 250 litre water. Shirale and Bidgire (2009) reported that spraying triazophos 40 EC @ 800 ml/ha twice, recorded less number of whiteflies at three, seven and ten days after spraying. The treatment *B. bassiana* @ 1 l/ha followed by Triazophos 40 EC @ 800 ml/ha was at par with treatment triazophos 40 EC @ 800 ml/ha followed by *B. bassiana* @ 1 l/ha at three days after spraying. Purwar and Yadav (2006) observed that Dimlin (diflubenzuron) 25 WP and triazophos 40 EC were effective to combat whitefly population and to manage yellow mosaic in soybean.

They are not well controlled by any of the available management practices such as cultural, mechanical, botanical and chemical. Among the various methods of whitefly management, chemical control is the primary method adopted. Systemic chemical insecticides *viz.*, acetamiprid, ethion, imidachlorpid, triazophos, provide better control of whiteflies; they also kill on contact, but are also taken inside the plant where they go onto protect against further attack for more than a few weeks (Wang *et al.*, 2009). Younger stages are generally more sensitive to insecticides as compared to older stages and higher toxic levels of certain insecticides work well against the first and second instar nymph stages. Nymphs of whitefly are more susceptible to insecticides; failure in control in crop fields may be due to the incomplete exposure on the underside of leaves (to the insecticides) where nymphs are present. Ghosh (2008) showed that imidacloprid reduced the whitefly populations to significant levels, if the insecticidal treatments are directed on the underside of the leaves, preventing the spread of MYMV and achieved more seed yield. Basal application of botanical insecticides controls the whitefly population in plants. Application of neem seed kernel extract (NSKE) and foliar spray of neem oil had a major impact by preventing the "nymphal" stage from developing into an adult; the nymphs tend to disappear from the treated plants (Dubey *et al.*, 2011).

Aphids

Aphids commonly known as green flies or black flies are the largest group of insect vectors. It is found all over the world and has large host range including cultivated crops, fruit trees, vegetables and ornamental plants. Among the legumes it damages beans, mungbean, urdbean, chickpea, clover, lentil, peas, cowpea, *etc.* (Hariri, 1981and Lal, 1992).

Nature of Damage

Aphids damage the plant by sucking the sap from tender parts, *i.e.*, top shoots, flowers, buds, developing pods and usually from the underside of the leaves. The mouth parts are modified for sap feeding in these sucking insects. The mandible and maxillae, which are jaw like structure in chewing insects, form two pairs of long bristle-like stylets. Due to its feeding, there is heavy drain of vital plant sap and the depletion of assimilates. As a result the rate of transpiration is increased (van Emden, 1973). The leaves, shoots and flowers deform and do not develop properly. The pods become shrivelled and deformed. The insect secretes honey dew, due to which black mould develops on infested plants, which interferes in photosynthesis. Under favourable conditions the damage caused is very high and there is substantial reduction in yield (Thakur *et al.*, 1984; Lal 1992; Ahmad and Lal, 1998).

Moreover, it has long been known that aphids are significantly attracted by yellow (Doring *et al.*, 2009) and that some plant viruses induce symptoms resulting in chlorosis, mosaics or other modifications that lead to yellowing of the leaves. Aphids have been shown to prefer to land and settle on virus-infected plants with yellowing leaves, possibly owing to the fact that insect vectors use visual cues to choose suitable plants (Fereres *et al.*, 1999; Hodge and Powell, 2008). Insect vectors also choose plants in response to olfactory cues, and the composition of the volatile organic compounds emitted by the leaves of some plants has been shown to change when the plant is infected by viruses (Ngumbi *et al.*, 2007; Medina-Ortega *et al.*, 2009). These changes can be more or less pronounced. For instance, plants infected by a circulative virus tend to attract their aphid vectors more than plants infected either by non-circulative viruses or by non-vectored viruses or healthy plants (Eigenbrode *et al.*, 2002). However, this pattern seems to depend on the host– virus interaction, since no such differences were observed on other plants and virus species (Castle *et al.*, 1998; Hodge and Powell, 2008). For several plant virus vectors reared on infected plants, studies have shown an increase in components of vector fitness such as survival or fecundity (Hunt and Nault, 1990; Blua *et al.*, 1994; Mayer *et al.*, 2002; Belliure *et al.*, 2005; Maris *et al.*, 2007 and Sisterson, 2008). Enhanced plant penetration by aphid stylets was observed on symptomatic virus-infected plants compared with healthy plants (Alvarez *et al.*, 2007).

Aphis craccivora is also a potent vector of some viral diseases. It transmits tobacco streak virus and cucumber mosaic virus in lentil. The insect prefers dry weather conditions, therefore, the infestation is high when winter rainfall fails or low (Lal, 1992). This pest is vector of pea leaf roll virus that causes chickpea stunt disease. This disease is very damaging in some areas of India, West Asia and Northern Africa. Besides, aphid is vector of alfalfa mosaic, bean common mosaic, urdbean leaf crinkle and bean yellow mosaic disease (Nene and Reddy, 1976).

Biology

Aphids are small (2-3 mm) and often inconspicuous, although their powers of reproduction are such that 0.4 hectares of plants may contain 2000 million individuals and roots may support a further 260 million (Dixon, 1973). The biology, host preferences and bionomics of *A. craccivora* have been studied extensively (Dorges *et al.*, 1966 and Bakhetia and Sidhu, 1977). Detailed biology of the pest was studied by Srikanth and Lakkundi (1988) on mungbean and urdbean at Bangalore, India. Fecundity and development vary to a great extent with host plant, soil fertility, soil moisture and temperature. Drought helps in mass multiplication and spread of aphids over large areas. The adults are small-sized insects, about 2 mm long, pear-shaped, green or greenish-black in colour. They have black bodies with prominent cauda and brown to yellow appendages. Males are rare. The common mode of reproduction is parthenogenic –producing young without fertilization and also viviparous, the young being born alive. Sexual reproduction with the production of eggs also occurs. The apterous (wingless) females are shining dark brown or black while the alate (winged) forms are greenish-black with transparent wings. The colour of newly laid nymphs is translucent except the reddish-brown compound eye. After a couple of minutes the transparency of body gradually disappears and changes to light yellow. After about half an hour the nymphs start moving and within an hour start feeding. There are generally four nymphal instars and the duration of each instar is 1-3 days. The apterous female starts laying the young ones within 24 hours of attaining that stage. A female can reproduce up to a maximum of 12 days. Breeding occurs almost throughout the year on one crop or the other and both alate and apterous forms are present (Chhabra *et al.*, 1986; Ahmad and Lal, 1998). A life cycle may take 6 to 8 days and an adult can lay 2 or 3 nymphs per day.

Management for Aphid

Host Plant Resistance

Some genotypes resistant to *A. craccivora* have been identified in mung bean, urd bean and cowpea but limited work seems to be undertaken in chickpea. Cultivars H 75-35 and 2184 of chickpea support minimum aphid population (Lal *et al.*, 1989). Mustaque (1977) tested performance of 12 chickpea cultivars and observed some of these as less susceptible. Chandra *et al.* (1992) reported that the variation in resistance to aphids infestation was correlated mainly with the colour of foliage of the genotypes; it has been observed that *A. craccivora* is quite attracted towards yellow colour.

Cultural Control

Aphids multiply and spread very fast under dry conditions. In endemic area, efforts should be made to keep the crop free from moisture stress. Wider spacing, *i.e.*, 60 x 20 cm is more favourable to aphids on chickpea than narrow spacing, *i.e.* 30 x 10 cm. The activity of aphid as a vector was slow down on paraffin treated plots which inhibited virus transmission (Kiss *et al.*, 2003). Proper weeding and destruction of weed hosts from the field area help to check the aphid population and diseases like leaf crinkle and mosaic mottle (Singh *et al.*, 1997).

Biological Control

Fifteen predators (*Coccinella septumpuncata* Linn, *C. transversalis* (Fabr), *C. nigritus, Menochiles sexmaculatus* (Fabr), *Brumus suturalis* (Fabr), *Harmonia arcuata, Vernia* sp., *Ischiodan javana* (Wied), *I. scutellaris* (Fabr), *Xanthogramma scutellare* (F.), *Chrysopa camea* Steph, *Cyclosa insulana* (Costa), *Luecangs celebesiana* (W), *Theridion* sp. and *Leucopis* sp. and three parasites (*Trioxys indicus, Praon* sp. and *Diaerecella rapaecurs*) have been recorded as natural enemies on *A. craccivora* (Ahmad and Lal, 1998; Yadava and Lal, 1998).

Chemical Control

A large number of insecticides including Acephate (0.02 per cent), bromophos (0.05 per cent), carbaryl (0.01 per cent), cypermethrin (0.004 per cent), oxydemeton methyl (0.05 per cent), dimethoate (0.03 per cent), endosulfan (0.07 per cent), fenvalerate (0.01 per cent), lindane (0.1 per cent), malathion (0.05 per cent), menazon (0.05 per cent), methomyl (0.03 per cent), phosdrin (0.02 per cent), primicarb (0.03 per cent), primiphos-methyl (0.03 per cent), thiometon (0.05 per cent) have been evaluated for the management of *A. craccivora* on pulse crops and most of them have been found quite effective to control this aphid (Singh and Sircar, 1980; Thakur *et al.*, 1984). Chhabra *et al.* (1993b) reported that the aphid can be controlled by spraying 625 ml of Malathion 50 EC (malathion) or 375 ml of Rogor 30 EC (dimethoate) or 375 ml of Metasystox 25 EC (oxydemeton methyl) per hectare.

Mites

Mites are closely related to the spiders and ticks. Eriophyid mites (Eriophyidae) and flat mites (Tenuipalpidae) are transmitters of plant viruses (Kitajima *et al.*, 2000, Paliwal 1980, Skare *et al.*, 2006 and Stenger *et al.*, 2005). They have needle-like stylet structures that can puncture cells at the surface inducing leaf deformations (Childers *et al.*, 2003 and Rancic *et al.*, 2006). The mite has assumed a pest status on a number of crops and fruit trees throughout India. The pest has been recorded on mungbean, urdbean, common bean, horse gram, cowpea, potato, cotton, sesame, citrus, hollyhock, marigold, tea, weeds and ornamental plants (Gupta and Dhooria, 1973; Dhooria, 1984; David, 1991 and Singh, 1996). About a dozen of eriophyid mites were reported to be vectors for important plant viruses and several other pathogenic agents of unknown etiology (Hiruki, 1992; Mararnorosch, 1994, Oldfield and Proeseler, 1996).The relationship between eriophyid vector and transmitted agent is highly specific. The plant pathogens transmitted by eriophyid mites are not known to be transmitted by other members of any other taxa or usually by more than one species of eriophyid (Oldfield, 1970).

Seth (1962) showed that under natural conditions SMD is transmitted by an eriophyid mite vector, *Aceria cajani*. Kumar *et al.* (1999 and 2000) established that SMD is caused by a virus referred to as pigeonpea sterility mosaic virus (PPSMV). This was further confirmed by Nene (1972) Nene and Reddy (1976a), Ramakrishnan and Kandaswamy, (1972) and Reddy *et al.* (1989). Report on nematode transmission of PPSMV by Arayanaswamy *et al.* (1963) was not confirmed (Mc Rae 1932, Nene, 1972, Ramakrishnan and Kandaswamy, 1972 and Mall *et al.*, 1977, Ghanekar, *et al.*, 1992).

PPSMV is transmitted by the eriophyid mite, *A. cajani*. This mite is highly host-specific and is largely confined to pigeonpea and its wild relatives, *C. scarabaeoides* and *C. cajanifolius*. Adult *A. cajani* measure 200 to 250 µm and have a very short life cycle of about 2 weeks comprising egg (30 × 40 µm) and two nymphal stages (Oldfield, 1981). Mites inhabit the lower surface of leaflets and are found predominantly on symptomatic leaves of PPSMV-infected plants (Kumar *et al.*, 2001, Muniyappa and Nangia 1982 and Reddy and Nene 1980). Their feeding causes no obvious damage to the host. Once established on PPSMV-susceptible genotypes, mites can multiply to high densities within a few weeks. Their dispersal is passive, assisted mainly by wind currents. A single *A. cajani* transmits PPSMV, but the maximum transmission achieved with single mites was about 50 per cent, which is high compared with the efficiency reported for the few other mite-transmitted viruses studied in detail (Gispert *et al.*, 1998 and Orlob, 1966). To transmit PPSMV, *A. cajani* required a minimum 15 minutes acquisition access period and 90-min inoculation access period, but these times were decreased to 10 and 60 minutes, respectively, when mites were starved prior to feeding. Viruliferous mites lose the ability to transmit PPSMV after feeding for 2 to 10 h on healthy plants, and there is no apparent latent period. As with other eriophyid mite– transmitted viruses, there is no evidence for transovarial transmission (Oldfield and Proeseler, 1996). *A. cajani* retains PPSMV for up to 6 h when feeding and for more than 13 h without access to a susceptible host, explaining the ability of *A. cajani* to transmit PPSMV after being carried in wind currents to new plants. Although *A. cajani* remain alive without feeding for up to 30 h in a moist chamber, they do not survive when transferred to plants, so that in nature it is unlikely that mites survive for many hours without feeding.

Eriophyid mites are amongst the smallest arthropods and are obligate plant pests in all active stages of the life cycle. Several of them cause direct damage by affecting plant growth and some indirectly by acting as vectors of plant viruses (Keifer *et al.*, 1982; Oldfield and Proeseler, 1996). *A. cajani* measures about 200-250 pm in length and can be seen under a stereomicroscope. These mites have short life cycle of less than 2 weeks that comprise an egg, two nymphal stages, and an adult (Janarthan, 1973 and Ramakrishnan and Kandaswamy, 1972; Oldfield *et al.*, 1981). *A. cajani* are distributed on all stages of leaves with their number more on young leaves (Reddy *et al.*, 1989) Their population density is more on SMD infected pigeonpea plants than on healthy. Even on infected plants, mite populations are concentrated towards the petiole end of young leaves and more than 90 per cent of the mites occur on the lower surface of the leaves (Dhar and Rathore, 1994). Several studies have shown that *A. cajani* populations on pigeonpea are uniform throughout the year in cooler parts of India, where as in semi-arid zones at higher temperatures mite populations decreased (Reddy and Raju, 1993; Lakshmikantha *et al.*, 1997). Dispersal and spread of mites in nature is passive and depend on wind currents, incidence and spread of the disease depends on vector population in the field (Reddy *et al.*, 1989). Efficiency of eriophyid mite vectors in transmitting viruses vary from species to species, and also depends on the host and type of virus it transmits. Efficency of *A. cajani*, of single mite was reported to range between 20 to 60 per cent in transmitting PPSMV (Reddy *et al.*, 1989). *A. cajani* required 5 min to acquire SMV (Reddy *et al.*, 1993). Nene and Reddy (1976) showed that *A. cajani* would not retain the PPSMV untill the death of the mite vector.

Nature of Damage

Both the adults and nymphs of plant mites feed by sucking the contents from the plant cell. In case of serious infestation, it causes downward curling of leaves and necrosis of young leaves and flowers. Initially the infested leaves appear pale to light brown. Severely infested leaves first turn light brown to brown and then to dark reddish brown. Bean mite has been recorded as a vector of viral disease mosaic motile in mungbean and urdbean (Beniwal and Chaudhary, 1979). Two families, the Eriophyidae and Tetranychidae transmit viruses. Eriophyidae mites are considered as the most important phalanx of vectors of plant viruses. These mites are about 0.2 mm long, four legged and elongated. Their main method of dispersion is by wind currents. They swarm on leaves, then leap into the wind to be carried away. The Tetranychidae mites are eight legged, up to 0.88 mm long, oval or pear shaped. Two species, *Tetranychus urticae* (Koch) and *T. telarius* (L.) have been reported to transmit PVY in the non persistent manner. At ICRISAT Kumar *et al.*, had reported that, in pigeonpea the transmission efficiency of single *A. cajani* was up to 50 per cent but it was 100 per cent when more than five mites per plant were used. *A. cajani* acquired PPSMV after a minimum acquisition access period (AAP) of 15 minutes and inoculated virus after a minimum inoculation access period (IAP) of 90 min. No latent period was observed.

Biology

Adult mites are hyaline in colour, oval in shape and have four pairs of legs. The females are larger than the males and are broader at abdominal end. The eggs are laid on the ventral surface of the new leaves in the depressions and are firmly stick to the plant substrate. The eggs are oval, elongate, covered with whitesh longitudinal rows of tubercles with flat base and are transparent shiny. Dhooria (1984) reported that Egg stage lasts for 1-3 days. Hatched out larvae have 3 pairs of legs and resemble the adults. The larvae are pear-shaped, milky-white with a dorsal white stripe in the middle region. Larval period ranges from 3 to11 days. Under optimal conditions a complete life cycle can be completed in about one week and 20-30 generations may actually be raised under tropical and subtropical regions. Various climatic factors such as temperature, relative humidity and precipitation affect field populations of mite. Mite population is low during winter due to a combination of low temperature and heavy rains (Gerson, 1992).

Management for Mite

Cultural Control

Proper weeding and destruction of weed hosts from the field area help to check the vector population and hence the mosaic mottle disease. The time of sowing may be adjusted to coincide with low vector population (Singh *et al.*, 1997).

Biological Control

About 11 natural enemies have been reported to predate on mite. *Amblyseius laryoensis, Typhlodromus annectens, T. porresi, Euseius victoriensis, T. rickeri* and *T. spulatus* are some of the major phytoseid mites which help to control mite in different parts of the world on different hosts.

Chemical Control

The mite can be controlled by spraying 375 ml of Rogor 30 EC (dimethoate) in 200-250 liters of water per hectare (Annonymous, 2010).

Thrips

Thrips are minute, slender-bodied insects (around 1/16 inch long, when fully grown) that have become an important agricultural pest because of their ability to transmit plant viruses. It is distributed in Japan, Korea and throughout India. The important host plants include mungbean, urdbean, pigeonpea, cowpea, beans, peas, sunhemp, groundnut, *Medicago sativa, Dolichos lablab, etc.* (Kooner *et al.*, 1983; Chakraborty and Dutia, 1999; Irulandi and Balasubramanian, 2000).

Nature of Damage

The mouthparts of thrips are composed of two maxillary stylets and one mandibular stylet that form a feeding apparatus (Cranston and Gullan, 2003). Thrips have been found to be vectors of at least four plant virus groups (families), bunyaviruses, ilarviruses, sobemoviruses, and caroviruses (Ullman *et al.*, 1997). Previously, eight species of thrips were reported to transmit TSWV (Wijkamp *et al.*, 1995). *Thrips tabaci* Lindeman, *T. setosus* Moulton, *T. palmi* Karny, *Frankliniella schultzei* Trybom, *F. occidentalis* (Pergande), *F. fusca* (Hinds) and *F. intonsa* Trybom were reported to be vector of TSWV (Wijkamp *et al.*, 1995, Ullman *et al.*, 1997). Webb *et al.* (1997) also reported *F. bispinosa* (Morgan) as a vector of TSWV. *Frankliniella tenuicornis* (Uzel) and *Scirtothrips dorsalis* Hood had been previously reported to be vector of TSWV, but experimental verification had not been done for all species as has been done for *F. occidentalis* and *T. tabaci* (Ullman *et al.*, 1997). Clearly, *F. occidentalis* and *T. tabaci* are common and important vectors of multiple plant viruses in many regions of the world (Ullman *et al.*, 1997). Thrips are generally transmit the viruses in propogative manner *i.e.*, virus multiplies in the vectors and posses through molting (Reddy and Wightman, 1988) Once vector acquires the virus no need to fresh feeding and the vector remain viruliferous throughour their life cycle. However, only thrips larvae are able to acquire the virus and then they transmit virus during larval stage and also as adult.

Thrips can attack green beans from the moment the seedlings emerge from the ground through to flowering and that has the greatest impact on damage. The feeding apparatus of thrips is unique amongst insects. Thrips have only one mandible, which they use to puncture the plant surface before inserting their stylets. Using their stylets, thrips suck the sap from the damaged cells, inducing a range of symptoms. Silvering is the most common symptom. It is readily seen on leaf tissue and occurs as a result of the cell contents being removed. Although of concern, leaf silvering does not cause significant yield reduction. Significant reduction in saleable yield does occur when thrips damage green bean pods. Scarring of the pods causes twisting and curling as well as mild rough patches on the skin. Kooner *et al.* (1983) reported that the thrips are found in flowers and due to their damage, the flowers shed before opening. Infested inflorescence becomes abnormal, showing symptoms of flower drop. Even if any such flower opens and pod is formed, that is abnormal; the grains in the matured pods are shrivelled and small in size. Morphology of such plants is entirely different from that of the healthy ones. Regeneration of vegetative growth of damaged plants occurs on axillary buds, the new leaves remain diminutive and consequently the plants attain a bushy growth. The crop looks dark green in colour. In case of severe incidence there is 100 per cent loss in grain yield. Chhabra and Kooner (1985a) estimated yield losses up to 54.3 per cent and deterioration in grain quality by this pest. Thrips are known vectors of tospoviruses such as tomato spotted wilt virus (TSWV) and groundnut bud necrosis virus (GBNV) .

Biology

Thrips belong to the insect order Thysanoptera (fringe-winged insects, their wings consisting of fine hairs rather than a membrane) Thrips are very small, dark brown insects. The body of the insect is quite shining. Adult thrips are small, cylindrical or cigar-shaped insects up to 2 mm long and range in colour from pale yellow to black. They have two pairs of narrow wings that are fringed with long hairs and rest along the length of their back. The young nymphs are similar in shape to the adults but smaller and wingless, and usually pale yellow to almost white. The life cycle of thrips includes an egg, two nymphal stages that actively feed, followed by two non-feeding stages, a pre-pupal and pupal stage, which generally takes place in the soil or leaf litter. The emerging adults are generally winged, but depending on the sex and species, some have short wings and others are wingless. Most thrips insert their eggs into plant tissue. The length of the life cycle depends on the temperature and the quality of the food source. In warm conditions, around 30°C, the life cycle can be completed in as little as 10–12 days. At 20°C the life cycle takes around 19 days (Persley and Sharman, 2007).

Management for Thrips

Host Plant Resistance

Malik (1990) observed that summer mung bean genotypes SML 77, UPM 82-4 and Pusa 107 were resistant to *M. distalis* under natural as well as screen house conditions. A number of mung bean and urd bean genotypes have been found resistant to the thrips (Chhabra and Kooner 1988, 1992, 1993). Urdbean varieties, N 212 and Khargone 3 and mungbean selections, P 15, P 56, P 101, P 135-2, P 382, P 386, P 435-1, P 435-2 and P 435-3 were found free from infection under natural conditions. Shanmugasundram (1988) reported variety Pant mung 3 to be moderately resistant to leaf curl virus. It was further investigated that low content of amino acids, total phenols, total minerals, total sugars, non-reducing sugars, calcium, potassium and high content of total carbohydrates were responsible in contributing resistance in SML 99 and SML 100 (Chhabra *et al.*, 1994).

Cultural Control

Intercropping of mungbean with cotton minimizes incidence of thrips as compared to sole crop.

Biological Control

Satyanarayan *et al.* (1995) reported *Brumoides suturalis, Loccinella septumpunctata, Menochilus sexmaculatus* (Coccinellids), *Chrysopa scelestes, Polistes hebraeus, Geocoris tricolor, Acantholepis* sp. (spiders) to be the primary natural enemies of *M. distalis.*

Chemical control

The bean thrips can be effectively controlled by spraying the crop at bud initiation stage with 250 ml of Rogor 30 EC (dimethoate) or Malathion 50 EC (malathion) or 300 ml of Metasystox 25 EC (oxydemeton methyl) or 90 ml of Dimecron 85 WSC (phosphamidon) in 200- 250 litres of water per hectare (Kooner *et al.,* 1983; Chhabra and Kooner, 1985b and 1986). Damage caused by thrips gets reduced considerably with the application of a combination of aldicarb at 1 kg a.i. per ha and monocrotophos 0.04 per cent (Chander and Singh, 1989). Gupta and Singh (1992) reported that one application of granular formulations of phorate, mephasulfan, disulfoton, carbofuran and aldicarb at 2.0 kg a.i. per ha were effective in controlling this pest. Irulandi and Balasubramanian (2000) reported neem seed kernal extract at 5 per cent and neem oil at 2 per cent to be as effective as monocrotophos 0.05 per cent.

Conclusions

Various methods are used for the management of the vectors. Some concern the elimination of the vector, but many others, the prevention of infection by changing cultural practices or areas of crop production. Methods have to be applied successively or at the same time and even then it may be difficult to completely stop the spread of virus. Great progress has been made in insect control by the development of several potent insecticides. Although they provide good control in many cases but good control in all the vectors and all the crops is many times not possible. In such cases, a number of methods of vector management are required which must be applied in combination with one another, so that the maximum control of the vector and the disease is achieved through a proper strategy.

Therefore, disease management has to focus on indirect measures such as reduction of insect vector populations through cultural or chemical methods, through natural enemies and induction of systemic resistance in host plants by exogenous application of biotic/abiotic inducers. Epidemics of existing plant virus diseases and emergence of novel virus diseases have become a serious threat to subsistence and commercial agriculture. Rapidly expanding global climatic change also creates favourable conditions for development and increased spread of plant virus diseases due to direct or indirect impacts on population dynamics of virus-transmitting insect vectors (Pautasso *et al.,* 2012). The important measures to combat the vectors associated problems should be followed as under:

1. Minimize the inoculums (virus source) and colonizing vectors by using cultural practices.
2. Follow the integrated pest management strategies for the management of insect vectors.
3. Tactics for integration are resistant plants, trap crops, barrier crops, sticky traps and insecticide sprays.

Manipulation of plant populations especially at relatively high plant densities also showed to affect the dynamics of virus vectors that transmit viruses to crop plants (Ogenga-Latigo *et al.*, 1992a, b). Similarly, according to Mukankusi *et al.* (1999) on groundnut rosette disease and Orawu *et al.* (2001) on cowpea viral diseases, also reported low virus incidences under high plant populations, suggesting that unfavourable microclimate disfavours virus vectors to colonize plants.

REFERENCES

Acharya, S.S. 1985. Prices and Price Policy for pulses and cereals. Sukhadia University, Udaipur XVIII+439+A 135 pp.

Agrios GN (2014). "Plant pathology," 5th Ed. pgs. 751 – 752, and Sanat printers Academic press: Elsevier, India.

Ahmad, R. and Lal, S.S. (1998). Insect pests of len per cent l and their management. In IPM System in Agriculture. Vol. IV. Pulses (Eds R.K. Upadhyay, K.G. Mukerji and R.L. Rajak), pp. 413- 426. New Delhi : Aditya Books.

Alvarez A. E., Garzo E., Verbeek M., Volman B., Dicke M., Tjallingii W. F. 2007Infection of potato plants with *Potato leafroll virus* changes attraction and feeding behaviour of *Myzus persicae*. *Entomol. Exp. Appl.* 125, 135–144.

Amin, P. (1985). Apparent resistance of groundnut cultivar Robut 33-1 to bud necrosis disease. *Plant Disease* 69, 718-719.

Amin, P. W., D.V.R. Reddy, and A. M. Ghanekar. 1981. Transmission of Tomato spotted wilt virus, the causal agent of bud necrosis of peanut, by Scirtothrips *dorsalis* and *Frankliniella schultzei*. Plant Disease 65: 663– 665.

Amin, P.W., 1985. Apparant resistance of groundnut cultivar Robut 33-1 to bud necrosis disease. *Plant Disease*, **69**: 718-719.

Ammar E-D, Jarlfors U, Pirone TP. 1994. Association of potyvirus helper component protein with virions and the cuticle lining the maxillary food canal and foregut of an aphid vector. *Phytopathology* 84: 1054–60.

Amos, J., Hatton, R.G., Knight, R.C. and Massee, A.M., 1927, Experiments in rhe transmisson of reversion of black currants. Ann. *Rep.* East Malling Res Sta., Kent, 1925, 11 Suppl., pp. 126-150.

Anderson, C.W. 1955. *Vigna* and *Crotalaria viruses* in Florida II. Notations concerning cowpea mosaic *virus* (*Marmor vignae*). *Plant Dis. Reptr* **39**: 349-353.

Andret-Link P, Fuchs M (2005).Transmission Specificity of Plant Viruses by Vectors. J. Plant Pathol. (2005), 87 (3), 153-165 Edizioni ETS Pisa, 2005 153.

Anno-Nyako, F.O. 1984. Identification and partial characterization of a mild mottle disease in soybean (*Glycine max* (L) Merril) in Nigeria. Ph.D. Thesis, University of Science and Technology, Kumasi, Ghana.

Anno-Nyako, F.O. 1984. Identification and partial characterization of a mild mottle disease in soybean (*Glycine max* (L) Merril) in Nigeria. Ph.D. Thesis, University of Science and Technology, Kumasi, Ghana.

Annonymous (2010) Package of practices for Kharif crops, pp. 58- 63, Punjab Agricultural University, Ludhiana.

Astier S, Albouy J, Maury Y and Lecoq H. 2001. Principes de Virologie Végétale. INRA, Paris, France.

Aykroyd, W. R., and J. Doughty. 1964. Legumes in Human Nutrition. FAO Nutritional Studies No. 19. Food and Agriculture Organization of the United Nations, Rome.

Azab, A.K., M.M. Megahed, and H.D. EI-Mirsawi. 1970. On the range of host-plants of *Bemisia tabaci* (Genn.). *Bulletin of the Entomological Society of Egypt* 54: 319· 326.

Bakhetia, D.R.C. and Sidhu, A.S. (1977). Biology and seasonal ac per cent vity of the groundnut aphid, Aphis craccivora Koch. *Journal of Research Punjab Agricultural University* 14: 299-303.

Belliure B., Janssen A., Maris P. C., Peters D., Sabelis M. W. 2005Herbivore arthropods benefit from vectoring plant viruses. *Ecol.Lett.* 8,70–79.

Beniwal SPS, Chaubey SN, 1979. Urdbean leaf crinkle virus: Effect on yield contributing factors, total yield and seed characters of urdbean (*Vigna mungo*). *Seed Res.* 7: 125- 181.

Beniwal, S.P.S. 1983. Urdbean leaf crinkle virus. pp. 85-86. *Viruses of legumes* 1983. Eds. K.F. Boswell and A.J. Gibbs. Descriptions and keys from VIDE. Austral. Natl Univ., Canberra.

Blua, M. J., Perring, T. M. and Madore, M. A. 1994 Plant virus-induced changes in aphid population development and temporal fluctuations in plant nutrients. *J. Chem. Ecol.* 20, 691–707.

Bock, K.R. 1971. Notes on East African plant virus diseases. I. Cowpea mosaic virus. *E.A. Agric. Forestry* 1. **37**: 60-62.

Bock, K.R. 1973. East African strains of cowpea aphid-borne mosaic virus. *Ann. appl. Biol.* **74**: 75-83.

BOS, L. and Want, J.P.H. van der. 1963. Early browning of pea, a disease caused by a soil and seed-borne virus. *Tijdschr. Pl.ziekten* **68**: 368-390.

Brunt, A.A. and Kenten, R.H. 1973. Cowpea mild mottle, a newly recognized virus infecting cowpeas (*Vigna unguiculata*) in Ghana. *Ann. appl. Biol.* **74**: 67-74.

Castle, S. J., Mowry, T. M. and Berger, P. H. 1998 Differential settling by Myzus persicae (Homoptera: Aphididae) on various virus infected host plants. *Ann. Entomol. Soc. Am.* 91, 661–667.

Chander, S and Singh, Y. (1989). Control of thrips, Megalurothrips distalis and pod borer *viz.* Euchrysops cnejus Fab. and Maruca testulalis Gayer infes per cent ng greengram, *Vigna radiata* (L.) Wilczek. *Indian Journal of Entomology* 51: 131-135.

Chandra, S., Ahmad, R. and Asthana, A.N. (1992). Field tolerance to aphid in len per cent l. LENS Newsletter 19: 45.

Chant, S.R. (1959). Viruses of cowpea, *Vigna unguiculata* (L.) Walp. in Nigeria. *Ann. appl. Biol.* **47**: 565-572.

Chapman RF. 1998. *The Insects: Structure and Function*. Cambridge, UK: Cambridge Univ. Press.

Chaudhary, J.P., Singh, A.K. and Yadav, L.S. (1981). Effect of disulfoton granules on pest complex, crop growth and yield of green-gram. *Indian Journal of Entomology* 43: 369-372.

Cheema, H.K., Taggar G.K., Singh R. and Kooner B.S. (2009). Evalua per cent on of insec per cent des against B. tabaci (Gennadius)on urdbean[*Vigna mungo* (L.) Hepper]. *J.Insect Sci* 22(4): 388-392

Chenulu VV, Verma A. 1988. Virus and virus-like diseases of pulse crops commonly grown in India. In: Baldev, B, Ramajunam, S, Jain, HK, editors. Pulse crops. New Delhi: Oxford and IBH; p. 338- 370.

Chhabra, K.S. and Kooner, B.S. (1985). Losses in summer mungbean Punjab. *Indian Journal of Entomology* 47: 103-105.

Chhabra, K.S. and Kooner, B.S. (1986). Comparative efficacy of different insecticides in controlling thrips, Megalurothrips distalis (Karny) on summer mungbean, Vigna radiata (L.) Wilczek. Legume Research 9: 39-42.

Chhabra, K.S. and Kooner, B.S. (1988). Varietal resistance in mungbean (*Vigna radiata* (L.) Wilczek) to thrips, *Megalurothrips disatalis*(Karny). The Tropical Vegetable Informaon Service News 3: 10-11.

Chhabra, K.S. and Kooner, B.S. (1992). Sources of resistance in mungbean against major insectpests and yellow mosaic virus. *Legume Research* 14: 175-184.

Chhabra, K.S. and Kooner, B.S. (1993). Screening of mungbean germplasm against insect pests and yellow mosaic virus. *Indian Journal of Pulses Research* 6: 69-75.

Chhabra, K.S., Kooner, B.S. and Cheema, H.S. (1986). Incidence of *Aphis craccivora* Koch. on grain legumes in the Punjab. In Aphidology in India, Proceedings of the Naonal Symposium(Ed B.K. Agarwala.), pp. 23-28. Agartala : ASI.

Chhabra, K.S., Kooner, B.S., Saxena, A.K. and Sharma, A.K. (1981). Effect of biochemical components on the incidence of insect pest complex and yellow mosaic virus in mungbean. *Crop Improvement* 8: 56-59.

Chhabra, K.S., Kooner, B.S., Saxena, A.K. and Sharma, A.K. (1994). Role of allomones in varietal resistance in summer mungbean against thrips, Megalurothrips distalis (Karny). *In Proceeding of Internaonal Symposium on Pulses Research*, p. 161. New Delhi.

Chhabra, K.S., Lal, S., Kooner, B.S. and Verma, M.M. (1993). Insect Pests of Pulses – Idenficaon and Control Manual. Ludhiana : Punjab Agricultural University, and Kanpur : Directorate of Pulses Research.

Childers CC, French JV, Rodrigues JC. 2003. *Brevipalpus californicus, B. obovatus, B. phoenicis,* and *B. lewisi* (Acari: Tenuipalpidae): a review of their biology, feeding injury and economic importance. *Exp. Appl. Acarol.* 30: 5–28.

Cockbain, A.J., Woods, R.D. and Calilung, V.C.J. 1983. Necrosis in field beans (*Vicia faba*) induced by interactions between bean leaf roll, pea early-browning and pea. enation mosaic viruses. *Ann. appl. Biol.* **102**: 495-499.

Cooper, J. I. and Jones, A. T. 1983 Responses of plants to viruses: proposals for the use of terms. *Phytopathology* 73, 127–128.

Costa, A.S. and Carvalho, A.M.B. 1961. Studies on Brazilian tobacco streak. *Phytopath. Z.* **42**: 113-138.

Cranston PS, Gullan PJ. 2003. Phylogeny of insects. In *Encyclopedia of Insects,* ed. VH Resh, RT Carde. Amsterdam: Academic.

Culbreath, A.K., J.W. Todd, and J.W. Demski. 1992. Comparison of hidden and apparent spotted wilt epidemics in peanut. *Proc. Am Peanut Res. Ed. Soc.* 24: 39 (Abstract.).

Culbreath, A.K.; Branch, W.D.; Brown, S.L.; Demski, J.W.; Beasley Junior, J.P. Effect of peanut cultivar Georgia Browne on epidemics of spotted wilt. *Plant Disease,* v.78, p.1185-1189, 1994.

Das gupta,B. and Chowdhury, A.K. (1985). Use of intervarietal intercropping to minimize the yellow mosaic virus of urd and mungbeans. *Indian Jour Plant Pathol.* 3 : 100-101.

David, P.M. (1991). Resurgence of yellow mite, *Polyphagotarsonemus latus* (Acari : Tarsonemidae) on chilli following applica per cent on of insec per cent cides. *Madras Agricultural Journal* 7: , 20-25.

Dhar, V. and Rathore, Y.S. 1994, Pattern of population distribution of Acerla cajani on pigeonpea plant. *Indian J. Pulse Res.,* 7: 137-143.

Dhar,V., Chaudhary, R.G. and Nairnuddin. 1998, Population studies on sterility mosaic vector (Aceria cajani). In, Annual report 1996-97. Indian Insititute of Pulses Research, Kanpur, pp 28.

Dhooria, M.S. (1984). Preliminary observa per cent on on the biology and host range of the tarsonemid mite, *Polyphagotarsonemus latus* (Acari : Tarsonemidae). Acaricide Newsle8er 4: 7-9.

Dickinson M (2005). Molecular Plant Pathology. Advanced Text, Edition Published by BIOS Scientific Publishers, Member of the Taylor and Francis Group.

Dijkstra, J., Bos, L., Bouwmeester, H.J., Hadiastono, Tutung and Lohuis, H. 1987. Identification of blackeye cowpea mosaic virus from germplasm of yard-long bean and from soybean, and the relationships between blackeye cowpea mosaic virus and cowpea aphid-borne mosaic virus. *Neth. J. Plant Path.* **93**: 115-133

Dixon, A.F.G. 1973. Biology of aphids. Institute of Biology's Studies *In* Biology No. 44. E. Arnold Limited, London.

Dorges, S.K., Dalaya, V.P. and Kabi, T.D. (1966). Studies on bionomics and control of groundnut aphid, Aphis craccivora Koch. *Labdev Journal of Science and Technology* 4 : 165-167.

Doring, T. F., Archetti, M. and Hardie, J. 2009 Autumn leaves seen through herbivore eyes. Proc. Biol. Sci. 276, 121– 127.

Dubey NK, Shukla R, Kumar A, Singh P, Prakash B. 2011. Global scenario on the application of natural products in integrated pest management programmes. In: Dubey NK, editor. Natural products in plant pest management, Vol. 1. Wallingford: CAB International; p. 1 - 20.

Duffy, S., Shackelton, L. A. and Holmes, E. C. 2008 Rates of evolutionary change in viruses: patterns and determinants. Nat. Rev. Genet. 9, 267–276.

Eigenbrode, S. D., Ding, H., Shiel, P. and Berger, P. H. 2002 Volatiles from potato plants infected with potato leafroll virus attract and arrest the virus vector, Myzus persicae (Homoptera: Aphididae). *Proc. R. Soc.* B 269, 455–460.

FAOSTAAT. 2011. Statistical Data Base of Food and Agriculture Organisation.

Fauquet, C., Lamy, D. and Thouvenel, J.C. 1979. Viral diseases of winged bean in the Ivory Coast. *FAO Plant Prot. Bull.* **27**: 81-87.

Feakin, S. D., 19 7 3 Pest control in ground nuts. Pest Articles and news summaries manual no 2. Centre for Overseas Pest Research, London. 197 pp.

Fereres, A., Kampmeier, G. E. and Irwin, M. E. 1999 Aphid attraction and preference for soybean and pepper plants infected with Potyviridae. *Ann. Entomol. Soc. Am.* 92, 542–548.

Fiedorow, Z.G. 1980. Some virus diseases of horse bean. Tag.-Ber. Akad. Landw. Wissensch. (DDR, Berlin) 184: 361-366.

Fiedorow, Z.G. 1980. Some virus diseases of horse bean. *Tag.-Ber. Akad. Landw. Wissensch.* (DDR, Berlin) **184**: 361-366.

Fiedorow, Z.G. 1983. Pea early-browning virus on horse bean (*Vicia faba* L. *ssp. minor*). *Zesz. Probl. Pst. Nauk Roln.* **291**: 97 110.

Ford R, Evans T (2003). Tobacco mosaic virus. The Plant Health Instructor. DOI: 10.1094/PHI-K-2003-0528-01.http: //www.apsnet.org.

Frencel, I. and Pospieszny, H. 1979. Viruses in natural infections of yellow lupin (*Lupinus luteus* L.) in Poland. IV. Bean common mosaic virus (BCMV). *Phytopathol. Acad. Sci. Hung.* **14**: 279-284.

Froissart R., Michalakis Y., Blanc S., 2002. Helper component- transcomplementation in the vector transmission of plant viruses. *Phytopathology* **92**: 576-579.

Froissart, R., Roze, D., Uzest, M., Galibert, L., Blanc, S. and Michalakis, Y. 2005 Recombination every day: abundant recombination in a virus during a single multi-cellular host infection. *PLoS Biol.* 3, e89.

Gerhardson, B. and Ryden, K. 1979. An isolate of pea early browning virus from field grown *Phaseolus vulgaris*. *Phytopath. Z.* **95**: 93-96.

Gerson, U. (1992). Biology and control of the broad mite, *Polyphagotarsonemus latus* (Banks) (Acari : Tarsonemidae). Experimental and Applied Acaricide 13: 163-178.

Ghanekar. A. M., Sheila, V. K., Beniwal, S. P. S., Reddy, M. V. and Nene, Y. L. 1992. Steril~ty mosaic of pigeonpea. Pp 415-428. In plant Diseases of International Importance: Diseases of Cereals and Pulses vol I (Singh, U S., Mukhopadhyay, A N., Kumar, J. and Chaubey, H S) New Jersey, USA. Prentice Hall.

Ghosh A. 2008. Management of yellow mosaic virus by chemical control of its vector, Whitefly (*Bemisia tabaci*) and its impact on performance of green gram (*Phaseolus aureus*) under rainfed lowland rice fallow. *Arch Phytopathol Plant Prot.* 41(1): 75-78.

Gibbs, A. J. and Harrison, B. (197 6) Plant Virology, the principles" John Wiley and sons, New York.

Gibbs, A.J. and Harrison, B.D. 1964. A form of pea early-browning virus found in Britain *Ann. appl. Biol.* **54**: 1-11.

Gibbs, A.J. and Paul, H.L. 1970. Echtes Ackerbohnemosaik-Virus. CMI/AAB Descriptions of Plant Viruses No. 20. Commonwealth Agricultural Bureaux, Slough.

Gispert, C., Oldfield, G. N., Perring, T. M., and Creamer, R. 1998. Biology of the transmission of peach mosaic virus by *Eriophyes insidiosus* (Acari: Eriophyidae). Plant Dis. 82: 1371-1374.

Gray, S. M. and Banerjee, N. 1999 Mechanisms of arthropod transmission of plant and animal viruses. *Microbiol. Mol. Biol. Rev.* 63, 128–148.

Greber, R.S. 1971. Some characteristics of tobacco streak virus isolated from Queensland. *Qld J. Agric. Anim. Sci.* **28**: 105-114.

Green, S.K. 1985. Seed transmission of a new mungbean virus and screening for resistance. *Progress Report AVRDC* 1985: 168-171.

Grewal JS. 1988. Diseases of pulse crops _ an overview. *Indian Phytopathol.* 41(1): 1-14.

Gullan PJ. 2001. Why the taxon Homoptera does not exist. *Entomologica* 33: 101–4

Gupta, P.K. and Singh, J. (1992). Effec per cent ve control of thrips by soil applica per cent on of granular insec per cent des in greengram. *Indian Journal of Entomology* 54: 465-470.

Gupta, S.K. and Dhooria, M.S. (1973). Occurrence of Hemitatorsonemus latus (Banks) on blackgram and its control in Punjab. *Science and Culture* 37: 484-485.

Haq, QMI. Arif A, Malathi VG. 2011a. Infectivity analysis of a blackgram isolate of mungbean yellow mosaic virus and genetic assortment with MYMIV in selective hosts. *Virus Genes.* 42(3): 429-439.

Hariri, G. (1981). Insect and other pests. In Lenls (Eds G. Webs and G. Haw per cent n), pp. 173-189. London : Commonwealth Agricultural Bureau.

Harrison, B.D. 1973. Pea early-browning virus. CMI/AAB Descriptions of Plant Viruses No. 120. Commonwealth Agricultural Bureaux, Slough.

Hiruki, C. 1992, Plant diseases associated with mites as vector of known viruses and unknown etiological agents. In Plant Diseases of Viral, Viroid, Mycoplasma and uncertain etiology. (Ed. Maramorosch). Westview Publication, 127pp.

Hodge, S. and Powell, G. 2008 Do plant viruses facilitate their aphid vectors by inducing symptoms that alter behavior and performance? Environ. Entomol. 37, 1573–1581.

Honda Y, Iwaki M, Saito Y. 1983. Mechanical transmission, purification and some properties of whitefly-borne mungbean yellow mosaic virus in Thailand. Plant Dis. 67: 801_804.

Hull R., 2002. Matthew's Plant Virology. Academic Press, New York, NY, USA.

Hull, R. 2001. Matthews' plant virology (4th ed.). Academic Press, San Diego, California.

Hunt, R. E. and Nault, L. R. 1990 Influence of life history of grasses and Maize chlorotic dwarf virus on the biotic potential of the leafhopper Graminella nigrifrons (Homoptera: Cicadellidae). *Environ. Entomol.* 19, 76–84.

ICTV (2012) Virus Taxonomy: Classification and Nomenclature of Viruses: Ninth Report of the International Committee on Taxonomy of Viruses. Elsevier Academic Press, San Diego, CA.

Iizuka, N., Rajeshwari, R., Reddy, D.V.R., Goto, F., Munyappa, V., Bharathan, N. and Ghanekar, A.M. 1984. Natural occurrence of a strain of cowpea mild mottle virus on groundnut (*Arachis hypogaea*) in India. *Phytopath. Z.* **109** : 245-253.

Irulandi, S. and Balasubramanian, G. (2000). Efficacy of insec per cent cides against pod borer, *Lampides boecus* L. (Lycaenidae : Lepidoptera) on green gram. *Insect Environment* 6: 108-109.

Janarthan, R, 1973, Ann Rept 1972-73 T N A U Coimbatore, India 138pp

Ji, Liang, 1987. Recent advances in legume virus diseases research in China. Soybean Science 6 : 157-166.

Jones DR. 2003. Plant viruses transmitted by whiteflies. *Eur J Plant Pathol.* 109: 195-219.

Kaiser, J.W., Danesh, D., Okhovat, M. and Mossahebi, H. 1968. Diseases of pulse crops (edible legumes) in Iran. *Plant Dis. Reptr* **52**: 687-691.

KaKaiser, W.J., Wyatt, S.D. and Pesho, G.R. 1982. Natural hosts and vectors of tobacco streak virus in eastern Washington. *Phytopathology* **72**: 1508-1512.iser *et al.*, 1982.

Kelfer, H. H., Baker, E. W., Kono, T., Delfinado, M. and Styer, W. E., 1982, An illustrated guide to plant abnormalities caused by eriophyid mites in North America. USDA Agricultural Handbook No. 573.

Kennedy JS, Day MF, Eastop VF. 1962. *A Conspectus of Aphids as Vectors of Plant Viruses*. London: Commonw. *Inst. Entomol.*

Kiss E F, Huszka T, Ockro I and Vidal S (2003) Pepper pathogen viruses and their control by resistance breeding and light summer oils. Bullen OILB-SROP 26: 201-10.

Kitajima EW, Rodrigues JCV, de Moraes GJ, Childers CC. 2000. *Brevipalpus* mite-borne viruses. *Virus Rev. Res.* 5: 44–45.

Kolte, S. J. and Nene, Y.L. 1975. Host range and properties of urd bean leaf crinkle virus. *Indian Phytopath.* **28**: 430-431.

Kooner, B. S. and Cheema, H. K.(2007a). Screening of mungbean germplasm against whitefly, Bemisia tabaciGenn. and MYMV. *Journal of Food Legumes* 20: 100-102.

Kooner, B. S. and Cheema, H. K.(2007b). Screening of mungbean germplasm against whitefly, Bemisia tabaci and MYMV. *Acta Horculturae* 752: 307-310.

Kooner, B.S. (1998). Identification of sources of resistance to whitefly, jassids and MYMV in kharif mungbean. *In Proceedings of National Symposium on Management of Bioiotic and Abiotic Stresses in Pulse Crops*, p. 40. Kanpur : Indian Society of Pulses Research and Development.

Kooner, B.S., Chhabra, K.S., Sekhon, H.S., Dhingra, K.K. and Cheema, H.S. (1983). A new deformity in summer mungbean, *Vigna radiata* (L.) Wilczek. Pulse Newsle8er 3: 23-26.

Kooner, B.S., Chhabra, K.S., Sharma, A.K. and Saxena, A.K. (1994). Phytochemical resistance in urdbean genotypes against major insect-pests. *In Proceedings of International Symposium on Pulses Research*, pp. 162. Kanpur : Indian Society of Pulses Research and Development.

Kowalska, Cz. 1979. Viruses infecting pea (*Pisum L.*) in Poland. *Genet. Pol.* **202**: 212-215.

Kumar, P. L, 2000, A novel virus associated with pigeonpea sterility mosaic disease in India. In Association of Applied Biologist Conference on Advanmces in Plant Virology. September *20-22*, 2000. Dundee University, Dundee, Scotland.

Kumar, P. L., Fenton, B., Duncan, G., Jones, A. T., Sreenivasulu, P., and Reddy, D. V. R. 2001. Assessment of variation in *Aceria cajani* (Acari: Eriophyidae) using analysis of nuclear rDNA ITS regions and scanning electron microscopy: Implications for the variability observed in host plant resistance to pigeonpea sterility mosaic disease. *Ann. Appl. Biol.* 139: 61-73.

Kumar, P. L., Jones, A. T., Sreenivasulu, P. and Reddy. D V. R, 1999a, Unfolding the mystery of the causal agent of pigeonpea sterility mosaic disease. In *Abstracts, International Conference on Life Sciences in Next Milleniurn* University of Hyderbad, India, pp-65-66.

Kumar, P. L., Jones, A. T., Waliyar, F., Sreenivasulu, P., Muniyappa, V., Latha, T. K. S., *et al.* (2007). Pigeonpea sterility mosaic disease. In P. L. Kumar, and F. Waliyar (Eds.), Diagnosis and detection of viruses infecting ICRISAT mandate crops (pp. 15–21). Patancheru, India: ICRISAT.

Ladipo, J.L. 1988. Viruses associated with a mosaic disease of *Crotalaria juncea* in Nigeria. *J. Phytopathol.* **121**: 8-18.

Lakshmi, K. V., J. A. Wightman, D.V.R. Reddy, G.V.R. Rao, A.A.M. Buiel, and D.D.R. Reddy. 1995. Transmission of Peanut bud necrosis virus by *Thrips palmi* in India, pp. 179 –184. In B. L. Parker, M. Skinner, and T. Lewis (eds), Thrips Biology and Management. Plenum, New York.

Lakshmikantha, B.P., Prabhuswamy, H.P. and Jagadish, K.S., 1997, Effect of staggered sowing on the incidence of sterlity mosaic and its mite vector, *Aceria cajani* ChannaBasavanna. *Mysore J. Agric. Sci.,* 31 : 29-32.

Lal, S.S. (1987). Insect pests of mung, urd, cowpea and pea and their management. In Plant Protection in Field Crops (Eds M. Veerabhadra Rao and S. Sithanantham), pp. 185-201. Hyderabad: Plant Protection Association of India.

Lal, S.S. (1992). Insect pest of lentil and their management : A review. *Agricultural Review.* 15: 225-232.

Lal, S.S., Yadava, C.P., Dias, C.A.R. and Nigam, R. (1989). Effect of planting density and chickpea cultivars on the infestation of black aphid, *Aphis craccivora* Koch. *Madras Agricultural Journal* 76: 461-462.

Lewis, G., Schrire, B., MacKinder, B., and Lock, M. (Eds.), (2005). Legumes of the world. UK: Royal Botanical Gardens.

Lima, J.A.A., Purcifull, D. and Hiebert, E. 1979. Purification, partial characterization, and serology of blackeye cowpea mosaic virus. *Phytopathology* **69**: 1252-1258.

Lockhart, B.E. and Fischer, H.U. 1976. Some properties of an isolate of pea early browning virus occurring in Morocco. *Phytopathology* **66**: 1391-1394.

Loebenstein, G., and Thottappilly, G. (Eds.). (2003). Virus and virus-like diseases of major crops in developing countries (p. p. 800). Dordrecht: Kluwer Academic Publishers.

Lovisolo, O. and Conti, M. 1966. Identification of an aphid-transmitted cowpea mosaic virus. *Neth. J. Plant Path.* **72**: 265-269.

Makkouk, K.M. and Azzam, O.I. 1986. Detection of broad bean stain virus in lentil seed groups. *LENS Newsl.* **13**: 37-38.

Makkouk, K.M., Bos, L., Azzam, O.I., Katul, L. and Rizkallah, A. 1987. Broad bean stain virus: identification, detectability with ELISA in faba bean leaves and seeds, occurrence in West Asia and North Africa, and possible wild hosts. *Neth. J. Plant Path.* **93**: 97-106.

Malathi, V. G., and John, P. (2008). Gemini viruses infecting legumes. In G. P. Rao, P. L. Kumar, and R. J. Holguin–Pen˜a (Eds.), Vegetable and pulse crops: Vol. 3. Characterization, diagnosis and management of plant viruses (pp. 97–123). USA: Studium Press LLC.

Mali, V.R., Shirashta, A.M. and Godbole, G M., 1977, Occurrence of PSMV In Marathawada, *Res. Bull. Marathawada Agnc UI~IV1.* : , 148-14 9.

Malik, S.P.S. (1990). Compara per cent ve resistance of summer mungbean genotype to the thrips, Megalurothrips distalis(Karny.). M.Sc. thesis, Punjab Agricultural University, Ludhiana.

Maramorosch, K. 1994, Current status of mite transmitted plant disease agents. In Proceedings of the international Symposium: Rose Rosette and other Eriophyid mite transmitted Plant Disease Agents of unknown Etiology. Lowa State University, pp. 7-9.

Maris P. C., Joosten N. N., Goldbach R. W., Peters D. 2007 *Tomato spotted wilt virus* infection improves host suitability for its vector *Frankliniella occidentalis*. *Phytopathology* 94, 706–711.

Mayer, R. T., Inbar, M., Mckenzie, C. L., Shatters, R., Borowicz, V., Albrecht, U., Powell, C. A. and Doostdar, H. 2002 Multitrophic interactions of the silverleaf whitefly, host plants, competing herbivores, and phytopathogens. *Arch. Insect. Biochem. Physiol.* 51, 151–169.

Mc Rae, W. 1932, Report of Imperial mycologist. Sci. Rept. Imp. Instt. *Agric.* Res Pusa 1930-31, pp. 73-81.

Medina-Ortega, K.J., Bosque-Perez, N.A., Ngumbi, E., Jim ´ enez- ´ Martínez, E.S. and Eigenbrode, S.D. (2009) Rhopalosiphum padi (Hemiptera: Aphididae) responses to volatile cues from barley yellow dwarf virus-infected wheat. Environmental Entomology, 38, 836–845.

Mehrotra RS, Aggarwal A (2010). Plant pathology. Second Ed. publisher. Tata McGraw Hill Education Private Limited. Printer: Gopaljee Enterprises. Delhi. Cover: SDR Printers, pp. 703 -748.

Mink, G.I. and Keswani, C.L. 1987. First report of cowpea mild mottle virus on bean and mung bean in Tanzania. *Plant Dis.* **71**: 557.

Mukankusi, C., Adipala, E., Kyamanywa, S., Epieru, G., Odeke, V., Warren, H.L. and Wilson, H.R. 1999. Effect of host genotype, time of planting and spacing on epidemics of groundnut rosette and Cercospora leaf spot diseases in eastern Uganda. *African Journal of Plant Protection* 9, 37-53

Muniyappa, V., and Nangia, N. 1982. Pigeonpea cultivars and selections for resistance to sterility mosaic in relation to prevalence of eriophyid mite *Aceria cajani* Channabasavanna. Trop. *Grain Legume Bull.* 25: 28-30.

Murugesan, S. and Chelliah, S., 1977, Influence of sowing time on the incidence of the vector Bemisia tabaci (Genn.) and the yellow mosaic disease of greengram. *Madras Agric. J.*, 64(2): 128-130.

Mushtaque, A. (1977). Preliminary studies on the infestation of Aphis craccivora Koch to chickpea and lentil. *Journal of Agricultural Research* 15 : 31-35.

Musil, M., Valenta, V., Kowalska, Cz., Wiatroszak, I. and Beczner, L. 1983. Serological properties of some comoviruses. *Biologia* (Bratislava) **38**: 231-236.

Narayanaswamy, P., Sheshdri, A.R. and Ramakrishnan, K., 1963, Perliminary notes on the suspected nematode transmission of redgram sterility mosaic. *Madras Agric. J.*, 50: 1 09-1 1 0.

Nath PD. 1994. Effect of sowing time on the incidence of yellow mosaic virus disease and whitefly population on greengram. Ann Agric Res. 15(2): 174-17.

Nath, P.D. and Saikia, A.K., 1995, Effect of time of sowing on the incidence of mungbean yellow mosaic virus disease and whitefly (*Bemisia tabaci* Genn.) population in greengram. *Ann. Agri. Res.*, 16(4): 483-484.

Nault LR, Ammar E-D. 1989. Leafhopper and planthopper transmission of plant-viruses. *Annu. Rev. Entomol.* 34: 503–29.

Nault LR. 1997. Arthropod transmission of plant viruses: a new synthesis. *Ann. Entomol. Soc. Am.* 90: 521–41.

Nene, Y. L, and Reddy, M. V 1976a. Screening for resistance to ster~lity mosaic of pigeonpea. *Plant Dis. Reptr.*, 60: 1034-1 036.

Nene, Y. L. 2006. Indian pulses through the millennia. Asian Agri-history 10: 179-202.

Nene, Y.L. and Kolte, S.J. 1972. Leaf curl. In: Viral diseases of pulse crops in Uttar Pradesh. Ed. Y.L. Nene. *Govind Ballabh Pant Univ. Agric. and Technol. Pantnagar Res. Bull.* **4** : 109-134.

Nene, Y.L. and Reddy, M.V., 1976. Preliminary information on chickpea stunt. *Tropical Grain Legumes Bulletin* 5: 31-32.

Nene, Y.L. 1973. Viral diseases of some warm weather pulse crops in India. *Plant Disease Reporter*, 57: 463-467.

Nene, Y.L., 1972, A survey of viral diseases of pulse crops in Uttar Pradesh. Research Bulletin No. 4, Uttar Pradesh Agri. Univ., Pantnagar, p. 191.

Ng JC, Falk BW. 2006. Virus-vector interactions mediating nonpersistent and semipersistent transmission of plant viruses. *Annu. Rev. Phytopathol.* 44: 183–212.

Ng JCK and Perry KL. 2004. Transmission of plant viruses by aphid vectors. *Molecular Plant Pathology* 5: 505-511Ng and Perry, 2004.

Ngumbi, E., Eigenbrode, S. D., Bosque-Perez, N. A., Ding, H. and Rodriguez, A. 2007 Myzus persicae is arrested more by blends than by individual compounds elevated in headspace of PLRV-infected potato. *J. Chem. Ecol.* 33, 1733–1747. (doi: 10.1007/s10886-007-9340-z).

Ogenga-Latigo, M.W., Ampofo, J.K.O. and Baliddawa, C.W. 1992a. Influence of maize row spacing on infestation and damage of intercropped beans. I. Incidence of aphids. *Field Crops Research* 30, 111-121.

Ogenga-Latigo, M.W., Baliddawa, C.W. and Ampofo, J.K.O 1992b. Influence of maize row spacing on infestation and damage of intercropped beans by the bean aphid (Aphis fabae Scop.). II. Reduction on bean yields. *Field Crops Research* 30, 123-130.

Oldfield, G. N., Reddy, M. V., Nene, Y. L., and Reed, W. 1981. Preliminary studies of the eriophyid vector of sterility mosaic. *Int. Pigeonpea Newsl.* 1: 25-27.

Oldfield, G., and Proeseler, G. 1996. Eriophiod mites as vectors of plant pathogens. Pages 259-275 in: Eriophyoid Mites - Their Biology, Natural Enemies and

Control. E. E. Lindquist, M. W. Sabelis, and J. Bruin, eds. Elsevier Science B. V., The Netherlands.

Oldfield, G.N, and Proeseler, G. 1996, Eriophyid mites as vectors of plant pathogens. In Eriophyid mites, their biology, natural enemies and control. (Eds. E.E. Lindquist, M.W. Sabelis and J. Bruin). Elsevier, pp. 259- 275.

Oldfield, G.N, and Proeseler, G. 1996, Eriophyid mites as vectors of plant pathogens. In Eriophyid mites, their biology, natural enemies and control. (Eds. E.E. Lindquist, M.W. Sabelis and J. Bruin). Elsevier, pp. 259- 275.

Oldfield, G.N., 1970, Mite transmission of plant viruses. *Annual Review of Entomology,* **15:** 343-380.

Oldfield, G.N., Reddy, M.V., Nene, Y.L. and Reed, W., 1981, Preliminary studies of the eriophyid vector of sterility mosaic. *Int. Pigeonpea Newsl.,* 1. 25-27.

Orawu, M., Adipala, E. and Warren, H. 2001. Integration of fungicide and insecticide usage for management of cowpea diseases and insect pests in Eastern Uganda. African Crop Science Conference Proceedings 5, 417- 422.

Orlob, G. B. 1966. Feeding and transmission characteristics of *Aceria tulipae* Kiefer as vector of wheat streak mosaic virus. *Phytopathol.* Z. 55: 218-238.

Paliwal,Y.C. 1980. Relationship of Wheat streak mosaic and barley stripe mosaic viruses to vector and non-vector Eriophyid mites. *Archives of Virology,* **63:** 123-132.

Paliwal YC. 1980. Relationship of wheat streak mosaic and barley stripe mosaic viruses to vector and nonvector eriophyid mites. *Arch. Virol.* 63: 123–32.

Pandey BP (2012). "Plant pathology", pathogen and plant diseases: Revised, updated and expanded. Ed., pp. 55- 56.

Patel, P.N. and Kuwite, C. 1982. Prevalence of cowpea aphid-borne mosaic virus and two strains of cowpea mosaic virus in Tanzania. *Indian Phytopath.* **35**: 467-472.

Patino, G. and Zaumeyer, W.J. 1959. A new strain of tobacco streak virus from peas. *Phytopathology* **49**: 43-48.

Pautasso, M., Doring, T.F., Garbelotto, M., Pellis, L. and Jeger, M.J. (2012). Impacts of climate change on plant diseases-opinions and trends. *European Journal of Plant Pathology*. doi: 10.1007/s10658-012-9936-1.

Persley, D and Sharman, M (2007). Thrips and Tospovirus: A Management Guide, CRC for Tropical Plant Protection, Brisbane, Department of Primary Industries and Fisheries, Queensland.

Phatak, H.C. 1974. Seed-borne plant viruses - identification and diagnosis in seed health testing. *Seed Sci. and Technol.* **2**: 3-155.

Pirone T.P., Blanc S., 1996. Helper-dependent vector transmission of plant viruses. *Annual Review of Phytopathology* **34**: 227-247.

Pospieszny, H. and Frencel, I. 1985. Viruses in natural infections of yellow lupin (*Lupinus luteus L.*) in Poland. VI. Pea early-browning virus. *Acta Phytopath. Acad. Sci. Hung.* **20**: 91-95.

Purseglove, J. W. 1968 Tropical Crops: dicotyledons. Volume 1 Jhon Wiley and sons, New York, 332 p.

Purwar J P and yadav S R (2006).Evaluation of biorational insecticides against whitefly (*Bemisia tabaci* Genn.): a vector of soybean yellow mosaic. *Journal Experimental Zoology* 9(2): 499-503.

Ramakrishnan, K. and Kandaswarny, T. K, 1972, Investigation on virus diseases of pulse crops in Tamil Nadu. *Final Technical Report,* T.N.A.U. Coimbattore, Tamil Nadu, India. Pp.53.

Ramusi M, Flett B (2012). Transmission of maize streak virus from grasses to maize.

Rancic D, Stevanovic B, Petanovic R, Magud B, Tosevski I, Gassmann A. 2006. Anatomical injury induced by the eriophyid mite *Aceria anthocoptes* on the leaves of *Cirsium arvense. Exp. Appl. Acarol.* 38: 243–53.

Rao, G. P., Kumar, P. L., and Holguin- Pen˜a, R. J. H. (Eds.). (2008). In Vegetable and pulse crops: Vol. 3. Characterization diagnosis and management of plant viruses. USA: Studium Press LLC, 408 p.

Ravindrababu, R. (1987). Studies on mungbean yellow mosaic virus disease. Ph.D. thesis, Tamil Nadu Agricultural University, Coimbatore.

Reddy, D. V. R., Mayo, M. A., Naidu, R. A. and Sherwood, J. L., 1995 (eds), Laboratory exercises for a Training Course on the detection and identification of Viruses. ICRISAT, Patancheru, India, pp-74.

Reddy, D.V.R., and Wightman, J.A. 1988. Tomato spotted wilt virus, thrips transmission and control (Harris K.R., ed.). Advances in Disease Vector Research 5: 203-220.

Reddy, M. V, and Raju, T. N., 1993. Some clues on increased ~nc~dencaen d seasonal variation of pigeonpea sterilty mosaic in Peninsular India. *Int Pigeonpea Newsl.,* **18:** 22-24.

Reddy, M. V., and Nene, Y. L. 1980. Influence of sterility mosaic resistant pigeonpeas on multiplication of the mite vector. *Indian Phytopathol.* 33: 61-63.

Reddy, M. V., Beniwal, S. P. S., Sheila, V. K., S~thananthan, S and Nene. Y. L 1989. Rloe of eriophyid mite(*Aceria cajani*) in transmission and spread of sterility mosaic of pigeonpea. Pp121-127 in Progress in Acarology (Channbasavanna, G.P. and Virakthmath, C.A. eds), Oxfard and IEH Publishing Co. New Dellhi, India.

Reddy, M. V., Sharma, S.B. and Nene, Y. L., 1990, Pigeonpea: Disease management. In The Pigeonpea (Eds., Y.L. Nene, S.D. Hall and V.K. Sheila). CAB International-ICRISA T, Pp. 349-374.

Reddy, M.V., Raju, T.N., Nene, Y.L., Ghanekar, A.M. Amin, K.S., Arjunan, G., Astaputra, J.V. Sinha, B.K., Muniyappa, V. Reddy, S.V., Gupta, R.P. and Kausalya, G., 1993a, Variability in sterility mosaic pathogen in Ind~al,n dian Phtytopathology, **46:** 206-21 2.

Robertson, D.G. 1966. Seed-borne viruses of cowpea in Nigeria. B.Sc. Thesis, University of Oxford, Oxford. Rossel, 1977

Sastry S. K., Zitter T. A. (2014). "Management of virus and viroid diseases of crops in the tropics," in*Plant Virus and Viroid Diseases in the Tropics* Vol. 2*Epidemiology and Management* (Dordrecht: Springer;) 149–480 10.1007/978-94-007-7820-7-2

Satyanarayana, J., Singh, K.M. and Singh, R.N. (1995). Insect pest succession in rice bean. *Bullen of Entomology* 36: 78-83.

Satyanarayana, J., Singh, K.M., Singh, R.N. and Singh, S. (1997). Influence of intercropping and insecticide. Incidence of major pests in ricebean, Vigna umbellata. *Indian Journal of Entomology* 59: 316-325.

Scholthof, K.-B. G. 2001. 1898 – the beginning of virology_time marches on. The Plant Health Instructor. DOI: 10.1094/PHI-I-2001-0129 01.

Seth, M.L. 1962, Transmission of pigeonpea sterility mosaic by an eriophyid mite *Indian Phytopathol.,* **15.** 225-227.

Shanmugasundaram, S.(1988). A catalog of mungbean cultivators released around the world. Asian vegetable research and development centre, Shanhua, Taiwan.

Sheila, V.K., Manohar, D.K. and Nene, Y.L., 1988, Biology and morphology of *Aceria cajani. Int. Pigeonpea Newsl.,* **7:** 28-29.

Shirale D. and Bidgire U. (2009). Integration of bioagents and synthetic insecticide in the management of whitefly and its effect on yield in soybean. Karnataka *Journal of Agricultural Science* 22(3): 631-632.

Shoyinka, S., Bozarth, R.F., Reese, J. and Rossel, H.W. 1978. Cowpea mottle virus, a seed-borne virus with distinctive properties infecting cowpeas in Nigeria. *Phytopathology* **6 8:** 6 9 3 - 6 9 9.

Singh RS (2005). Introduction to principles of Plant padhology. Fourth Ed. Publisher; Rajuprimlari for oxford 1 Bit publishing Co. Pvt. Ltd printed at chama Enterprises, New Delhi.

Singh, D.S. and Sircar, P. (1980). Relative toxicity of pyrethroids to Lipaphis erysimi Kalt and *Aphis craccivora* K. *Indian Journal of Entomology* 42: 597-605.

Singh, M. (1996). Studies on the biology of leaf curl mite Polyphagotarsonemus latus (Banks) (Acari : Tarsonemidae) on chilli (Capsicum annum). M.Sc. thesis, Punjab Agricultural University, Ludhiana.

Singh, R A, Gurha S N and Ghosh A (1997) Diseases of mungbean and urdbean and their management. In: Diseases of Field Crops and their Management (Eds T S Thind), pp. 179-204, Daya Publishing House, Delhi.

Singh, R.A. and Gurha, S.N., 1994, Influence of cropping seasons on the incidence of yellow mosaic disease in mungbean genotypes. *Indian J. Pulses Res.,* 7(12): 206-208.

Sisterson M. S. 2008Effects of insect–vector preference for healthy or infected plants on pathogen spread: insights from a model. *J. Econ. Entomol.* 101, 1–8.

Skare JM,Wijkamp I, Denham I, Rezende JA, Kitajima EW, *et al.,* 2006. A new eriophyid mite-borne membrane-enveloped virus-like complex isolated from plants.*Virology* 347: 343– 53.

Spence NJ (2001). Biotic interactions in plant-pathogen associations, pp. 15-26. Oxford University Press, New York, USA.

Srikanth, J. and Lakkundi, N.H. (1998). Instar period, fecundity and longevity of cowpea aphid, *Aphis craccivora* Koch. on seven leguminous hosts. *Journal of Aphidology* 2: 18-19.

Strange RN (2003). Introduction to Plant Pathology. Prnters: TJ International Ltd., Padstow, Cornwall. Great Britain, pp. 75-80.

Sylvester ES. 1958. *Aphid transmission of plant viruses.* Presented at Int. Congr. Entomol., 10th, Montreal.

Tapio, E. 1970. Virus diseases of legumes in Finland and in the Scandinavian countries. *Annls Agric. Fenn.* **9**: 1-97.

Thakur, B.S., Verma, R., Ptitunda, A. and Rawat, R.R. (1984) Chemical control of aphid, Aphis craccivora Koch on lentil. *Indian Journal of Entomology* 46: 103-105.

The International Committee on Taxonomy of Viruses (ICTV) (2012). The viruses. In A. M. Q. King, M. J. Adams, E. B. Carstens, and E. J. Lefkowitz (Eds.), Virus taxonomy—Ninth report of the international committee on taxonomy of viruses (pp. 21–1197). United Kingdom: Elsevier/Academic Press.

Thomas, H.R. and Zaumeyer, W.J. 1960. Red node, a virus disease of beans. *Phytopathology* **40**: 832-846.

Thongmeearkom P, Honda Y, Saito Y, Syamananda R. 1981. Nuclear ultra structural changes and aggregates of virus like particles in mungbean cells affected by mungbean yellow mosaic disease. *Phytopathology.* 71: 41_44.

Thouvenel, J.C., Montsarrat, A. and Fauquet, C. 1982. Isolation of cowpea mild mottle virus from diseased soybeans in Ivory Coast. *Plant Dis.* **66**: 336-337.

Tsuchizaki, T., Senboku, T., Pholauporn, S., Srithongchai, W., Deema, N. and Ching, A.O. 1984. Blackeye cowpea mosaic virus from asparagus bean (*Vigna sesquipedalis*) in Thailand and Malaysia, and their relationships to a Japanese isolate. *Ann. Phytopath. Soc. Japan* **50**: 461-468.

Ullman, D. E., J. L. Sherwood, and T. L. German. 1997. Thrips as vectors of plant pathogens, pp. 539 –565. In T. Lewis (ed), Thrips as Crop Pests. CAB International, New York.

Vadodaria, M.P. and Vyas, H.N. (1987). Control of whitefly, *Bemisia tabaci* (Gennadius) and its impact on yellow mosaic virus (YMV) in greengram, (Vigna radiata (L.) Wilczek) and the grain yield. *Indian Journal of Agriultural Research* 21: 21-26.

Van Emden, H.F. (1973). Aphid host plant rela per cent onship : some recent studies. In Perspectives in Aphid Biology (Ed A.D. Lowe), pp. 54-64. Auckland Entomology society of New Zealand.

Varma A, Dhar AK, Mandal B. 1992. MYMV transmission and control in India. In: Green SK, Kim D, editors. Mungbean yellow mosaic disease. Taipei: Asian Vegetable Research and Development Centre; p. 8-27.

Varma A, Malathi VG. 2003. Emerging geminivirus problems: a serious threat to crop production. *Ann Appl Biol.* 142: 145-164.

Verma, P. M. 1963, Transmission of plant viruses by white flies, Bulletin of the National Institute of Sciences of India., 24: 11-33.

Vijaya Lakshmi, K. 1994. Transmission and ecology of Thrips palmi Karny, the vector of peanut bud necrosis virus. PhD thesis. Andhra Pradesh Agricultural University, Rajendranagar, Hyderabad, Andhra Pradesh, India. 99 pp.

Vroon, C.W., Pietersen. C. and Tonder, H.J. van. 1988. Seed transmission of soybean mosaic virus in *Lupinus albus L. Phytophylactica* **20**: 169-175.

Walkey, D. 1991. Applied Plant Virology. 2nd edition. Chapman and Hall, London.

Wang ZY, Yao MD, Wu YD. 2009. Cross-resistance, inheritance and biochemical mechanisms of imidacloprid resistance in B-biotype Bemisia tabaci. Pest Manag Sci. 65: 1189-1194.

Watson MA, Roberts FM. 1939. A comparative study of the transmission of Hyocyamus virus 3, potato virus Y and cucumber virus 1 by the vectors *Myzus persicae* (Sulz.), *M. circumflexus* (Buckton), and *Macrosiphum gu* (Koch). *Proc. R. Soc. London Ser. B* 127: 543–76.

Webb, S.E.; Kok-Yokomi, M.L.; Tsai, J.H. 1997. Evaluation of *Frankliniella bispinosa* as a potential vector of tomato spotted wilt virus. *Phytopathology* 87(6 SUPPL.): S102.

Whitney, W.K. and Gilmer, R.M. 1974. Insect vectors of cowpea mosaic virus in Nigeria. *Ann. appl. Biol.* **77**: 12-21.

Wijkamp I., Almarza N., Peters D., 1995. Median latent period and transmission of tospoviruses vectored by thrips. In: Wiljkamp I., Almarza N., Peters D. (eds.). Thrips Biology and Management, pp. 153-156. NATO ASI Series, Plenum Press, New York, USA.

Xu, Z.G., Pu, Z.Q., Cao, Q., Fang, Z.D. and Cockbain, A.J. 1986. Virus diseases of the faba bean (Vicia faba L.) found in the Yangtze river valley. pp. 116-122. In: The selected work of virus diseases in leguminous crops (1968 - 1985) to celebrate 60 years teaching career of professor Yu Dafu. Nanjing Agric. Univ. Publ., Nanjing.

Yadava, C.P. and Lal, S.S. (1998). Major insect pests of chickpea and their management. In IPM System in Agriculture. Vol. IV. Pulses (Eds R.K. Upadhyay, K.G. Mukerji and R.L. Rajak), pp. 197-231. New Delhi : Aditya Book.

Yein, B.R. (1983). Efficacy of certain insec per cent cides against B. tabaci (Genn.) and Pagria signata (Motsch) on blackgram. *Journal of Research Assam Agricultural University* 4: 45-48.

Zaitlin, M. 1998. The discovery of the causal agent of tobacco mosaic disease. Pages 105-110 in: Discoveries in Plant Biology. S.D. Kung and S.F. Yang, eds. World Publishing Co. Ltd. Hong Kong.

Chapter 16

Diseases of Arid Legumes: Extent and Management

Ritu Mawar, Manish Mathur & A.S. Tomer

Legumes play an important role in improving livelihood, nutritional security of farmers and populations in less developed countries as well as sustainability of agriculture in dry areas worldwide. Grain legumes are important rotational crops in the Pacific Northwest and Northern Great Plains of the United States and central California. The introduction of pulse crops into the agricultural systems in Western Australia is relatively recent and has been supported by considerable research to adapt and develop varieties and production systems to suit this Mediterranean environment. Traditionally grain legume or pulses have been an important constituent of predominantly vegetarian diet of the people of India. These are cultivated over a vast area-nearly 25 million hectares constituting 18 per cent of the total cultivated area in this country, the largest in the world. Legume crops, which include pulses are important source of protein for a large population of vegetarian people of India. The important grain and seed legumes of India-Chickpea (*Cicer arietinum* L.), lentil (*Lens culinaris* Medik.), faba bean (*Vicia faba* L.), grasspea (*Lathyrus sativus* L.) and fieldpea (*Pisum sativum* L.) are grown in cool season, while Clusterbean (*Cyamopsis tetragonoloba* (L.) Taub.), pigeonpea (*Cajanus cajan* (L.) Millsp.), blackgram (*Vigna mungo* (L.) Hepper), mungbean (*V. radiata* (L.) Wilczek), horsegram (*Macrotyloma uniflorum* (Lam.) Verdc.), cowpea (*V. unguiculata* (L.) Walp) and soybean (*Glycine max* (L.) Merr.) are the warm season legumes. These grain legumes grown worldwide are prone to attack by a large number of plant pathogens, from fungi, bacteria, phytoplasmas, and viruses to nematodes and parasitic angiosperms, which result in severe economic losses globally. Among these, fungi and viruses are the largest and perhaps the most important groups affecting all parts of the plant at all stages of growth (Table 16.1). In the present chapter the diseases of guar, cowpea, mothbean and horse gram have been reviewed.

Table 16.1. Important Foliar Diseases of Legumes and their Distribution

Legumes	*Diseases*	*Pathogen*	*Distribution*	*Losses*
Chickpea (*Cicer arietinum* L.)	Ascochyta blight	*Ascochyta rabiei*	West Asia, northern Africa, Mediterranean region	> 50 per cent
	Botrytis gray mold	*Botrytis cinerea*	India, Nepal, Bangladesh, Pakistan, North Africa, Australia, America	50-100 per cent
	Stunt	Bean leaf roll luteovirus (BLRV)	North Africa, Middle East, India, Spain, Turkey and the USA.	20-30 per cent
Lentil (*Lens culinaris* Medik.),	Rust	*Uromyces viciae-fabae*	Bangladesh, Chile, Ecuador, Ethiopia, India, Morocco, Nepal and Pakistan	50-100 per cent
	Ascochyta blight	*Ascochyta lentis*	Argentina, Australia, Brazil, Canada, Chile, Cyprus, Ethiopia, Greece, Iran, Jordan, New Zealand, Pakistan, Russia, Spain, Syria, and USA.	Up to 70 per cent
	Stemphylium blight	*Stemphylium botryosum*	Bangladesh, Egypt, Syria, and USA	Up to 70 per cent
Faba bean (*Vicia faba* L.),	Ascochyta blight	*Ascochyta fabae*	Mediterranean countries	5-50 per cent
	Chocolate leaf spot	*Botrytis cinerea/Botrytis fabae*	Mediterranean countries	Up to 50 per cent
	Rust	*Uromyces viciae-fabae*	Mediterranean countries	Up to 50 per cent
	Necrotic yellows	Faba bean necrotic yellows virus	West Asia and North Africa	Up to 80 per cent
Pea (*Pisum sativum*, L.)	Powdery mildew	*Erysiphe polygoni*	India, Nepal	10 per cent
	Downy mildew	*Peronospora viciae*	India, Nepal	30 per cent
Clusterbean (*Cyamopsis tetragonoloba* (L.) Taub.)	Dry root rot	*Macrophomina phaseolina*	Asian countries, USA, Australia and Africa	30-40 per cent
Pigeonpea (*Cajanus cajan* (L.) Millsp.)	Sterility mosaic	Pigeonpea sterility mosaic virus	Bangladesh, India, Myanmar, Nepal, Sri Lanka and Thialand	Annual loss of 205, 000 tonnes in India
Blackgram (*Vigna mungo* (L.) Hepper) and Mungbean (*V. radiata* (L.) Wilczek)	Yellow vein mosaic	Yellow mosaic virus	Bangladesh, India	10-100 per cent
	Cercospora leaf spot	*Cercospora cruenta, C. canescens*	Bangladesh, India, Indonesia, Taiwan, Thailand, Philippines, Malaysia	Upto 50 per cent

Legumes	*Diseases*	*Pathogen*	*Distribution*	*Losses*
	Powdery mildew	*Erysiphe polygoni*	India, southeast Asian countries	9-50 per cent
Cowpea (*Vigna ungiculata* [L.] Walp.)	Cercospora leaf spot	*Cercospora canescens* and *Pseudocercospora cruenta*	Fiji, Brajil, Kenya, Nigeria, Zimbabwe, India, Bangladesh, Egypt, Iran, Japan, Malaysia, Thailand	18-42 per cent
	Cowpea golden mosaic	Cowpea golden mosaic virus	Kenya, Nigeria, Tanzania, Cuba, Surinam, USA	60-100 per cent
	Cowpea aphid-borne mosaic	Cowpea aphid-borne mosaic virus	Europe, Africa, Mediterranean basin, Turkey, Iran, India, Indonesia, China, Japan, Australia, Brazil, USA	13-87 per cent

Source: Pande *et al.,* 2007.

Besides source of proteinacious diet in Indian desert, these crops also providing ecosystem services mainly cultural and provisional (Table 16.2).

Table 16.2. Arid Legumes and their Ecosystem Services (Cultural and Provisional)

Arid Legume	*Ecosystem Services*
Cyamopsis tetragonoloba	Food, fodder, vegetable, manure, industrial, medicinal.
Vigna aconitifolia	Green-small vegetable, source of vegetable protein, green manure, confectionery items.
Vigna ungiculata	Confectionery items, vegetable, medicinal (dissolving stone, gynaecological problems, whooping cough, purgative).
Macrotyloma uniflorum	Food item, medicinal.
Vigna radiata	Main staple source of protein for vegetarians.

In general, legumes are mostly grown in two seasons: (i) rainy season or kharif (June-October), and (ii) rabi (October-April). Chickpea, lentil, and dry peas are grown in the rabi season, while clusterbean, pigeonpea, urdbean, mungbean, and cowpea are grown during the kharif season. Kumar (2005) had recognized arid legumes for taming drought, sustaining soil health and diversifying agriculture and for providing food, feed and variety of products of industrial use. Garg and Burman (2009) have reported drought physiology of arid legumes and their drought related traits (stomatal conductance, photosynthetic capacity, timing of phonological phases, starch availability, rooting depth, osmotic adjustment, anti-oxidative stress and accumulation of stress related protein). Arid legumes are adjudged for sustaining growth and production, especially in drier eco-systems encountered with harsh and hostile growing environments. Sustained breeding efforts have been made to improve these crops so as to make them more productive (Kumar and Rodge, 2012).

Arid Legumes and their Major Biotic Constraints

Major arid legumes likes cluster bean/guar (*Cyamposis tetragonoloba*), cowpea (*Vigna unguiculata*), moth bean (*Vigna aconitifolia*) and horsegram (*Macrotyloma uniflorum*) suffer from many biotic constraints and subsequently their performance and yield are reduce at considerable level. In subsequent part of the chapter major disease of these arid legumes, their pathometery (casual organism, symptoms, predisposing factors, epidemic threshold and epidemic injury levels), losses assessment/measurement and their control measures (chemical, physical, biological along with integrated disease management) are discussed in details.

Guar (*Cyamopsiss tetragonoloba* L.)

Clusterbean, (*Cyamposis tetragomoloba* (L.) Taub) commonly known as Guar, is a drought and high temperature tolerant deep rooted summer annual legume of high social and economic significance. The qualities of the crop like high adaptation towards erratic rainfall, multiple industrial uses and its importance in cropping system for factors such as soil enrichment properties, low input requirement, *etc.* have made the guar one of the most significant crops for farmers in arid areas in India. The crop is mainly grown in the dry habitats of Rajasthan, Haryana, Gujarat and Punjab and to limited extent in Uttar Pradesh and Madhya Pradesh. India is the largest producer of Guar and contributes 80 per cent of total Guar production in the world. In India, Guar crop is cultivated mainly during *Kharif* season, with an annual production of around 2.7 million metric tons. Guar crop has experienced a remarkable journey from a traditional crop grown on marginal lands mainly for food, animal feed and fodder to a crop with various industrial usages. Guar gum is an important ingredient in producing food emulsifier, food additive, food thickener and other Guar gum products. Guar gum is derived from endosperm which contains significant amounts of galactomannan gum (19 to 43 per cent of the whole seed). India is the largest producer of Guar gum and its derivatives. Guar gum is largely an export oriented commodity with about 75-80 per cent of total output being exported from the country. India was the leading exporter of mucilages and thickeners in the world with a share of more than 73 per cent in value terms during 2011. India earns a foreign exchange of approx. 1000 crores from guar gum. Like other legumes, guar is an excellent soil-building crop with respect to availability of nitrogen. Root nodules contain nitrogen-fixing bacteria and crop residues, when ploughed under, improves yields of succeeding crops. The yield of guar has been very low and sometimes uneconomical mainly due to erratic and uncertain rains, and cultivation of traditional cultivars which are late maturing, poor yielders and highly susceptible to a number of diseases. Increasing guar yields to a maximum and to improve the quality it is imperative that we increase our knowledge of many diseases, which affect the quality and yield of the crop.

Fungal Diseases

Alternaria Leaf Spot

Pathogen

Alternaria cucumerma var. *cyamopsidis*

It is seed borne disease affecting the Guar crop across the country. The fungus survives through seeds and mycelium in infected crop debris. The seed and soil borne inoculums infects the seedlings. Such infected plants serve as foci of infection. On availability of favourable environmental condition of the disease development the pathogen infects the large population of the crop. Excessive rainfall and humidity favour the disease incidence (Saharan *et al.*, 2003).

Symptoms

The disease is visible primarily on leaf blade as dark brown, round to irregular spots varying from 2 to 10 mm in diameter. In the early stage of infection, water soaked spots appear on the leaf blade which later on turn on greyish to dark brown with concentric zonation, demarked with light brown lines. On the under surface the spots are light to grayish brown. In severe infection, several spots merge together involving a major part of the leaf blade. The maximum severity take place between bloom and pod set. If the plants are infected in the early stage of growth, they may not flower, thereby causing severe losses. High losses (43-87 per cent) occur when it attacks at seedling and reproduction stage. Predisposing factors include excessive rainfall, humidity (80 per cent) and 25-31°C temperature and subsequently cumulative effects of these factors favors maximum disease intensity. Biochemical studies indicate that polyphenol oxidase and peroxidase enzyme increase the intensity of this disease, while catalase enzyme activities may reduce its intensity (Joshi *et al.*, 2004).

Management

Resistant Cultivars

Sowing of pathogen free seeds or cultivars resistant to pathogen. There is dilemma regarding its resistant cultivar, according to Yadav and Shalendra, (2014), however, varieties HG 182. CAZG 9007, GAUG 9008 and RGC 1014 were resistant (Saharan *et al.* 2001).

Cultural

In view of seed born nature of pathogen thus, cultural practice like crop rotation with non-host crop can be a better management practice.

Chemical

Thiram was the most effective fungicide tested against seed borne fungi when used individually or in mixture (Shivanna *et al.*, 1992). Two sprays of zineb (0.2 per cent) and copper oxychloride at an interval of 15 days are also proven chemical approach. A combination of zineb with streptocycline is recommended for simultaneously controlling bacterial as well as Alternaria blight. Crop weather disease interactions in *Cyamopsis tetragonoloba* in semi arid environment were carried out by Gupta *et al.* (1998), Saharan *et al.* (2001), Chanderia (2008) and Chanderia and Singh (2009). These researchers have concluded (a) positive correlation of disease index of *Alternaria* blight of clusterbean with temperature, relative humidity and rainfall, whereas its negatively related with sunshine duration and rainfall events (b) minimum temperature significantly correlated with the lesion size and (c) early sowing is best cultural practice to avoid this disease.

Powdery Mildew

Pathogen

Odio taurica, Erysphe polygoni

The disease manifests only on leaves. Causal organisms are many and includes *Odio taurica, Erysphe taurica, Leveilula taurica, Erysiphe polygoni* and *Sphaerotheca fuliginea.*

Symptoms

The disease appears as scattered white powdery mass on the adaxial leaf surface. The affected leaves are covered with mycelia patches dotted with the fruiting bodies (perithecia). Severely affected plants are defoliated and weakened by premature drying and death of infected leaves. The incidence of powdery mildew is more observed in the regions where crop season is prolonged. Warm temperature (33 per cent or above), high humidity (more than 80 per cent) and bright sunshine are congenial conditions for disease development. The perithecial stage of the fungus first appears on the dried leaves. Under favorable condition the mildew causes considerable defoliation.

Management

Resistant Cultivars

GAUG 63, an unbranched type, was the best performing variety showing moderately resistant to bacterial blight and highly resistant reaction to powdery mildew (Pandey *et al.*, 1993).

Cultural

Cultural practices like early sowing particularly first fortnight of July is reported as best management strategy.

Chemical

Application of benlate, wetttable sulphur compounds *etc.* inhibits the germination of conidia of *Erysiphe*. Seed treatment with streptocycline (0.025 per cent) and spraying a combination of streptocycline and dinocap (0.01 per cent) is very effective for controlling powdery mildew as well as bacterial blight.

Charcoal Rot or Dry Root Rot

Pathogen

Macrophomina phaseolina (Tassi) Goid (Syn: *Rhizoctonia bataticola*)

Symptoms

Erratic rainfall and short and long duration of moisture stress favour development of charcoal rot or dry root rot caused by *M. phaseolina*, a ubiquitous pathogen, which has wide range of hosts in commercially valuable crops of the region *e.g.*, Sorghum, sesame, cowpea, moth bean, jojoba, Euphorbia besides clusterbean (Dhingra and Sinclair, 1978, Lodha *et al.*, 1986). The disease may occur at any stage of the crop from pre-emergence to maturity and can appear as seedling blight or dry root rot. Seedling blight occurs primarily on the cotyledons in the form of damping off or as elongated black cankers on growing seedlings. Infected seedlings can be easily uprooted. In grown up plants, the disease appears as bronzing of leaves on one or more branching followed by drooping of the upper tender parts of the shoots. Affected plants continue to grow without having clusters or pods, but may break easily in strong winds. Nodule formation in these plants also reduces drastically. Highly affected plants rarely produce any grain or produce infected seeds. Lodha *et al.* (1986) estimated 32 per cent losses in the seed yield of guar only due to dry root rot under years of frequent moisture stresses. The inoculums build up of the pathogen is more in infested soil than seed borne inoculums or plant inoculations. Inoculums and disease development correlate with moisture and the native population of total bacteria and actinomycetes in soil (Lodha and Burman, 2004).

Management

Resistant Cultivar

Many workers have evaluated the germplasm for resistance to the pathogen under natural and artificially inoculated conditions (Gupta *et al.*, 2003). The genotypes *i.e.* HG-75 and RGC-1030 were free from the disease. Similarly, Kutch-8 and RGC-471 were found to be moderately resistant (Lodha and Solanki, 1998). Out of 146 lines 40 lines are resistant (Gupta *et al.*, 2003). High yielding variety CAZRI 27-1 possessing high gum content (32.8 per cent) was having field tolerance to the disease (Lodha and Burman, 2004).

Cultural

Crop rotation with less susceptible crop like moth bean and pearl millet or keeping the land fallow reduced the pathogen population (Lodha *et al.*, 1990).

Organic Amendments

Mulching of the soil with pearl millet stover (3.5 ton ha^{-1}), low plant density (1.6 lakh ha^{-1}) and farm yard manure (10 ton ha^{-1}) singly or in combination effectively conserved the soil moisture and reduced the population densities of the pathogen (Lodha, 1996). A significant reduction in the severity of dry root rot for two successive years in all the weed amended plots except Polycarpaea signified that organic amendments were important in the control of *M. phaseolina* (Mawar and Lodha, 2006). Efficacy of mustard oil-cake (4 t ha^{-1}) and cauliflowers leaf residues (5 t ha^{-1}) combined with summer irrigation and/or solarization was ascertained on the population of *M. phaseolina* (Lodha *et al.*, 1997). One summer irrigation of the dry non amended plots caused 40 per cent reduction in *M. phaseolina* counts at 0-30 cm depth. Amending soil with cruciferous residues augmented the efficiency of irrigation by eliminating a sizeable proportion of *M. phaseolina* in non-solarized plots (Israel *et al.*, 2005). Soil solarization of amended and irrigated plots elevated soil temperature by 4-6°C compared to non-solarized plots. Combined effect of moisture, amendments and temperature completely eliminated viable propagules of *M. phaseolina*, irrespective of soil depth. In yet another study (1998-2000) combined effects of *Brassica* amendments (mustard oil - cake or pod straw -2.5 t ha^{-1}) and summer irrigation on survival of *M. phaseolina* and *Foc* in soil, and charcoal rot intensity on cluster bean (July-Oct) and wilt of cumin (November-March) was studied in the same field. Both the amendments were significantly superior in reducing incidence of both the diseases (Mawar and Lodha, 2002). Of the residues, mustard oil-cake was significantly more effective than pod straw with a 34 per cent greater reduction in wilt incidence.

Effectiveness of different doses of *Brassica* amendments in reducing viable propagules of *M. phaseolina* was ascertained in field (Lodha and Sharma 2002). Mustard oil-cake was significantly better where complete reduction in viable propagules was achieved at 0.9 per cent (2 t ha^{-1}) compared to 0.22 per cent (5 t ha^{-1}) with mustard pod residues. Under shade, magnitude of reduction in *M. phaseolina* propagules was low but significant improvement in the reduction was estimated with increased concentration of amendments. Since mustard oil- cake was considered expensive, a need was felt to improve the efficiency of mustard pod residue by integrating it with other easily available, cost-effective practical management strategies. Therefore, effects of soil solarization, residue incorporation, summer irrigation and biocontrol agents alone or in combination on survival of *M. phaseolina* and *Foc* were ascertained (Israel *et al.*, 2005). Combining amendments and soil solarization elevated the soil temperatures by 0.5-5°C and 2.5-13.0°C compared to non-amended solarized and non-solarized plots, respectively. These treatment combinations significantly reduced *M. phaseolina* and *Foc* propagules compared to control. Suppressive effect of composts on dry root rot was studied at CAZRI, Jodhpur in a series of experiments. After summer exposure 4 week old composts of pearl millet, clusterbean, weeds and cauliflower residues (each @ 4 t/ha) were incorporated to a depth of 0-30 cm in separate plots and cluster bean crop was grown. Dry root rot severity was significantly reduced and seed yield of cluster bean was enhanced by amendment with different composts. Amendment of soil with different composts or FYM or weed residues resulted in significant reduction in plant mortality (Table 16.3) due to charcoal rot compared to non-amended control (Mawar and Lodha, 2008, 2009; Bareja *et al.*, 2010, 2013).

Table 16.3: Effect of different Compost Amendments on Dry Root Rot Mortality Induced by *M. phaseolina* and Seed Yield of Cluster Bean

Compost	*Mortality (per cent)*	*Seed Yield (Kg ha^{-1})*
Pearl millet	3.2	403
Clusterbean	4.7	436
Weeds	5.5	449
Cauliflower leaves	4.1	461
Control	10.7	313

Soil Solarization

Soil mulching with transparent polyethylene (LLDPE) sheet during summer effectively increases the soil temperatures and decreased the soil saprophyte (*M. phaseolina* from 350 to 7 sclerotia g^{-1}) after 15 days of application (Lodha,1995).

Biological Control

Bacillus firmus, a native strain isolated from Brassica residues amended soil was found antagonistic to *Macrophomina phaseolina* (Lodha *et al.*, 2013). Soil microbial population was increased when biocontrol agents and soil amendments were used together (Singh *et al.*, 2012). Further, biological control agents like *Glomus mosseae, G. fasciculatum, G. constrictum, Gigaspora gilmorei* efficiently control the pathogen through various mechanisms. *Aspergillus versicolar* produces an antagonistic substance mycoversilin (Samanta *et al.*, 1983) that reduces the pathogen inoculums. *Bacillus firmus, Trichodemra* spp. and *Pseudomonas fluorescens* are other promising bio-control agents (Dhoke and Kurundkar, 2006). Incorporation of Zinc or Copper (5kg/ha) or Sulphur (10kg/ha) coupled with FYM also bring down the dry root rot incidence. Bavistin seed treatment @ 2 g per kg seed and spray with the same @

0.1 per cent is suggestive measure to control the disease. Lodha and Burman (2004) reported CAZRI 27-1 as a prominent resistant variety.

Bacterial Diseases

Bacterial Blight

Pathogen

Xathomonas axonopodis pv cyamopsidis

This disease was first recorded in India from Patna (Patel *et al.*, 1953), the disease is widespread in all the cluster bean growing regions and occurs in sever forms causing yield losses of 32 – 68 per cent.

Symptoms

The bacterium is internally seed borne (Shrivastava and Rao, 1963). Symptoms on the leaves first appear as water soaked spot. The spots enlarge into initially flaccid but then brown and necrotic lesions with lemon-yellow margins. Lesions often coalesce to cause such extensive tissue damage that defoliation occurs. Wilting is evident if the pathogen invades the plant vascular system. Yellow exudates may form on lessons. Disease is favored by spattering rains, high humidity (80-90 per cent) and moderate temperature (27-30°C). Further, 4-6 weeks old crop is most susceptible. Light and nitrogen conditions are also associated with disease intensity in reciprocal manner. According to Yadav *et al.* (2005) crop sown on last week of July showed highest disease incidence (58.3 per cent) while Khatri (2009) recommended that sowing on mid of the July is most suitable for clusterbean in the Bikaner region. High nitrogen doses (20-40 kg/ha) may also result in higher disease index. Plants at early growth (20-30 days) may be more susceptible, leading to their drying due to stem rotting.

Management

In India, under arid and semi arid conditions the National Network Research Project on Arid Legumes (2014) provided the management of bacterial blight of cluster bean under epiphytotic conditions which are as follows:

1. Under arid conditions (Bikaner), treatment of streptocycline 500 ppm+ streptocycline 250 ppm + copper oxychloride (0.2 per cent) sprays at 15 days interval, which is further followed by streptocycline @500 ppm and 2 spray of copper hydroxide at 15 days interval provided minimum disease intensity and higher grain yield
2. Under semi arid condition (Hanumangarh) treatment of streptocycline 250 ppm + copper oxychloride (0.1 per cent) 2 spray at fifteen days interval. Streptocycline 500 ppm + streptocycline 250 ppm + copper oxychloride (0.2 per cent) at 15 days interval provided minimum disease intensity and higher grain yield.

Seed dip for two hours in streptocycline (0.025 per cent) was highly effective (Lodha and Anantha Ram, 1993). Even hot water treatment at 56°C for 10 min was found to eradicate the seed borne inoculums (Singh and Swarup 1986). Among the biological control agents *Penicillium citrinum, Bacillus subtilis* and *Pencillium citrinum* are most promising and environmental friendly management tool.

Bacterial Brown Spot

Pathogen

Pseduomonas cyamopsicola and *P. syringae* pv. *syringae*

Bacterial brown spot is the most economically significant disease and it occurs in many warmer parts of the world. The bacterium *P. syringae* pv. *Syringae* primarily survives as an epiphytes on weed host, and in plant debris it can survive over one year. Pathogen spread through windblown rain or overhead sprinkler irrigation. Due to infection necrotic yellow green zone appears on leaves. Necrosis followed by bacterial oozing and water soaking. The leaf tissue around the lesion may be wrinkled and on pods, dark brown lesions are generally small and cause pod malformation by inciting cessation of growth of nearby tissue. Overcast, cloudy conditions and high humidity are the major predisposing factors, especially if such conditions immediately follow rain or irrigation. Moderate and warm temperatures are conducive to disease development. High wind velocity correlates with plant susceptibility. Control measures like crop rotation is highly recommended. Sprays with copper containing chemicals and use of resistant cultivars are also recommended.

Virus Diseases

In general clusterbean is regarded as a local lesion host for many viruses such as bean top necrosis, rose mosaic vines, sun hemp mosaic virus, strawberry mottle virus, tobacco necrosis and top necrosis virus. Thus this arid legume is regarded as a potential local lesion indicator by Hansen and Lesemann (1978). They had reported occurrence and characteristics of a seed transmitted potyvirus from Indian African and North American Guar. Mosaic disease of guar is characterized by mottling and crinkling of leaves with intermediate stunting. Subsequently this reduced the plant height and other morphological feature of plant along with phytomass (Begum and Madhusudan, 1989). Ben-Moshe *et al.* (1991) reported a new green-sterile disease of guar in South Africa, which is transmitted through *Myzus persicae*. In this disease the infected plants produces smaller leaves and had fewer, often sterile, inflorescences' along the stem than did healthy plants. In an advance stage of the disease, about 50 per cent seeds infected. A leaf curl disease of guar with Geminivirus like symptoms has been consistently reported from various parts of India (Khan *et al.*, 2003). The symptoms consisted of curling, shortening and malformation of leaves, shortening of internodes and stems with overall stunting of diseased plant. Previous studies demonstrated the association of an unknown begomovirus with this disease. The disease agent was transmitted by whitefly (*Bemisia tabaci*) from diseased to healthy guar.

Integrated Management of Guar Diseases

Integrated disease management approaches for Guar are as follows:

- ☆ Growing short duration varieties like HG-365, HG 563, RGC-936 to escape the disease incidence
- ☆ Using hot water treatment of seed (56 °C) for10 minutes before sowing
- ☆ Soil solarization in the month of May is found most effective for managing soil borne plant pathogens
- ☆ Spraying Streptocycline @ 150 ppm twice at 15 days interval at the appearance of BLB
- ☆ Spraying Dianthane M-45 @ 0.2 per cent thrice at an interval of 15 days at the appearance of Alternaria Blight, also spray imidacloprid once to check insect pests
- ☆ Use kitazin @ 2 ml/kg seed along with Chloropyriphos @ 4 ml/kg and streptocyline @ 1g/kg seed to check the presence of insect pests, fungi and bacterial during storage
- ☆ Using mustard residue (2.5 t/ha) + One summer irrigation in crop free period before sowing to check pre and post emergence seedling mortality
- ☆ Spraying the crop with wettable cream and copper @ 1 kg/ha at an interval of 15 days at the appearance of PM disease (Yadav and Shalendra, 2014)

Cowpea (*Vigna unguiculata*)

Cowpea is mainly grown in tropical and sub-tropical regions in the world for vegetable and grains and to lesser extent as a fodder crop. It is a most versatile pulse crop because of its smothering nature, drought tolerant characters, soil restoring properties and multi-purpose uses (Oyewale and Bamaiyi, 2013). It is found to be suitable for both normal and late sown condition during rainy seasons. The total area in Rajasthan is about 1.0 lakh hectare with an estimated production of 0.50 lakh tons of grains. It is also suitable for inter and mixed cropping. Rust, blight, charcoal rot, anthracnose and mosaic are the most important diseases occurring in this state.

Fungal Diseases

The fungal diseases of cowpea are numerous that includes: Damping-off (*Pythium* spp.) may occur on seedlings under moist conditions and in dense plantings. Root rot (*Rhizactonia* spp.) and stem rot (*Fusarium* spp.) may also be a problem. Cowpea is susceptible to powdery mildew (*Erysiphe polyqoni*) during wet winter months and under humid conditions. Other diseases that affect cowpea include anthracnose (*Colletotrichum lindemuthianum*), charcoal rot (*Rhizactonia bataticola*), and fusarium wilt (*Fusarium oxysporum vr. tracheiphilum*). Major fungal diseases, their pathogen and plant part affected are summarized in Table 16.4.

Anthracnose

Pathogen

Four species of *Colletothrichum i.e. C. lindemuthuanum, C. capsici, C. truncatum* and *C. dematium* have been reported to cause the disease in cowpea crop. However *C. destructivum* has also been reported to cause the infection (Shen-Song *et al.*, 2001).

Table 16.4. Fungal Disease of Cowpea and their Casual Organisms

Disease Name	Pathogen	Plant Part Affected
Alternaria leaf spot	*Alternaria cassia*	Leaf (Grangeand and Aveling, 1998)
Cercospora leaf spot	*Cercospora canescens*	Leaf (Amadi, 1995)
Choanephora pod rot	*Choanephora cucurbitarum*	Pods and fruits (Munoz and Tamayo, 1994)
Colletotrichum stem disease	*Collectotrichum dematium*	Stem (Smith *et al.*, 1999)
Anthracnose	*Colletotrichum destructivum*	Whole plant (Akinbode and Ikotun, 2008)
Brown blotch	*Colletotrichum capsici*	Whole parts Latunde-Dada *et al.*, 1999 and Latunde-Dada and Lucas, 2007)
Fusarium wilt	*Fusarium oxysporum* f. sp.*tracheiphilum*	Seedling (Smith *et al.*, 1999)
Fusarium wilt	*Fusarium oxysporum* f. sp. *Vasinfectum*	Seedling (Ushamalini *et al.*, 1997b)
Blight	*Macrophomna phaseolina*	Seedling (Ratnoo *et al.*, 1997)
Leaf spots	*Mycospherella cruenta*	Leaf (Emechebe and Lagoke, 2002)
Phomopsis leaf spot	*Phmopsis longicola*	Pods and fruits (Roy and Ratnayake, 1997)
Leaf smut	*Protomcopsis phaseoli*	Folia (Santos *et al.*, 1997)
Phytium soft rot	*Phythium aphanidermatum*	Stem (Bankole and Adebanjo,1998)
Damping off	*Pythium ultimum*	Seedling (Aveling and Adandonon, 2002)
Stem canker	*Phytopthora cactoorum*	Stem or root (Fernando and Linderman, 1997)
Phytopthora stem rot	*Phytopthora vignae*	Stem or root (Fernando and Linderman, 1997)
Scab	*Sphaceloma* species	Hypocotyls and epicotyles (Emechebe and Lagoke, 2002)
Basal stem rot/wilt	*Sclerotium rolfsii Sacc*	Stem (Muqit *et al.*, 1996)
Brown rust	*Uromyces appendiculatus*	Foliar (Heath, 1998)
Web blight rot	*Thanatephorus cucumeris*	Root or seedling (Emechebeand Lagoke, 2002)

Anthracnose disease remains a devastating health problem to cowpea crop and an equated hindrance to its economic cultivation. The major afflicting pathogen *C. destructivum* has the virulence of one hundred per cent (100 per cent) infection on a single crop stand (that is every part of the crop is subject to attack and infection by *C. destructivum* at a given pathogenic situation, (Obi and Barriuso-Vargas, 2014). *C. destructivum* is a polycyclic, hemibiotrophic and produces anthracnose symptoms within 96 hours of inoculation (Latunde- Dada *et al.*, 1999), in the form of lesions as small angular brown spot on the leaf petiole, the lower surface of leaves and leaf vein of cowpea (Adebitan and Ikotun, 1996). These various spots created later coalesced to produce a brick red to brown discoloration of the entire leaf. Symptoms are usually delayed until production of flowers/pods (Obi and Barriuso-Vargas, 2014). Infected seeds are the primary source of infection. Damp, warm and high humidity are the major predisposing factors for cow pea. Dispersal takes place through agencies like air-current, water and contact.

Management

Control strategies include cultural, use of resistant cultivar, use of botanical extracts, chemical and biological agents. Adegbite and Amusa (2008) have reported that anthracnose incidence and severity can be reducing by applying intercropping practices. Ogu and Owoeye (2013) have reported the potential of leaf extract of *Cyathula prostra* and *Diodia scandens* for controlling anthracnose disease of cowpea in Africa. Genotypes like Sel-263, COVY2, C15, V26, RC25 and RC48 are resistant for this pathogen. Plant extracts of *Piper nigrum, P.betle, Cymbopogon citrates, Ocimum gratissiumum* and *O. sanctum* are the best botanicals tested and evaluated at filed levels (Adegbite and Amusa, 2008). Biological control agents like *Trichoderma viride* and *T. harzianum, Bacillus subtilis* significantly controls the *C.lindemuthianum.* Further chemicals like carbendazim, mancozeb, benomyl and thiram are also recommended.

Charcoal Rot (Damping Off)

Pathogen

Charcoal rot caused by *Macrophomna phaseolina* (Tassi) Gold.

It is prevalent in all cowpea growing areas of Haryana and Rajasthan. Symptoms of the disease are first noticed at the collar region of the plant as dark brown to black lesions encircling the stem partly or wholly which later enlarge upward and downward and after a few days cover the whole plant giving affected tissues a charcoal appearance. Charcoal colouration may or may not be contiguous on stems or branches. Production of sclcrotia a hard mass of mycelium is common while fungus was cultured artificially and these bodies were also present in affected tissues of root and stem. The sclcrotia remain in soil on plant debris and constitute main source of infection. Pycnidia were also noticed on some of the varieties as dot like bodies on stem surface formed at the maturity of plant which are sub-merged under the epidermis. High temperature (30-35°C) and low soil moisture are the major predisposing factors. Seed the primary source of infection where pathogen survive with the seed (both internal as well as external) while, the secondary infection take place through air borne condia or scleortia.

Management

Soil solarisation (Lodha *et al.,* 1997), planting later-maturing cultivars (Pearson *et al.,* 1984) and crop rotation (Singh *et al.,* 1990) and use of, antagonistic bacteria and fungi (Pal *et al.,* 2001; Singh *et al.,* 2002; Baird *et al.,* 2003) are the major management practices. Ndiaye *et al.* (2007) suggested combination of solarisation and organic amendment for better management of this disease. Among the fungal antagonists, *Trichoderma* spp. are generally the most frequently reported. Application of *T. harzianum* as seed treatment, suspension for soil drenching or wheat husk bran culture reduced infection of *Rhizoctonia bataticola* to 18 per cent, 28 per cent and 14 per cent, respectively, as compared to 70 per cent in the control variant (*Parakhia et Vaishnav,* 1986). Afouda *et al.* (2012) had identified *Bacillus subtilis, B. sphaericus* and *Paenibacillus polymixa* as an effective antagonists. They reported that *B. subtilis* strain A-11 reduced the disease incidence by 89.92 per cent. Lodha *et al.* (2013) reported that *Bacillus firmus* a bio control agent is specially showed the antagonistic activity against *M. Phaseolina*. Among the plant species, extracts of *Vitex negunda, Adenocalymma alliaceum* and *Ocimum sanctum* effectively controls the sclerotial germination of *M. phaseolinai* (Ushamalini *et al.*, 1997a and b).

Cercospora and *Pseudocercospora* Leaf Spot

Pathogen

Cercospora leaf spot is caused by pathogen *Cercospora canescens* while Psedocercospora leaf spot is induced by *Mycosphaerella cruenta* in the form of its anamorph, *Pseudocercospora cruneta. Cercospora* leaf spot is characterized mostly by circular to irregular cherry red to reddish-brown lesion on both surfaces, while, *Pseudocercospora* leaf spot is characterized by chlorotic or necrotic spots on the upper leaf surface and profuse masses of conidiophores and conidia, appearing as downy gray to black mats, on the lower leaf surface (Adegbite and Amusa, 2008). During non-crop period both pathogens survive on infected crop residue and in infected seed. Disease intensity linearly associated with humid temperature and dissemination of spore carried out through wind and rain splash. Management strategies involve sanitation, roguing, seed treatment with thiram or captan, benomyl, mancozeb or with zineb.

Brown Rust

Pathogen

It is induced by *Uromyces phaseoli var. vignae* and *U. appendiculatus*.

In this disease the particular rust symptoms appear 56 days after the crop plantation. High humidity, cloudy weather, heavy rain with temperature ranges of 22-28°C are the major predisposing factors. Rust disease physiology was studied by Heath (1998), he reported the involvement of reactive oxygen species in hypersensitive response of cowpea to infection by the rust. He proved that rust fungi initially negate nonspecific defensive responses in both resistant and susceptible cells and part of the establishment of biotrophy and he also suggested that the hypersensitive response in the cowpea-cowpea rust fungus pathogen system is not triggered by an oxidative burst. In India, Rangaiah (1997) investigated the inheritance of resistance in resistant cowpea genotype and showed that a minimum of two genes control resistance. Brown rust control with the fungicides was also studied in India by Kale and Anahosur (1996) who found that triadimefon and mancozeb sprays effectively controlled brown rust in cowpea. From the management perspective, Cherian *et al.* (1995) identified B-766, 774, 1484 and 1485 as highly resistant lines to *U. appendiculatus*. Further, according to Chandrasekhar *et al.* (1989) adjustment in sowing dates can minimize the disease incidence, early sown crop (July) shows less susceptibility compared to late sown (September). Foliar sprays of 1000 ppm oxycarboxin or mancozeb, diclobutrazol are also recommended.

Cowpea Leaf Smut

Pathogen

Protomycopsis phaseoli (Ramak and Subram) is the casual agent of the cowpea leaf smut. This disease is frequently present and reported from Nigeria. Pathogen formed dark ash-grey to sooty-black lesion. False smut occurs mostly in humid and fertile soil causing yield losses between 23 and 48 per cent (Adejumo and Ilotun, 2003). 26-27°C temperature and 70-82 per cent humidity are the major predisposing factors for this disease. Cowpea cultivars like IT85F-584-1 and TVu 4031 and TVu 11067 were found resistant to leaf smut (Adejumo *et al.*, 2001 and Adegbite and Amusa, 2008). *Bacillus* sp., *Aspergillus fumigates, Fusarium oxysporum, Trichoderma harzianum, T. koningii* and Yeast are the potential biological control agents of *P. phaseoli* (Adejumo *et al.*, 1999). Cultural practices like roguing, crop rotation and zero tillage are highly recommended disease management tools.

Web Blight and Root Rot Seedling Diseases

Pathogen

It is caused by *Thanatephorus cucmeris* (Frank) Donk and its anamorphic state, *Rhizoctonia solani* Kuhn

The disease is spread by airborne basidiospores, mycelia bridges between plants, rain splashed sclerotia and infected soil debris (Gody-Lutz *et al.*, 1996). Web blight is induced by aerial types, usually belonging to AG-1, while the strains that induce root rots/seedling diseases are soil borne in nature, in contrast to the aerial strain. The two phases of the disease complex are seed transmitted (Adegbite and Amusa, 2008). Web blight pathogens infect leaves and many other young stem tissues. Initial symptoms are small circular brown spot which enlarge and often show concentric banding and become surrounded with irregular shaped water soaked areas. Under humid conditions the lesions develop rapidly and coalesces leading to extensive blighting and defoliation (Allen and Lenne, 1998). From India web blight was reported by Verma and Mishra, (1989). Practices like use of clean seed, avoidance of dense planting and adjustment in sowing time (avoid peak rainfall period) can minimize the disease incidence. Seed treatment with chloroneb (2g/kg seed) and thiabendazol was found most effective in controlling the seedling rot and spray of oxycarboxin (0.16kg a.i./ha) has been found effective in reducing the foliar damage (Gupta *et al.*, 1998). The phylloplane phase of disease can also be managed by using bio-pesticides. It has been observed that the use of *Trichoderma koningii* spore suspension twice on artificially inoculated foliage with *R. solani* can restrict the extent of disease by causing antagonistic effect and results have been found to be significantly better over mixture of chemical protecting (Gupta *et al.*, 1998). Organic management with *Trichoderma harzianum, Pseudomonas fluorescens* was reported by Vavilapalll *et al.* (2014).

Powdery Mildew

Pathogen

It is caused by *Erysiphe polygoni*. The disease is worldwide in distribution and prevailing more in tropical America and India. In this disease mycelium forms superficial, scattered spreading patches, at first white turning greyish on leave and other plant parts. The affected plants are dwarfed; leaves turn yellow and curl. Cool night temperature and low humidity are major predisposing factors. Pathogen survives as mycelium or as active infections on volunteer plants. The disease can be managed by spraying the crop with wettable sulphur, application of benomyl, tridemorph and dinocap.

Fusarium Wilt

Pathogen

Disease is caused by *Fusarium oxysporum* f. sp. *tracheiphilum*.

Fairly wide-spread, reported from North and South America, Asia and Australia. Fusarium wilt usually causes the lower leaves on one side of the plant to turn yellow. Infected plants usually are stunted and wilted as the organism develops in the food and water conducting tissues. Brick red tissue can be observed in the stem when it is split lengthwise. The pathogen is soil-borne and probably also seed transmitted.

Management

Use of resistance varieties like IT84S-2049 and 524B; Cultural practices like long crop rotation, roguing and sanitation, soil nutrient management; Chemicals like thiram, capton and Use of biological agents like *Glomus mosseae, Glomus intraradices, Trichoderma harzianum* and *Trichoderma viride* (Ramezani and Khaniki, 2014).

Sclerotium Blight and Rot

Pathogen

Sclerotium rolfsii is responsible for seedling decay, seedling bight, stem rot and crown rot in cow pea.

The wide spread nature of this pathogen has been reported by several workers (Saxena *et al.*, 1998; Richard *et al.*, 2002; Adandonon *et al.*, 2004 and 2006). The fungus infects the bases of stems producing a fan of silky white mycelium and large round sclerotia which are initially white and gradually darken. Infected plants wilt and die. The mycelium and presence of sclerotia serve to distinguish this disease from Pythium stem rot which it otherwise resembles. This pathogen is an unspecialised parasite capable of extensive saprophytic growth in surface layers of soil, persisting on crop residues and weed hosts. The sclerotia are disseminated by cultivation, wind and water, and occasionally as contaminants amongst seed.

Management

Use of healthy seeds, clean cultivation and two applications of Captafol @ 0.3 kg/100 1 at weekly intervals may also be effective. Adandonon *et al.* (2006) have utilized various biocontrol agents in combination with *Moringa oleifera* extract and they had reported that *Moringa* extract concentration of 15 kg leaves 10^{-1} water (W/V) showed highest disease control. Further seed treatments using *Trichoderma*

Kd 63 and soil sprinkle using *Trichoderma* IITA 508 had a significant effect on a disease incidence.

Septoria Leaf Spot

Pathogen

Septoria leaf spot of cowpea is induced by three species of *Septoria*, namely *S. vignae, S. vignicola*, and *S. kozopolzanskii* (Emechebe and Shoyinka 1985 and Emechebe and Florini, 1997).

The most prevalent and most economically important across Africa is *S. vignae*, with reports of *S. vignicola* in East Africa and of *S. kozopolzanskii* in Zimbabwe (Mariga *et al.*, 1985). By contrast, *S. vignicola* has been consistently reported as the pathogen of the disease in India (Rawal and Sohi 1981, 1984, 1986), while *S. vignae* is a minor pathogen in Nicaragua (Lin and Rios 1985). Lesions caused by *S. viginicola* are dark red, circular to irregular, 2-4 mm diameter, appearing similar on both leaf surfaces. Spots are often concentric ringed and sometime raised, giving the leaf a freckled appearance. Black fruiting bodies (pycnidia) on the lesions contain septate, threadlike conidia. Pathogen is seed borne in nature and high humidity is the major predisposing factor for it. Management practices include crop rotation, spraying with thiabendazole @225-315 g a.i/ha or benomyl.

Bacterial Diseases

In cowpea two major bacterial diseases are bacterial blight and bacterial pustule. The former one is induced by *Xanthomonas campestris* pv. *vignicola* while X. *campestris* pv. *vignicola* is responsible for latter one. Their biochemical characteristics, phylogenetics studies and management approaches have been addressed (Vauterin *et al.*, 2000; Allen *et al.*, 1998; Sikirou, 1999 and Emechebe and Lagoke 2002).

Bacterial Blight

This disease is widely affecting the crop in the Africa, America and India. Seedling mortality resulting from seed borne infection may be up to 60 per cent. Appearance of tiny water soaked dots on leaves is the first symptoms of this disease. These dots remain small and the surrounding tissue dies, developing a tan to orange coloration with a yellow halo. Pathogen also infects the stem, causing cracking (stem canker) and causes water soaking of pods from where the pathogen enters the seed. The disease is primarily seed borne, which are the primary source of inoculums, while, the secondary spread of disease is usually from the infected parts of the plant. 70-95 per cent relative humidity and 25-30°C are the major predisposing factors for this disease. For management, use of hot water treatment for eliminating this seed borne pathogen is most effective. Seed treatment with streptomycin sulphate and copper oxychloride are also recommended.

Viral Diseases

Viruses are known to be a major constraint to production wherever cowpeas are grown in the world. Hampton *et al.* (1997) have listed nine viruses considered most damaging to cowpea, seven of which that are seedborne being the following:

Blackeye cowpea mosaic potyvirus (BlCMV), *Cowpea aphid-borne mosaic potyvirus* (CABMV), *Cucumber mosaic virus* (CMV), *Cowpea mosaic comovirus* (CPMV), *Cowpea severe mosaic comovirus* (CPSMV), *Southern bean mosaic sobemovirus* (SBMV), and *Cowpea mottle carmovirus* (CPMoV). The two non-seedborne viruses considered important by Hampton *et al.* (1997) are *Cowpea golden mosaic begomovirus* (CGMV) and *Cowpea chlorotic mottle bromovirus* (CCMV). They listed about eight other viruses considered to be of minor or undetermined importance. In India, cowpeas are known to be infected spontaneously at least by 11 different viruses belonging to the taxonomic group of *Alfalfa mosaic virus* (AMV), *Cowpea mild mottle carlavirus* (CpMMV), *Cowpea mosaic comovirus* (CPMV), *Southern bean mosaic sobemovirus* (SBMV), cowpea cucumoviruses, cowpea tobamoviruses, cowpea potyviruses, cowpea nepovirus, cowpea golden mosaic geminivirus, urdebean leaf crinkle virus (ULCV) (Ojuederie *et al.*, 2009).

Alfalfa Mosaic Virus (AMV)

This virus reported from Iran and India with little disease incidence, however, seed yield reduces 20-50 per cent upon artificial inoculation. In cowpea this virus is transmited through aphid in a non-persistent manner and other than cowpea its seed borne in nature. Infected plant shows typical yellow mosaic foliage symptoms. From the management perspective no resistance to AMV in cowpea has been found: however, two cowpea lines PI-228920 and PI-293474 have been reported from European countries.

Cowpea Potyviruses

Potyviruses that infect cowpea includes blackeye cowpea mosaic (BICMV), cowpea aphid borne mosaic (CAMV), cowpea green vein banding (CGVBM) and Peanut mottle (PMV). In India BICMV and CAMV are frequently reported but not PMV (Mali 1986, Mali and Thattappilly, 1986 and Ojuederie *et al.*, 2009). BlCMV occurs wherever the crop is grown. It is particularly damaging when it occurs in combination with other viruses (Hampton *et al.*, 1997). The virus contains single standard RNA. It causes cytoplsmic cylindrical inclusions and associated scrolls in its hosts. Disease caused by BICMV and CAMV can be controlled by the use of resistant varieties and virus free seeds. Both potyviruses are transmitted through seed, therefore, establishment of seed certification practices are prerequisite. In India S-288 cultivar possessed immunity to BICMV and hypersensitive resistance to CAMV.

Cowpea Mosaic Comovirus (CPMV)

Two strain of CPMV namely CPMV-SB and CPMV-Vu have been reported from India (Khurana, 1998). CPMV is transmitted through mechanically, grafting as well as by seed. Insect transmission is carried out by thrips (*Sericothrips occipitalis* and *Taeniothrips sjostedti*) and by grasshopper's species. In insect vector the optimum acquisition and inoculation feed periods are 10 minutes and 24 hours, respectively. Virus is characterized by isometric, non-envelop particles about 20-24 nm in diameter. It has two kinds of nucleoprotein particles with single stranded RNA. Its thermal inactivation point ranges between 55-60°C, dilution end point at 10^{-5}.

Sainsbury *et al.* (2014) have conducted its genetic engineering and characterization of cowpea mosaic virus empty virus like particles. Vector-control with insecticides and the use of resistant cultivars are two promising control methods of CPMV. As the virus is seedborne, care should be taken with importation of seed and these preferably, should be quarantined. Resistant cultivar includes Victor, K-798, VITA-3, Braham, K-892, Groit, and Arlington.

Cowpea Cucumoviruses

Cucumoviruses causing cowpea diseases are variously described from different countries. Cowpea infecting cucmovirus so far reported include cucumber mosaic virus -cowpea strain (CMV-CS), CMV bean seed borne strain (CMV-BS), CMV-Cowpea or ring spot isolate (CMV-RS), CMV-banding mosaic isolate (CMV-BM) and cowpea ring spot virus (CpRSV). In India, CMV BM is found more pathogenic than CNV. All the cowpea cucumoviruses are characterized by isometric particles about 25 to 30 nm in diameter. Serologically CpBMV and CpNV are distinct. General symptoms includes crinkling and leaf size reduction, vein banding. The infected plants show concentric and necrotic ring-spots, venial necrosis, necrotic streaking of stem and stunting.

All these cucmovirueses are seed-borne in cowpea except CpNV. However, the level of seed transmission has been found to vary among cowpea cucumorviruses, which ranges 4-26 per cent for CMV-CS, 15-30 per cent for CMV-BM, 15-20 per cent for CpRSV and 3.5-18.5 per cent for CMV-RS. It has been established that the level of seed transmission varies between cucumoviruses and is dependent upon virus-cowpea cultivar reactions. All the cowpea cucmoviruses except CpRSV are transmitted by aphid vectors in a non-persistent manner. Aphid vectors includes *Aphis craccivora, A. euonymie, A fabae.* Because non-persistent virus -vector relationship, use of insecticides is lesser effective compare to use of virus free seed. Physical treatment like application of dry heat treatment for 15 minutes at 65°C followed by 2,4 or 8 days incubation at 30°C eliminates seed transmission of the virus without affecting seed viability. Chemical treatments includes, soaking of seeds in maleic-hydrazide at 40, 100 or 400 ppm for 90 minutes, 2-thiouracil at 500 or 700 ppm for 60 minutes, Naphthalene acetic acid (NAA) at 40 ppm for 240 minutes and teepol (5-10 per cent) for 240 minutes eliminates seed transmission of the virus (Khurana, 1998).

Cowpea Tobamoviruses

This viral disease of cowpea described and reported from many parts of the world with different vernacular names. In general TMV-CS causes only a mild green mottle on trifoliate leaves and leaf malformation ad reduction and dwarfing of vines in highly susceptible cowpea lines. Thermal inactivation point of this virus ranges from 80-85°C, while, dilution end point between 10^{-5} to 10^{-7}. Host range of this virus includes *Glycine max, Phaseolus lathyroides, Vigna vexillata, Chenopodium amaranticolor, C. tetragonoloba* and *Physalis floridana.* Control method includes uses of virus free seeds and resistant cultivar.

Southern Bean Mosaic Sobemovirus (SBMV)

This disease was reported from India, Pakistan USA, Ghana and Nigeria. Infected seedling shows mottling chlortic lesions and dark green vein banding along the major veins. Host range includes *Cassia tora,* and *Vigna sinensis*. This virus is isometric seed-borne as well as vector born (aphid). Use of insecticides, virus free planting and resistant cultivar like RC-80, HC-2 and CH-86.

In brief we can categorize the viral diseases of cowpea into three categories

Seed Borne Viruses	*Non-Seed Borne Viruses*	*Combination of Viruses Causing Diseases*
☆ Blackeye cowpea mosaic potyvirus (BICMV) ☆ Cowpea aphid borne mosaic potyvirus (CABMV) ☆ Cowpea severe mosaic comovirus (CPSMV) ☆ Southern bean mosaic sobemovirus (SBMV) ☆ Cowpea mosaic virus (CPMV) ☆ Cowpea mottle carmovirus (CPMoV) ☆ Cucumber mosaic cucumovirus (CMV)	☆ Cowpea golden mosaic geminivirus (CPGMV) ☆ Cowpea chlorotic mottle bromo virus (CCMV)	☆ BICMV+CMV ☆ BICMV+CPSMV ☆ CMV+CPSMV+SBMV

Moth Bean (*Vigna aconitifolia*)

Moth bean also called dew gram or mat bean, is one of the most drought-tolerant grain legume grown during the monsoon either as sole or mixed with pearlmillet. The seed is used, either whole or split, as a pulse. It is also used in making a salted snack called Bikaneri Bhujia. It is a quality forage in arid and semi arid areas. In India, moth bean is grown in Rajasthan, Gujarat, Maharashtra, Haryana and Punjab on 1.3 million hectares. Rajasthan covers about 90 per cent of the total area and accounts for 80 per cent of the total production. The crop occupies an important place in the grain-legume economy of the state. The area and production of the moth bean fluctuates from year to year mainly because of erratic rainfall and occurrence of several diseases in its growing tracts. Several fungal plant pathogens such as *S. rolfsii, A. alternata, R. solani, M. phaseolina, C. capsici* are involved in causing serious damage at any stage of the Moth bean.

Fungal Disease

From Indian arid and semi arid regions 6 fungal diseases on *Vigna aconitifolia* in Jodhpur, Rajasthan, caused by *Alternaria alternata* (leaf spot), *Colletotrichum truncatum* (causing leaf spot and pod infection), *Curvularia lunata* (*Cochliobolus lunatus*) (leaf spot), *Fusarium moniliforme* (*Gibberella fujikuroi*) (seedling rot) and *Myrothecium roridum* (pod infection) are reported by Singh and Srivastava (1989) and *Macrophomina phaseolina* by Garg and Gupta (2010). Mishra *et al.* (2011) and Surekha *et al.* (2013) have quantified the biocontrol efficacy of *Trichoderma viride* against moth bean fungal pathogen like *Rhizoctonia solani, Sclerotium rolfsii, Macrophomina phaseolina, Alternaria alternata, Fusarium solani* and *Colletotrichum capsici.* Many of these fungal pathogens are seed borne in nature and affect various traits of seeds like colour, germ inability, size, yield *etc.*

Dry Root Rot

Pathogen

Root rot incited by *Macrophomina phaseolina* (Tassi) Goid has been rated as most destructive disease of mothbean. The pathogen attacks on all parts of plant *i.e.* root, stem, branches, petioles, leaves, pods and seeds. Moreover, seed infection of *Rhizoctonia bataticola* (*M. phaseolina*) ranges from 2.2-15.7 per cent which causes 10.8 per cent in grain yield and 12.3 per cent in protein content of seed in mungbean. The infected seeds act as an important source of primary inoculums for new areas (Kumari *et al.*, 2012). Hot and dry weather are the major predisposing factors for this disease. Infected seedling has black sclerotia of fungus. Pre-emergence plant mortality varied from 2.5-24.6 per cent. Integrated management includes use of FYM, seed treatment with carbendazim and use of genotypes like IPCMO-138 and IPCMO-522. Javid and Saddique (2011) have reported that *M. phaseolina* root rot in mungbean can effectively be managed by amending the soil with 1.5 per cent dry leaf manure of *D. metel.* Kumari *et al.* (2012) have studied the integrated management by using five different types of herbal oils, three plants products and seven fungicides. According to them all herbal oils showed complete inhibition of mycelial growth of pathogen at 2 per cent concentrations. *Asafoetida* was found least effective. The relative efficacy of herbal oils and plant products under pots house condition exhibited palmarosa oil to be the most effective as seed dresser in reducing root rot incidence. Among fungicides, bavistin was reported as the most effective to inhibit mycelial growth of pathogen as well as reducing root rot incidence, followed by captan or thiram, indofil M-45 and vitavax or raxil, while copper sulphate was the least effective treatment in both conditions. In the case of organic manures, vermin-compost was the most effective in reducing the root rot incidence under pots conditions. FYM and goat manure was found moderately effective in controlling root rot incidence. Integrated management approach showed that vermin-compost and carbendazim in combination was more effective in reducing the root rot incidence in pots conditions. Experiment conducted by National network Research Project on Arid Legumes (ICAR-Indian Institute of Pulses Research, Kanpur) at Bikaner station (2014-15) have reported the management of root rot through local isolated bio agents under epihpytotic condition at Bikaner. According to their trial, treatment combination of *T. harzianum* + *P. fluorescens* seed treatment (4+4 gm/kg seed) + soil application of *T. harzianum* + *P. fluroescens* (1.25 +1.25 kg in 50 kg FYM for each//ha) was found most effective with disease incidence (19.0 per cent over 41.25 per cent)

Pod Rot

Pathogen

The disease is initiated by pathogen *Choanephora cucurbitarum*. Pathogen mostly attacks tissues that have been damaged by insects or mechanical means, or crops that are poorly adapted to a hot humid climate. In the initial stage of disease progress, water soaked lesion appear on green pods which turned in brownish black in color. No data is available about the yield reduction. However, for management point of view spraying of mancozeb, 0.2 per cent or copper oxychloride, 0.2 per cent can be recommended.

Brown Web Blight

Pathogen

Caused by *Rhizoctonia bataticola* that survive in soil and in plant debris. Heavy rains and high temperature are the major predisposing factors and temperature of 25-30°C and RH of 85 per cent is most congenial for the development of this disease. Disease occurs through soil, seeds and naturally infected hosts. Losses due to this disease are more severe at the seedling stage. Light small round web like patches can be seen on both surfaces of the leaves. Every plant part but flower can bear the patches. Inclusion of non-leguminous crop in the rotation is advised. Close planting may be avoided. Seed treatment with Benlate and Brassicol @ 2g/kg seed may result in complete elimination of seed-born infection.

Bacterial Disease

Bacterial Leaf Spot

Pathogen

Xanthomomas phaseoli was identified as the causal bacterium of this disease. This disease is most common in Rajasthan. It spreads rapidly during high humidity. Many small, large and irregular brown necrotic spots appear on the leaves proving very severe on the leaves. These patches are more prominent on upper than on lower leaf surface. Minute water-soaked round irregular spots appear in group on leaf surface, their sizes increase and turn brown to black. In the extreme cases, leaf may fall down. Ultimately petioles, stems and pods may show extended brown spots.

Management

For management perspective, the genotypes *viz.*, IC-8833, Amravati local and PLM-11 were found resistant under artificial inoculation against Bacterial leaf spot. Other cultivars *viz.*, RDM~63, RDM-168 and RDM-182 also showed resistance to Bacterial leaf spot under artificial inoculation. Seed treatment with Streptocyline (0.01 per cent) + Captan (2 g kg^{-1} seed) was found better in reducing primary infection of Bacterial leaf spot/blight. The intensity of Bacterial leaf spot blight could be reduced by three sprays of Blitox (0.3 per cent)/Streptocycline (0.01 per cent) + Blitox (0.3 per cent)/Bavistin (0.05 per cent) + Blitox (0.3 per cent) (Kumar, 2002).

Viral Diseases

Yellow Mosaic Disease

Yellow mosaic is regarded as the most serious and devastating viral disease of moth bean, it starts appearing on the young leaves and considerably affects growth, pod set and yield of moth bean plants. Yellow mosaic causing virus is transmitted through white flies (*Bemisia tabaci*). It may occur at any stage of plant growth, however, if occurs at the initial stage may result in slow growth, delayed flowering, poor pods, great loss to the canopy chlorophyll, causing yield reduction even upto 90 per cent (Kumar, 2002). Reports also indicate that yield losses depending on the intensity and severity of disease may reach upto 100 per cent. Initial visible signs are the appearance of yellow spots scattered over the leaf, expanding rapidly. Leaves depict yellow patches alternatively with green areas, the latter may turn yellow. Completely yellowish leaves may give whitish look and ultimately may become necrotic. Plants may show decreased growth. Plants bear few, small and curled pods which bear few and shrivelled seeds. Plants show distinct stunted growth, golden yellow look of the leaves and their curling behaviour. Local varieties are highly susceptible to YMV infection; these may be replaced by YMV resistant varieties, CAZRl Moth-l, Jawala and early maturing ones, like RMO-40, RMO-257 and RMO-225. These varieties have shown field resistance against YMV. (Population build-up of vector (white flies) may help disease spread rapidly. Studies from Fatehpur revealed that vector of YMV could be controlled by single spray of monocrotophos (0.04 per cent) at 35 DAS. Rogor (0.02 per cent) sprayed twice at the interval of 15 days has also proved most effective. Guar is the most effective trap crop for checking white flies and, therefore, spread of YMV.

Horsegram

Horsegram is scientifically known as *Macrotyloma. uniflorum* Lam. It also goes by the name *Dolichos uniflorus* due to a lot of confusion in the *Dolichos* category. Horsegram is native to the old world tropics. It was probably domesticated in India where its cultivation is known since prehistoric times. Now a day's horse gram is cultivated as a low grade pulse crop in Southern Asia mainly from India to Myanmar, it is also grown as a forage and green manure in many tropical countries especially Africa. Horse gram is recorded to occur wild or neutralized in Central, East and Southern Africa. It is a popular pulse, locally known as Gaheth belongs to the family Fabaceae that still remain an under exploited legume crop (Mehra and Upadhyaya, 2013). Traditionally, it has been widely used in the treatment of kidney stones, inflamed joints, fever, musculoskeletal disorders, sinus wounds and localized abdominal tumours. Experimentally, the seeds are reported as hepato-protective, diuretic and antioxidant (Das *et al.*, 2014).

Fungal Disease

Aspergillus flavus, Aspergillus niger, Alternaria alternata, Fusarium equiseti, Helminthosporium sp., *Mycelia sterilia, Sclerotium* sp. and *Scopularopsis* sp. are the major fungal pathogen infecting horsegram. Most of them are seed born in nature (Narayan and Krishirasagar, 2013).

Anthracnose

Pathogen

Colletrotrichum lindemuthianum and *C. truncatum*.

The fungus causes dark brown elongated, more or less angular spots along the veins on the petioles, stem and lamina. When infection occurs on the hypocotyl, the plant collapses. Seeds when infected turn brown or black. Crop rotation, sanitation, roughing as well as seed treatments with captan, carbendazim, mancozeb are recommended for its managment. Udaya *et al.* (2012) have evaluated 21 genotypes and 4 released varieties (AK-21, AK-26, Palem-1 and Palem 2) against anthracnose, they reported that out of 21 genotypes, 11 were moderately resistant, eight were susceptible and two were resistant against *C. lindemuthianum*.

Cercospora Leaf Spot

Pathogen

The disease is initiated by *Cercospora dolichi* which is seed and air borne in nature. Water soaked lesions on the leaf blade appear followed with powdery growth of fungus. Seed treatment with captan @3 g/kg seed and spraying the infected plant with mancozeb can reduce the disease intensity. National Network Research Project on Arid Legumes (ICAR-Indian Institute of Pulse Research, Kanpur, 2014-15) has evaluated nine genotypes of horesegram against *Cercospora* leaf spot at S. K. Nagar. They have found that all the nine genotypes (CRHG-23, CRHG-19, CRHG-22,BGHG-1, VLG-39,BHG-13-2,VLG-15 (C), VLG-38 and BHG-13-1 were moderately resistant to this disease with disease intensity ranged between 0.067-3.67 per cent.

Powdery Mildew

Pathogen

Levillula taurica (Lev.) Arnaud is the casual agent, which is endophytic and remain sub-cuticle within host.

The pathogen is air borne in nature and infected host shows white powdery growth on the leaves which spreads to cover the whole plant. Management practices include the cultural practices like adjustment in sowing date, crop sown during 1st and 2nd fortnight of the August yielded higher compared to crop sown during the 1st fortnight of September (Naik and Rangaswamy, 1994), use of resistant variety like KBHG-1, chemical like calixin, dinocap and tridemorph and biological control (hyperparasitism) with *Amelomycis quisqualis*.

Other lesser extent fungal disease of horsegram includes dry root rot (*Macrophomina phaseolina*), rust (*Uromyces phaseoli*), rhizoctonia aerial blight (*Rhizactonia solani*).

Viral Disease

Yellow Mosaic

Caused by Mung Bean Yellow Mosaic virus that belongs to Gemini viruses. In initial stage of disease small spots on young leaf appear followed by discoloration of infected tissue. Infected plant produces small and distorted pods. Virus transmitted through insect vector *Bemisia tabaci*. Disease management involves: roguing of infected plants, adjustment in sowing time (first fortnight of August and September); use of resistant cultivar like PHG-9 and PHG-11; vector management with systemic insecticide like monocrotpohos.

Future Concerns

If we go for a brief introspection into the work done on legume diseases in Rajasthan, we would find some of the original research of the highest order. A number of new diseases of these crops have been reported, sources of resistance identified and management strategies worked out. Diseases are one of the major biotic factors, which limiting production in dry regions. Such biotic constraints may be soil, seed or vector born. Considering the present review of diseases of arid legumes and their subsequent research revealed that in last two decades many efforts have been put forth for development of resistant genotypes/varieties, their integrated management approaches (includes, biological, cultural and chemical) as well as enhancing their productivity. In addition to these agronomic practices, efforts have also been conducted with physiological, biochemical and molecular aspects. Despite of all these efforts, our researches and farmers still not satisfying the growing demands of arid legumes. The gap between demand and production have been summarized by many agencies like National Institute of Agricultural Marketing (2013, 2014 and 2015), ICRISAT, NBPGR, ICAR *etc.* Singh *et al.* (2015) have provided three different approaches *viz.*, vertical, horizontal and policy intervention for improving legume production. In Vertical approach, possible methods and technique related to achieve more production without expanding crop area, here focus is on to improve productivity of crop per unit area, while, horizontal approach includes the efficient utilization of fallow/wastelands for arid legume production. Considering their recommendations following research efforts at laboratory, field and end-user level are needed for prioritization

- ☆ Modeling of crop loss estimation with different modules that can provide the reference values between theoretical, attainable and actual yield. Hot spot of each disease should be identified.
- ☆ Trait identification (disease resistant, drought and salinity tolerance *etc.*) isolation and their incorporation in desire arid legume.
- ☆ Legumes are mostly grown in water deficient farming system but the chemical sprays require consumption of this scarce commodity. Research efforts are therefore required to integrate resistance genotype, moisture conservation practices that reduce disease incidence, inter and mixed cropping, soil amendments, biological control, use of trap or decoy crop, early and late sowing varieties *etc.* Dedicated efforts of multi disciplinary team of scientists in different agro-climatic zones can meet this requirement.
- ☆ Epidemiological studies at present have not received adequate attention. These studies should be undertaken for diseases that are prevalent all over the state. The information generated will help in the development of sound disease forecasting systems.
- ☆ Development of multiline varieties and gene deployment.
- ☆ Identification of predisposing factors to avoid possible epidemic.
- ☆ Identification of cropping combination and cropping sequence that may enhance productivity and reduce the disease incidence.

- ✰ Encouraging accelerated adoption of present technology for bridging the yield gap.
- ✰ Institutional support to boost seed replacement rate and quality production as well.
- ✰ Strengthening of life-saving irrigation in pulse growing pockets.
- ✰ Guaranteeing availability of critical inputs *viz.*, seed fertilizer, pesticides.
- ✰ Gradual mechanization for arid legume production.
- ✰ Supply of quality seed material (breeder, founder and certified seed).
- ✰ Public–Private partnership for sustaining chain and to minimizing post-harvest losses and policy support for value chain for pulse.
- ✰ Solid government policies related with minimum support price, specifically when crop failure occur due to biotic epidemic and/or biotic constraints.

REFERENCES

Adandonon, A. Aveling, T. A. S., Labuschagne, N. and Tamo, M. 2006. Biocontrol agents in combination of *Moringa oleifera* extract for integrated control of *Sclerotium-* caused cowpea damping off and stem rot. *European Journal of Plant Pathology*, 115: 409-418.

Adandonon, A., Aveling, T. A.S. and Tamo, M. 2004. Occurrence and distribution of cowpea damping-off and stem rot and associated fungi in Benin. *Journal of Agricultural Science*, 142: 561-566.

Adebitan, S.A. and Ikotun, T. 1996. Effect of Plant Spacing and Cropping Pattern on Anthracnose *Colletotrichum lindemuthianum* of Cowpea. *Fitopatología Brasileira*, 21: 5-12.

Adegbite, A. A. and Amusa, N. A. 2008. The major economic field diseases of cowpea in the humid agro-ecologies of south-western Nigeria. *African Journal of Biotechnology*, 7 (25): 4706-4712.

Adejumo, T.O. and Ikotun, T. 2003. Effect of planting date on incidence and severity of leaf smut of cowpea in northern Nigeria. *Moor Journal of Agricultural Research*, 4: 106-110.

Adejumo, T.O., Florini, D.A. and Ikotun, T. 2001. Screening of cowpea cultivars for resistance to leaf smut. *Crop Protection*, 20: 303-309

Adejumo, T.O., Ikotun, T. and Florini, D.A. 1999). Biological control of *Protomycopsis phaseoli*, the causal agent of leaf smut of cowpea. *Journal of Phytopathology*, 147: 371-375.

Afouda, L. C., Schulz, D., Wolf, G. and Wydra, K. 2012. Biological control of *Macrophomina phaseolina* on cowpea (*Vigna unguiculata*) under dry conditions by bacterial antagonists. *International Journal of Biological and Chemical Sciences*, 6 (6): 5068-5077.

Afouda, L., Wolf, G. and Wydra, K. 2009. Development of a sensitive serological method for specific detection of latent infection of *Macrophomina phaseolina* in cowpea. *Journal of. Phytopathology*, 157: 15–23

Akinbode, O.A. and Ikotun, T. 2008. Evaluation of Some Bio agents and Botanicals in *In Vitro* Control of *Colletotrichum destructivum*. *African Journal of Biotechnology*, 7: 868-872.

Allen, D.J. and Lenne, J.M. 1998. Diseases as constraints to production of legumes in agriculture. In Pathology of Food and Pasture Legumes. Allen DJ, Lenne JM (Eds.). CAB International, Wallingford, UK. pp. 1- 61.

Amadi, J.E. 1995. Chemical Control of *Cercospora* Leaf Spot Disease of Cowpea (*Vigna unguiculata* (L.) Walp.). *Agrosearch*, 1: 101-107.

Annual Report, 2014-15. National Network Research Project on Arid Legumes. ICAR-ICAR-Indian Institute of Pulses Research, Kanpur. p. 171.

Aveling, T.A.S. and Adandonon, A. 2002. First Report of Pre and Post Emergence Damping-Off of Cowpea Caused by *Pythium ultimum* in South Africa. *Plant Disease Reporter*, 84: 922-926.

Baird, R.E., Watson, C.E. and Scruggs, M. 2003. Relative longevity of Macrophomina phaseolina and associated mycobiota on residual soybean roots in soil. *Plant Diseases*, 87: 563-566.

Bankole, S.A. and Adebanjo, A. 1998. Efficacy of Some Fungal and Bacteria Isolates in Controlling Wet Rot Disease of Cowpea Caused by *Pythium aphanidermatum*. *Journal of Plant Protection* 11: 37-43.

Bareja, M, Kumar, P and Lodha, S. 2010. Effect of composts on microbial dynamics and activity, dry root rot severity and seed yield of cowpea in the Indian arid region. *Phytopathologia Mediterranea* 49 381-392.

Bareja, M., Mawar, R., Mathur, M and Lodha, S. 2013. On –farm waste-based composts suppressive to *Macrophomina phaseolina* inciting dry root rot of guar in an arid environment. *Australasian Journal of Plant Pathology* 42: 9-16.

Begum, H. and Madhusudan, T. 1989. Effect of mosaic virus infection on growth and yield of clusterbean. *Indian Journal of Mycological Research* 27: 191-194.

Ben-Moshe, H., Rey, M. E. C., Sibara, M. and Garnett, H. M. 1991. A new green sterile disease of Guar in South Africa. *Plant Diseases*, 75 (6): 638-643.

Chandaeria, U. K. 2008. Studies on Alternaria blight of Guar, Ph.D. Thesis. Jiwaji University, Gwalior, India.

Chanderia, U. K. and Singh, R. 2009. Crop-weather disease interaction in *Cyamopsis tetragonobloba* L. (Taub.) in semi arid environment. In: Legumes in Dry Ares (Eds. D. Kumar, A. Henry, and K. P. R. Vittal). Indian Arid Legumes Society and Scientific Publishers (India, Jodhpur, India. 502-505

Chandershekhar, M., Kumar, T. B. A., Saifulla, M. and Yadahilly, Y. H. 1989. Effect of different dates of sowing cowpea on the severity of leaf rust caused by *Uromyces phaseoli* var. *vignaei*. *Tropical Agriculture*, 66: 149-152.

Cheian, S., Anil, Kumar, T. B. and Rawal, R. D. 1995. Evaluation of vegetable cowpeas for slow rusting resistance. *Plant Disease Research*, 10: 22-27.

Das, S., Vasudeva, N. and Sharma, S. 2014. Chemical composition of ethanol extract of *Macrtyoloma uniflorum* (Lam.) Verdc. Using GC-MS spectroscopy. Organic and Medicinal Chemistry Letters, 4 (13): 1-4

Dhingra O D and Sinclair J B 1978. Biology and pathology of *Macrophomina phaseolina.* Universrsidade Federal de vicosa, MG, Brasil, pp. 166.

Dhoke, P. K. and Kurundkar, B. P. 2006. *In vitro* efficacy of fluorescent *Pseudomonas* isolates against *Macrophomina phaseolina. Indian Journal of Pulses Research*, 19: 131-132.

Emechebe, A. M. and Florini, 1997. Shoot and pod diseases of cowpea induced by fungi and bacteria. *Advances in cowpea research* In: Singh, B.B., D.R. Mohan Raj, K.E. Dashiell, and L.E.N. Jackai (eds.). Co-publication of International Institute of Tropical Agriculture (UTA) and Japan International Research Centre for Agricultural Sciences (JIRCAS). UTA, Ibadan, Nigeria. Pp. 176-192.

Emechebe, A.M. and Lagoke, S.T.O. 2002. Recent Advances in Research on Cowpea Diseases in Challenges and Opportunities for Enhancing Sustainable Cowpea Production. *Proceeding of the World Cowpea Conference III*, the International Institute of Tropical Agriculture (IITA), Ibadan, 4-8 September 2000, 94-123.

Emechebe, A.M., and S.A. Shoyinka. 1985. Fungal and bacterial diseases of cowpeas in Africa. Pages 173-192 in Cowpea research, production and utilization, edited by S.R Singh and K.O. Rachie. John Wiley and Sons, Chichester, UK

Fernando, W.G.D. and Linderman, R.G. (1997) The Effect of Mycorrhizal (*Glomus intraradisces*) Colonization on the Development of Root and Stem Rot (*Phytophthora vignae*) of Cowpea. *Journal of the Natural Sciences*, 25: 39-47.

Garg, B. K. and Uday, B. 2009. Drought physiology of arid legumes. In: Legumes in Dry Ares (Eds. D. Kumar, A. Henry, and K. P. R. Vittal). Indian Arid Legumes Society and Scientific Publishers (India, Jodhpur, India. 94-109.

Garg, N. and Gupta, H. 2010. Isolation and purification of fungal pathogen (*Macrophomina phaseolina*) induced chitinase from moth beans (*Phaseolus aconitifolius*) Journal of Pharmaacy and Bioallied Sciences, 2 (1): 38-43.

Gody-Lutz, G., Arias, J., Steadman, J. R. and Eskridge, K. M. 1996. Role of natural seed infection by the web blight pathogen in common bean seed, damage, seedling emergence and early disease development. *Plant Disease*, 80 (8): 887-890

Grangeand, N.L.A. and Aveling, T.A.S. 1998. First Report of *Alternaria cassiae* on Cowpea. *Plant Disease*, 82: 1171.

Gupta, P. P. Arora, R. N. and Gandhi, S. K. 1998. Disease of kharif fodder crops and their management. In: *Diseases of Field Crops and their Management* Thind, T. S. (Ed.). National Agricultural Technology Information Centre, Ludhiana, India. pp 373-430.

Gupta, P. P., Gupta, V., Singh, J. V. and Singh, V. P. 1998. Alternaria blight disease management strategy in clusterbean, proceeding of 7th International Congress of Plant Pathology, Edinburgh, Scotland.

Gupta P P., Gupta V., Kumar R., Singh J V and Jatasara D S 2003. Effect of single and multiple inoculation of cluster bean on growth, nutrition and tolerance to infection by *Macrophomina phaseolina*. Human impact on desert environment. 305-315.

Gupta P P., Gupta V., Singh J V., Punia A and Kapoor R 2003. Screening of cluster bean germ plasm against root rot and bacterial leaf blight. *Advances in Arid Legume Research*, 415-418.

Hampton, R.O., Thottappilly, G. and H.W. Rossel. 1997. Viral diseases of cowpea and their control by resistance-conferring genes. In: *Advances in cowpea research*, B.B. Singh, D.R. Mohan Raj, K.E. Dashiell, and L.E.N. Jackai (Eds.). Co-publication of International Institute of Tropical Agriculture (IITA) and Japan International Research Centre for Agricultural Science (JIRCAS). IITA, Ibadan, Nigeria P. 159–175.

Hansen, A. J. and Lesemann, D. E. 1978. Occurrence and characteristics of a seed transmitted potyvirus from India, Africa and North America Guar. American Phytopathological Society, 68: 841-847.

Heath, M.C. 1998. Involvement of reactive oxygen species in the response of resistant (hypersensitive) or susceptible cowpeas to the cowpea rust fungus. *New Phytologist* 138(2): 251–263.

Israel, S., Mawar, R. and Lodha, S. 2005. Soil solarization, amendments and bio-control agents for the control of *Macrophomina phaseolina* and *Fusarium oxysporum* f. sp. *cumini* in aridisols. *Annals of Applied Biology* 146: 300-310

Joshi, H. N., Gupta, P. P., Gupta, V. and Kumar, S. 2004. Biochemical factors in clusterbean that impart *Alternaria* blight resistance. *Journal of Mycology and Plant Pathology*, 34: 581-583.

Kale, J. K. and Anahosur, K. H. 1996. Chemical control of cowpea rust. Karnataka *Journal of Agricultural Sciences*, 9 (1): 179-181.

Khan, J. A., Sohrab, S. S. and Aminuddin. 2003. Guar leaf curl disease from India is caused by Tomato leaf curl virus. *Plant Pathology*, 52: 796.

Khatri, N. K. 2009. Effect of dates of sowing and varieties on development of bacterial blight on clusterban at Bikaner. In: Legumes in Dry Ares (Eds. D. Kumar, A. Henry, and K. P. R. Vittal). Indian Arid Legumes Society and Scientific Publishers (India, Jodhpur, India. 463-466.

Khurana, S. M. P. 1998. *Pathological Problems of Economic Crop Plants and their Management*, Scientific Publisher, Jodhpur, P. 655.

Kumar, D. 2002. Production Technology for Moth bean in India. Indian Turisum Publication. P. 41.

Kumar, D. 2005. Status and direction of arid legumes research in India. *The Indian Journal of Agricultural Sciences*, 75 (7): 375-391.

Kumar, D. and Rodge, A. B. 2012. Status, scope and strategies of arid legumes research in India- A review. *Journal of Food legumes*, 25 (4): 255-272.

Kumari, R., Shekhawat, K. S., Gupta, R. and Khokhar, M. K. 2012. Integrated management against root rot of Mungbean (*Vigna radiate* L.) incited by *Macrophomina phaseolina*. *Plant Pathology and Microbiology*, 3 (5): 1-5

Latunde-Dada, A.O. and Lucas, J.A. 2007. Localized Hemibiotrophy in *Colletotrichum*: *Cytological* and Molecular Taxonomic Similarities among *C. destructivum*, *C. linicola* and *C. truncatum*. *Plant Pathology*, 56: 437-447.

Latunde-Dada, A.O., O'Connell, R.J., Bowwyer, P. and Lucas, J.A. (1999) Cultivar Resistance to Anthracnose Disease of Cowpea (*Vigna unguiculata* (L.)Walp.) Caused by *Colletotrichum destructivum* O'Gara. *European Journal of Plant Pathology*, 105: 445-451.

Lin, M.T. and G.P. Rios. 1985. Cowpea diseases and their prevalence in Latin America. Pages 199-204 in Cowpea research, production and utilization, edited by S.R. Singh and K.O. Rachie. John Wiley and Sons, Chichester, UK

Lodha, S 1995. Soil solarization, summer irrigation and amendments for the control of Fusarium oxysporum f. Sp. cumini and Macrophomina phaseolina in arid soils. Crop Protection, 14: 215-219.

Lodha, S 1996. Influence of moisture conservation techniques on *Macrophomina phaseolina* population, dry root rot and yield of cluster bean. *Indian Phytopathology*, 49: 342-349.

Lodha, S and Anantha, Ram K 1993. Effectiveness of streptocycline spray schedule on the bacterial blight intensity and yield of clusterbean. *Annals of Arid Zone*, 32: 237-240.

Lodha, S and Sharma, SK 2002. Effect of natural heating and different concentration of *Brassica* amendments on survival of *Macrophomina phaseolina*. *Indian Phytopathology* 55: 303-305

Lodha S and Solanki K R 1993. Inheritance of dry root rot resistance in clusterbean. *Indian Phytopathology*, 45: 430-433.

Lodha S., Gupta G K and Singh S 1986. Crop disease situation and some new records in Indian arid zone. *Annals of Arid Zone* 25: 311-320.

Lodha S., Mathur B K and Solanki K R 1990. Factors influencing population dynamics of *Macrophomina phaseolina* in arid soils. *Plant and Soil* 125: 75-80.

Lodha, S., Sharma, S K., Mathur, B K., Aggarwal, R K 2003. Integrating sub-lethal heating with Brassica amendments and summer irrigation for control of *Macrophomina phaseolina*. *Plant Soil* 256: 423-430

Lodha S, Sharma S. K, Aggarwal R. K. 1997. Natural and solar heating of irrigated soil amended with cruficerous residue and improved control of *Macrophomina phaseolina*. *Plant Pathology* 46: 186-190

Lodha, S., Mawar, R., Chakarbarty, P K and Bhagvan Singh 2013. Managing *Macrophomina phaseolina* causing dry root rot of legumes by a native strain of *Bacillus firmus*. *Indian phytopathology* 66 (4) : 356-360.

Lodha, S. and Burman, U. 2004. Towards managing soil-borne plant diseases in arid region. Central Arid Zone Research Institute, Jodhpur, India, *Technical Bulletin*, 84 pp.

Mali, V. R. 1998. Cowpea virus diseases-the Indian scene. In: *Pathological Problems of Economic Crop Plants and their Management*, (Ed.) S. M. Paul. Scientific Publishers, India, pp. 133-157.

Mali, V.R. and Thottappilly, G. 1986. Virus diseases of Cowpea in the Tropics. In: Raychadhuri, S.P. and Verma, J.P. (Eds.). *Reviews of Tropical Plant Disease* 3: 361-403

Mariga, I.K., Giga, D. and P. Maramba. 1985. Cowpea production constraints and research in Zimbabwe. *Tropical Grain Legume Bulletin* 30: 9-14.

Mawar, R and Lodha, S. 2006. Relative efficacy of on-farm weeds as soil amendment for managing dry root rot of clusterbean in an arid environment. *Phytopathology Mediterranea*, 45, 215-224.

Mawar, R. and Lodha, S. 2002. *Brassica* amendments and summer irrigation for the control of *Macreophomina phaseolina* and *Fusarium oxysprum* f. sp. *Cumini* in hot arid region. *Phytopathology Mediterranea*, 41: 45-54.

Mawar, R. and Lodha, S. 2008. Efficacy of weed residues vis-a-vis their composts on survival of *Macrophomina phaseolina* and dry root rot incidence on clusterbean. *Indian Phytopathology* 62: 119-121.

Mawar, R. and Lodha, S. 2009. Prior weakening of *Macrophomina phaseolina* and *Fusarium* propagules for enhancing efficiency of *Brassica* amendments. *Crop Protection* 28: 812-817

Mehra, A. Upadhyaya, M. 2013. *Macrotyloma uniflorum* Lam. A traditional crop of kumaun Himalaya and etthnobotanical perspectives. *International Journal of Agricultural and Food Science*, 3 (4): 148-150.

Mishra, B., Mishra, K., Kumar, R. Mishra, R. Tiwari, C., Kumar, A., Singh, R. and Dikshit, A. 2011. Biocontrol efficacy of *Trichoderma viride* isolates against fungal plant pathogens causing disease in *Vigna radiate* L. *Archives of Applied Science Research*, 3 (2): p. 361.

Munoz, M.A. and Tamayo, M.P.J. 1994. The Process of *Choanephora cucurbitarum* in Cowpea [*Vigna unguiculata* (L.) Walp.]. ASCOLF Informa, 20: 4-5.

Muqit, A., Haque, M.S. and Hossain, M.M. 1996. Reaction of Cowpea Lines against *Sclerotiumrolfii*. *Bangladesh Journal of Plant Pathology*, 12: 63.

Naik, R. G. and Rangaswamy, S. D. 1994. Effect of date of sowing and application of chemicals on the incidence and control of powdery mildew in horsegram. *Legume Research*, 3 (4): 203-206.

Narayan, M. G. and Krishirasagar, A. D. 2013. Study of seed borne fungi of different legumes. *Trends in Life Sciences*, 2 (1): 32-34.

Ndiaye, M., Termorshuizen, A. J. and Van Bruggen, A. H. C. 2007. Combined effects of solarization and organic amendment of charcoal rot caused by *Macrophomina phaseolina* in the Sahel. Phytoparasitica, 35 (4): 392-400.

Obi, V. I. and Varguas-Barriuso, J. J. 2014. Situation of biofungicides reconnaissance, a case of anthracnose disease of cowpea. *American Journal of Plant Sciences*, 5: 1202-1211.

Ogu, G. I. and Owoeye, A. E. 2013. *In vitro* control of anthracnose disease of cowpea (*Vigna unguiculata* L. Walp) caused by *Colletotrichum destructivum*. *Journal of Intercultural Ethnopharmacology*. 2 (1): 29-36.

Ojuederie, O. B., Odu, B. O. and Ilori, C. O. 2009. Serological detection of seed borne viruses in cowpea regenerated germplasm using protein a sandwich enzyme linked immunorsorbent assay. *African Crop Science Journal*, 17 (3): 125-132.

Oyewale, R. O. and Bamaiyi, L. J. 2013. Management of cowpea pests. *Scholars Academic Journal of Biosciences*, 1(5): 217-226.

Pal, K.K., Tilak, K.V.B.R., Saxena, A.K., Dey, R. and Singh, C.S. 2001. Suppression of maize root diseases caused by M. phaseolina, Fusarium moniliforme and Fusarium graminearum by plant growth promoting rhizobacteria. *Microbiology Research*, 156: 209- 223.

Pandey R N., Pathak A R., Kher H R and Bharodia P S 1993. Evaluation of clusterbean (*Cyamposis tetragonoloba*) germplasm for resistance to bacterial blight (*Xanthomonas campestris* pv. *Cyamopsidis*) and powdery mildew (*Leveillala taurica*). *Indian Journal of agricultural Science*, 63: 68-70.

Patel M K., Dhande G W and Kulkarni Y S 1953. Bacterial leaf spot of *Cyamopsis tetragonoloba* (L.) Taub. *Current Science*, 22: 183.

Patel, D.S., Patel, S.I., Desai, B.G., and Khandar, R.R. 1998. *Neocosmospora vasinfecta* an incident of wilt of clusterbean. *Indian Phytopathology*, 51 (3): 305-309.

Pearson, C.A.S., Schwenk, F.W., Crowe, F.J. and Kelly, K. 1984. Colonization of soybean roots by Macrophomina phaseolina. *Plant Diseases*, 68: 1086-1088.

Ramezani, H., Khaniki, G. B. 2014. Evaluation effect of bio-agents on *Fusarium* wilt of cowpea. International Journal of Developmental Research, 4 (10): 2100-2103.

Rangaiah, S. 1997. Inheritance of resistance to *Uromyces phaseolus* in *Vigna unguiculata* (L.) Walp. *Crop Improvement*, 24 (2): 251-252.

Rao, A. N. S. 1989. Seed treatment of cowpea with macozeb against seed and soil borne diseases. *Current Research. University Agricultural Science*, Bangalore, 18: 7-8.

Ratnoo, R.S., Jain, K.L. and Bhatnagar, M.K. 1997. Effects of Atmospheric Temperature on the Development of Ashy Stem Blight of Cowpea. *Journal of Mycology and Plant Pathology*, 27: 90-91.

Roy, K.W and Ratnayake, S. 1997. First report of *Phomopsis lengicolla* infection of cowpea pods and seeds in Mississippi. *Plant Diseases*, 81: 693.

Rawal, R.D and H.S. Sohi. 1981. Sources of resistance in cowpea to *Septoria vignicola*. *Indian Phytopathology*, 34(4): 504-505.

Rawal, R.D and H.S. Sohi. 1986. Studies on the control of leaf spot disease of cowpea caused by *Septoria vignicola*. *Pesticides* 20(9): 40-43

Rawal, RD. and H.S. Sohi. 1984. Assessment of losses in pod yield of cowpea due to *Septoria vignicola* Rao. *Indian Journal of Plant Pathology* 2(1): 59-63.

Richard, L. F., Philip, D. and Duke, Sr. 2002. Southern blight (*Sclerotium rolfsii* Sacc.) of cowpea: yield-loss estimates and sources of resistance. *Crop Protection*, 21: 403-408.

Saharan, M. S., Sahara, G. S. and Ram Niwas. 2001. Progress of Alternaria blight on different varieties of clusterbean in relation to environmental conditions. *Journal of Mycology and Plant Pathology*, 31: 220-226.

Saharan, M. S., Saharan, G. S. and Gupta, P. P. 2003. Conidial germination of *Alternaria cucumerina* var. *cyamopsidis* causing Alternaria blight of cluterbean. *Journal of Mycology and Plant Pathology*, 33: 444-445.

Sainsbury, F., Saxena, P., Aljabali, A. A., Saunders, K., Evans, D. J. and Lomonossoff, G. P. 2014. Genetic engineering and characterization of Cowpea mosaic virus empty virus-like particles. *Methods of Molecular Biology*, 1108: 139-153.

Samanta, A. K., Kole, H. K., Goswami, S. K. and Bose, S. K. 1983. Mycoversilin a new antifungal antibiotic from a mutant derivative of *Aspergillus versicolor*. *Indian Journal of Experimental Biology*, 21: 577-578.

Santos, A.A., Quindere, M.A.W. and Melo, M.B. 1997. Evaluation of Cowpea Genotypes for Resistance to Cowpea Smut (*Entyloma vignae*). *Fitopatologia Brasileira*, 22: 77-78.

Saxena, M., Saxena, D.R., Bhale, M.S. and Khare, M.N. 1998. Diseases of cowpea and their management. Pages 239-252. In: Diseases of Field Crops and their Management (ed. Thind T.S.), National Agricultural Technology Information Centre, Ludhiana, India.

Shen Song Hai, Goodwin P and Hsiang T 2001. Hemibiotrophic infection and identity of the fungus. *Mycological Research* 105: 1340-1347.

Shivanna, M B., Meera M S and Shetty H S 1992. Evaluation of seed dressing fungicides against the diseases of cluster bean. *Indian Phytopathology*, 45: 373-376.

Sikirou, R. 1999. Epidemiological investigations and development of integrated control methods of bacterial blight of cowpea caused by *Xanthomonas campestris* pv. *vignicola*. Ph.D. thesis. University of Gottingen, Germany. 218 pp.

Singh J. and Swarup J 1986. Control of bacterial blight of cluster bean by seed treatment and foliar spray. Pesticides 20: 23-24.

Singh, A. K., Singh, S. S., Prakash, V., Kumar, S. and Dwivedi, S.K. 2015. Pulse production in India: Present status and bottleneck and way forward. Journal of Agrisearch, 2 (2): 75-83.

Singh, S. K. and Srivastava, H. P. 1989. Some new fungal disease of moth bean. Indian Phytopathology, 42 (1): 164-167.

Singh, S.K., Nene, Y.L. and Reddy, M.V. 1990. Influence of cropping systems on *Macrophomina phaseolina* populations in soil. Plant Diseases, 74: 812-814

Singh, T., Srivastava, A.K. and Arora, D.K. 2002. Horizontal and vertical movement of Pseudomonas fluorescence towards exudates of *Macrophomina phaseolina* in soil: influence of motility and soil properties. Microbiology Research, 157: 139-148.

Singh, V., Mawar, R and Lodha, S. 2012. Combined effect of biocontrol agents and soil amendments on soil microbial populations, plant growth and incidence of charcoal rot on cowpea and wilt on cumin. Phytopathologia Mediterranea 51 (2): 307-316

Smith, J. E.,. Korsten, L. and Aveling, T. A. S 1999. Infection process of *Colletotrichum dematium* on cowpea stems. Mycological Research 103: 230-234.

Srivastava, D. N. and Roa, Y. P. 1963. Seed transmission and control of bacterial blight of guar *Cyamposis tetragonoloba* (L) Taub. Indian Phytopathology, 16: 390-391.

Surekha, C. H., Neelapu, N. R. R., Kamala, G., Prasad, S. B., Ganesh, P. S. 2013. Efficacy of *Trichoderma viride* to induce disease resistance and antioxidant responses in legume *Vigna mungo* infested by *Fusarium oxysporium* and *Altenraria alternate*. International Journal of Agricultural Sciences and Research, 3 (2): 285-294.

Udaya, S. A., Meena, J. K. V.S., Rajasri, M., Sunil, N., Kumar, S., Prasanthi, Y. and Anitha, K. 2012. Evaluation of horsegram (*Macrotyloma uniflorum*) genotypes against anthracnose disease). Indian Journal of Plant Protection, 40 (4): 301-305.

Ushamalini, C., Rajappan, K. and Gangadharan, K. 1997a. Inhibition of *Macrophomina phaseolina* and *Fusarium oxysporum* f. sp. *tracheiphilum* by Antagonists under *in Vitro* Conditions. Plant Disease Research, 12: 168-170.

Ushamalini, C., Tajappan, K and Gangadharan, K. 1997b. Control of cowpea wilt by non-chemical means. Plant Disease Research, 12: 122-129.

Vauterin, L., Rademaker, J. and Swings, J. 2000. Synopsis on the taxonomy of the genus *Xanthomonas*. Phytopathology 90(7): 677–682.

Vavilapalll, S., Celine, V. A. and Girja, V. K. 2014. Collor rot and web blight caused by *Rhizoctonia solani* Kuhn in vegetable cowpea (*Vigna unguiculata* (L.) Wapl.) and its organic management. 2nd International Conference on Agricultural and Horticultural Sciences. Feb. 3-5. Pp. 2-4.

Verma, J.S. and Mishra, S.N. 1989. Evaluation of improved lines from IITA in humid-subtropical India. Trop. Grain Legume Bulletin, 36: 38-39

Yadav, H. and Shalendra, 2014. *An Analysis Of Performance Of Guar Crop In India*. National Institute of Agricultural Marketing, Jaipur, India. Pp.98.

Yadav, S. C., Nath, R. and Yadav, R. K. 2005. Occurrence of bacterial blight (*Xanthomonas axonopodis* pv. *Cyamosidis*) on clusterbean. Journal of Mycology and Plant Pathology, 35 (3): 496-499.

Chapter 17

Cultural Practices: A Sustainable Option for Management of Grain Legumes Diseases

S.S. Vaish

Grain legumes, also called **pulses**, are plants belonging to the family *Leguminosae* (alternatively *Fabaceae*) which are grown primarily for their edible seeds. India is the largest producer, largest consumer and the largest importer of pulses in the world. For majority of vegetarian population in India, pulses are the major source of protein. Pulses and pulse crop residues are also major sources of high quality livestock feed in India. In India pulses are grown in around 24-26 million hectares of area producing 17-19 million tonnes of pulses annually. India accounts for over one third of the total world area and over 20 per cent of total world production. India primarily produces several grain legumes or pulses, *viz.*, chickpeas (*Cicer arietinum*), Pigeonpea (*Cajanus cajan*), lentil (*Lens culinaris*), peas (*Pisum sativum*), cowpea (*Vigna unguiculata*), mung bean/green gram (*Vigna radiata*), black gram/urd (*Vigna mungo*), cluster bean (*Cyamopsis tetragonoloba*), lima bean (*Phaseolus lunatus*), faba bean (*Vicia faba*), groundnut (*Arachis hypogaea*), soybean (*Glycine max*) and hyacinth bean, (*Dolichos lablab*). India is the top world producer of chickpeas and Pigeonpea.

Sustainable agriculture is the management and utilization of the agricultural ecosystem in a way that maintains its biological diversity, productivity, regeneration capacity, vitality and ability to function, so that it can fulfil- today and in the future - significant ecological, economic and social functions at the local, national and global levels, and that does not harm other ecosystems (Lewandowski *et al.*, 1999). The sustainability of agriculture has faced some of the most significant challenges in recent years (Hanson *et al.*, 2007; Oborn *et al.*, 2003). In addition, plant diseases continue to play a major limiting role in agricultural production.

The production of grain legumes is affected by several diseases such as root rot, wilt, mildews, blights and many bacterial and viral diseases. These diseases under favourable environment for pathogen cause huge losses and affect grain legume production seriously. These diseases are caused by seed, soil and airborne pathogens. However, soil borne diseases caused by soil-inhabitant long persistent pathogen, *viz., Rhizoctonia, Sclerotinia, Sclerotium, Pythium, Plasmodiophora etc.*, are difficult to manage. Effective management of diseases depends on over all integrated crop management system (ICMS). El-Zick and Frisbie (1985) defined ICMS as "a system whereby all interacting crop production and pest control tactics aimed at maintaining and protecting plant's health are harmonized in the appropriate sequence to achieve optimum crop yield, quality and maximum net profit, in addition benefiting society and mankind". The main objective of ICMS is to maintain plant's health throughout the growing season. "Plant's health" means relative freedom from the biotic and abiotic stresses that limit the plant's productivity.

Cropping system refers to a set of crop systems, which are making up the cropping activities of a farm system. Further, cropping system comprises all components required for the production of a particular crop and the interrelationships between them and environment. In other words, a cropping system usually refers to a combination of crops in time and space. Combination in time occurs when crops occupy different growing period and combinations in space occur when crops are inter planted. When annual crops are considered, a cropping system usually means the combination of crops within a given year.

In cropping system cropping pattern, the yearly sequence and spatial arrangement of crops and fallow on a given area plays its role on dynamics of diseases. Adequate adjustment in crop production techniques can modify the environment in such manner that it becomes unfavourable for the pathogen and pathogenesis. Cultural practices which include manipulation and/or adjustment of crop production techniques have been as old as possibly agriculture itself. The present chapter deals with various types of cultural practices, *viz.*, crop rotation, fallowing, mixed cropping, monoculture, adjustment of date of sowing, plant spacing *i.e.*, rate of sowing and density of stand, method of sowing/planting, irrigation water management, selection of seed and seed material (use of seeds of resistant varieties from certified sources), levelling of field and provision of drainage facilities, barrier cropping, yellow sticky traps, field and plant sanitation, roguing, decoy cropping, soil solarization, organic amendment of soil, management of plant nutrition and their effectiveness to manage diseases of pulses especially of seed and soil borne nature. Root rot and wilt has been the major problem of food legumes. Chickpea, lentil, and pea are susceptible to a number of soilborne fungal diseases. Seed and seedling rots are caused primarily by *Pythium* spp. and *Rhizoctonia solani.* Full season root rots and other root diseases are caused by *Aphanomyces euteiches, Fusarium solani, Sclerotium rolfsii, Rhizoctonia solani,* and *Pleiochaeta setosa.* Wilt of these crops is caused primarily by various form a special is of *Fusarium oxysporum. Fusarium udum* is a serious problem of Pigeonpea.

The most economical and durable control of *Fusarium* wilt is to grow resistant varieties. New races of a wilt pathogen have arisen due to increased selection pressure from growing resistant varieties in short rotations but have not outpaced the development of resistant cultivars. Thus, cultural practices which has also been component of biological control in its broader sense has their great importance in management of soil borne diseases of grain legumes by encouraging development of such microorganisms in the environment (especially soil) which destroy or suppress pathogens through antagonism.

The definition of cultural practices in the context of disease control is complicated. We suggest classifying cultural practices into three categories (Katan, 1996):

a) Cultural practices which are usually applied for agricultural purposes that are not related to crop protection, such as fertilization and irrigation. They may or may not have positive or negative side effects on disease incidence.

b) Cultural practices which are used solely or mainly for disease control, such as sanitation for the eradication of plant residues and flooding.

c) Cultural practices which are used for both agricultural purposes and disease control, such as crop rotation, grafting, composting and flooding.

When analysing the potential of cultural practices (CPs) as tools for the management of soil borne diseases, the following should be determined (Palti and Katan, 1997):

a) Level of disease reduction

b) Economic and technological feasibility of the method

c) Epidemiological aspects

d) Reproducibility of disease reduction by the CP

e) Possible long-term effects

f) Mechanisms of disease reduction, which should be investigated since they may provide tools for improving the method.

Crop Rotation

Crop rotations play very effective role in suppression of soil borne root inhabitant pathogens those having limited survival ability, dependent on their host debris, highly specialized and have low competitive saprophytic ability *e.g., Fusarium, Verticillium, Aphanomyces.*

The advantageous effect of crop rotation against various soil borne pathogens are found to be owing to their restricted host range. The host-dependent reproduction of most pathogens limits inoculum build-up (Garrett and Cox 2006). Alternations of dicotyledonous with monocotyledonous crops are effective in limiting the inoculum levels of many soil borne plant pathogens (Agrios 1997).

Crop rotation reduces diseases by means of starvation, contribution of antagonistic microorganisms and introduction of chemically different residues. There are two important considerations, firstly, the crops grown between susceptible crops should not favour the survival of the pathogen directly or indirectly *i.e.*, it should be immune; secondly, one should know about the survival structures of the pathogen and its longevity in soil. These considerations will help in selection of the crops as well as determining the length of the crop rotation. Crop rotation will not be helpful to manage diseases that are wind-blown or insect vectored from outside the area. Nor will it help control pathogens that can survive long periods in the soil without a host *i.e.*, soil inhabitant with high competitive saprophytic ability and long persistent in large proportion *e.g.*, *Rhizoctonia*, *Sclerotium rolfsii*, *Pythium etc.* Rotation, by itself, is only effective on pathogens that can overwinter in the field or be introduced on infected seeds or transplants. Of course, disease-free transplants or seed should be used in combination with crop rotation. The period of time between susceptible crops is highly variable, depending on the disease. The crop rotation of Pigeonpea – Sorghum was found effective against wilt disease caused by *Fusarium udum*. It is important to keep at least three years between the same types of pulse crop to allow for the breakdown of crop residue on which disease pathogens survive. Longer rotations may be required for *Aphanomyces euteiches*, due to long-lived resting spores in the soil. Since there are diseases that affect more than one type of pulse crops, it is still important to maintain at least two years between different pulse crops. Crotalaria, velvet bean, and grasses like rye are usually resistant to root-knot nematodes (Peet, 1996). Rotations like these will not only help prevent nematode populations from reaching economic levels, they will also help control plant diseases and insect pests.

Bean anthracnose fungus *Colletotrichum lindemuthianum*, which lives in soil for 1 to 2 years. They can be eliminated from soil by adopting 3 or 4 year rotation with non-host crops. Crop rotation is effective in the control of brown stem rot of soybean (*Cephalosporium gregatum*). The disease incidence can be reduced to a great extent by rotating with corn for 4 to 5 years between two soybean crops. Soybean seed infection by *Phomopsis* sp. can be reduced by rotating soybean with maize. For various other soilborne pathogens, *e.g.*, *Verticillium dahliae*, *Rhizoctionia solani*, root knot nematodes (*Meloidogyne* spp.) and root lesion nematodes (*Pratylenchus* spp.), the design of a proper rotation can be difficult because these pathogens are capable of infecting and/or surviving on multiple hosts.

Green manure or cover crops cultivated in wintertime can be part of the crop rotation. The main reason to grow a green manure crop is to protect soil from erosion and to prevent leaching of mineralized nitrogen. In narrow rotations with a high pressure of soil borne pathogens, the choice of the optimal green manure crop can be a challenge. For example, to reduce nitrate leaching in sandy soils in wintertime in the Netherlands, it is now obligatory to grow a green manure crop following maize cultivation.

Incorporation of several *Brassica* species has been shown to reduce disease incidence caused by *Rhizoctonia solani*, *Phytophthora erythroseptica*, *Pythium ultimum*, *Sclerotinia sclerotiorum*, or *Fusarium sambucinum* in potato. The underlying mechanism involves the production of toxic volatiles during decomposition of the cruciferous organic matter. Marigold (*Tagetes* spp.) is grown as a green manure to specifically suppress *Pratylenchus penetrans*, which is likely due to toxic plant exudates (Hiddink *et al.*, 2010). Crop rotation is a flexible disease management system that is capable of reducing disease losses caused by many soil borne pathogens. However, the need for rotating high-value crops with lower-value crops and the relatively high risk of losing a complete crop make this system often less attractive to farmers.

Fallowing

It is of three type *viz.*, dry, wet and flooding. Dry fallowing acts by means of starvation without crop while flooding by keeping soil flooded with excessive water. These are found effective against diseases caused by plant parasitic nematodes. Flooding the soil for seven to nine months kills nematodes by reducing the amount of oxygen available for respiration and increasing concentrations of naturally occurring substances such as organic acids, methane, and hydrogen sulfide that are toxic to nematodes (MacGuidwin, 1993). However, it may take two years to kill all the nematode egg masses (Yepsen, 1984). Flooding works best if both soil and air temperatures remain warm. An alternative to continuous flooding is several cycles of flooding (minimum two weeks) alternating with drying and disking (MacGuidwin, 1993).

Wet fallowing can be practiced by giving frequent irrigation during the fellow period to make the pathogen propagule susceptible to attack of saprophytes. Sclerotia of *Sclerotium rolfsii*, *Verticillium dehliae* and *Sclerotinia sclerotiorum* (that cause root rot and wilt diseases in pulses) get germinated quickly under the influence of alternate drying and wetting of soil and lysed by saprophytes. Diseases like *Macrophomina* root rot on different crop plants is controlled by following this method.

Mixed Cropping

Mixed cropping, *i.e.*, cultivation of multiple crops in the same field at the same time helps in reduction of disease incidence and development due to reduction in number of host plants, creation of a barrier by roots of non-host plants for obstructing the movement of pathogen in soil and release of toxic substances.

Mixed crops have been grown for ages and because of their yield stability mixed cropping is still practiced for this reason in tropical regions (Vandermeer 1990). Co-occurring crops compensate for failure of one of the crops due to soil and airborne pathogens, weeds, temperature-, and water stress (Vandermeer 1990). This kind of growth compensation is an important reason for mixed cropping.

Mixed cropping of pegionpea and sorghum shows its effectiveness against wilt disease of pegion pea owing to release of cynide (HCN) by sorghum roots which interferes with mitochondrial respiration of the fungus. In addition to this, sorghum roots act as a barrier for obstructing the movement of the pathogen in soil and also favour the growth of *Trichoderma* that antagonises *Fusarium*. Blight of pulse crop (*Phyllosticta phaseolina*) has been successfully overcome by growing pulses as a mixed crop with cereals like sorghum and pearl millet. Hiddink *et al.* (2010) also discussed various possible mechanisms involved in suppression of soil borne diseases or pathogens in mixed-cropping, *viz.*, host dilution, allelopathy, physical barrier, enhanced antagonistic populations, root camouflage, change in microclimate and can bring about ISR (induced by non-pathogenic microorganisms) or SAR (stress inducers like water stress, salinity, allelopathic substances, or pathogens).

The effect of host dilution will likely be a reduction in disease incidence rather than disease severity on infected plants. Host dilution might have direct *i.e.*, an effect on the pathogen itself as well as indirect (influencing other factors than the pathogen) effects on disease suppression in mixed crops. Pathogens with wide host range such as *R. solani*, slightly or moderately susceptible plants may serve as nutrient source without expressing striking disease symptoms, therefore, reduces the host dilution effect. Further, the intensity of root intermingling in mixed cropping may be an important factor for the interference processes and the level of disease suppression possibly determined by the crops or cultivars grown and their root architectures. Wu *et al.* (2001) defined allelopathy as any biochemical interaction among plants, including those mediated by microorganisms, resulting in either detrimental or beneficial effects on the interacting plants. Allelopathy plays a vital role in suppression of many soil borne diseases in soil environment. For example, delayed germination of spores of *F. udum*, causing wilt in Pigeonpea, has been credited to allelopathic substances exuded from sorghum roots. For the effectiveness of allelopathy in inhibiting rhizosphere-inhabiting pathogens, allelopathic substances should be there at adequately high concentrations in the micro sites close to roots where the pathogen is located and roots of mixed crops need to be in vicinity.

The antagonistic activity governed by the enhanced antagonistic populations has also been proposed as a key mechanism for disease reduction in mixed-cropping systems. Rhizosphere microbial communities, including pathogens, antagonists, and plant-growth-promoting bacteria are crop- and cultivar-specific and it might be worthwhile to investigate if these communities can be influenced judiciously by the choice of cultivars in a mixed-crop setting.

The efficacy of the biological control agent *Trichoderma harzianum* was tested on various levels of *Fusarium udum*, the pigeonpea wilt pathogen. Field plots infested with the pathogen at three inocula levels (log 3.04, log 4.98 and log 5.34 colony-forming units (cfu)/g of soil) were identified (Prasad, *et al.*, 2002). *Trichoderma harzianum* was applied as a seed treatment (10 and 20 g/kg seed) and as a soil amendment (10 and 20 g/9 m^2). The Trichoderma population increased to more than 10^8 cfu/g soil by 60 days in treated plots, whereas for seed treatments, fungal population reached a maximum of $10^{4.62}$ cfu/g soil within 45 days, and thereafter started to decline. At log 3.04 pathogen concentration, soil amendment, with *T. harzianum* at 10 and 20 g gave 42.9 per cent and 61.5 per cent disease control, respectively. Seed treatments resulted in less than 30 per cent disease control when compared to the check. The disease control achieved with fungicide seed treatment was insignificant. At log 4.98 and log 5.34 pathogen concentrations, disease control ranged between 22 per cent and 35.3 per cent for soil treatments. Even at the highest pathogen density (log 5.34), soil amendment with *T. harzianum* at 10 g gave about 30 per cent disease reduction. In seed treatment plots, disease control ranged between 4.36 per cent and 13.7 per cent. In general, soil application of *T. harzianum* was found to be more effective than seed treatment for disease suppression. These results suggest the need to augment soil applications of *T. harzianum* for obtaining effective control of Pigeonpea wilt.

Monoculture

Continuous cropping or monoculturing provides opportunity for perpetuation of pathogenic organisms in the soil when the same crop is raised year after year in the same field. In modern agriculture, cultivation of single crops in a rotation is the most common cropping system for a vast range of crop species worldwide. However, suppression of pathogens does occur with respect to continuous single-crop production.

Cultivation of single or closely allied crop species in annual or seasonal succession with interruption only by fallow or intermittent growing of soil amendment not necessarily after each crop is known as monoculture (Palti, 1981). Most of soil borne diseases are favoured by monoculturing due lack of diversity as the host is available periodically which provides the good disease perpetuating conditions. Two types of disease and pathogen development has been observed with respect to monoculturing as mentioned below:

Irreversible

Disease incidence increases until a certain level is reached and maintained due to high competitive saprophytic ability and or persistent survival structures and their tolerance to other soil organisms. For example, *Verticillium alboatrum* or *V. dahlia*, *Meloidogyne* spp.

Reversible

The of disease incidence first raising but subsequently declining to more or less fixed level due to their low competitive saprophytic ability. For example, *Phymatotrichum omnivorum*, *Gaumannomyces graminis etc.*

Certain diseases show natural suppression during continuous cultivation. Induction of disease suppression can take many years and generally it is vanished after growing other crops. The mechanisms involved have been studied extensively and are linked to the microbial community in soil or the rhizosphere. The best-known mechanisms include antibiotic production (*e.g.*, by strains of *Pseudomonas fluorescens*), competition by closely related non-pathogenic strain s (*e.g.*, competition for carbon by non-pathogenic *Fusarium oxysporum*), and parasitism (*e.g.*, by *Trichoderma* spp.). For these types of disease suppression to develop and to sustain, both the pathogen and a susceptible host plant need to be present and a certain level of damage has to be accepted.

Overall, adequate disease suppression in continuous mono-cropping system can be induced in several pathogen–crop combinations. However, other pathogens on the same crop can become problematic. Moreover, the unpredictable time span needed for induction of specific disease suppression and the inflexibility of the cropping system, result in limited applicability of this system for soil borne disease management.

Adjustment of Date of Sowing

Its main principle is to reduce to a minimum the period over which infective agents (propagule, vectors) meet the susceptible stage of the host, so that the susceptible stage of plant growth does not coincide with the environment favourable for the pathogen. When peas and chick pea are sown late, *i.e.*, after October, root rot and wilt incidence (a complex of *Fusarium*, *Rhizoctonia* and *Sclerotium*) get reduced as the delayed sowing utilizes the lower temperature and moisture in soil that do not favour the pathogens. Few crops are even attacked by a single pathogen and sowing of crops on a specific date may help to reduce the risk of one disease, but heighten that of another disease. For example, late sowing peas helps in reducing losses from root rot and wilt, it exposes the crop to powdery mildew and rust at a time when pods are forming. Under such circumstances, the farmers have to choose the suitable date for planting, according to which disease is more important in particular area and for what purpose the crop has been grown.

Plant Spacing *i.e.*, Rate of Sowing and Density of Stand

Virus diseases are a limiting factor in cowpea production in all agro-ecological zones of Nigeria and yield reductions of between 80-100 per cent have been reported. Planting density at 25x75 cm consistently produced the lowest viral disease incidence (Aliyu and Balogun, 2011). The incidence of brown rot (*Cephalosporium gragatum*) of soybean is also higher in widely spaced planting than in closer rows.

Method of Sowing/Planting

In places where water accumulation is a problem to the crop growth sowing of seeds on the sides or ridges is found effective in reducing the incidence of *Sclerotium rolfsii* on groundnut and *Sclerotinia sclerotiorum* and *Rhizoctonia solani* on various crops and *Phytophthora* blight of Pigeonpea. Ridging is disadvantageous in water deficit areas where it encourages pathogens like *Macrophomina phaseolina*.

Irrigation Water Management

Irrigation especially at seed-development stage, may favour seed infection. Irrigation time and amount of water should be controlled so that the relative humidity is not raised to such an extent that it becomes conducive for seed infection. Control of seed-borne diseases favoured by wet climate can be achieved by raising the seed crop in dry areas. Some examples are anthracnose of bean and cucurbits (*Colletotrichum* spp.), *Ascochyta* blight of pea (*Ascochyta* spp.) and bacterial blight of legumes. Such crops can be grown in dry areas with the help of irrigation so that these aerial parts remain dry and do not contact infection. Virus-free potato seed tubers can be produced more successfully in areas where temperature and moisture conditions do not favour build-up of populations of the insect vectors. *Macrophomina* root rots of many crops spread mainly through irrigation and drainage water.

Sclerotia, smut spores, chlamydospores, oospores and mycelium found in the soil are carried from one field to another through irrigation and drainage water. Stem rot, sheath blight and bacterial blight diseases of rice, damping off of vegetables and Macrophomina root rots of many crops spread mainly through irrigation and drainage water. Hence care should be taken not to irrigate a healthy field using drainage/irrigation water from a diseased field. The practice of flooding somewhat resembles soil disinfestation. The harmful effect on soil borne pests may be related to lack of O_2, increased CO or various microbial activities under anaerobic conditions, *e.g.* production of substances that are toxic to the pathogen (Bruehl, 1987). This practice was known in ancient civilizations in the Near and Far East. Flooding apparently destroys *Pseudomonas solanacearum*, and *Radopholus similis* and plant parasitic nematodes (Stover, 1962).

Irrigation strongly affects both foliar and soilborne diseases (Rotem and Palti, 1969). It alters the moisture content of the soil and consequently influences its aeration and temperature, and these, in turn, affect the incidence of diseases through their impact on biotic and abiotic processes in the soil or foliage. Irrigation also affects disease incidence indirectly due to changes in the agricultural regime, such as intensification of cropping, and changes in the date of sowing and growing seasons (Katan,1996). Irrigation alleviates the water stress which predisposes the plants to certain diseases. Water stress had a predisposing effect on the severity of Phytophthora root rot in safflower and on disease caused by *Macrophomina phaseolina* (Ghaffar and Erwin, 1969), and an adequate irrigation regime reduced disease incidence.

Selection of Seed and Seed Material

Selection of seeds and seed materials is very effective preventive measure against many diseases caused by fungi, bacteria, viruses and phytoplasmas and they are primary source of infection. Hence, seed and seed materials like cuttings, tubers, grafts, setts *etc.*, should be disease free, from a certified source, uninjured and have a high germinating capacity. The practice reduces initial inoculum subsequently definitely helpful in delaying or suppressing the incidence of the disease. The diseases like *Ascochyta* blight (chick pea, lentil), seed rot and seedling blight (*Botrytis, Sclerotinia, Rhizoctonia,* and *Fusarium* species), seed rot and damping off (*Pythium* and *Phytophthora* species) of chick pea and fieldpea and diseases caused by viruses and mycoplasma, leaf crinkle of black gram *etc.*, are transmitted through seeds. Peanut mosaic and peanut stripe diseases are seed borne. Some of the important fungal diseases of faba bean reported to be seed borne are chocolate spot (*Botrytis fabae*), blight (*Ascochyta fabae*), and rust (*Uromyces vicia-fabae*). *Ascochyta pisi* causes leaf and pod spot; *Mycosphaerella pinodes*, the perfect stage of *Ascochyta pinodes*, results in blight; and *Ascochyta pinodella* (now called *Phoma medicaginis* var. *pinodella*) causes foot rot of peas and known to survive through seed and soil. Downy and powdery mildew of pea are also seed borne. Fusarium wilt, however soil borne, but seed play important role in its transmission.

Leveling of Field and Provision of Drainage Facilities

Water stagnation in different patches of field favours the fungi like *Pythium, Phytophthora, Rhizoctonia solani, etc.*, for which proper levelling of field before sowing or planting is very essential. Further improving the drainage is necessary in the control of diseases favoured by excess of water.

Barrier Cropping

Taller crops can be grown to protect a crop of lesser height from virus vectors. The insects may land at the taller crops (barrier crops) and the dwarf crop may escape from virus diseases by those insects. Barrier cropping with 3 rows of maize or sorghum or pearlmillet around the main crop namely blackgram or greengram is effective in reducing the vector population and incidence of yellow mosaic.

Yellow Sticky Traps

Sticky, yellow polythene sheets erected vertically on the windward side of black gram and green gram are in use to attract and kill the whitefly vectors which spread yellow mosaic of blackgram and greengram.

Field and Plant Sanitation

Field and plant sanitation is an important method of disease control through cultural practices. The inoculum present on field plants in the field may multiply on the plant or in the soil and in due course of time may be sufficient to nullify or reduce the effect of control practices. Many pathogens overwinter or over summer on plant debris during the off-seasons and become active when the crop is again grown in the field. Hence plants bearing pathogens or plant debris introducing inoculum into the soil should be removed as early as possible. In most of the soil borne diseases like wilt and root rot, it has been reported that as long as the dead roots and other roots and other affected parts are present in the soil, the fungus continue its growth. When such diseased plant materials are removed, there is quick decline in the population of pathogens in the soil.

Avoidance of the transfer of inoculum from one field to another by man, machine or water is one of the ground rules of cultural control. Where soil-borne diseases are concerned, anything that carries soil is suspect, this includes wheels, boots and water flowing either from adjacent fields, or through drainage ditches from distant fields. As regards sap-borne.viruses, attention must be paid to disinfection of wheels and of the hands of labourers, as they pass from one field to another. Where such virus can also be carried on clothing, the work should be planned so that the labourers do not go from older to younger fields on the same day.

Roguing

Roguing entails about complete removing or uprooting the diseased plants to prevent further spread of the disease. This method is widely used in management of virus diseases spread by insects (yellow mosaic of blackgram and greengram,). Root rot and wilt attached plants after their death should be as and when noticed in the field uprooted and burnt to check the inoculum build up in the soil.

Decoy Cropping

It involves use of non-host crops to waste the infection potential of soil borne pathogens by destructing the crop before the pathogen can reproduce or complete its life cycle. For example, *Tagetes minuta* against *Rotylenchus* spp. and *Pratylenchus* spp. in soybean.

Soil Solarization

Katan and his co-workers (1976, 1981, 1983, and 1984) were the first to extensively use this simple approach to manage soil borne diseases. It is hydrothermal soil disinfestation method for the management of heat sensitive soil borne pathogens which involves use of solar heat as lethal agent by capturing the solar energy through the use of thin transparent polyethylene traps and keeping the soil wet during the period of intense solar irradiation say during the month of May-June in plains of northern India where temperature goes up to 45°c. Solar irradiation intensity, air temperature, humidity, polyethylene gaze, soil properties, soil heat flux are important factors governing the processes of soil solarization.

Solarization is an integrated method of increasing plant health, growth, and yield. It appears to be adaptable to a wide range of agricultural applications, alone and in conjunction with agricultural chemicals and biological control agents. The possible uses of soil solarization both pre-plant and post-plant, are being explored in field, orchard, nursery, greenhouse and garden situations, and in environmental and landscape improvement. Under the limitations of its applicability, soil solarization is a safe and effective method for disease and pest control that may reduce the necessity for chemical applications to soil: it represents a significant advance in soil disinfestation mulching technology (Stapleton and Devay, 1986).

The most pronounced effect of soil solarization is physical one *i.e.*, an increase in soil temperature (52°c) for several hours of the day. However, other processes include shift in biological equilibrium, changes in chemical and gas composition, high soil moisture level also play vital role in suppression of the pathogens of several crops *e.g.*, *Verticillium*, *Fusarium*, *Rhizoctonia solani*, *Sclerotium rolfsii*, *Plasmodiophora brassicae*, *Orobanche*, *Pratylenchus*, *Heterodera*, *Meloidogyne* and other plant parasitic nematodes.

Soil solarization is a special mulching process which causes hydrothermal disinfestation and other physical and biological changes in soil which are beneficial to plant health and growth.

Plastic film laid over moist soil during periods of high air temperature, usually for 1-2 months, can greatly reduce or eradicate a number of pathogens and pests including fungi, bacteria, nematodes, arthropods and weeds. Following soil solarization, growth of microflora beneficial to plant growth or antagonistic to pathogens and pests may slow the reinfestation of soil by these organisms for more than one growing season. Increased plant growth and yield of annual and perennial field, row, and nursery crops usually occur following soil solarization. In addition, the availability of increased mineral nutrients following solarization may reduce crop fertilization requirements. Soil solarization has been effective as a pre-plant and as a post-plant treatment, and has been compatible with chemical soil treatments and also biological soil amendments after solarization. Soil solarization is a significant advance in the non-chemical control of many pathogens and pests (Stapleton and Devay, 1986).

In Japan, solarization of soil, further insulated in closed plastic greenhouses (increasing the greenhouse effect) effectively controlled *Fusarium oxysporum* f. sp. *fragariae, Sclerotium rolfsii, Rhizoctonia solani* and *Verticillium albo-atrum* (Kodama and Fukui, 1979, 1982). Population reductions, varying from 42 per cent to 100 per cent, were achieved by soil solarization for species of plant- parasitic nematodes in at least 10 genera including *Meloidogyne, Heterodera, Globodera, Pratylenchus, Ditylenchus, Paratrichodorus, Criconemella, Xiphinema, Helicotylenchus and Paratylenchus* in Israel (Hadar *et al.*, 1983; Katan, 1984), in California (Stapleton and DeVay, 1983) and in New York. *Verticillium* and *Fusarium* wilts of several crops have been successfully controlled by solarization, as well as diseases caused by *Bipolaris sorokiniana, Didymella lycopersicii; Phytophthora cinnamomi, Plasmodiophora brassicae, Pyrenochaeta lycopersici, Pyrenochaeta terrestris, Pythium myrothecium, Pythium ultimum, Rhizoctonia solani, Sclerotium oryzae, Sclerotium rolfsii,* and *Thielaviopsis basicola*. Pathogenic fungi including Pythium *irregulare, Sclerotium cepivorum* and *Sclerotinia minor* were reduced in artificially inoculated soils.

Organic Amendment of Soil

Decomposition of organic substrates (Farm yard manure, compost, oil cakes, decomposed leaves, animal litter compost *etc.*) is a biological/microbial phenomenon and has a multidirectional impact. It improves soil health, soil structure, enhances microbial population which may directly act as antagonist or indirectly their metabolites affects biological equilibrium. For example, compost is effective because it nurtures a more diverse soil environment in which innumerable soil organisms exist. Further, compost acts as a food source and shelter for the antagonists that compete with plant pathogens, for those organisms that prey on and parasitize pathogens, and for those antagonists that produce antibiotics. Root rots caused by *Pythium* and *Phytophthora* are generally suppressed by the high numbers and diversity of beneficial microbes found in the compost. Such antagonists prevent the germination of spores and infection of plants growing on the amended soil (Harrison *et al.*, 1999). To get more reliable results from compost, the compost itself needs to be stable and of consistent quality.

It has become evident that a "one size fits all" approach to composting used in disease management will not work. Depending on feed stock, inoculum, and composting process, composts have different characteristics affecting disease management potential. For example, high carbon to nitrogen ratio (C:N) tree bark compost generally works well to suppress Fusarium wilts. With lower C:N ratio composts, Fusarium wilts may become more severe as a result of the excess nitrogen, which favors Fusarium (Hoitink *et al.*, 1997). Compost from sewage sludge typically has a low C:N ratio.

Some of the beneficial microorganisms that re-inhabit compost from the outside edges after heating has subsided include several bacteria (*Bacillus* species, *Flavobacterium balustinum* and various *Pseudomanas* species) and several fungi (*Streptomyces, Penicillium, Trichoderma* and *Gliocladium verens*). The moisture content following peak heating of compost is critical to the range of organisms inhabiting the finished compost. Three approaches can be used to increase the suppressiveness of compost. First, curing the compost for four months or more; second, incorporating the compost in the field soil several months before planting; and third, inoculating the compost with specific biocontrol agents (Hoitink *et al.*, 1997). Two of the more common antagonists used to inoculate compost are strains of *Trichoderma* and *Flavobacterium,* added to suppress *Rhizoctonia solani. Trichoderma harzianum* acts against a broad range of soil-borne fungal crop pathogens, including *R. solani,* by production of anti-fungal exudates.

Soil-borne diseases result from a reduction in the biodiversity of soil organisms. Restoring important beneficial organisms that attack, repel, or otherwise antagonize disease-causing soil organisms will reduce their populations to a manageable level. Beneficial organisms can be added directly, or the soil environment can be made more favourable for them with compost and other organic amendments. Compost quality determines its effectiveness at suppressing soil-born plant diseases. Compost quality can be determined through laboratory testing.

Management of Plant Nutrition

Management of plant nutrition helps to protect crops from pathogens by avoidance of nutrient strain which may directly lower the resistance of the crops to pathogens and by manipulation of nutrients to the relative disadvantage of the pathogen. The deficiency of nutrients *viz.*, nitrogen, and micronutrients (zinc, molybdenum, magnesium, manganese, copper, Iron, boron), if noticed in standing crop can be eliminated by spraying the required nutrient as top dressing. However, phosphorus, potassium and calcium can be applied as basal dressing during the field preparation.

All the essential nutrients can affect disease severity (Huber and Graham, 1999). However, there is no general rule, as a particular nutrient can decrease the severity of a disease but can also increase the severity of the disease incidence of other diseases or have a completely opposite effect in a different environment (Marschner, 1995; Graham and Webb 1991; Huber, 1980). A particular element may decrease the severity of some diseases, but increase others, and some have an opposite effect in different environments. Some forms of biological disease control and 'suppressive soils' are manifestations of microbial activity that influence nutrient availability (Huber, 1989; Huber and Graham 1999). In general, the greatest benefit to the plant is provided when full nutrient sufficiency is provided; however, the response to a particular nutrient may be different when going from deficiency to sufficiency than from sufficiency to excess. Since each nutrient functions as part of a delicately balanced interdependent system with the plants genetics and the environment, it is important to establish a nutrient balance for optimum crop response. Through an understanding of the disease interactions with each specific nutrient, the effects on the plant, pathogen, and environment can be effectively modified to improve disease control, enhance production efficiency, and increase crop quality.

High Ca in tissues is correlated with resistance to macerating diseases caused by *Erwinia carotovora, Fusarium solani, Pythium myriotylum, Rhizoctonia solani, Sclerotinia minor*, and *Sclerotium rolfsii* (Bateman and Basham, 1976, Kelman *et al.*, 1989, and Huber, 1994). The availability of nutrient elements such as nitrogen and its form, potassium, phosphorus, sulfur, and silicon will influence disease development directly or indirectly (*e.g.*, Walters and Bingham 2007) in mixed crops. Nitrogen in ammonia form is toxic to *Sclerotium rolfsii* and *Plasmodiophora brassicae*. High nitrogen is also recommended to reduce *Sclerotium rolfsii*.Calcium cynide is toxic to *S. rolfsii*, *Pythium* and *Fusarium*.

Conclusion

The ultimate objective of integrated crop management system can be fulfilled by the judicious application of the various cultural practices suited against the pathogens by the modification of environment. To accomplish good management, every and each method of cultural practices should be assessed for tits use as a potential tool for disease management. A thorough knowledge of ecology and plant husbandry/crop cultivation of a particular crop is absolutely essential for disease management (Garrett, 1970). Integrated management is less appropriate, but it decreases the risks and hazards stemming from the use of a single method of management. There should be an increasing interest in cultural practices for disease management that can replace the use of harmful pesticides, and that can be integrated into pest/disease-management programmes.

REFERENCES

Agrios, G.N. (1997) Plant pathology. Academic, London, UK, pp. 635

Aliyu, T.H. and Balogun,O.S. (2011). Effects of Variety and Planting Density on the Incidence of Common Viral Diseases of Cowpea (*Vigna unguiculata*) in a Southern Guinea Savannah Agro-ecology. *Asian Journal of Plant* Pathology. 5: 126-133.

Bateman., D.F., and Basham, H.G. (1976) Degradation of plant cell walls and membranes by microbial enzymes. In: Heitefuss R, Williams PH (Eds.) Encyclopedia of plant physiology : vol. 4, physiological plant pathology. New York : Springer, pp 316-355

Bruehl, G.W. (1987). Soil-Borne Plant Pathogens. MacMillan, New York, NY, USA.

Dunniway, J.M. (1977). Predisposing effect of water stress on the severity of Phytophthora root rot in safflower. *Phytopathology*. 67: 884-889.

El-Zik, K. M. and Frisbie. R. E. (1985). Integrated management systems for pest control and plant protection, in N. B. Mandava (ed.), CRC Handbook of Natural Pesticides: Methods. Vol. I. Theory, Practice, and Detection. CRC Press, Inc., Boca Raton, FL. pp. 21-122. Focus Publishing, Newburyport, MA. p.75-77.

Garrett, K.A. and Cox, C.M. (2006) Applied biodiversity science: managing emerging diseases in agriculture and linked natural systems using ecological principles. In: Ostfeld R, Keesing F, Eviner V (eds) Infectious disease ecology: the effects of ecosystems on disease and of disease on ecosystems. Princeton University Press, Princeton, NJ, pp 368–386

Garrett, S.D. (1970). Pathogenic Root-Infecting Fungi. Cambridge University Press, Cambridge, UK.

Ghaffar, A. and Erwin, D.C. (1969). Effect of soil water stress on root rot of cotton caused by *Macrophomina phaseolina*. *Phytopathoogy*. 59: 795-797.

Graham, R.D. and Webb, M.J. (1991) Micronutrients and disease resistance and tolerance in plants. In: Mortvedt, J.J., Cox, F.R., Schuman, L.M. and Welch, R.M. (eds) Micronutrients in agriculture. Madison Wis : Soil Sci Soc America, pp 329-370

Hadar, E., Sofer, S., Brosh, S., Mordechai, M., Cohn, E. and Katan, Y. (1983). Control of clover cyst nematode on carnation. *Hadasseh* 63: 1698-1700.

Hanson, J.D., Liebig, M.A., Merrill, S.D., Tanaka, D.L., Krupinsky, J.M. and Stott, D.E. (2007) Dynamic cropping systems: increasing adaptability amid an uncertain future. *Agronomy. J*. 99: 939–943.

Harrison, Una J. and Frank J. Louws. (1999). Disease management through suppressive soils. Department of Plant Pathology, North Carolina State University (draft document). September 23. 14 p.

Hiddink, G.A., Termorshuizen, J. and van Bruggen, A.H.C. (2010) Mixed Cropping and Suppression of Soil borne Diseases. In: Eric Lichtfouse (Ed.) pp119-146. Sustainable Agriculture Reviews Volume 4¸ Genetic Engineering, Bio fertilisation, Soil Quality and Organic Farming, Springer Dordrecht Heidelberg London New York. DOI 10.1007/978-90-481-8741-6

Hoitink, H.A.J., Stone, A.G. and Han. D.Y. (1997). Suppression of plant diseases by composts. Hort Science. Vol. 32, No. 2. p. 184–187.

Huber, D. M. (1980). The role of mineral nutrition in defense. In J. G. Horsfall and E. B. Cowling (eds.), *Plant Pathology: An Advanced Treatise*. Academic Press, New York, pp. 381–406.

Huber, D.M. (1989) Introduction. In: Engelhard AW (ed) Soilborne plant pathogens : management of diseases with macro-and microelements. St Paul, Minn : APS Press, pp 1-8

Huber, D.M. (1994) The influence of mineral nutrition on vegetable diseases. Horticultura Brasileira 12: 206-214.

Huber, D.M. and Graham, R.D. (1999) The role of nutrition in crop resistance and tolerance to diseases. In: Rengel, Z. (Ed.) Mineral nutrition of crops: fundamental mechanisms and implications. New York : Food Products Press, pp 169-206.

Katan, J. (1981). Solar heating (solarization) of soil for control of soilborne pests. Annual Review of Phytopathology 19,21 l-236.

Katan, J. (1984). Soil solarization. In: Second International Symposium on management of plant pathogens and pests. *Crop Protection*. 5(3): 190-198.

Katan, J. (1996). Cultural practices and soil-borne disease management. In: Utkhede R.S., Gupta V.K. (eds). Management of Soil Borne Diseases. pp. 100-122. Kalyani

Publishers, New Delhi, India.

Katan, J., Fishler, G. and Grinstein, A. (1983). Short- and long-term effects of soil solarization and crop sequence on Fusarium wilt and yield of cotton in Israel. Phytopathology 73,1215-1219.

Katan, J., Greenberger, A., Alon, A. and Grinstein, A. (1976). Solar heating by polyethylene mulching for the control of diseases caused by soil-borne pathogens. *Phytopathology*. 76: 683-688.

Kelman, A., McGuire, R.G. and Tzeng K.C. (1989) Reducing the severity of bacterial soft rot by increasing the concentration of calcium in potato tubers. In: Engelhard, A.W. (Ed) Soilborne plant pathogens : management of diseases with macro- and microelements. St Paul, Minn : APS Press, pp 102-123.

Kodama, T. and Fukui, T. (1979). Solar heating sterilization in the closed vinyl house against soil-borne diseases. I. The movement of soil temperature and determination of thermal lethal conditions for some soil-borne pathogens. *Bulletin of the Nara Prefecture Agriculture Experiment Station*. 10: 71-82.

Lewandowski, I., Hardtlein, M. and Kaltschmitt M. (1999) Sustainable crop production: Definition and methodological approach for assessing and implementing sustainability. *Crop Sci*.39: 184-193.

MacGuidwin, A.E. (1993). Management of Nematodes. p. 159ñ166. In: Randell C. Rowe (ed.) Potato Health Management. APS Press, St. Paul, MN.

Marschner, H. (1995). Relationship between mineral nutrition and plant diseases and pests. In *Mineral Nutrition of Higher Plants*, 2nd Edn. Academic Press, New York, pp. 436–460.

Oborn I., Edwards A.C., Witter E., Oenema O., Ivarsson K., Withers P.J.A., Nilsson S.I. and Richert Stinzing A. (2003) Element balances as a toll for sustainable nutrient management: a critical appraisal of their merits and limitations within an agronomic and environmental context, *Eur. J. Agron*.20: 211-225.

Palti J. (1981). Cultural Practices and Infections Crop Diseases. Springer-Verlag, Berlin, Germany.

Palti J. and Katan J. (1997). Effect of cultivation practices and cropping systems on soilborne diseases. In: Hillocks R.J., Waller J.M. (eds). Soilborne Disease of Tropical Crops, pp.

Peet, M. (1996). Sustainable Practices for Vegetable Production in the South. 377-396. CAB International, Wallingford, UK.

Prasad, R.D., Rangeshwaran, S.S.V., Hegde, S.V. and Anuroop, C.P. (2002). Effect of soil and seed application of Trichoderma harzianum on pigeonpea wilt caused by Fusarium udum under field conditions . *Crop Protection*. 21(4): 293-297.

Rotem J. and Palti, J. (1980). Irrigation and plant disease. *Annual Review of Phytopathology* 7: 267-288.Soil Disinfestation. Acta Horticulturae No. 152, p. 227 (ed. by C. Van Assche). Leuven, Belgium: International Society for Horticultural Science.

Stapleton, J.J. and De Vay, J. E. (1986).Soil solarization: a non-chemical approach for

Stover, R.H. (1962). Fusarium wilt of Panama diseases of bananas and other Musa species. *Phytopathological Paper No. 4*,Commonwealth Mycological Institute, Kew,UK.

Vandermeer, J.H. (1990) Intercropping. In: Carrol CR, Vandermeer JH, Rosset P (eds) Agroecology. Mcgraw Hill, New York, pp 481–516

Walters, D.R. and Bingham, I.J. (2007). Influence of nutrition on disease development caused by fungal pathogens: implications for plant disease control. *Annals of Applied Biology*. 151: 307-324.

Wu. H., Pratley, J., Lemerle, D. and Haig, T. (2001) Allelopathy in wheat (*Triticum aestivum*). *Ann Appl Biol*. 139: 1-9.

Yepsen, Roger B. Jr. (1984). The Encyclopedia of Natural Insect and Disease Control. Rev. ed. Rodale Press, Emmaus, PA. p. 267-271.

Chapter 18

Role of Plant Products in the Management of Diseases of Pulses

J.N. Srivastava, A.K. Singh & S.K. Biswas

During ancient times man evolved many cultural and physical practices to protect crops from ravages of insect and pathogens. This was followed by the use of botanicals. The earliest recommendation for the use of plant products was given by Democritus in 470 B.C. He recommended the control of plant blights by sprinkling plants with the olive grounds left after extraction of olive oil. Later Surapala in 9th century suggested the use of vidanga (*Embelia ribis*), mustard and sesame for curing tree diseases in his book "Vrikshayurveda"

As agriculture advanced man invented many new techniques for the control of pathogen. These included the use of inorganic and organic chemicals. The chemical control was so efficient in control of diseases that man started using it indiscriminately. This indiscriminate use continued until 1962 when Rachel Carson for the first time attracted attention towards the bad effect of chemicals in her book "Silent Spring"

Excessive use of chemicals causes soil, air, surface and ground water pollution besides affecting the crop produce. Chemical pesticides have been used for a long time, but have serious drawbacks (Sharaby, 1988), such as direct toxicity to beneficial pathogens and insects, fishes and human (Goodland *et al.*, 1985), pesticide induced resistance (Georghiou and Taylor, 1977), health hazard (Bhaduri *et al.*, 1989) and increased environmental and social costs (Pimental *et al.*, 1980). The chemicals have entered our food chain also. Judging the seriousness of this problem, attention has been paid into search for an effective alternative with practically no residual or toxic effects on ecosystem. Botanicals can serve as an alternative approach in this situation.

The natural plant products have been derived from plants which played a significant role in discovery of a new germicide, nematicide and to certain extent of viricide, either by their direct application to diseases or through their exploitation in the resilienting plant with optimum biological and physical properties. Biopesticides are natural plant products that belong to the so called secondary metabolites, which include alkaloids, terpenoids, phenolics, and minor secondary chemicals. Plants are rich source of bioactive organic chemicals. It is estimated that the plants may contain as many as 4000,000 secondary metabolites (Mamun, 2011b). It is estimated that as many as 2121 plant species have been reported to posses' pest control properties. Botanicals like neem, ghora-neem, mahogoni, karanja, adathoda, sweet flag, tobacco, derris, annona, smart weed, bar weed, datura, calotropis, bidens, lantana, chrysanthemum, artemisia, marigold, clerodendrum, wild sunflower and many others may be grown by planters with minimum expense and extracted by indigenous methods. These botanical materials can be used as an alternative to chemical pesticides. The use of botanical pesticides is now emerging as one of the prime means to protect crops and their products and the environment from pesticides (Sanjay and Tiku, 2009). Botanicals degrade more rapidly than most chemical pesticides, and therefore are considered to be ecofriendly and less likely to kill benefecial pests than synthetic pesticides with longer environmental retention. Most of the botanical pesticides generally degrade within a few days, and sometimes even within a few hours (Siddiqui and Gulzar, 2003).

This paves the way for the healthy and pollution free sustainable environment. In prospects of good safety and public health, plant products satisfy following aims:

a) They are harmless and environmental friendly
b) Biodegradable
c) They can increase host metabolism
d) They are slow in action of limited persistence
e) Have a shelf life of 2-3 years
f) Target specific
g) Low mammalian toxicity
h) Increase crop quality
i) Systemic in nature
j) Optimum bioactivity
k) Increase benefit: cost ratio
l) Novel mode of action
m) Low toxicity
n) Phytotoxicity of ecotolerance
o) Less likely development of resistance
p) Cost efficacy

However, there are some drawbacks of botanicals control:

a) Loss of immediate killing action

b) Higher dosage requirement

c) Some plant oils show phytotoxicity of lighter doses, for *e.g. Azadirachta indica,Cymbopogon sp.*

d) Problems in preparation of formulation of spraying

e) Some oils are non-edible, so their equipped toxicity problem also needs investigation

Plant extracts used to control of the phytopathogens have been obtained mainly from tree species such as eucalyptus and neem (24 per cent of the studies with extracts) and herbaceous species like garlic, citronella, mint, rue, yarrow, ginger, basil, camphor, turmeric and ocimum (54 per cent). Besides these there are many other plants flora whose antimicrobial potential was tested by researchers. With respect to groups of pathogens, the majority of the work is with those that cause disease in the plant canopy (30 per cent of the works with extracts), like the genus *Alternaria, Bipolaris, Crinipellis, Corynespora* and *Colletotrichum*, which respond alone for 15 per cent of the works. The soil-born pathogens represent 20 per cent of the researches, especially *Rhizoctonia, Sclerotium, Sclerotinia, Fusarium* and *Phytophthora*. Post-harvest pathogens like *Penicillium, Aspergillus* and *Rhizopus* are in 9 per cent of the works and *Meloidogyne* nematode in 9.5 per cent. For the host plants, 30 per cent of the work are with crops like beans, soybeans, coffee, wheat, cotton and cassava; 20 per cent with vegetables like cucumber and tomato, this later representing alone 15 per cent of all the researches with extract; and 10 per cent with the fruits like papaya, strawberry and cocoa (Stangarlin *et al.*, 2011)

Table 18.1. Some Plants and their Part Used as Botanicals

Common Name	Botanical Name	Family	Plant Part Used
Neem	*Azadirachta indica*	Meliaceae	Leaf, oil cake, bark, fruit pulp.
Garlic	*Allium sativa*	Amaryllidaceae	Cloves
Onion	*Allium cepa*	Amaryllidaceae	Bulbs
Ghrit Kumari	*Aloe vera*	Liliaceae	Leaves
Custered apple	*Annona squamosa*	Annonaceae	Leaves
Satavar	*Asparagus racemosus*	Liliaceae	Root
Kalmegh	*Andrographis paniculata*	Acanthaceae	Leaves, Stem, roots
Bael	*Aegle Eegle mamrelos*	Rutaceae	Leaves and Fruit
Babool	*Acacia nilotica*	Mimosaceae	Leaves, stem, bark
Adusa/vasaka	*Adhatoda vasica*	Acanthaceae	Leaves
Apamarga	*Achyranthus aspersa*	Amranthaceae	Leaves and seeds
Anethium	*Anethum graveolens*	Apiaceae	
Brahmi	*Bacopa monnieri*	Scrophulariaceae	Leaves, whole plant
Palas/Dhak/Tesu	*Butea monosperma*	Leguminaceae	Stem/bark
Bougainvillea	*Bougainvillea spectabilis*	Nyctaginaceae	Leaves
Turmeric	*Curcuma longa*	Zingiberaceae	Leaves and rhizomes

Common Name	Botanical Name	Family	Plant Part Used
Amahaldi	*Curcuma amada*	Zingiberaceae	Leaves and rhizomes
Chenopodium	*Chenopodium ambrosioides*	Chenopodiaceae	Leaves
Lemon grass	*Cymbopogon flexuosus*	Poaceae	Leaves
Sadabahar	*Cartharanthus pusillium*	Apocynaceae	Leaves
Citronella	*Cymbopogan winterianus*	Poaceae	Leaves
Cumin	*Cuminum cyminum*	Apiaceae	seeds
Periwinkle	*Catharanthus roseus*	Apocynaceae	Leves, flowers
Cannabis	*Cannabis sativa*	Cannabinaceae	Leves, flowers
Papaya	*Carica papaya*	Carcicaceae	Leves, flowers
Amaltas	*Cassia fistula/C.occidentalis*	Caesalpineaceae	Leaves, pods and seeds
Cleome	*Cleome viscose*	Capridaceae	Leaves
Coleus	*Colius aromaticus*	Lamiceae	Foliage
Callistemon	*Callistemon lanceolatus*	Myrteaceae	Leaves
Cissus	*Cissus quadrangularis*	Vitaceae	Stem
Datura	*Datura metel,D.stromonium*	Solanaceae	Leaves
Yam	*Dioscoria bulbifera*	Dioscoreaceae	Root tuber rhizomes.
Desmostachya	*Desmostachya bipinnata*	Graminae	Leaves, flower
Eucalyptus	*Eucalyptus citridoa*	Myrtaceae	Leaves and bark
Aonla	*Emblica officinalis*	Euphorbiaceae	Leaves and fruits
Chotidudhi, baridudhi	*Euphorbia microphyla,E. pulcherima*	Euphobiaceae	Leaves
Jatropa	*Jatropa gossypifolia, J. laminarioides*	*Euphorbiacceae*	Leaves
Bhasmpatti	*Kalanchoe heterophylla*	*Labiataeae*	Leaves
Lantana	*Lantana camara*	*Verbinaceae*	Leaves
Pundina	*Mentha arvensis*	*Labiateae*	Leaves
Murraya/mitha neem	*Murraya konengii*	*Rutaceae*	Leaves
Melia	*Melia azadirachta*	*Meliaceae*	Leaves
Shajana	*Moringa olifera*	*Moringaceae*	Leaves, pods, root, Flower
Majorana	*Majorana hortensi*	*Labiateae*	
Mint	*Mentha spp.*	*Labiateae*	Leaves
Tulsi	*Ocimum sanctum*	*Lambiaceae*	Leaves
Ocimum	*O.canum giatisimum*	*Lambiaceae*	Leaves
Oxalis	*Oxalis quatosella*	*Oxalidaceae*	Leaves
Solanum	*Solanum nigrum*	*Solanaceae*	Leaves, flower
Sage	*Salvia futicosa*	*Labiaceae*	Leaves
Rumex	*Rumex japonicas*	*Polygonaceae*	Leaves
Nyctanthes	*Nyctanthes arbortistis*	*Oleaceae*	Leaves flower
Marigold	*Tagetues erecta, T. patula*	*Compositae*	Whole plant
Thuja	*Thuja oriantalis*		Leaves and bark

Common Name	Botanical Name	Family	Plant Part Used
Tamrind	*Tamrindus indicus*	*Caesalpineaceae*	Seed
Arjun	*Terminalia arjuna*	*Combretaceae*	Leaves, bark
Fenugreek	*Trigonella foenum granatum*	*Papionaceae*	seeds
Tiliacora	*Tiliacora racemosa*		Roots
Thevitia	*Thevetia peruvina*	*Apocynaceae*	Leaves seed, roots
Khas	*Viterveria zizanoides*	Poaceae	roots
Verbina	*Verbena offcinalis*		Leaves
Ashwagandha	*Withamia somnifera*	Solanaceae	Leaves, green berries, seeds and roots.
Patchauli	*Pogastemon patchaulii*	Labiateae	Leaves
Bachochii	*Psoralea corylifolia*		Leaves, seeds
Carrot grass	*Parthenium hysterophorus*	Astoaceae	Leaves, stems, root, inflorescence
Polyalthia	*Polyalthia longifolia,P.suberosa*	Anonaceae	Leaves
Isabgol	*Plantago ovate*	Plantoginaceae	Seeds, husk
Sarpagandha	*Rauwolfia serpentine*	Apocynaceae	Root, stem, leaves, bark
Castor	*Ricinus communis*	Euphorbiaceae	Seeds and leaves
Kava	*Piper methysticum*	Piperaceae	Roots
Sunflower	*Helianthus annus*	Asteracaea/ Compositae	receptacle
Hyptis	*Hyptis suveolens*	Lamiaceae	Roots
Clove	*Syzygium aromaticum*	Myrtaceae	Flower buds
Ginger	*Zingibar officinalis*	Zingberaceae	rhizome
Eelgrass	*Zosetra marina*	Poaceae	leaves
Sida	*Sida cordifolia*	Malvaceae	leaves
Ber	*Zizyphus jujuba*	Rhamnaceae	Leaves and flower
Nerium	*Nerium oleander*	Apocynaceae	leaves

Antimicrobial Compounds Present in Plants

The presence of antifungal compounds in higher plants has long been recognized as an important factor to diseases resistance (Mahadevan, 1982). Such compounds being biodegradable and selective in their toxicity are considered valuable for controlling some plant diseases (Singh and Dwivedi, 1987). A large number of active secondary metabolic compounds including alkaloids, cyanogenic glycosides, terpenoids, diterpenoids, triterpenoids, lipids, fattyacids, steroids, polyacetylenes, quassinoids, glucosinolates, isothiocyanates, flavonoids, simple and complex phenolics *etc.* have been reported in many plants and these secondary metabolites have antimicrobial properties that protect the plant against attack of several pathogens (Chitwood, 2002; Blum, 1996). Some of the important antimicrobial secondary metabolites are:

Glycosides

On wounding of plants, cynogenic glycosides and glycosinolates and their enzyme mingle and interact producing cyanide, isocyanates, nitriles and thiocynates. All these compounds are toxic to all organisms and also to fungi (Agios, 2006)

Detection of Glycosides

1 ml of water extracts of plants are taken in a test tube and 10 ml of 50 per cent H_2SO_4 added. The mixture is heated in boiling water for 15 minutes. After this 2 ml of Fehling solution is added and mixture is boiled. Brick red precipitate indicates glycosides.

Terpenoids

It is a 10 carbon compound. It is a major component of essential plant oils and posses activity against predators and pathogens.

Detection of Terpenoids (Salkovski test)

A small amount of extract solution is taken and 5 drops of H_2SO_4 and 1 ml chloroform is added. Change of yellow color to red indicates presence of terpenoids (REF).

Flavonoides

These are phenolic compounds present in plant tissues especially the vacuoles. They are also present in various bark and heart wood. Flavonoides are inhibitory/ toxic to pathogens and some of them eg: mediacarpin act as phytoallexin in induding defense against fungi (Agrios, 2005)

Detection of Flavonoids (Shinoda Test)

1 gram $MgSO_4$ powder and two drops of conc. HCl are added to 3 ml of plant extracts. Red colorations indicate the presence of flavonoides.

Saponins

Constitutive substances/preformed substances made from glycosylated steroidal triterpenoid compounds have antifungal membranolytic activity. Saponins form complexes with cell membrane, form pores and lead to loss of membrane integrity of pathogens. *e.g.* tomatins in tomato, avenacin in oats.

Detection of Saponins (Frothing test)

2 ml of plant extracts are taken in separate test tubes and are vigourously shaken for 2 minutes. An observation of frothing indicates presence of saponins.

Tannins

A group of compounds containing phenol, hydroxy acids and glucosides. They are phenolic oligomers. Condensed tannins are proanthocyanadins are group of tannin formed by polymerization of flavonoids.

Detection of Tannin

2 drops of $FeCl_3$ are added to 1 ml extract. A dirty green precipitate showed presence of tannins.

Phenols

Toxic compounds accumulate at faster rate after infection. *e.g.* Chlorogenic acid, caffeic acid and ferulic acid.

Detection of Phenols

Materials are extracted in ethanol and evaporated to dryness. The residue is then dissolved in distilled water and 0.5 ml Folin Coalteau Reagent (FCR) is added followed by 2 ml of 20 per cent Na_2CO_3 solution. Development of bluish colour indicates phenols.

Alkaloids

Detection of Alkaloids

20 mg extract and 2 ml distilled water are taken and filtered. To the filtrate 2-4 drops of 1 per cent HCl is added and steam is passed through it. Now 1 ml of this solution is mixed with 6 drops of Wagner's reagent. Brown red precipitate indicates presence of alkaloids.

Management of Fungal Disease by Plant Products

Many types of antifungal compounds are present in plants which are produced either constitutively or produced in response to attack by pathogen. Compound produced constitutively are called phytoanticipins while those produced in response to attack by pathogen are called phytoallexins.

The constitutive antifungal compounds belong to following major classes of secondary metabolites

1. Terpenoids- *e.g.* Iridoides, sesquiterpines, saponins *etc.*
2. Nitrogen and Sulphur containing compounds- *e.g.* alkaloids, amides amines *etc.*
3. Aliphatic- long chain alkanes and fatty acids.
4. Aromatics- *e.g.* phenol, flavinoids, stilbinoids, biobonyls, xanthones, bimoguimones *etc.*

Neem and Fungal Disease Management

Although man is using neem since ancient times for its multifarious uses, its scientific renaissance regarding insecticidal, fungicidal, nematicidal, antiviral properties came into existence quite recently.

Table 18.2. Effect of Various Neem Products on different Pathogens

Sl.No.	*Neem Product*	*Effective against Pathogen*
1	Neem oil cake	*Sclerospora sacchari*
2	Fruit pulp	*Rhizoctonia solani* (supress formation of sclerotia)
3	Extract oil	Inhibition of germ tube growth of *Erysiphe polygoni* (Singh and Singh, 1981)
4	Extract	Mycelial growth and spore germination of *Curvularia lunata, Alternaria alterneta.*
5	Neem oil	*Fusarium monoliforme, M.phaseolina, D. rostrtatum,* powdery mildew, Black spots.

Neem oil formulation called "Trilogy" has been approved by the EPA for use on food while "Rose Defence" and "Triact" (for control of Powdery mildew, rust, black spot, *Botrytis,* Downy mildew and other common diseases) are designed for use on ornamental.

Table 18.3. Fungal Diseases of Pulses Managed by Plant Products

Name of Fungal Diseases	*Causal Agent*	*Botanical Control*	*References*
Aphanomyces root rot of pea	*Aphanomyces euteiches*	Green manuring with oat, rape and sweet corn, soil application of paper mill residues.	Williums Woodward (1997)
Powdery mildew of pea	*Erysiphe pisi*	Nemadole a neem product, Extracts of *Allium cepa, Allium sativum*, rhizome of ginger and neem.	Sindhar *et al.* (1999); Singh and Prithviraj (1997)
Anthracnose of bean	*C.lindemuthianum*	Oil of eelgrass	Stanley *et al.* (2002)
Charcoal rot of soyabean/cluster bean	*Macrophomina phaseolina*	Pearlmillet, residue composet, cauliflower leaf compost	Lodha and Aggarwal (2002)
Web blight of urdbean	*Rhizoctonia solani*	Extract of *Canabis* and *Datura*	Kumar and Tripathi (2012)
Wilt of lentil	*Fusarium oxysporum* f. sp. *lentis*	Extract of ginger, garlic and neem	Garkoti *et al.* (2013)
Root rot of French Bean	*Rhizoctonia solani*	Extracts of *Artemisia vulgaris, coix lacryma jobi, Lantana camera*	Mangang and Chhetry (2012)
Wilt of chickpea	*Fusarium oxysporum* f. sp. *ciceri*	Aqueous leaf extracts *Azadirachta indica* and *Lantana camara*	Kamdi *et al.* (2012)
Root rot of mungbean	Macrophomina phaseolina	Oil of palmarosa, lemongrass, citronella, mentha and tulsi	Kumari *et al.* (2012)
Wilt of pegion pea	*Fusarium oxysporum* f. sp. *udam*	Leaf extract of *Azadirachta indica, Datura festilosa, Tagetes erecta, Eucalyptus citridora, Aegle marmelos* and *Mimusops elengi*	Singh *et al.* (2010)

Viral Disease Management

Viral disease are transmitted by many means like- mechanical transmission through sap, through vegetative propagating materials, seeds, pollens or through vector like insect, mites, nematodes, and fungi. When virus is transmitted mechanically through sap, intense, sanitation methods are carried out for the control but when it is transmitted by vectors, the best method is to control the vector. Several antiviral proteins obtained from various plant parts have been found effective in various virus host system as shown in the Table 18.4.

Table 18.4

Sl.No.	*Antiviral Protein*	*Source*	*Virus-Host System*
1	Abrin	*Abrus precatorius*	TMV/*N.glutinosa* seeds
2	Agrotins	*Agrostemma githago*	TMV/*N.glutinosa* seeds
3	Bryodin	*Bryonia dioica* leaves and roots	TMV/*N.glutinosa*
4	Ca-SRI	*Cleodendrum aculeatum* leaves	TMV/*N.glutinosa* TMV/*N.tabaccum/ Cymopsis tetragonoloba.*
5	Dianthins	*Dianthus cryophyllus*	TMV/*N.glutinosa*
6	EHL(Eranthis Hyemalis Lectin)	*Eranthis hyemalis* bulb	AMV/*Vigna vigna*
7	Gelonin	*Gelonium multiflorum*	TMV/*N.glutinosa*
8	MAP	*Mirabilis jalapa* roots	TMV/*N.glutinosa*
9	Modeccins	*Adenia digitata* roots	TMV/*N.glutinosa*
10	Momordin	*Momordica charantia*	TMV/*N.glutinosa* seeds
11	PAP	*Phytolacca americana* seeds leaves	TMV/*N.glutinosa*
12	:995	*Phaseolous vulgaris*	TMV/*P. vulgaris*
13	AIMV	*Chenopodium amranticolar*	PVX/*Gmphrema globosa*
14	RICIN	*Ricinus communis*	ACMV/*N. benthaminana* CaMV/ *Brassica compertris*
15	Soponins	*Saponaria officinalis*	TMV/*N.glutinosa* leaves,root
16	YLP(yucca recurvifolia protein)	*Yucca recurvifolia* leaves	TMV/*C. amranticolour*

Source: Sendhilvel *et al.*, 2004.

Table 18.5. Use of Botanicals in Viral Disease Management of Pulses

Name of Viral Diseases	*Name of Virus*	*Botanical Control*	*References*
Yellow mosaic disease of mungbean and urdbean	Mungbean yellow mosaic virus	Root extract of *Boerhaavia diffusa* and leaf extract of *Azadirachta indica* Leaf extract of *Clerodendrum aculeatum*	Singh and Awasthi (2009) Verma and Singh (1994)
Urdbean leaf crinkle disease	Urd bean leaf crinkle	Leaf extract of *Datura metal*	Chaudhury and Saha (1985)
	Yellow mosaic disease	Neem oil and mustard oil	Trivedi *et al.* (2014)

Induced Resistance

Antiviral proteins identified in many plants species can stimulate the active defense mechanism against viruses. Activation of such mechanism results in plants becoming highly resistant to subsequent infection. As for example flower extracts of *Azadirachta indica, Euphorbia milli* induce resistance against PVX infection in hypersensitive host *Chenopodium amranticolor*

Table 18.6. Some Plant Products and their Efficacy

Sl.No.	Plant Product	Virus	Vectors
1	Neem oil(5 per cent spray)	RTV	N.Virecens
2	Neem cake	"	"
3	Custard apple oil	"	"
4	Custerd + neem oil	"	"
5	Achras sapota	Ragi mosaic	"
6	Basella	Nubra	"
7	*Mirabilis jalapa*	"	"

Nematode Management by Plant Products

Phytoparasitic nematodes are very difficult to control because most of them spend their lives in soil or within plant roots and the cuticle and other surface structure of nematodes are impermeable to many organic molecules. Therefore, nematologists are not optimistic about the chemical ways of nematode control. So, attempts are made in the direction of plant products and phytochemical based strategies.

Research with crude plant extract has lead to the development of "Sincocin", the trade name of recently developed product containing a mixture of extracts from prickly pear (*Opuntia engelmanni lindheimeri*, Cacteaceae), southern red oak (*Quercus falcota*, Fegaceae), the sumac *Rus aromatica* (Anacardiaceae) and the mangrove *Rhizophora mangle* (Rhizophoraceae). In the field situation "Sincocin" has been found effective in controlling *Tylenchulus semipenetrans* (citrus nematode) on orange, *Rotylenchulus reniformis* (reniform nematode) on sunflower and cyst nematode *Heterodera schactti* on sugarbeet. Rizvi *et al.* (2012) has observed that greatest reduction of *Meloidogyne incognita, R. reniformis*, and *T. brassicae* in chickpea by the use of extract of *Argemone mexicana, Calotropis procera, Solanum xanthocarpum* and *Eichhornia echinulata*. Chickpea plant treated with higher concentrations of leaf extracts of Persian lilac showed the least impact of *M. incognita* (Rehman *et al.*, 2012). Mojumder and Mishra (1993) reported that soaking of *Vigna radiata* Linn. Seed with aqueous extracts of Neem reduced the *M. incognita* juveniles in soil. Soil application of Neem and Calotropis leaf extract reduced the root knot nematode infestation in cowpea (Umamaheswari and Sundarababu, 2001a, 2001b). The nematicidal activity of plant extracts may be due to the presence of active toxic principles like azadirachtin and nimbin in *Azadirachta indica* and thiophene and α-terthienyl in *Tagetes erecta*. Such toxic principles might be absorbed by the plant roots which changes the chemical composition of plants and such roots excerted some influence on the pathogenesis of *M. incognita* as well as improved plant growth (Rao and Parmar, 1984). Due to the exorbitant cost of nematicial chemicals and the environmental hazards they cause, it is high time to modify the nematode management options like botanicals which are safer and cost-effective.

Mechanism: A number of nematode antagonistic compounds are obtained from many plants. They inhibit or control the nematode population by following ways:

Inhibition of hatching of eggs: For *e.g.* isothiocyanates of black mustard seed inhibit hatching of *Globodera rostochenensis* eggs.

Inhibition of motility: A combination of linalool and methyl chavicol isolated from essential oil of *Pelargonium graveolens* (Geraniaceae) inhibited motility of *H. cajani* and *M. incognita.*

Inhibiton of enzyme action: All members of Brassicaceae produce thioglucose conjugate called glucosinolates which are hydrolysed either in soil or by mammal to form isothiocyanates.These isothiocynates react with sulphahydryl group of protein or enzymes and metabolism is disturbed.

Repellant: For *e.g.,* Heartwood extract from *Pinus massoniana* (Pinaceae) containing sesquiterpene humulene repelled *B. xylophilus. M. incognita* is repelled by high cucurbitacin produced by cucumber.

Induction of mortality of juveniles/direct toxicity: Linalool, geraniol and citronellol isolated from essential oil of *Pelargonium gravealens* (Geraniaceae) induced mortality in *M.incognita* juveniles. A triterpenoid caumeric acid from *Lantana camara* can directly kill *M.incognita* juvenile at 1 per cent concentration.

Inhibiton of penetration of host: Three Quassinoids from seeds of *Hannoa undulata* (Simarubaceae)-chaparrinone, Klaineanone and glaucaubolone inhibited penetration in tomato roots by *M. incognita* at juvenile stage.

Inhibiton of host finding juvenile ability: α-terthienyl from marigold inhibit host finding and induce mortality in entamopathogenic nematode *Steinernema glaseri.*

Some Commercial Safe and Systemic Fungicides

a. Funga stop — Commercial formulation of mint oil

b. Sun Spray — Refined horticultural oil

c. Triology — Neem oil formulation

d. Rose defence — Neem oil formulation

e. Triact — Neem oil formulation

f. Milsana — Knotweed (*Reynoutria sachalinensis*)

g. Spiroxanine

h. Ac-90085

i. 3 preventol

j. 7 L-378257

k. KBR 27- 38 exprodionil.

Conclusion and Future Prospect

The pesticidal compounds of plant origin, are much more effective and have little or no side effects on human, still a lot of work needs to be done before large scale utilization of botanical pesticides. Once a potentially useful plant species is identified, intensive breeding and selection work will have to be undertaken so that economic production, from such plant of antagonistic compounds is possible. The safety and selectivity of such botanicals needs to be tested. People are not interested at large in mode of action than in development of environmentally safe, inexpensive ergonomically useful compounds. The international agronomical industry is well aware of both public perception of chemical hazards and customers deires for alternative methods to protect their livelihood. This has stimulated a wide range of technologist to complement synthetic chemicals with other cost effective natural products for all types of farmers in future. It is likely that current global thrust on botanical pesticide may yield new leads and prototype modules to guide future research programme in plant protection.

REFERENCES

Agrios, G.N (2005) Plant Pathology, 5th edition. p295.

Bhaduri, N., Gupta, D.P. and Ram, S. 1989. Effect of vegetable oils on the ovipositional behaviour of *Callosobruchus chinensis* Fab. pp. 81-84. *In:* Proc. 2nd Int. Symp. On Bruchids and Legumes (ISBL-2). Okayama, Japan.

Blum, U. 1996. Allelopathic interactions involving phenolic acids. *J. Nematol.*28: 259-267.

Chowdhury, A.K. and Saha, N.K. 1985. Inhibition of urdbean leaf crinkle virus by different plant extract. *Indian Phytopathology*, 38: 566-568.

Chitwood, D.J. 2002. Phytochemical based strategies for nematode control. *Ann. Rev. Phytopathol.* 40: 221-249.

Garkoti, A. Kumar, V. and Tripathi, H.S. 2013. management of vascular wilt of lentil through aqueous plant extracts in tarai region of uttarakhand state. *The bioscan*, 8(2): 473-476, 2013.

Georghiou, G.P. and Taylor, C.E. 1977. Pesticide resistance as an evolutionary phenomenon. *In*: Proc. 14th Int. Cong. Entomol. p759.

Goodland, R., Watson, C. and Ledec, G. 1985. Biocides bring poisoning and pollution to 3rd world. The Bangladesh Observer, 16th and 17th January, 1995. p3.

Kamdi, D.R., Mondhe, M.K., Jadesha, G., Kshirsagar, D.N. and Thakur, K. D. 2012. Efficacy of botanicals, bio-agents and fungicides against *Fusarium Oxysporum F. Sp. Ciceri*, in chickpea wilt sick plot. *Annals of Biological Research* 3 (11): 5390-5392.

Kumar, S. and Tripathi, H.S. 2012. Evaluation of plant extracts against *Rhizoctonia solani* Kuhn, the incitant of web blight of Urdbean. *Plant Disease Research*, 27 (2): 190-193.

Kumari, R., Shekhawat, K.S., Gupta, R., Khokhar, M.K. 2012. Integrated Management against Root-rot of Mungbean [*Vigna radiate* (L.) Wilczek] incited by *Macrophomina phaseolina*. *J Plant Pathol. Microb.*, 3: 136.

Lodha, S. and Aggarwal, R.K. 2002. Inactivation of *Macrophomina phaseolina* propagule during composting and effect of compost on dry root rot severity and on seed yield of cluster bean. *Eur.J.Plant Pathol.* 108 (3): 253.

Mahadevan, A. 1982. Biochemical aspects of plant disease resistance part I. Preferred inhibitory substance prohibition. Today and Tomorrow Printer and Publishers, New Delhi.

Mamun, M.S.A. 2011b. Development of tea science and tea industry in Bangladesh and advances of plant extracts in tea pest management. *Int. J. Sustain. Agril. Tech.*,*7*(5): 40-46.

Mangang, H.C. and Chhetry, G.K.N. 2012. Antifungal properties of certain plant extracts against *Rhizoctonia solani* causing root rot of French been in organic soils of Manipur. *International Journal of Scientific and Research Publications*, 2 (5): 1-4.

Pimental, D., Andow, D., Dyson-Hudson, D., Gallahan, D., Jacobson, S., Irish, M., Croop, S., Moss, A., Schreiner, I., Shepard, M., Thompson, T. and Vinzant, B. 1980. Environmental and social cost of pesticides. A preliminary assessment. *Oikos*, 34: 125- 140.

Rao, K.N. and Parmar, B.S. 1984. A compendium of chemical constituents of neem. Neem News letter 1: 39-46.

Rehman B., Ganai, M A., Mansoor, K.P. Siddiqui, A. and Usman, A. 2012. Management of Root Knot Nematode, Meloidogyne incognita Affecting Chickpea (Cicer arietinum) for Sustainable Production. *Bioscience International, 1 (1): 01-05.*

Rizvi, R., Mahmood, I., Tiyagi, S. A. and Khan Z. 2012. Effect of some botanicals for the management of plant-parasitic nematodes and soil-inhabiting fungi infesting chickpea. *Turk J Agric*, 36: 710-719.

Sanjay, G. and Tiku, A.K. 2009. Botanicals in Pest Management Current Status andnFuture Perspectives, *Biomed. Life Sci.*, p317.

Sendhilvel, V., Kavitha, K., Nakkeeran, S., Raguchander, T. and Marimuthu, T. 2004. Bioagents and Mycoherbicides. *Glimses of Plant Pathology*.

Sharaby, A. 1988. Evaluation of some Mytraceae plant leaves as protectants against the infestation by *Sitophilus oryzae* L. and *Sitophilus granarius* L. *Insect Sci. Appl.*, 9: 465-468.

Siddiqui, F.A., Gulzar, T. 2003. Tetra cyclic triterpenoids from the leaves ofn*Azadirachta indica* and their insecticidal activities. *Chem Pharm Bull.* (Tokyo); (51): 415-417.

Singh, P.K., Khan, A., Gogoi, R. and Jaiswal, R.K. 2010. Plant leaf extracts and bioagents for ecofriendly management of wilt of Pigeonpea caused by *Fusarium udum*. *Indian Phytopath.*, 63 (3): 343-344.

Singh, R.K. and Dwivedi, R.S. 1987. Effects of oils on *Sclerotium rolfsii* causing root rot of barley. *Indian Phytopath.*, 40: 531-533.

Singh, S. and Awasthi, L.P. 2009. Plant products for the management of yellow mosaic disease of mungbean and Urdbean. *The Journal of Plant Protection Sciences*, 1(1) : 87-91.

Singh, U.P. and Prithviraj, B. 1997. Neemazol-a product of neem (*Azadiracta indica*) induces resistance in pea against *Erisiphe pisi. Physiol. Mol. Plant Pathol.* 51: 181.

Stanley, M.S. Callow, M.E., Perry *et al.* 2002. Inhibition of fungal spore spore adhesion by zosteric acid as the basis for the novel non toxic crop protection technology. *Phytopathology*. 92(4): 378.

Trivedi, A., Sharma, S.K., Ameta, O.P. and Sharma, S.K. 2014. Management of viral and leaf spot disease complexes in organic farming of blackgram. *Indian Phytopath*. 67 (1): 97-101.

Umamaheswari, R. and Sundarababu, R. 2001a. Efficacy of neem leaf extract on root knot nematode *Meloidogyne incognita* infecting cowpea. *Indian J.Nematol.*, 31(2): 126-128.

Umamaheswari, R. and Sundarababu, R. 2001b. Management of root knot nematode *Meloidogyne incognita* infecting cowpea by Calotropis leaf extract. *Indian J. Nematol.*, 31(2): 133-135.

Verma, A. Singh, R.B. 1994. *Clerodendrum aculeatum* a possible prophylactic agent against natural viral infection in mungbean. *Annals of Plant Protection Sciences* 2: 60- 63.

Williams–Woodward, J.L. 1997. Green manures of oat, rape and sweet corn for reducing common root rot in pea (*Pisum sativum*) caused by *Aphanomyces euteiches. Plant Soil*. 188 1: 43.

Stangarlin, J. R., Kuhn, O. J., Assil, L., Schwan-Estrada K. R. F. 2011. Control of plant diseases using extracts from medicinal plants and fungi. Science against microbial pathogens: Communicating current research and technological advances. A. Mandez-Vilas (Ed.) © Formatex. p1033-1042.

Chapter 19

Bio-control Agents (BCAs): Eco-friendly Approaches for Management of Pulses Pathogens

R.K. Mishra, Sonika Pandey, Monika Mishra & Naimuddin

Pulses play an important role in maintaining soil fertility and supplying protein to the large vegetarian population of the country. Thus, they provide nutritional security to the people, and also to the soil. Their cultivation has remained an integral component of subsistence farming systems since time immemorial due to ability to thrive under harsh fragile ecosystem of dry lands in India. Importance of pulses in producing protein rich food is more in India as majority of the population is religiously dependent on pulse protein and follows the religious taboos in eating meat. Many pulse crops are grown in different part of the country. Several biotic stresses synergistically hamper the pulse production and productivity in the country (Table 19.1 and Figure 19.1). Biocontrol agents are common soil inhabitants that are well known for their antagonistic and plant growth promotion potential. They inhibit the growth of soil borne phytopathogens, through various mycoparasitic mechanisms such as antibiosis, secondary metabolite production, enzyme secretion, competition for food and space *etc.* In rhizosphere the exchange and recognition of signaling molecules between BCAs and plants alter the physiology and biochemistry of both. BCAs in association with plant roots can trigger systemic resistance and improve plant health. These are also colonizers of cellulosic substances and thus they can be easily found on decaying materials (Kubicek *et al.*, 2008; Jaklitsch, 2009). BCAs also colonize their habitats very efficiently, because they utililize the substrate efficiently and secrete metabolites, antibiotics and enzymes. Among the BCAs, *Trichoderma* is well-established promising biological control agents widely acceptable worldwide and they also produce several secondary metabolites with potential applications as novel antibiotics. The aim of this chapter is to discuss the details scope and mechanisms of biocontrol agents and their exploitation for effective management of pathogens of pulse crops.

Table 19.1. List of Major Diseases of Pulses that can Effectively Managed by Biocontrol Agents

Sl.No.	*Name of Crop*	*Name of Pests*	*Causative Agents*
1	Pigeonpea	Phytophthora stem blight	*Phytophthora drechsleri* f. sp. *cajani.*
		Wilt	*Fusarium oxysporum* f. sp. *udum* (Butler) Snyd. and Hans.
		Leaf spot	*Cercospora* spp.
		Powdery mildew	*Erysiphe polygoni* DC
		Stem rot	*Sclerotinia sclerotiorum*
2	Chickpea/Gram	Wilt	*F. oxysporum* f. sp. *ciceri* Matuo and Sato
		Dry Root Rot	*Rhizoctonia bataticola*
		Stem Rot	*Sclerotinia sclerotiorum*
		Rust	*Uromyces ciceris-arietinii* (*Grogn.*) Jacz. and Boy.
		Wet Root Rot	*Rhizoctinia solani*
		Botrytis gray mold	*Botrytis cinerea*
3	Lentil	Wilt	*F. oxysporum* f. sp. *lentis*
		Root rot	*Macrophomina phaseolina*
		Rust	*Uromyces fabae*
		Ascochyta blight	*Ascochyta fabae* f. sp. *lentis*
4	Fieldpea	Rust	*U. fabae*
		Powdery mildew	*E. polygoni*
		Root rot	*R. solani*
5	Mungbean/ Urdbean	Root rot	*M. phaseolina*
		Leaf Spot	*Cercospora* spp.

Biocontrol Agents (BCAs) Research and Development

During the past two decades, an urgent need was realized for management strategies that are safe for environment and agriculture. Farmers are shifting towards eco-friendly technology for the management of diseases through BCAs or BCAs based formulations, referred to as bio-pesticides. Most important BCAs are *Trichoderma* spp., *Pseudomonas* spp., *Bacillus* spp., *Agrobacterium radiobacter*, nonpathogenic *Fusarium* spp., *Coniothyrium.* spp and *Aspergillus niger*, *Bacillus thuringiensis* (Bt), *Metarhizium* spp., *Beauveria bassiana* and nuclear polyhedrosis virus (NPVs), which are popularly used in plant protection. According to a recent report (NAAS, 2013), nearly 1400 BCAs products were sold and 175 biopesticide active ingredients and 700 products were registered worldwide for their commercialization. A growing

Figure 19.1. Symptoms of Major Diseases Affecting Pulses.

(a) Dry root rot infected chickpea field, (b) Chickpea rust, (c) Chickpea stem rot, (d) Chickpea wilt, (e) *Aschochyta* blight of chickpea, (f) *Alternaria* blight on pigeonpea, (g) Pigeonpea wilt affected field, (h) *Phytophthora* blight of pigeonpea, (i) Pigeonpea stem rot, (j) Pea rust, (k) Lentil rust, (l) Anthracnose affected bean.

body of research articles report on the identification and efficacy of different BCAs against a number of pests and pathogens, however their slow embrace is evident from the fact that only 2 per cent bio-pesticides are currently used for crop protection worldwide. Various carrier-based formulations available worldwide, alginate pellet- and talc-based formulations of BCAs have emerged as most important carrier for the application in the management of crop diseases. This, however, has yielded inconsistent performance given that a single agent might not remain active in all soil conditions. Further, enabling mass production with high level of microbial count and viability also assumes greater significance.

Management of Pulse Pathogens through Biocontrol Agents (BCAs)

Several efforts have been made to manage different diseases of pulses by incorporating bioagents colonized natural substrates into the soil. Among the antagonistic fungi *Trichoderma* species are the most studied bioagents in pulse ecosystem. They are successfully used to control wilts, root rots, collar rot and stem rot diseases incited by different *Fusarium* spp., *Rhizoctonia solani*, *R. bataticola*, *Sclerotium rolfsii*, *Sclerotinia sclerotiorum*, *Ascochyta*, *Cercospora*, *Alternaria* spp. *Phytophthora* spp. and *Pythium* spp., in different pulses and other field crops (Table 19.2). Indian Institute of Pulses Research (IIPR), India has identified several native potential strains of different biocontrol agents *i.e. Trichoderma* spp. (*Trichoderma harzianum*, *T. asperellum*, *T. afroharzianum*, *T. longibrachiatum*, *T. asperelloides*, *T. brevicompactum*, *T. atrobrunneum and T. aureoviride*) and plant growth promoting rhizobacteria (*Pseudomonas* spp, *Bacillus* spp, *Serretia* spp) from rhizospheres in major pulse growing areas in India and evaluated these for their antagonistic potential against a variety of pathogens. Accordingly, mass production technology has been developed and popularized among the pulse-growing farmers in different agro-ecosystems (Chaudhary *et al.*, 2004; Mishra *et al.*, 2015, 2016, 2018).

Table 19.2. Biocontrol Efficacy of *Trichoderma/PGPRs* against Pulses Pathogens

Crops	*Pests/pathogens*	*Effective Trichoderma spp/BCAs*
Chickpea	*F. oxysporum* f. sp. *ciceri* *Sclerotium rolfsii* *R. bataticola, R. solani* *A. rabei*	*Trichoderma harzianum, T. asperellum, T. koningii, T. virens, Gliocladium virens, Pseudomonas* sp., *Bacilus* sp.
Pigeonpea	*F. udum* *P. drechsleri* f. sp. *cajani*	*T. harzianum, T. asperellum, G. virens, Bacillus subtilis, T. koningii, T. hamatum, A. niger, Penicillium* sp, *Glomus mosseae, Verticillium chlamydosporum, Glomus mosseae, P, fluorescens, HaNPV, Bt,*
Lentil	*F. oxysporum* f. sp. *lentis* *M. phaseolina* *M. phaseolina* *A. fabae* f. sp. *lentis*	*T. asperellum, T. harzianum, G. virens, Pseudomonas fluorescens* *Rhizobium leguminoserum*
Pea	*U. fabae* *E. polygoni* *R. solani*	*T. harzianum, T. viride, T. virens, T. hamatum* *Rhizobium leguminoserum*
Mungbean/ Urd bean	*M. phaseolina* *Cercospora* spp.	*T. asperellum, T. harzianum, T. konninghii, T. longibrachiatum, P. fluorescens, Bacillus subtilis, A. niger* and *Pencillium citrinum*

Chickpea Diseases

Chickpea wilt caused by *Fusarium oxysporum* f. sp. *ciceri*, the major factor for limiting chickpea production worldwide is widely distributed in chickpea growing areas. Many species of *Trichoderma* and PGPRs isolates from different locations were screened against races of *F. oxysporum* f. sp. *ciceri* commonly prevalent in India. Ranchi isolates of *T. viride* followed by T. *harzianum* inhibited maximum mycelial growth of the pathogen. The integrated application of *T. harzianum* (10g/kg seed) and carboxin (2 g/kg seed) as a seed treatment reduced wilt incidence (44.1-60.3 per cent) and increased seed germination by 12.0-14.0 per cent and grain yields by

42.6-72.9 per cent. Application of Tea leaves formulation of *T. harzianum* by seed treatment at 10gm per kg seeds was found effective for the management of chickpea wilt followed by wheat bran saw dust formulation (Singh *et al.*, 2007). *T. virens* based seed dressing formulation, *viz.*, Pusa 5SD and soil application formulations, *viz.*, Pusa Biopellet 16G (PBP 16G) and Pusa Biogranule 6 (PBG 6) were used for management of wet root rot caused by *R. solani*. Other diseases like Damping-off, root and/or stem rot caused by various organisms can be effectively managed by integrating *T. viride* (isolate TVM2).

Pigeonpea Diseases

Among the available biocontrol agents, *Trichoderma* is most effective and predominantly used against pigeonpea wilt (Mishra *et al.*, 2015, 2016, 2018a, 2018b). Mandhare and Suryawanshi (2005) reported that 63.25 per cent wilt reduction was achieved by seed and soil application of *Trichoderma* spp. Shukla and Chaudhary (2007) studied the efficacy of 30 different *Trichoderma* isolates against *F. udum* and found variation in the antagonistic property of these isolates. Two isolates showed highest colony growth reduction of 65.5 to 67.2 per cent, seven between 50 to 60 per cent and 21 isolates between 41-50 per cent. They further observed that eleven isolates reduced conidia production by >90 per cent. The culture filtrates of *Trichoderma* spp. also reduced > 90 per cent colony growth of *F. udum*. Phytophthora stem blight (PSB), caused by *Phytophthora drechsleri* f. sp. *cajani*, is another economically important disease of pigeonpea (*Cajanus cajan*), especially when excessive rains fall within a short span of time and hot and humid weather persists during the crop season. Several isolates of *Trichoderma* species and PGPRs (*Pseudomonas* spp and *Bacillus* spp) were characterized and evaluated against *Phytophthora* of pigeonpea (Mishra *et al.*, 2018).

Lentil Diseases

Fusarium oxysporum f. sp. *lentis* causes wilt disease in lentil worldwide and is major limiting factor its production. In India it causes yield loss up to 50 per cent (Bhat *et al.*, 2006). The fungal bio agent, *T. hamatum* was reported to reduce up to 33 per cent per cent of mortality due to wilt in lentil (El-Hasan *et al.*, 2013). Kumar *et al.* (2001) suggested that the chemical spray of Bayleton + *T. harzianum* showed significant response with respect to the disease severity, 1000-grain weight (g), and yield (kg/ha) followed by Mancozeb + *P. fluorescens* and Captaf + *T. harzianum*.

Mungbean and Urdbean Diseases

Root rots are caused by various fungal pathogens *viz.*, *Rhizoctonia bataticola*, *R. solani*, *S. rolfsii*, *M. phaseolina* and *F. solani*. Soil application of *Trichoderma virens* based formulations, Pusa Biogranule 6 and Pusa Biopellet16G (PBP 16G) and seed dressing formulation, Pusa 5SD (PBG6) were found to be superior in reducing wet rot disease incidence and increasing seed germination and shoot and root lengths in mungbean. Combined application of PBP 16G (*T. virens*) and Pusa 5SD (*T. virens*) + Carboxin was superior compared to individual treatments (Dubey ef al., 2011). However, *T. harzianum* based formulations, Pusa 5SD for seed dressing and Pusa Biopellet 10G effectively combined in integrated management of dry root rot of mungbean (Dubey *et al.*, 2009). The bioagent, *T. viride* is reported to be effective against various soil borne pathogens of mungbean *viz.*, *R. solani*, *S. rolfsii*, *M. phaseolina*, and *F. solani* by confrontation assay (Mishra *et al.*, 2011). Cercospora leaf spot of mungben can be effectively managed by integration of *T. viride* along with fungicides.

Pea and Beans Diseases

Root rot disease complex is considered as the most destructive in pea and bean as it affects its initial plant stand and more than 20 different pathogens have been reported to be associated with the disease from different parts of the world. Among these, *Fusarium oxysporum* f. sp. *pisi*, and *Pythium* spp. are highly destructive. Different strains of *Trichoderma* effectively manage the root rot disease of fieldpea caused by *F. solani* f. sp *pisi* (Di Pietro *et al.*, 1992, Kapoor *et al.*, 2006).

Delivery System of Biocontrol Agents in Pulses

Biocontrol formulations are delivered through several means based on survival nature and mode of infection of the pathogen. It is delivered through seed treatment, soil application, and foliar application or through combination of several methods.

Seed Treatment

This is the most effective method for management of phytopathogens. In seed treatment, seed priming is the process in which hydration of seed is controlled to a level that permits pre germinative metabolic activity to take place without emergence of the radical. Treating Pigeonpea and chickpea seeds with talc-based formulation of *T. harzianum*, *Trichoderma viride*, *T. hamatum*, *T. virens*, *Bacillus subtilis*, and *Pseudomonas fluorescens* facilitates management of Fusarium wilt in both crops. On-farm demonstrations have evidenced that the seed treatment with 2 per cent talc-based formulation of *Trichoderma harzianum* (IPT-31) led up to 32.1 and 14.3 per cent decrease in root rot incidence in chickpea and lentil, respectively, thus correspondingly improving crop yield by 16.6 and 12.6 per cent (Purushottam *et al.*, 2014).

Soil Application

Fully active growing population of bioagents is applied in the soil at the time of sowing. *Trichoderma* can be applied as powder form as well as drench at initial stage of the crops. According to Vidhyasekaran and Muthamilan (1995), soil application of peat-based formulation with *P. fluorescens* (Pf1) at 2.5 kg of formulation mixed with 25 kg of well-decomposed farm yard manure improved management of chickpea wilt. Combining *P. fluorescens* with safer fungicides reduced the wilt complex in Pigeonpea.

Foliar Spray

Several researchers have reported that the application of *Trichoderma* sp., in bean leaves reduces the incidence of bean rust (*Uromyces phaseoli*). In a similar manner, seed treatment and foliar application of *P. fluorescens* (Pf1) reduced the severity of *Puccinia arachidis* of groundnut and rust (*U. fabae*) of fieldpea under field conditions.

Conclusions

Chemical fungicides which are normally used to control plant disease are very harmful for the environment. They pollute air, water and soil causing harm to human health. Use of biocontrol agents makes the environment safe. *Trichoderma* is a well known biocontrol agent and it is effective against a variety of pathogens. There are many biocontrol genes in *Trichoderma* which help the plant in developing resistance against plant pathogens. *Trichoderma* based bioformulation are cheap, safe and environmental friendly. However, there is a need of more work to be done to develop stable, improved and cost effective *Trichoderma* formulations which can control pathogens more effectively and efficiently. In present agriculture scenario, use of biocontrol agents (BCAs) is of utmost importance in pulses for biotic stresses management, but their potential is yet to be exploited fully mainly because the research in this area is still confined to the laboratory and very little attention has been paid to produce the commercial formulations of biocontrol agents. Moreover, whatever has been commercially produced has not been used efficiently by the farmers owing to the lack of information regarding its application and importance. So, the concept of biological control and their different formulations need to be popularized at field level.

REFERENCES

Babychan M and Simon S (2017). Efficacy of *Trichoderma* spp. against *Fusarium oxysporum f. sp. lycopersici*. (FOL) infecting pre-and post-seedling of tomato. *Journal of Pharmacognosy and Phytochemistry*, *6:* 616-619.

Bhat NA, Beig MA, Maheshwari SK, and Masoodi SD(2006). Screening and yield of lentil germplasm as influenced by Fusarium wilt. *Ann PI Protec Sci* 14: 139-141.

Bjorkman T, Blanchard LM and Harman GE (1998). Growth enhancement of shrunken-2 sweet corn when colonized with *Trichoderma harzianum* 1295-22 effect of environmental stress. *Journal of American Society of Horticultural Sciences*, 123: 35-40.

Brotman Y, Landau U, Cuadros-Inostoza A, Takayuki T, Fernie A R *et al.* (2013). *Trichoderma*-Plant Root Colonization: Escaping early plant defense responses and activation of the antioxidant machinery for saline stress tolerance. https: //doi.org/10.1371/journal.ppat.1003221.

Buensanteai N, Mukherjee PM, Horwitz BA, Cheng C (2010). Expression and purification of biologically active *Trichoderma virens* proteinaceous elicitor Sm1 in *Pichia pastoris*. *Protein Express*. 72: 131-138.

Carpenter MA, Ridgway HJ, Stringer AM, Hay AJ, Stewart A (2008). Characterization of a *Trichoderma hamatum* monooxygenase gene involved in antagonistic activity against fungal plant pathogens. *Curr. Genet*. 53: 193-205.

Chaudhary RG and Prajapati RK (2004). Comparative efficacy of fungal, biological agents *Fusarium udum*. *Annals of Plant Protection Sciences*. 12(1): 75-79

Chet I (1987). *Trichoderma* application, mode of action, and potential as biocontrol agent of soil-borne pathogenic fungi. Pages 137-160.in: Innovative Approaches to Plant Disease Control. I. Chet, ed., John Wiley, New York.

Chowdappa P, Kumar SPM, Lakshmi MJ, Upreti KK (2013). Growth stimulation and induction of systemic resistance in tomato against early and late blight by *Bacillus subtilis* OTPB1 or *Trichoderma harzianum* OTPB3. *Biological Control,* 65: 109–117.

Claydown KL, Emerson OH, and Sauthwell RJ (1987). The isolation of a toxic substance from the culture filtrate of *Trichoderma. Phytopathology* 36: 1068.

Contreras-Cornejo HA, Macías-Rodríguez L and Cortés-Penagos C (2009). *Trichoderma virens,* a plant beneficial fungus enhances biomass production and promotes lateral root growth through an auxin dependent mechanism in *Arabidopsis. Plant Physiology,* 149: 1579-1592.

Deshmukh S, Hueckelhoven R, Schaefer P, Imani J and Sharma M (2006). The root endophytic fungus *Piriformospora indica* requires host cell death for proliferation during mutualistic symbiosis with barley. *Proc Nat Acad Sci* USA 103: 18450—18457

Di Pietro A, Gut-Rella M, Pachlatko JP and Schwinn FJ (1992). Role of antibiotics produced by *Chaetomium globosum* in biocontrol of *Pythium ultimum,* a causal agent of damping-off. *Phytopathology* 82: 131- 135.

Doni F, Isahak A, Radziah C, Zain CM, Ariffin SM, Nurashiqin W, Mohamad Wan, Mohtar W, and Yusoff Wan. (2014). Formulation of *Trichoderma* sp. SL2 inoculants using different carriers for soil treatment in rice seedling growth. *Springer Plus,* 3: 1-5.

Donzelli BGG, Lorito M, Scala F, Harman GE (2001). Cloning, sequence and structure of a gene encoding an antifungal glucan 1,3- β-glucosidase from *Trichoderma atroviride* (*T. harzianum*). *Gene,* 277: 199-208.

Druzhinina IS, Kopchinskiy AG, Komon M, Bissett J, Szakacs G, Kubicek CP (2005) An oligonucleotide barcode for species identification in *Trichoderm*a and *Hypocrea. Fungal Genet Biol* 42: 813–828.

Dubey SC, Bhavani R and Singh B(2009). Development of Pusa 5SD for seed dressing and Pusa Biopellet 10G for soil application formulation of *T. harzianum* and their evaluation for integrated management of dry root rot of mungbean (*Vigna radiata*). Bio] Control 50: 231-242.

Dunlop RW, Simon A, Siwasithamparam D and Ghisalberti EL(1989). An antibiotic from *Trichoderma koningii* active against soil-borne plant pathogens. *Journal of Natural Products,* 52: 67-74.

El -Hassan, $A and Gowen SR (2006). Formulation and delivery of the bacterial antagonist *Bacillus subtilis* for management of lentil vascular wilt caused by *Fusarium oxysporum f. sp. Lentis.* J Phytopathol 154: 148-155.

Gams W, Bissett J (1998) Morphology and identification of *Trichoderma.* In: Harmann GE, Kubicek CP (eds) *Trichoderma* and *Gliocladium.* Taylor and Francis, London, pp 3–34.

Haqu, MM, Ilias GNM. and Molla AH (2012). Impact of *Trichoderma*-enriched Biofertilizer on the Growth and Yield of Mustard (*Brassica rapa* L.) and Tomato (*Solanum lycopersicum* Mill.) *The Ariculturists.* 10: *109-119.*

Harman G E, Howell CR, Viterbo A, Chet I and Lorito M (2004). *Trichoderma* species opportunistic, avirulent plant symbionts. *Nature Reviews Microbiology,* 2: 43–56.

Harman GE, Lorito M, Lynch JM (2004) Uses of *Trichoderma* spp. to alleviate or remediate soil and water pollution. *Adv Appl Microbiol* 56: 313–330.

Hermosa R, Botella L, Keck M, Jimenez JA, Montero Barrientos M, Arbona V, GomezCadenas A, Monte E, Nicolas C (2011). The overexpression in *Arabidopsis thaliana* of a *Trichoderma harzianum* gene that modulates glucosidase activity, and enhances tolerance to salt and osmotic stresses. *J. Plant Physiol.* 168: 1295-1302.

Howell CR, Hanson RD, Stipanovic and L.S. Puckhaber(2000). Induction of terpenoid synthesis in cotton roots and control of *Rhizoctonia solani* by seed treatment with *Trichoderma virens*. *Phytopathol.,* 90: 248-252.

Howell CR (2003) Mechanisms employed by *Trichoderma* species in the biological control of plant diseases: the history and evolution of current concepts. *Plant Dis* 87: 4-10

Howell CR, Hanson LE, Stipanovic RD and Puckhaber LS (2000). Induction of terpenoid synthesis in cotton roots and control of *Rhizoctonia solani* by seed treatment with *Trichoderma* virens. *Phytopathology* 90: 248-252.

Idowu OO, Olawole OI, Idumu OO Salami AO (2016). Bio-control Effect of *Trichoderma asperellum* (Samuels) Lieckf. and *Glomus intraradices* Schenk on Okra Seedlings Infected with *Pythium aphanidermatum* (Edson) Fitzp and *Erwinia carotovora* (Jones). *American Journal of Experimental Agriculture.* 10: 1-12

Jaklitsch WM (2009) European species of Hypocrea. Part I. The green-spored species. *Stud Mycol* 63: 1–91

Jin X, Harman GE and Taylor AG (1996). Conidial biomass and desiccation tolerance of Trichoderma harzianum produced at different medium water potentials. *Biol Control* 1: 237-243.

Kapoor AS, Paul YS and Singh A (2006). Integrated management of white rot and root rot-wilt disease complex of pea. *Indian Phytopath* 59: 467-474.

Katatney and Idris (2014). Effects of Single and Combined Inoculations with *Azospirillum brasilense* and *Trichoderma harzianum* on Seedling Growth or Yield Parameters of Wheat (*Triticum vulgaris L.,* Giza 168) and Corn (*Zea mays L.,* Hybrid 310). *Journal of Plant Nutrition* 37: 1913-36.

Kubicek CP, Komon-Zelazowska M, Druzhinina IS (2008) Fungal genus *Hypocrea/ Trichoderma*: from barcodes to biodiversity. J Zhejiang Univ Sci B 9: 753–763.

Kuhls K, Lieckfeldt E, Samuels GJ, Kovacs W, Meyer W, Petrini O,Gams W, Borner T, Kubicek CP (1996) Molecular evidence that the asexual industrial fungus Trichoderma reesei is a clonalderivative of the ascomycete Hypocrea jecorina. *Proc Natl Acad Sci* U S A 93: 7755–7760

Kumar D and Dubey SC (2001). Management of collar rot of pea by the integration of biological and chemical methods. *Indian Phytopath* 54: 62-66.

Kumar V, Shahid M, Singh A, Srivastava M, Mishra A, Srivastava YK, Pandey S and Shrarma A (2014). Effect of Biopriming with Biocontrol Agents *Trichoderma harzianum* (Th.Azad) and *Trichoderma* viride on Chickpea Genotype (Radhey). *Journal of Plant Pathology and Microbiology*, 5: 1-4.

Lalitha P, Srujana and Arunalakshmi K (2012). Effect of *Trichoderma viride* on germination of mustard and survival of mustard seedlings. *International Journal of Life Sciences Biotechnology and Pharma Research*.1: 137-140.

Leong j (1986). Siderophores: Their biochemistry and possible role in biocontrol of plant pathogen. *Annual Review of Phytopathology*, 24: 187-209.

Li M, Yang Q (2007). Isolation and characterization of a β-tubulin gene from *Trichoderma harzianum*. *Biochem. Genet*. 45: 529-534.

Mandhare VK and Suryawanshi AV(2005). Standarization of storage conditions to increase the shelf life of Trichoderma formulations. *Agri Sci Digest* 25: 71-73.

ManteroBarrientos M, Hermosa R, Nicolas C, Cardoza RE, Gutierrez S, Monte E (2008). Overexpression of a *Trichoderma* HSP70 gene increases fungal resistance to heat and other abiotic stresses. *Fungal Genet. Biol*. 45: 1506-1513.

Marcello CM, Steindorff AS, Silva SP, Silva RN, Bataus LAM (2010).Expression analysis of the exo-β-1,3-glucanase from the mycoparasitic fungus *Trichoderma asperellum*. *Microbiol. Res*. 165: 75-81.

Martinez C, Blanc F, Le CE, Besnard O, Nicole M and Baccou JC(2001). Salicylic acid and ethylene pathways are differentially activated in melon cotyledons byactive or heat-denatured cellulase from Trichoderma longibrachiatum. *Plant Physiol* 127: 334-344.

Matsouri F, Björkman T, Harman GE (2010). Seed treatment with *Trichoderma harzianum* alleviates biotic, abiotic and physiological stresses in germinating seeds and seedlings. *Phytopathology*, 100: 1213-1221.

Migheli Q, GonzalezCandelas L, Dealessi L, Camponogara A, RamonVidal D (1998). Transformants of *Trichoderma longibrachiatum* overexpressing the β-1,4-endoglucanase gene *egl*1 show enhanced biocontrol of *Pythium ultimum* on cucumber. *Biol. Control*, 88 (7): 673-677.

Mishra A and Salokhe VM (2011) Rice growth and physiological responses to SRI water management and implications for crop productivity. *Paddy Water Environment*, 9: 41–52.

Mishra BK, Mishra RK, Mishra RC, Tiwari AK, Yadav RS and Dikshit A (2011). Biocontrol efficacy of *Trichoderma viride* isolates against fungal plant pathogens causing disease in *Vigna radiata* L. *Arch Appl Sci Res* 3: 361-369.

Mishra RK, Bohra A, Naimuddin, Kumar Krishna, Sujayanand GK, Saabale PR, Naik SJ Satheesh, Sarma BK, Kumar D, Mishra Monika, Srivastava DK and Singh NP. (2018a). Utilization of biopesticides as sustainable solutions for management of pests in legume crops: achievements and prospects. *Egyptian Jr. Biological Pest Control*.DOI 10.1186/s41938-017-0004-1.

Mishra RK, Mishra Monika, Naimuddin and Kumar Krishna (2018b). *Trichoderma asperellum:* A potential biocontrol agents against wilt of Pigeonpea caused by *Fusarium udum* Butler. *Journal of Food Legumes* 31(1): 50-53

Mishra RK, Naimuddin, Monika Mishra and Sandeep Sharma. (2016). *Trichoderma*: Potential biocontrol agents for pulses. *Dalhan Alok* 14: 80-86.

Mishra RK, Naumuddin, Akram Mohd., Sachan DK, Mishra Monika and Kumar Krishna.(2016). Characterization of Indigenous *Trichoderma* spp. through ITS based sequences. *Pulses News Letter*.27 (2): 5.

Mishra RK, Saabale PR, Naimuddin, Jagadeeswaran R, Mishra O. (2015). Potential *Trichoderma* sp. from Pulses Rhizosphere. *Pulses News Letter*. April-June pp: 3.

Mukherjee M, Mukherjee PK, Kale PS (2007). cAMP signalling is involved in growth, germination, mycoparasitism and secondary metabolism in *Trichoderma virens*. *Microbiology*, 153: 1734-1742.

Mukherjee PM, Latha J, Hardar R, Horwitz BA (2003). *Tmk*A, Mitogen activated Protein Kinase of *Trichoderma virens* is involved in biocontrol properties and repression of conidiation in the dark. *Eukaryot. Cell* 2: 446-455.

Mukhtar I (2008). Influence of *Trichoderma* species on seed germination in okra. *Mycopathology*. 6: 47-50.

Naznin A, Hossain MM, Anju-Man Ara K, Hoque A, Islam M Hasan T (2015). Influence of Organic Amendments and Bio-Control Agent on Yield and Quality of Tuberose, *Journal of Horticulture*, 2: 4 http: //dx.doi.org/10.4172/2376-0354.1000156.

Okoth SA, Otadoh, J.A. and Ochanda OJ (2011). Improved seedling emergence and growth of maize and beans by *Trichoderma harziunum*. *Tropical and Subtropical Agroecosystems*, 13: 65 – 71.

Persoon CH (1794) Disposita methodica fungorum. Römer's Neues Mag Bot 1: 81–128.

Pieterse CM, Leon-Reyes A, Vander Ent S, Van Wees SC (2009). Networking bysmall-molecule hormones in plant immunity. *Nature Chemical Biology*. 5: 308-16.PLoSONE, 6(1): 1-15.

Pozo MJ, JongMin B, Garcia JM, Kenerley CM (2004). Functional analysis ot *tvsp1*, a serine protease-encoding gene in the biocontrol agent *Trichoderma virens*. *Fungal Genet. Biol.* 41: 336-348.

Rehman SU, Lawrence R, Kumar EJ and Badri ZA(2012). Comparative efficacy of *Trichoderma* viride, *T. harzianum* and carbendazim against damping-off disease of cauliflower caused by *Rhizoctonia solani* Kuehn *Journal of Biopesticides*. 5: 23-27.

Rifai MA (1969) A revision of the genus Trichoderma. Mycol Pap 116: 1–56.

Saba H, Vibhash D, Manisha M, Prashant KS, Farhan H and Tauseef A (2012): *Trichoderma* – a promising plant growth stimulator and biocontrol agent. *Mycosphere*, 3: 524–531.

Samuels GJ (2006) *Trichoderma*: systematics, the sexual state, and ecology. *Phytopathology* 96: 195–206.

Seidl V, Seibel C, Kubicek CP, Schmoll M (2009a) Sexual development in the industrial workhorse *Trichoderma reesei*. *Proc Natl Acad Sci* U S A 106: 13909–13914

Sharma K, Mishra, A K Misra RJ (2009). Morphological, biochemical and molecular characterization of Trichoderma harzianum isolates for their efficacy as biocontrol agents. *J. Phytopathol.* 157, 51–56. doi: 10.1111/j.1439-0434.2008.01451.x

Shi WL, Chen XL, Wang LXet al. (2016) Cellular and molecular insight into the inhibition of primary root growth of Arabidopsis induced by peptaibols, a class of linear peptide antibiotics mainly produced by *Trichoderma* spp *J Exp Bot* DOI: 10.1093/jxb/erw023.

Shukla N and Chaudhary RG (2007). Reduction of inoculums potential of Fusarium udum by Trichoderma spp. Abstract of Paper presented during National Symposium on "Plant Pathogens: Exploitation and Management" at R. D. University, Jabalpur (M.P.), India during January 16-18, 2007. *Indian Phytopath* 60: 392.

Shukla N, Awasthi RP, Rawat L and Kumar J (2012). Biochemical and physiological responses of rice (Oryza sativa L.) as influenced by *Trichoderma harzianum* under drought stress. *Plant Physiology and Biochemistry*, 54: 78-88.

Singh A, Srivastava S and Singh HB(2007). Effect of substrates on growth and shelf life of *Trichoderma harzianum* and its use in biocontrol of diseases. Bioresource Technol 98: 470-473.

Singh S, Dureja P, Tanwar RS and Singh A (2005). Production and antifungal activity of secondary metabolites of *Trichoderma* virens. *Pesticides Research Journal*. 17: 26-9.

Sriram S, Palanna KB and Ramanujam B(2010). Effect of chitin on the shelf-life of *Trichoderma harzianum* in talc formulation. *Indian J Agri Sci* 80: 930-932.

Sriram S, Roopa KP and Savitha MJ (2011). Extended shelf life of liquid formulations of *Trichoderma harzianum* with addition of glycerol. *Crop Prot* 30: 1334-1339.

Srivastava SN, Singh V and Awasthi SK (2006). *Trichoderma* induced improvement in growth, yield and quality of sugarcane. *Sugar Tech.* 8: 166. https: //doi.org/10.1007/BF02943654.

Stepanova AN, Yun J, Likhacheva AV, and Alonso JM (2007). Multilevel interactions between ethylene and auxin in Arabidopsis roots. *Plant Cell.* 19, 2169–2185.

Thakur AK, Uphoff N and Antony E (2010). An assessment of physiological effects of system of rice intensification (sri) practices compared with recommended rice cultivation practices in India. *Experimental Agriculture*, 46: 77–98.

Thakur M. and Sohal BS (2013). Role of elicitors in inducing resistance in plants against pathogen infection: A review. ISRN *Biochemistry*: 1–10. doi: 10.1155/2013/762412

Tijerino A, Cardoza RE, Moraga J, Malmierca MG, Vicente F, Aleu J,Collado IG, Gutierrez S, Monte E, Hermosa R (2011). Overexpression of the trichodiene synthase gene *tri5* increases trichodermin production and antimicrobial activity in *Trichoderma brevicompactum*. *Fungal Genet. Biol.* 48: 285-296.

Tucci M, Ruocco M, De Masi L, De Palma M, Lorito M (2011). The beneficial effect of *Trichoderma* spp. on tomato is modulated by the plant genotype. *Molecular Plant Pathology*. 12: 341-54.10.1111/j.1364-3703.2010.00674.

Vinale F, Flematti G, Sivasithamparam Ket al.(2009) Harzianic acid, an antifungal and plant growth promoting metabolite from *Trichoderma harzianum J Nat Prod* 72 2032 5.

Vinale F, Marra R. and Ruocco M (2014) *Trichoderma* Secondary Metabolites Active on Plants and Fungal Pathogens Article in *The Open Mycology Journal*, 8: 127-139.

Yoshioka Y, Ichikawa H, Naznin HA, Kogure A and Hyakumachi M (2011). Systemic resistance induced in Arabidopsis thaliana by *Trichoderma asperellum* SKT-1, a microbial pesticide of seedborne diseases of rice. *Pest Manag Sci* http: //dx.doi.org/10.1002/ps.2220

Zhong YH, Wang TH, Wang XL, Zhang GT, Yu HN (2009). Identification and characterization of a novel gene, *TrCCD1*, and its possible function in hyphal growth and conidiospore development of *Trichoderma reesei*. *Fungal Genet. Biol.* 46: 255-263.

Chapter 20

Role of Vesicular Arbuscular Mycorrhizae (VAM) Fungi in Leguminous Crops

Amar Bahadur

The term mycorrhiza (fungus root) was coined by Frank (1885). Harley and Smith (1983) defined mycorrhiza as an association between fungal hyphae and roots of higher plants concerned with absorption of mineral substances from the soil. The Greek *mycos,* meaning 'fungus' and *rhiza,* meaning 'root'. Mycorrhiza is a symbiotic mutualistic relationship between special soil fungi and fine plant roots.

Two main types of mycorrhizae may be found, depending on penetrates into the root cells. *Ectomycorrhizae* - this type causes a drastic change in the root shape. The ectomycorrhizal fungus penetrates between the cell walls of the cortex and forms a covering sheath, or mantle, of fungal hyphae around the entire root. Most ectomycorrhizal fungi produce mushrooms and can be cultivated in culture media in the laboratory. *Endomycorrhizae* or *Arbuscular Mycorrhizae* (AM) do not form a mantle over the root, and the fungus actually enters the cortex cells. This type is characterized by the presence of arbuscles in the region of the root cortex; vesicles may or may not be present; and they may function as reserve organs and also for fungal multiplication. Arbuscles absorb nutrients from the plant cells but also release mineral elements to the plant. AM fungi also produce spores outside the root. Endomycorrhizal fungi have to be multiplied through a host plant. Most of the plants form symbiotic relationship with a group of fungi called mycorrhiza, which function as a bridge for the flow of energy and matter between plants and soils (Cardon and Whitbeck, 2007). VA-mycorrhizae have long been known to enhance the growth/yield of various crops (Gerdemann 1968; Smith and Read

1997). Pulses are highly dependent on mycorrhizae due to their high P requirement (Bethlenfalvay 1993). The density of their spores in the field and/or the degree of root colonization are considered measures of plant performance (Fyson and Oaks 1990; Wani, McGill, and Tiwari 1991) as most of the plants are mycorrhizal in nature (Gerdemann 1968). Earlier researchers have documented that the inoculation of AMF is beneficial to crops by way of mobilizing nutrients especially P (Lambert, Baker, and Cole 1980; Sekar, Vanangamudi, and Suresh 1997). The symbiotic association benefits plants by improving nutrient uptake (Mosse 1973). Recent investigations reveal that mycorrhizal fungi enhance the growth and productivity of different crops (Khaliq, Gupta, and Kumar 2001; Rajeshwari, Latha, Vanangamudi *et al.*, 2001). Brundrett (2004) defined mycorrhiza as a symbiotic association for one or both partners between a fungus and a root of living plant that is primarily responsible for nutrient transfer.

Leguminous crops are the important protein source in human dietary systems particularly in developing countries. The legume productions have, however, been seriously hampered due to heavy metal contamination of soils. Arbuscular mycorrhizal fungi (AMF) establish mutual symbioses with the majority of higher plants and after colonization, contribute to the plant growth in metal-contaminated sites by increasing plant access to P, by improving soil quality, and by restricting the movement of metals within plant tissues. The heavy metals affect legumes and the incidence of AMF in metal-polluted sites. In addition, the role of AMF in restoring heavy metal-contaminated site. Maintenance of soil quality adopting various remediation strategies including biological approaches is therefore important. Mineral fertilizers which are required to raise crop production levels are too expensive and most smallholder farmers cannot afford them. Intercropping or rotation of cereals with grain legumes (due to their N2-fixing abilities) and crop residue management have been recommended as alternatives to improve cereal yields on these soils. The N2-fixing abilities of legumes can be enhanced not only by *Rhizobium* spp. but also by colonization of their roots by arbuscular mycorrhiza fungi (Asimi *et al.*, 1980; Bayne and Bethlenfalvay, 1987). Mycorrhiza is a mutualistic association between fungi and higher plants (Menge, 1983.). Different types of mycorrhizae occur, distinguished by their morphology and to a certain extent, in their physiology. The ectomycorrhizae characterized by an external sheath of fungal cells surrounding the root, often penetrates between the cells of epidermis and the first few cells of cortex and the fungal hyphae typically infect the roots of forest trees of the temperate region. While endomycorrhizae like vesicular arbuscular mycorrhizal (VA) fungi forms no sheath, the fungus infects the root system of most cultivated crops and usually it invades several layers of the outer root cortex. VA-fungal hyphae penetrate individual cells and form arbuscules within the cell and vesicles outside their host cells which led to their name (Tisdale, *et al.*, 1995.). They transport water and mineral nutrients from the soil to the plant while the fungus is benefiting from the carbon compounds provided by the host plant. Therefore VA-fungi have a pervasive effect upon plant form and function (Bethlenfalvay and Barea, 1994.). VA-fungi are associated with improved growth of many plant species due to increased nutrients uptake, production of growth promoting substances, tolerance to drought, salinity and transplant shock and synergistic interaction with other beneficial soil microorganisms such as N-fixers and P-solubilizer (Sreenivasa and Bagyaraj, 1989.). Symbiotic association of plant roots with VA-fungi often result in enhanced growth because of increased acquisition of phosphorus (P) and other low mobile mineral nutrients (Kwapata and Hall, 1985.). Effective nutrient acquisition by VA-fungi is generally attributed to the extensive hyphal growth beyond the nutrient depletion zone surrounding the root (Tisdale, *et al.*, 1995.). A better understanding of the mycorrhizae of agronomic crops is needed because of their potential involvement in systems of sustainable agriculture (Khalil, *et al.*, 1992). Sreenivasa and Bagyaraj (1989) studied the effect of fungicides on root colonization with VA mycorrhizal fungi and indicated that reduction from 10 to 20 per cent of root infection percentages recorded when the recommended level of fungicides were used. While some fungicides significantly increased the percentage root colonization at half the recommended level.

Water, an integral part of living systems, is ecologically important because it is a major force in shaping climatic patterns and is biochemically important because it is a necessary component in physiological processes. Plant acclimation to drought stress depends on the process like osmotic adjustment and osmoregulation. Among hydrophilic proteins, several proteins including dehydrins accumulate in various plant parts during the process of osmotic adjustment. Moisture stress has also been shown to induce phenolic accumulation through up-regulation of phenylalanine ammonia lyase (PAL) and proline accumulation. Reactive oxygen species (ROS) levels increase when plants are exposed to abiotic stress conditions. Carbon metabolism and the levels of specific sugars are severely affected by abiotic stress. Mycorrhizae can facilitate plant growth directly by facilitating the uptake of nutrients from the environment, by influencing phytohormone production and by enzymatic lowering of plant ethylene levels.

Mycorrhizal association benefits higher plants by improving water and nutrient uptake, helping in the development of roots and binding the soil, storing carbohydrates and oil, protecting plants from soil-borne diseases, and detoxifying soils contaminated by metals. Arbuscular mycorrhiza benefit plants by improving the supply of nutrients, especially phosphorus and minerals such as zinc, copper, sulphur, potassium, and calcium (Cooper and Tinker 1978), and the plant supplies the fungus with photosynthetic sugars (Verma and Schuepp 1995), increased tolerance of mycorrhizal plants to toxic heavy metals concentrations in the soil makes mycorrhiza significant

Mechanism of VAM can enhance resistance to drought stress in host plant and may include many possible aspects of VAM improves the properties of soil in rhisophere; enlarges root areas of host plants, and improves its efficiency of water absorption; enhances the absorption of P and other nutritional elements, and then improves nutritional status of host plant; activates defence system of host plant quickly; protects against oxidative damage generated by drought and affects the expression of genetic material.

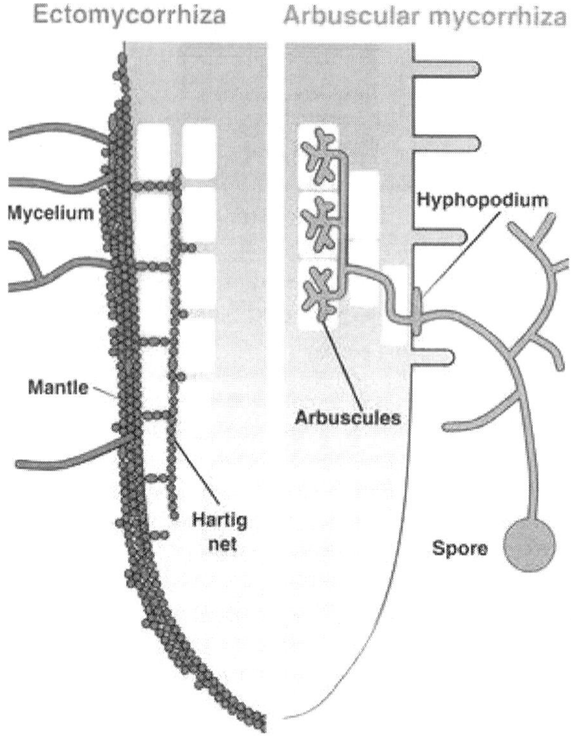

Figure 20.1. Ecto- and Endo-Mycorrhizae Attached with Root System.

Beneficial Effects of VAM

Mycorrhizal association benefits higher plants result from one or several of these mechanisms by improving water and nutrient uptake, helping in the development of roots, increased overall absorption capacity of roots due to morphological and physiological changes in the plant. There is increased absorption surface area, greater soil area explored (since the fungus acts as an extension of the root) greater longevity of absorbing roots, better utilization of low-availability nutrients, and better retention/storage of soluble nutrients, thus reducing reaction with soil colloids or leaching losses. Increased mobilization and transfer of nutrients (P, N, S, micronutrients Cu, Zn) from the soil to the plant. Mycorrhizal fungi have been estimated to "substitute" up to 500 lb/a of P for citrus and 170 lb/a for soybeans in tropical areas. Better development of P solubilizing bacteria in the myco-rhizosphere. Increased establishment, nodulation and atmospheric nitrogen fixation capacity in legumes. Modification of plant-pathogen relations: mycorrhizae influence the colonization of roots by other microorganisms, reduce the susceptibility (or increase the tolerance) of roots to soil-borne pathogens such as nematodes or phytopathogenic fungi such as *Fusarium oxysporum, Fusarium solani, Rhizoctonia solani* and *Macrophomina phaseolina*. Usually, plants of soybeans, cotton, tomato, oats, and cucumbers are less susceptible to nematode invasion when they are mycorrhizal. In studies with fungi such as Pythium, Phytophthora, Fusarium and

Verticillium, in most cases (53 per cent) the mycorrhizal interaction is beneficial for onions, cucumbers, and tomatoes. Secretion of antibiotics and support of a community that competes or antagonizes pathogenic microorganisms, thus aiding in disease suppression. Increased production of plant growth hormones such as cytokinins and gibberellins. Modification of soil-plant-water relations, promoting better adaptation of plant to adverse environment conditions (drought, metals). At elevated heavy metal concentrations in soils, mycorrhizal fungi have been shown to detoxify the environment for plant growth.

Significance of VA Mycorrhizae

VA-fungi are known to be effective in increasing nutrient uptake, particularly phosphorus and biomass accumulation of many crops in low phosphorus soil (Osonubi, 1991). Several investigators indicated that there is a beneficial effect of VA-fungi inoculation on nutrient uptake and on plant growth especially in sterilized soils (Menge, 1983., Abbott, 1984., Hayman, 1982., Powell, 1984, Turk, 2006). VA-fungi on the uptake of nitrogen and micronutrient uptake may be attributed to two situations. In the first one is mycorrhizal hyphae act as extension to plant root, increasing root surface area and exploring larger soil volume, which will increase the chance of more micronutrient uptake. Mycorrhizal association with plant root may also enhance translocation between root and shoot of the infected plant, hence enhancing the plant growth (Osonubi, 1991).

VA-fungi infection has been reported to increase nutrient uptake in water stressed plants (Busse, and. Ellis, 1985) enable plant to use water more efficiently and to increase root hydraulic conductivity (Graham, and Syversen, 1984.). VAM was able to alter water relations and played a great role in the growth of host plant in the condition of drought stress. Augé compiled the existing literature on plant water relations, drought and VAM symbiosis (2001). VAM plants would recover from wilting more quickly than NM plants upon relief of drought (Gemma *et al.*, 1997). VAM symbiosis has generally not affected stomatal density and guard cell size when comparing VAM and NM plants with similar leaf areas, even when transpiration or stomatal conductance differed (Henderson and Davies 1990). It has been proved that concentration of chlorophyll in VAM plants was higher than their control NM plants (Gemma *et al.*, 1997, Davies *et al.*, 1993., Mathur and Vyas, 1995.) and higher concentration of chlorophyll is associated with higher photosynthesis rate (Davies et, al., 1993). Colonization with VA fungi might change the architecture of root, which may be used to increase the interaction of root and soil (Atkinson, 1994). Endogenous hormone balance changed by VA fungi colonization contributed to the enhancement of plant drought resistance. In legume plants, there are evidences showing VAM can alleviate drought induced nodule senescence and reduce oxidative damage to lipids and proteins. Bartels proposed that both prevention of oxidative stress and the elimination of reactive oxygen species are the most effective approaches used by plants to gain tolerance against several abiotic stresses, including drought (2001).

AMF inoculation improved plant nutrition; compensated the root damage and changed mycorrhizosphere microbial populations and augmented activation of plant defense mechanisms (Hooker *et al.*, 1994; Linderman, 1994). AM inoculated plant generated various compounds such as phytoalexins, enzymes of the phenylpropanoid pathway, chitinases, β-1,3-glucanases, peroxidases, pathogenesis-related (PR) proteins, callose, hydroxyproline-rich glycoproteins (HRGP) and phenolics (Bowles, 1990; Gianinazzi-Pearson *et al.*, 1994). These compounds are related to plant defense mechanisms. AMF initiated host defense responses which were subsequently suppressed disease infestations. The enhanced lignification of root endodermal cells by AMF colonization has been recorded (Dehne, 1979). Nevertheless, AMF symbiosis-specific genes control the expression of the genes concerned to plant defense during the establishment of AMF (Gianinazzi-Pearson *et al.*, 1994; 1995; 1996). AMF are the main components of many plant rhizosphere and play an important role in decreasing plant disease (Akhtar and Siddiqui, 2008., Sharma *et al.*, 2014). The application of AMF as a biocontrol agent performed a vital role in plant resistance and display greater potential to protect bean plants against the infection with *F. solani* (Al-Askar and Rashad, 2010).

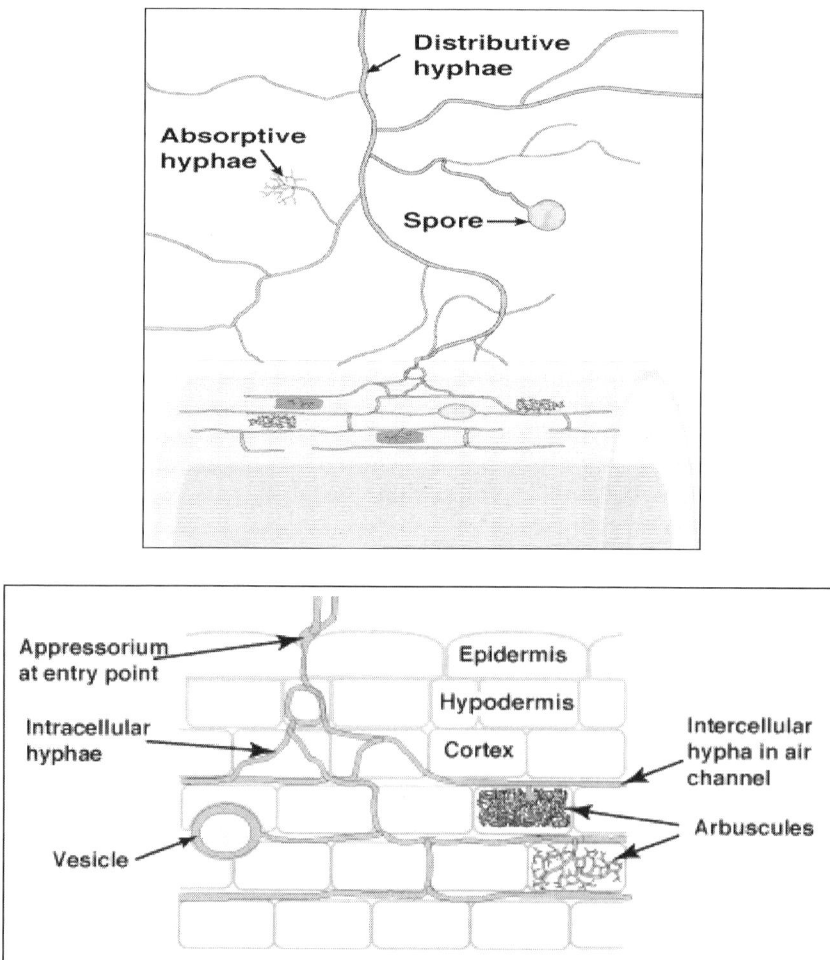

Figure 20.2. Structures in Soil and Structures in Roots.

Endomycorrhizal and ectomycorrhizal fungi in protecting host plants from phytopathogens and mineral elements toxicity. Still it was indicated that ectomycorrhizal fungi protect trees from high concentrations of toxic heavy metals, because these tend to be accumulated and immobilized in the mycorrhizal sheath (Jackson, and Mason, 1984). Arbuscular mycorrhizal (AM) symbiosis is the most important mutualistic association between arbuscular mycorrhizal fungi (AMF) from soils and the roots of terrestrial plants (Gadkar *et al.*, 2001). It is well documented that AM symbiosis can improve plant water relations or enhance drought resistance of host plants, thus protecting host plants against detrimental effects caused by drought stress (Augé 2001). It is now accepted that the contribution of AM symbiosis to plant drought resistance is the result of accumulative physical, nutritional, physiological and cellular effects (Aliasgharzad *et al.*, 2006). Mycorrhizal fungi is that the fungi induce plant tolerance to biotic stresses (Dehne, 1982; Barea *et al.*, 1996) and abiotic stresses (Parke *et al.*, 1983; Stahl *et al.*, 1998; Schreiner *et al.*, 2001; Borkowska, 2002; Shi *et al.*, 2002; Aleksandrowicz-Trzciñska, 2004; Swaty *et al.*, 2004).

Mass Production of Mycorrhiza

Ectomycorrhizal fungi are facultative parasites/saprophyte and can be cultured on artificial/synthetic media whereas Arbuscular mycorrhizal fungi are obligate parasite and require living cells (host) to grow. EM fungi can be cultured either on synthetic/laboratory media or natural media (most common PDA) (Thakur *et al.*, 2012). Specific nutrient requirement have also been observed in different fungi Invariably EM fungi rarely sporulate under culture conditions, therefore a grain culture (spawn) can be prepared by using locally available grains like wheat, maize, sorghum, bajra *etc.* similar to grain spawn prepared for production of mushroom fungi.

AMF require living host to grow. Fibrous roots of grasses like sudan, bahia, guinea, cenchrus and rhodes as well as white/red clover are highly susceptible to colonization by AMF. These hosts are also known as nurse host species and are utilized for mass multiplication of the inoculums. Asexual propagules-chlamydospores of known endophytes are collected from some known source like research organizations or extracted from the soil by wet sieving and decanting method (Gardemann and Nicolson 1963). These Chlamydospores are inoculated by mixing in the pot mixture 1:1 (Sterilized sand: soil mix for mass production) or making layers 2mm below the seeds of grasses/nurse hosts. When plants are 10-12 weeks old the root system colonized by the fungus could be observed by staining the infected roots (Phillips and Hayman 1970). The root mass is uprooted and chopped into small segments. The infected root segments along with the soil containing chlamydospores are used as the inoculum. The VAM species, *Glomus mosseae*, can be maintained on onion (*Allium cepa* L.) to prepare pot cultures following the procedure of Krishna *et al.* (1985). Once the pot culture was established, its soil along with the roots of onion was used as a source of inoculums.

Methods of Inoculation

Ectomycorrhiza

Grain inoculums can also be mixed with the soil in pot cultures. In case of seedlings/rooted suckers, grain inoculum is applied near the root zone/callus where root initiation takes place (Thakur and Sharma 2013). When inoculum is fruiting structure, mycelial segments obtained from such structure or spore mass released by fruiting bodies has to be mixed in pot mixture giving optimum organic matter. Grain inoculum can be placed in the form of column or a layer below the seed of a host to be inoculated in such a way that root after seed germination passes through the inoculum column.

Endomycorrhiza

Inoculation can be made by inoculum placement in layers, band placement or side dressing, mixing inoculum with soil, seed pelleting and pre-inoculation of transplanted seedlings. Seed/Seedling dressing can be done by mixing 1 per cent carboxy methyl cellulose for ensuring sticking of propagules with seed/seddlings. For inoculation of cuttings of multipurpose trees/fruit trees/other perennial host species AMF inoculum can be smeared at distal 2 cm area of cuttings to be buried in rooting mixture. In AMF inoculum consists of either chlamydospores extracted from AMF infected hosts or root segments of nurse host plants infected with AMF.

Determination of Mycorrhiza in Plant Roots

The details of the technique for extraction and quantification of AMF has reviewed by Utobo *et al.* (2011). Different species of AMF are identified on the basis of synoptic key consisting of spore size, colour, shape, wall layer characteristics and attachment of hyphae (Hall and Fish 1979; Trappe 1982; Benny *et al.*, 2001; Morton and Redecker 2001; Mukherji 1996; Sharma *et al.*, 2008). To determine the degree and intensity of root colonization by mycorrhizal fungi, procedure of Phillips and Hayman (1970) modified by Kormanick *et al.* (1980) is widely used for clearing and staining roots for rapid assay.

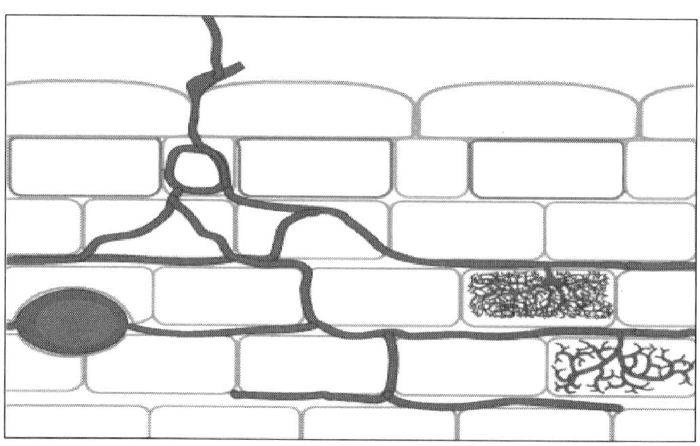

Figure 20.3 Mycorrhizas Produced by *Glomus* Species.

VA- Mycorrhizal Association with Legume Crops

Legume crops have a high (phosphorus) requirement for nodule formation nitrogen fixation and optimum growth. Mycorrhizal condition of legume crops found to increase its vegetative growth and seed yield in addition to improve nodulation on it's root system (Lambert and Weidensaul, 1991., Mathur and Vyas, 2000.). Nair *et al.* (1990) reported that higher level of VA mycorrhizal infection was beneficial for plant growth of cowpea (*Vigna unguiculata* L.) under field condition. Jackson and Mason (1984) found positive relationships among (P) availability, VA mycorrhizal infection and pod yield in groundnut (*Arachis hypogaea* L.). It was indicated that mycorrhizal colonization in several cowpea genotypes was host dependent and heritable (Molina, 1981). Alloush *et al.* (2000) found that chickpea plants inoculated with mycorrhizal fungus *Glomus versiforme* had higher number of nodules, shoot phosphorus content, shoot dry weight and grain yield than uninoculated chickpea plants. Inoculation with VA-fungus did not significantly affect seed yield of pea (*Pisum sativum* L.) plants in soil which is rich in organic matter (OM) amendment and phosphorus. On the contrary, seed yield was significantly enhanced with VA-fungi inoculation in soil which is poor in OM and phosphorus (Bethlenfalvay and Barea, 1994).

The association of arbuscular mycorrhizal (AM) fungi with different leguminous crops in five agro-ecological zones (AEZs) of Bangladesh was assessed through a survey. Among the selected crops grasspea, lentil, chickpea, mungbean, cowpea, blackgram, gardenpea, and soybean were highly mycotrophic. Variations in AM colonization and spore population were observed in different leguminous crops in different AEZs in Bangladesh. The AM fungal structures (hyphae, arbuscules and vesicles) in the root system of the leguminous crops varied. *Glomus* was the most common genus of AM while *Sclerocystis* was the least prevalent genus in the rhizosphere soil (Molla and Solaiman, 2009). Asimi *et al.* (1980) also indicated that stimulation of nodule activity in soybean grown at different soil P levels might be due to direct and preferential enhancement of nodule formation by vesicular-arbuscular mycorrhizal (VAM) fungi. Since cowpea had the highest AMF colonization than all the other legumes, the associated extensive AMF hyphal system may have facilitated the absorption of any ammonium that might have been released from nodules through their (AMF hyphae) production of glutamine synthetase, thereby, improving the N economy of the cowpea plant (Smith *et al.*, 1985). The improved shoot P uptake associated with mycorrhizal plants is largely the result of enhanced shoot growth of such mycorrhizal plants (Daft and El-Giahmi, 1974; Mosse *et al.*, 1976; Atayese *et al.*, 1993) as observed in cowpea in this study. In a study to investigate the response of three mycohhrizal legume species (cowpea, pigeonpea and groundnut) to the combined application of rock phosphate and a soluble phosphate, Ahiabor and Hirata (2003) observed that cowpea was the most responsive in terms of shoot growth and shoot P uptake but concluded that the positive shoot growth response of cowpea was not due to a favourable P nutrition but might have been the result of the effects of a greater mycorrhizal activity and an enhanced root growth on some unmeasured factor(s).

The deficiency is due to inherent low soil P, high fixation by aluminum (Al) and iron (Fe) oxides in soils (Buresh *et al.*, 1997), and insufficient use of fertilizers to replace P taken up in plant products or correct inherent low P (Obura *et al.*, 2001). Much of the semi- arid soils were derived from acidic parent materials low in P. In soils with low P availability, its uptake by diffusion alone is often exceeded by plant P demand (Marschner, 1995), thereby creating depletion zones in the rhizosphere (Junk and Claassen, 1989). Under such conditions, plant roots release a number of organic acids particularly malate and citrate in the rhizosphere which chelate Fe and Al thereby releasing P (Marschner, 1995). Despite the role of organic acids in P release in the rhizosphere, they have contributed to about 50 per cent total P depletion in the rhizosphere (Helal and Dressler 1989). In acid tropical soils, where P availability is low and the capacity to scavenge or to use added P with high efficiency is critical, mycorrhiza fungi are of particular significance (Swift *et al.*, 1994). Mycorrhizal associations are symbiotic between plant roots and fungi, such that the host plant receives mineral nutrients while the fungi obtains photosynthetically derived carbon compounds (Young, 1997). There are two types of mycorrhizas, ectomycorrhiza, which has fungal sheath surrounding the roots, and endomycorrhiza, which penetrates them. Endomycorrhiza cover three groups, with arbuscular mycorrhiza (AM) forms being the most common tropical agro ecosystems, (Dodd, 2000; Swift *et al.*, 1994). The AM are characterized by having vesicles and or arbuscules (haustoria like structures). It has four recognizable genera *viz., Acaulospora* (*A*), *Glomus* (*G*), *Gigaspora* (*Gi*) and *Sclerocystis* (*S.*) (Gerdermann and Trappe, 1974). They are primarily responsible for nutrient transfer from the soil to plants, but have other roles such as soil aggregation, plant protection against pathogens and drought stress. This is achieved by growth of their fungal mycelium within the host plant roots and out into the soil beyond (Dodd, 2000). The mechanism of P uptake by mycorrhiza is generally similar to P acquisition by plant roots, however they are able to absorb more P than the unassociated roots. The increased spatial availability of P, effective P acquisition by external mycorrhizal hyphae is related to a number of factors. Mycorrhizal hyphae extends beyond the P depletion zone around the roots to absorb P from areas where roots alone cannot reach. They are able to absorb P effectively from lower soil concentration than the unassociated roots because mycorrhizal hyphae have higher affinity for P therefore they maintain uptake from low soil P concentration. This is because soil exploitation by mycorrhizal hyphae involves smaller energy expenditure per unit absorbing area than by roots. The hyphae forms internal polyphosphotase and thus maintain low internal P concentration (Jungk and Claassen, 1989). The infected plants can exploit sources of soil phosphate not available to uninfected ones such as rock phosphate fertilizer or fixed organic or inorganic soil P (Harley and Smith, 1983). It is important to note that the nutritional benefits of mycorrhizas are not confined to P alone, but to increased uptake of other nutrients, such as N, Ca, Mg, Zn, Cu, Mn and K, have been demonstrated in a variety of plants (Clark *et al.*, 1999; Swift *et al.*, 1994).

Arbuscular mycorrhizal fungi (AMF) enhanced uptake of nutrients such as phosphorus, zinc, copper, boron and molybdenum, iron and manganese from the soil (Lambert *et al.*, 1979., Shibata and Yano,2003) Through the intensive exploitation of the soil volume by the root external mycelia, AM fungi can efficiently and intensively extract soluble nutrients from the soils. AM fungi enhanced the uptake of phosphorus which is important for nodules formation in legumes such as Pigeonpea. Apart from AM fungi contribution, legume/cereal crop interactions may occur when nodulating plant (Pigeonpea) are interplanted with non-nodulated arable crop such maize. This may result in transfer of fixed nitrogen from nodulating plant to non-nodulating crop (Lambert *et al.*, 1979., Shibata and Yano,2003) Farmers adopt intercrop of cereals such as maize and legumes (Pigeonpea) to enhance productivity and income (Egbe and Adeyemo, 2006). AM fungi will further improve the soil quality which will consequently increase crop yield for profitability. Pigeonpea is used in a wide variety of cropping system. Intercropping of grain legume and cereals results in the legume deriving a greater proportion of its nitrogen from nitrogen fixation than when grown alone. Pigeonpea enriched the soil with over 40 per cent nitrogen per hectare in fallen leaves during its growth. Large proportion of the nitrogen fixed by Pigeonpea remains in the field which can give a substantial benefit to subsequent and cereals intercrop (Kumwenda *et al.*, 1996). The inoculation of arbuscular mycorrhizal fungi to mycorrhizal crops improved their growth and yield (Fagbola *et al.*, 2005., Dania *et al.*, 2009). AMF can indirectly affect nitrogen availability because it enhances the uptake of phosphorus which increased nodulation of leguminous plants. The inoculation of mycorrhiza to Pigeonpea increased the nutrient uptake of Pigeonpea. The nutrient (nitrogen, phosphorus and potassium) uptake in Pigeonpea was influenced by mycorrhizal inoculation and this was in accordance with earlier work done (Kumar and Dart, 1987).

Conclusion

VAM of roots increase absorption capacity due to morphological and physiological changes in the plant. VAM symbiosis often results in altered rates of water movement into, through and out of host plants, with consequent effects on plant morphology and physiology. The mechanisms of VAM enhancing resistance to drought stress are still debated. Two aspects are necessary when VAM symbiosis enhance resistance of host plants to high stress, one is VAM symbiosis activating defence system of host plant quickly; and the other is some biochemical compounds that can resist high stress are synthesized by VAM symbiosis. VAM symbiosis to plant drought tolerance is the result of accumulative physical, nutritional, physiological and cellular effects. Phosphorus deficiency is a major limiting factor to food production in tropical agriculture and mycorrhiza has the potential to alleviate. It reduces the vulnerability of plant species to pathogens by various protective mechanisms like enhancing biochemical defense mechanisms, production of plant growth hormones/antibiotics and altering root exudates as well as microbial population of mycorrhizosphere. Mycorrhizae can be effectively used to enhance the fertilizer use efficiency particularly in low input crops like legumes. Considering the multi-dimensional protective role of mycorrhizae against diseases and pests, it can be utilized as a component of IPM strategies.

Future Strategies

Degradation and loss of arable lands have attracted attention of the scientists worldwide to protect lands from further declining. The multifaceted functions of legumes on the other hand in the improvement of natural and managed terrestrial ecosystems have necessitated their sustainable production. Understanding the role of mycorrhizae in metal detoxification is likely to improve the agronomic strategies in order to take full advantage of mycorrhizal association for legume production in disturbed cropping locations. Most authors report extensive colonization to occur mainly in plants growing in soils of low fertility. Field and greenhouse studies demonstrated that crops growing in nutrient-poor soils had higher levels of mycorrhizal colonization than crops growing in better soils. Phosphorus deficiency is a major limiting factor to food production in tropical agriculture and mycorrhiza has the potential to alleviate it if well managed. Its exploitation can tackle P deficiency. Adoption of cropping system such as agroforestry and inclusion of highly mycorrhizal crops in the systems are important. The possibility of achieving positive results through use of soil from areas with highly mycorrhizal plant as a source inoculant when applied in little quantities

REFERENCES

Atkinson D. (1994): Impact of mycorrhizal colonization on root architecture, root longevity and the formation of growth regulators. In: Gianinazzi S *et al.* (eds). *Impact of arbuscular mycorrhizas on sustainable agriculture and natural ecosystem*, pp. 89-99.

Augé R.M. (2001): Water relations, drought and vesicular-arbuscular mycorrhizal symbiosis. *Mycorrhiza*, 11: 3-42.

Aleksandrowicz-Trzciñska M. (2004): Kolonizacja mikoryzowa i wzrost sosny zwyczajnej (*Pinus silvestris* L.) w uprawie za³o¿onej z sadzonek wró¿nym stopniu zmikoryzowanych. *Acta Sci. Pol. Silv. Colendar. Rat. Ind. Lignar* 3(1): 5-15.

Augé R.M. (2001): Water relations, drought and vesicular- arbuscular mycorrhizal symbiosis. *Mycorrhiza, 11*: 3–42

Asimi S., Gianinazzi-Pearson V. and Gianinazzi S. (1980): Influence of increasing soil phosphorus levels on interactions between VA mycorrhizae and *Rhizobium* in soybeans. *Can. J. Bot.* 58: 2200–2205.

Atayese M. O., Awotoye O. O., Osonubi O. and Mulongoy K. (1993): Comparisons of the influence of vesiculararbuscular mycorrhiza on the productivity of hedgerow woody legumes and cassava at the top and base of a hillslope under alley cropping systems. *Biol. Fert. Soils.* 16: 198–204.

Ahiabor B. D. and Hirata H. (2003): Associative influence of soluble phosphate, rock phosphate and arbuscular mycorrhizal fungus on plant growth and phosphorus uptake of three tropical legumes. *W. Afr. J. appl. Ecol.* 4: 75– 90.

Aliasgharzad N., Neyshabouri M.R., and Salimi G. (2006): Effects of arbuscular mycorrhizal fungi and *Bradyrhizobium japonicum* on drought stress of soybean. *Biologia*, 61: 324–328

Alloush, G.A.Z., S.K. Zeto and Clark, R.B. (2000): Phosphorus source, organic matter and arbuscular mycorrhiza effects on growth and mineral acquisition of chickpea grown in acidic soil. *J. Plant Nutr.*, 23: 1351-1369.

Abbott, L.K., A.D. Robson and Boer, G. De. (1984): The effect of phosphorus on the formation of hyphae in soil by the vesicular-arbuscular mycorrhizal fungus *Glomus fasciculatum. New Phytol.*, 97: 437-446.

Akhtar, M.S. and Siddiqui, Z.A. (2008): Arbuscular mycorrhizal fungi as potential bioprtectants against plant pathogens. *In: Mycorrhizae: Sustainable Agriculture and Forestry.* Siddiqui, Z.A., M.S. Akhtar and K. Futai (eds.), Springer Netherlands, Dordrecht, The Netherland

Al-Askar, A.A. and Rashad, Y.M. (2010): Arbuscular mycorrhizal fungi: A biocontrol agent against common bean fusarium root rot disease. *Plant Pathol. J.*, 9: 31–38

Bethlenfalvay, G.J. and J.M. Barea, (1994): Mycorrhizae in sustainable agriculture. I. Effects on seed yield and soil aggregation. *Am. J. Altern. Agric.*, 9: 157-161.

Bayne H. G. and Bethlenfalvay G. J. (1987): The Glycine-Glomus-Rhizobium symbiosis. IV. Interactions between mycorrhizal and nitrogen-fixing Endophytes. *Pl. Cell Envir.* 10: 607–617.

Buresh R.J., Smithson P.C. and Hellums C. (1997): Building phosphorus capital in Africa p. 111-150. *In:* Buresh R.J., Sanchez P.A., and Calhoun. Replenishing soil fertility in Africa. Soil Sci. Soc. of America, Madison, Wisconsin, USA.

Bethlenfalvay G J. (1993): Vesicular-arbuscular mycorrhizal fungi in nitrogen fixing legumes: problems and prosects In *Methods in microbiology techniques for mycorrhizal research*, pp. 835–849, edited by J R Norris D J Read, A K Verma New York: Academic Press

Busse, M. D. and Ellis, J. R. (1985): Vesicular-arbuscular mycorrhizal (Glomus *fasiculatum*) influence on soybean drought tolerance in high phosphorus soil. *Can. J. Bot.*, 63: 2290-2294

Bartels D. (2001): Targeting detoxification pathways: an efficient approach to obtain plants with multiple stress tolerance. *Trends Plant Sci.*, 6: 284-286.

Borkowska B. 2002. Growth and photosynthetic activity of mikropropagated strawberry plants inoculated with endomicorrhizal fungi (AmF) and growing under drought strees. *Acta Physiol. Plantarum.* (4): 365-370.

Barea J.M., Calvet C., Estaun V., and Camprubi A. (1996): Biological control as key component in sustainable agriculturae. Plant Soil. 185: 171-172.

Bowles, D.J. (1990): Defense-related proteins in higher plants. *Ann. Rev. Biochem.*, 59: 873–907

Brundrett M C. (2004): Diversity and classification of mycorrhizal associations *Biological Review* 79: 473–496

Benny G.L, Humber, R. A. and Morton, J.B. (2001): Zygomycota: Zygomycetes, Mycota VII. In J *System Evol Microbiol* 7: 113-146.

Cardon, Z.G. and Whitbeck, J.L. (2007): *The Rhizosphere*, p: 235. Elsevier Academic Press.

Clark R.B., Zobel R.W. and Zeto S.K. (1999): The effects of mycorrhizal fungus isolates on mineral acquisition by Penicum virgatum in acid soils. Mycorrhiza 9(3): 167-176. Copyright CAB Internationals.

Cooper K M and Tinker P B. (1978): Translocation and transfer of nutrients in vesicular arbuscular mycorrhizae *New Phytologist* 81: 43–53

Daft M. J. and El-Giahmi A. A. (1974): Effect of endogone mycorrhiza on plant growth. VIII. Influence of infection on the growth and nodulation in French bean (*P. Vulgaris*). *New Phytol*. 73: 1139–1147.

Dodd J. C. (2000): The role of arbuscular mycorrhizal fungi in agro- and natural ecosystems. Outlook on Agriculture 29(1): 55-62 In: Elderidge D.J. and Stokes S. (Eds.). Copyright: CAB Internationals.

Dania S.O., Dania M.I., and Fagbola, O. (2009): Performances of maize as influenced by *Cajanus cajan* compost and mycorrhizal inoculation. *Nig J Mycol.;* 2(1): 117–122.

Dehne, H.W. and Schonbeck, F. (1979): Influence of endotrophic mycorrhiza on plant diseases. II. Phenol metabolism and lignification. *J. Phytopathol.,* 95: 210–216

Dehne H.W. 1982. Interactions between vesicular-abuscular mycorrhizal fungi and plant pathogens. *Phytopathology*. 72: 1115- 1119.

Davies, F.T., Potter, J.R. and Linderman, R.G. (1993): Drought resistance of mycorrhizal pepper plants independent of leaf P-concentration – response in gas exchange and water relations. *Physiol Plant,* 87: 45-53.

Egbe, O.M., and Adeyemo M.O. (2006): Estimation of the effects of intercropped Pigeonpea on the yield and yield components of maize in Southern Guinea Savannah of Nigeria. *J Sust Dev Environ.;* 2(1): 107-119

Fagbola, O, Oyetunji, O.J., Osonubi, O. and, Mulongoy K. (2005): Greenhouse evaluation of two woody hedgerows as affected by arbuscular mycorrhizal fungus. Soil quality and Moisture. *Arch Agron Soil Sc*. 51(3): 335–349.

Frank A B. (1885): Uber die auf Wurzelsymbiose beruhende Ehnahrung gewisser Baume durcha unterirdische Pilze *Berlin Deutsche Boanische Gesssalschaft* 3: 128–145

Fyson A and Oaks A. 1990 Growth promotion of maize by legume soils *Plant and Soil* 122: 259–266

Gianinazzi-Pearson, V., Gollotte, A., Cordier, C., and Gianinazzi, S. (1996): Root defense responses in relation to cell and tissue invasion by symbiotic microorganisms: cytological investigations. In: *'Histology, Ultrastructure and Molecular Cytology of Plant-Microorganism Interactions'*, pp: 177–191. Nicole M. and V. Gianinazzi-Pearson (eds.). Kluwer Academic Publishers, Dordrecht, The Netherlands

Gemma J.N., Koske R.E. and Roberts E.M. (1997): Mycorrhizal fungi improve drought resistance in creeping bentgrass. *J Turfgrass Sci.,* 73: 15-29.

Gianinazzi-Pearson V., Lemoine, M.C., Arnould, C., Gollotte, A., and Morton, J.B. (1994): Localization of β (1=>3) glucans in spore and hyphal walls of fungi in the Glomales. *Mycologia,* 86: 478–485.

Gianinazzi-Pearson, V., Gollotte, A., Lherminier, J., Tisserant, B., Franken, P., Dumas-Gaudot, E., Lemoine, M.C., van Tuinen, D., and Gianinazzi, S. (1995): Cellular and molecular approaches in the characterization of symbiotic events in functional arbuscular mycorrhizal associations. *Can. J. Bot., 73*: 526–532.

Graham, J.H. and Syversen, J.P. (1984): Influence of vesicular arbuscular mycorrhiza on the hydraulic conductivity of roots of two citrus rootstocks. *New Phytol.,* 97: 277-284

Gerdermann J.W. and Trappe J.M. (1974): The Endogonaceace in the Pacific Northwest, Mycologia, Mermoir 5. Gerdermann J.W., 1964. The effect of mycorrhiza on the growth of maize. *Mycologia,* 56: 342-349.

Gadkar V., David-Schwartz R., Kunik T., and Kapulnik Y. (2001): Arbuscular mycorrhizal fungal colonization. Factors involved in host recognition. *Plant Physiology,* 127: 1493–1499.

Gerdemann J W. (1968): Vesicular-arbuscular mycorrhiza and plant growth *Annual Review of Phytopathology* 6: 397–418

Gardemann, J.D., and Nicolson, T.H. (1963): Spores of mycorrhizal endogene extracted from soil by wet sieving and decanting. *Trans Br Mycol Soc*46: 235-240.

Harley, J. L., and Smith, S. E. (1983): Mycorrhiza symbiosis. Academic Press, London.

Hall, I.R. and Fish, B.J. 1979. A key to the Endogonaceae. *Trans Br Mycol Soc* 73: 261-270.

Hayman, D.S. (1982): Influence of soils and fertility on activity and survival of vesicular-Arbuscular Mycorrhizal Fungi. *Phytopathology.,* 72: 1119-1125.

Helal, H. M. and Dressler, A. (1989): Mobilization and turnover of phosphorus in the rhizosphere. Zeitschrift fur Pflanzenernahrung und Bodenkunde 152: 175-180.

Harley, J. L. and Smith, S. E. (1983): *Mycorrhizal Symbiosis* London: Academic Press

Henderson, J.C. and Davies, F.T. (1990) Drought acclimation and the morphology of mycorrhizal Rosa hybrida L. cv Ferdy is independent of leaf elemental content. *New Phytol,* 115: 503-510.

Hooker, J.E., Jaizme-Vega, M. and Atkinson, D. (1994): Biocontrol of plant pathogens using arbuscular mycorrhizal fungi. *In: Impact of Arbuscular Mycorrhizas on Sustainable Agriculture and Natural Ecosystems,* pp: 191–200. Gianinazzi, S. and H. Schüepp, (eds.). Birkhäuser Verlag, Basel, Switzerland.

Jackson, R.M. and Mason, P.A. (1984): Mycorrhiza. Edward Arnold, Ltd., London, pp: 60. ISBN 0-7131- 2876-3

Kormanick, D.P, Bryan, W.C. and Schultz, R.C. (1980): Procedures and equipment for staining large numbers of plant roots for endomycorrhizal assay. *Can J Microbiol* 26: 536-538.

Krishna K P, Shetty, K. G, Dart, D. J. and Andrews, D. J. (1985): Genotype dependent variation in mycorrhizal colonization and response to inoculation of pearl millet *Plant and Soil* 159: 89–102

Kumwenda, J.T., Waddington, S.R., Snapp, S.S., Jones, R.B. and Blackie, M.J. (1996): Soil fertility management research for the maize cropping system of smallholders in Southern Africa: A review Natural Resource Group Paper. 96-102. CIMMYT. Mexico. D.F

Kwapata, M.B. and Hall, A.E. (1985): Effects of moisture regime and phosphorus on mycorrhizal infection, nutrient uptake and growth of cowpeas (*Vigna unquiculata* L.). *Field Crops Res.*, 12: 241-250.

Kumar Rao, J.V.D.K. and Dart, P.J. (1987): Nodulation, nitrogen fixation and nutrient uptake in Pigeonpea of different maturity groups. *Plant and Soil.*;99: 255-266.

Khalil, S., Loynachan, T.E. and McNabb, H.S. (1992): Colonization of soybean by mycorrhizal fungi and spore populations in Iowa soils. *Agron. J.*, 84: 832-836.

Khaliq A, Gupta, M. L., and Kumar, S. (2001): The effect of VAM fungi on growth of peppermint *Indian Phytopathology* 54(1): 82–84

Lambert, D.H, Baker, D. E, Cole, H. J. R. (1979): The role of mycorrhizae in the interactions of phosphorus with zinc, copper and other elements. *SSS Am J.* 43: 975-980.

Lambert, D.H. and Weidensaul, T.C. (1991): Element uptake by Mycorrhizal soybean from sewage- sludge-treated soil. *Soil Sci. Soc. Am. J.*, 55: 393-398.

Lambert, D. H., Baker, D. E, and Cole, H. (1980): The role of mycorrhizae in the interactions of phosphorus with zinc and other elements. *Soil Science Society of American Journal* 43: 976–980

Linderman, R.G. (1994): Role of VAM fungi in biocontrol. *In: Mycorrhizae and Plant Health,* pp: 1–125. Pfleger, F.L. and R.G. Linderman, (eds.).*American Phytopathological Society*, St. Paul, Minnesota, USA

Morton, J.B. and Redecker, D. 2001. Two new families o f Glomales Archaeosporaceae and Paraglomaceae, with two new genera *Archaeospora* and *Paraglomus,* based on concordant molecular and morphological characters *Mycologia* 93: 181-195.

Mukerji, K.G. (1996): Taxonomy of endomycorrhizal fungi. In: *Advances in Botany* KG Mukerji, B Mathur, BP Chambola and P Chitralekha (Eds). A P H Publication Corporation, New Delhi. 213-221 pp.

Mathur, N. and Vyas, A. (1995): Influence of VAM on net photosynthesis and transpiration of Ziziphus mauritiana. *J Plant Physiol*, **147**: 328-330.

Menge, J.A. (1983): Utilization of vesicular-arbuscular mycorrhizal fungi in agriculture. *Can. J. Bot.*, 61: 1015-1024.

Mathur, N. and Vyas, A. (2000): Influence of arbuscular mycorrhizae on biomass production, nutrient uptake and physiological changes in Ziziphus mauritiana Lam. under water stress. *J. Arid Envir.*, 45: 191-195.

Molina, R. (1981): Ectomycorrhizal specificity in the genus *Alnus. Can. J. Bot.*, 59: 325-334.

Molla, M. N. and Solaiman, A. R.M. (2009): Association of arbuscular mycorrhizal fungi with leguminous crops grown in different agro-ecological zones of Bangladesh. *Archives of Agronomy and Soil Science*. 55, (3): 233-245

Marschner H. (1995): Mineral nutrition of higher plants. Academic Press, London. Jungk A. and Claaseen N. (1989): Availability in soil and acquisition by plants as the basis for phosphorus and potassium supply to plants. *Zeitschrift fur Pflanzenernahrung und Bodenkunde*, 152: 151-157

Mosse, B., Powell, C. L. I. and Hayman, D. S. (1976): Plant growth responses to vesicular-arbuscular mycorrhiza. IX. Interactions between VA mycorrhiza, rock phosphate and symbiotic nitrogen fixation. *New Phytol*. 76: 331-342.

Mosse B. (1973): Advances in the study of vesicular arbuscular mycorrhizae *Annual Review of Phytopathology* 11: 171–176

Menge, J.A. (1983): Utilization of vesicular-arbuscular mycorrhizal fungi in agriculture. *Can. J. Bot.*, 61: 1015-1024.

Nair, S.K., Peethambaran, C.K., Geetha, D., Nayar, K. and Wilson, K.I. (1990): Effect of soil solarization on nodulation, infection by mycorrhizal fungi and yield of cow pea. *Plant and Soil*, 125: 153-154.

Obura, P.A., Okalebo, J.R., Othieno, C.O. and Woomer, P.L. (2001): The effect of prep-pac product on maize-soybean intercrop in the acid soils of western Kenya. *African Crop Science Conference Proceedings*, Vo. 5: 889-896.

Osonubi, O., K. Mulongoy, O.O. Awotoye, M.O. Atayese and Okali, D.U. (1991): Effects of ectomycorrhizal and vesicular-arbuscular mycorrhizal fungi on drought tolerance of four leguminous woody seedlings. *Plant and Soil*, 136: 131-143

Parke, J.L., Linderman, R.G., Black, C.H. (1983): The role of ectomycorhizas in drought tolerance of Douglas-fir seedlings. *Newphytol*. 95: 83-95.

Powell, L., Conway, I. and Bagyaraj, D.J. (1984): VA Mycorrhiza, CRC Press, Boca Raton Fl., pp: 1-3, 205-222

Phillips, J.M. and Hayman, D.S. (1970): Improved procedures for clearing and staining parasitic and VAM fungi for rapid assessment of infection. *Tran Br Mycol Soc* 72: 204.

Rajeshwari, E., Latha, T. K. S., Vanangamudi, K. A., Selvan. K., and Narayanan, R. (2001): Effect of AM and phosphorous on seedling growth of *Casuarina equisetifolia Indian Phytopathology* 54(1): 85–87

Swift, M.J., Dvorak, K.A., Mulongoy, K., Musoko, M., Sanginga, N. and Tian, G. (1994): The role of soil organisms in the sustainability of tropical cropping systems. In: Syers J.K. and Rimmer D.L. Soil science and sustainable land management in the tropics pp155-170. CAB International, Cambridge University Press, UK.

Shibata, R. and Yano, K. (2003): Phosphorus acquisition from non-labile sources in peanut and Pigeonpea with mycorrhizal interaction. *Appl soil Ecol.*;24: 133-141.

Sekar L, Vanangamudi K, and Suresh K K. (1997): Influence of inoculation of biofertilizers on biomass production of seedlings of Shola tree species Van Vigyan 35(2): 57–62

Smith, S. E. and Read, D. J. (1997): *Mycorrhizal symbiosis* New York: Academic Press

Smith S. E., St. John B. J., Smith F. A. and Nicholas D. J. D. (1985): Activity of glutamine synthetase and glutamate dehydrogenase in *Trifolium subterraneum* L. and *Allium cepa* L.: Effects of mycorrhizal infection and phosphate nutrition. *New Phytol.* 99: 11–227.

Shi, L., Guttenberg, M., Kottke, I. and Hampp R. (2002): The effect drought on mycorrhizas of beech (*Fagus sylvatica* L.) changes in community structure and the content of carbohydrates and nitrogen storage bodies of the fungi. *Mycorrhoiza.* 12(6): 303-311

Sreenivasa, M.N. and Bagyaraj, D.J. (1989): Use of pesticide for mass production of vesicular-arbuscular mycorrhizal inoculum. Plant and Soil, 119: 127-132.

Swaty, R.L., Deckert, R.J., Whitham, T.G. and Gehring, C.A (2004): Ectomycorrhizal abundance and community composition shifts with drought: predictions from tree rings. *Ecology.* 85: 1072-1084

Schreiner, P.R., Ivors, K.L. and Pinkerton J.N. (2001): Soil solarization reduces arbuscular mycorrhizal fungi as a consequence of weed suppression. *Mycorrhiza.* 1(6), 273-277.

Sreenivasa, M.N. and Bagyaraj, D.J. (1989): Use of pesticide for mass production of vesicular-arbuscular mycorrhizal inoculum. *Plant and Soil,* 119: 127-132.

Stahl, P., Schuman, G.E., Frost, S.M and Williams S.E. (1998): Arbuscular mycorrhizae and water stress tolerance of Wyoming big sagebrush seedlings. *Soil Sci. Am. J.* 62: 1309-1313.

Sharma, S., Prakash, B. and Aggarwal, A. (2008): Glomales I: A monograph of *Glomus* spp. (Glomaceae) in the sunflower rhizosphere of Haryana, India *Helia* 31: 13-18.

Sharma, Y.P, Santosh, W. and Thakur, J.S. (2014): Role of Mycorrhizae: A Component of Integrated Disease Management Strategies. *J Mycol Plant Pathol*, 44: (1) 12-20.

Tisdale, S.L., Nelson, W.L. and Baton, J.D. (1995): Soil Fertility and Fertilizers. Macmillan Publishing Company, USA.

Turk, M.A., Assaf, T.A. Hameed, K.M. and Al-Tawaha, A.M. (2006): Significance of Mycorrhizae. *World J. of Agri. Sci.s.* 2 (1): 16-20.

Thakur, J.S., Sharma, Y.P. and Lakhanpal, T.N. (2012): Standardization of ectomycorrhizal inoculation technique for apple seeds *Ind Phytopath* 65: 190

Thakur, J.S. and Sharma, Y.P. (2013): Effect of ectomycorrhizal inoculation on the growth of apple seedlings. *Pl Dis Res* 28: 35-38.

Trappe, J.M. 1982. Synoptic keys to the genera and species of zygomycetous mycorrhizal fungi. *Phytopathology* 72: 1102-1108.

Utobo, E.B., Ogbodo, E.N. and Nwogbaga, A.C. (2011): Techniques for Extraction and Quantification of Arbuscular Mycorrhizal Fungi. *Libyan Agri Res Cen J Int* 2: 68-78.

Verma, A. and Schuepp, H. (1995): Mycorrhization of the commercially important micropropagated plants, pp. 313–323 In *Methods in Microbiology,* edited by Morris Jr, D J Read, and A K Verma London, UK: Academic Press

Wani, S. P., McGill, W. B. and Tiwari, J. P. (1991): Mycorrhizal and common root rot infection, and nutrient accumulation in barley grown on Breton loam using N from biological fixation or fertilizer *Biology and Fertilizer of Soils* 12: 46–54

Young, A. (1997): Agroforestry for soil management. 2nd ed. CBA International New York, USA.

Chapter 21

Indian Type Culture Collection: A Treasure for Disease Management Research of Grain Legumes

Deeba Kamil, T. Prameela Devi, Damaram Choudhary & R. Sudeep Toppo

Grain legumes, also called pulses, are plants belonging to the family Leguminosae (alternatively Fabaceae) which are grown primarily for their edible seeds (Table 21.1). These seeds are harvested mature and marketed dry to be used as food or feed or processed into various products. Besides these direct benefits, legumes generate cash and grain legumes fetch more cash when sold than cereals. Grain legumes play important roles on farms, in human diets and for the sustainability of agriculture. Legumes (green and dry beans, soybean, groundnuts, cowpeas and pigeonpea) are major sources of protein and vitamins in addition to carbohydrate, for the resource poor farmers in India, where chick pea and pigeonpea are the major legume grown. These crops are grown in a range of cropping rain fed systems ranging from pure stands to mixed/intercropping (Deshpande and Deshpande, 1991).

Table 21.1. Important Grain Legumes with their Common and Scientific Names

Sl.No.	Common Names	Scientific Names
1	Broad bean or faba bean	*Vicia faba*
2	Common bean or dry beans	*Phaseolus vulgaris*
3	Chick pea	*Cicer arietinum*
4	Cowpea	*Vigna unguiculata*
5	Lentil	*Lens culinaris*

Sl.No.	Common Names	Scientific Names
6	Mungbean	*Vigna radiata*
7	Pea	*Pisum sativum*
8	Groundnut	*Arachis hypogaea*
9	Pigeon pea	*Cajanus cajan*
10.	Soybean	*Glycine max*

Grain legumes contribute significantly towards reducing poverty, improving food security, improving nutrition and health, and sustaining the natural resource base. Grain legumes are a cost-effective option for improving the diets of low-income consumers who cannot easily afford meat, dairy products and fish. Grain legumes also generate substantial benefits to the well-being of small holder farm families.

Diseases of Grain Legumes

However, legumes are more susceptible to pests and diseases than cereals. The important fungal diseases are given below.

Chickpea

Ascochyta blight (*Ascochyta rabie* (Pass.) Labr.), *Fusarium* wilt (*Fusarium oxysporum* f. sp. *ciceri*), dry root rot (*Rhizoctonia bataticola*), *Botrytis* gray mold (*Botrytis cinerea*) and rust (*Uromyces cicerisarientini*) are major diseases of Chickpeaa.

Ascochyta Blight (*Ascochyta rabie* (Pass.)

Ascochyta blight is a major disease in West Asia, North Africa and North Europe. The disease was first reported in India in 1911 by Butler. The disease has been reported in Punjab, Bihar, Jammu, Gujarat, Maharashtra, Uttar Pradesh and Madhya Pradesh. The disease is usually seen around flowering and podding time as patches of blighted plants in the field. Plants are more vulnerable for infection at flowering and podding stage. Initially water soaked lesions on leaves are more prominent. On leaves, stem and pod spots become roundish with brown margin with yellowish grey center and coalesce to blighten the plant. In the center of spot, numerous minute black pycnidia were arranged in concentric rings.

Fusarium Wilt (*Fusarium oxysporum f. sp. ciceri*)

Fusarium wilt disease is widely spread in most Chickpeaa growing areas of Asia (India Burma), Africa, southern Europe and the Americas. The disease can be observed at any stage of the plant. In seedling stage infected seedling collapse and lie flat on ground disease. These seedlings retain their dull green color. Dark brown to black discoloration of the internal stem tissues is clearly visible on affected seedlings when split open vertically from the collar downwards or cut transversely. In grownup plants, leaves turn yellow to light brown colour and drop-off prematurely. On collar region necrosis and discoloration can observed. In the adult stage affected plants show typical wilting, results drooping of the petioles, rachis and leaflets.

Cowpea

Anthracnose (*Colletotrichum lindemuthianum*), Cercospora leaf spot (*Pseudocercopora cruenta*), Ascochyta blight (*Ascochyta phaseolorum*) are major diseases of cowpea.

Pigeonpea

Pigeonpea suffers from many diseases in the Indian sub-continent, where over 90 per cent of the world crop is grown. These are Fusarium wilt (*Fusarium udum*), Alternaria leaf spot (*Alternaria tenuissima*), Collar rot (*Sclerotium rolfsii*), Powdery mildew (*Oidiopsis taurica*) and Phytophthora blight (*Phytophthora dreschleri* (Pal *et al.*, 1970; Kannaiyan *et al.*, 1981).

Wilt (*Fusarium udum*)

In India disease was reported from Maharashtra, UP, Bihar, Tamil Nadu and Gujarat. Fusarium wilt is a worst disease of red gram in India causing severe damage wherever crop is grown. The most characteristic symptom of disease is gradual wilting of seedling plant, yellowing of leaves, drooping down of foliage within 3-4 days. The disease appears in patches and spread in concentric circle. Affected plant shows discolouration of roots and vascular bundles prominently. The pathogen produces toxic Fusaric acid which causes plugging of xylem and phloem vessels.

Stem Rot (*Phytophthora drechsleri* f. sp. *cajani*)

Pathogen produced necrotic lesion on basal portion of stem. Initially lesions are very small and later on enlarge showing slight depression. Infected tissues at collar region become soft and whole plants wilt. On infected collar portion bark become brown colour, tissues become soften as well as swelling of collar region. Upper portion of the twigs, branches shows wilting and pre-mature drying.

Dry Beans

Anthracnose (*Colletotrichum lindemuthianum*), rust (*Uromyces phaseoli*), angular leaf spot (*Isariopsis griseola*) and Fusarium root rot (*Fusarium solani*) are major diseases of dry beans.

Anthracnose (*Colletotrichum lindemuthianum*) Per Stage: *Glomerella cingulata*

The disease has been known since 1875 when it was reported in Italy and Germany. It occurs in Australia, Europe, India, Japan and New Zealand. All aerial parts of the plants are susceptible to infection. Diseases appear in the second or third week of June in Northern Indian hills. Black sunken lesion appear on pods, cotyledons and stems. Pinkish spore mass appear on lesion during wet weather condition. Infection may go deep into seed and seeds carry fungus from season to season. Stems of seriously affected plants may break causing them to topple. Symptoms also develop on petioles and veins.

Rust (*Uromyces phaseolitypica*)

The disease occurs throughout the country where beans are cultivated. With 3.0 per cent infection the yield loss of dry beans is 2.7 per cent and with 79.0 per cent infection it may be 36.7 per cent. Rust pustules appear on all above ground plant parts. They are mostly by on undersides of leaves, less abundant on pods and they occur sparingly on stems. Initial symptoms are minute, almost white, slightly raised which later become distinct, reddish brown, circular sori. The sori are relatively tall, especially on pods where the elevations may be 6 mm in height. On some varieties, the sori are surrounded by a yellow halo. When leaves become thoroughly rusted, they fall from the plant. During an epidemic, most of the disease vines are deformed and severe yield reduction occur.

Pea

Downy mildew (*Peronospora viciae*), powdery mildew (*Erysiphe polygoni*), rust (*Uromyces viciae-fabae* and U. *pisi*), and wilt (*Fusarium oxysporum* f. sp. *pisi*) are important diseases of pea.

Wilt (*Fusarium oxysporum* f.sp. *pisi*)

It is also known as "St. Johns Wilt". Firstly reported in Minnesota in 1918 but symptoms were described and fungus was identified in 1928 in Wisconsin. This disease appears at or before blossoming time. Yellowing of lower leaves and stunting of plants first appear. Leaf margin generally curl downwards with incomplete filling of pods. Stem tissue is discoloured turning to orange brown. At first, root appears as normal but when the lowest part of the stem is split, a yellow to orange and finally black discolouration is noted in the xylem. Stem near the soil line is thickened slightly. Stipules and leaflets also curved down wards. Disease is likely to appear in circular spots in a field which enlarge as the season progresses. The disease affects most of the plants in heavily infected soil.

Powdery Mildew (*Erysiphae polygoni*)

This disease was first appear on leaves and then other parts. White powdery patches were formed on both sides of the leaf as well as on tendrils, pods and stems. These patches originate as a minute discoloured specks from which a powdery mass covers on all sides. Transpiration and respiration increase during night and photosynthesis reduces.

Lentil

Ascochyta blight (*Ascochyta lentis*), rust (*Uromyces fabae*), and wilt (*Fusarium oxysporumf.* sp. *lentis*) are considered serious disease problems of lentil in India.

Faba Bean

Chocolate spot (*Botrytis fabae* and *B. cinerea*), Ascochyta blight (*Ascochy tafabae*), and rust (*Uromyces viciae-fabae*) are reported to be widespread and serious diseases of faba bean.

Groundnut

Rust (*Puccinia arachidis*), Early leaf spot (*Cercospora arachidicola*), Late leaf spot (*Cercosporidium personatum*) andStem rot/root rot (*Sclerotium rolfsii and Rhizoctonia solani*)

Rust (*Puccinia arachidis*)

The disease is usually found when the plants are over 6 weeks old. Small brown to chestnut brown dusty pustules occur on the lower surface of leaves. The epidermis ruptures and exposes a powdery mass of brown uredospores. Corresponding to the sori, small, brown, necrotic spots appear on the upper surface of leaves. The pustules are 0.3 to 2.0 mm in diameter and usually surrounded by a yellow halo. In severe infection lower leaves wither and drop up prematurely. The kernels formed in the affected plants are shriveled and small.

Leaf Spot (Tikka/Early and Late) Diseases

Tikka disease is the major disease of groundnut in India. It occurs in groundnut growing areas of the world, including United States of America, most countries of Africa, the Philippines, Indonesia, Sri-lanka, China, Malaysia and Australia. All the groundnut verities grown in India are susceptible to this disease.

Early Leaf Spot (*Cercosporaar achidicola*)

The early leaf spots caused by *C. arachidicola* are small circular to irregular in out line, 1-10 mm in diameter but larger than late leaf spot. They are surrounded by yellow halo from the very beginning. On the upper surface, the necrotic area is reddish brown to black while on the lower surface it is light brown. The yellow halo is indistinct or not present on lower leaf surface. The fungus do not produce haustoria.

Late Leaf Spot (*Cercosporidium personatum* (Revised: *Phaeoisariopsis personata*)

C. personatum produces smaller, more circular spots which are 1-6 mm in diameter and dark brown to black in color. These brown spots are not surrounded by yellow halo in early stages. The lower surface of the spots is carbon black in colour. Pathogen produced conidia are restricted to the lower surface and cushions of conidiophores can be seen in concentric rings. They occur mostly on leaf but may occur also on the petiole and stem. Severely infected leaves may drop off prematurely. This disease comes late in season. Minimum after 1 month of sowing and becomes severe in between flowering and harvesting stage under favorable climatic condition. Late spot develop 2-4 weeks later than the early spots. This disease is more damaging causing spots and spread faster and cause early defoliation.

Mungbean

Cercospora leaf spot (*Cercospora canescens*) is the major disease problem of mungbean.

Role of Culture Collection in Grain Legume Research

Fungal plant pathogens are potential biotic threat to crop production system. The reliable identification and documentation of Fungal pathogens requires type strains of the particular causal species or pathovars and subspecies. Generally, researchers working on Fungal pathogens procure the reference strains from abroad culture collection centers in order to do confirmation of their strains, which is time consuming as well as cost incurring. To address this issue, a repository of major microbial plant pathogens along with their identification and supply services has been established as Indian Type Culture Collection (ITCC) in the Division of Plant Pathology, Indian Agricultural Research Institute, New Delhi in 1936. At present 4,000 fungal cultures are available with ITCC. ITCC is an affiliate member of the World Federation for Culture Collections (WFCC) and is registered with the World Data Centre for Microorganisms (WDCM, registration number 430). Central Insecticide Board, Directorate of Plant Protection, Quarantine and Storage, Ministry of Agriculture, Government of India had made it mandatory to deposit the bio-agents at ITCC before they are released. Also the accession number of ITCC is a prerequisite for publication.

Culture collection centers are worldwide encouraged to create novel and better techniques for bio prospecting of novel microorganisms, carry out molecular sequence analysis, define phylogenetic relationships, execute taxon-based research on population structure of natural consortia, and pools of microorganisms, monitor and track genetically-engineered microorganisms that have been released for use in the environment and industry, and Many identified species/strains of Microbes related to grain legumes available with ITCC need to be explored for disease management research.

Table 21.2. Culture Available in ITCC with Especial Reference to Grain Legumes (Prameeladevi, 2012)

ITCC No.	*Sources*	*Year of Deposition*
	***ALTERNARIA ALTERNATA* (Fr.) Keissier**	
3467	Soybean seed, New Delhi, Vishwadhar	1985
5558*	Groundnut, Coimbatore, Ramaiah	2004
6622*	Green gram seed, Head, Navsari	2010
	***ALTERNARIA CUCUMERINA* (ELLIS and Everh) Elliot**	
6012*	Cluster bean, Bapatla, S. Reddy	2005
6121*	Cluster bean, Bapatla, S. Reddy	2005
	***ASCOCHYTA RABIEI* (Passerini) Labrousse**	
4634*	Chickpeaa, Ludhiana, M. Pal	1995
4638*	Chickpeaa 0-11, Delhi, M. Pal	1995
4785*	Chickpeaa, Jammu, S.K. Singh	1998
	***ASPERGILLUS FLAVUS* Link**	
1652	*Arachis hypogaea*, Hyderabad, A.S. Indulkar	1971
1653	*Arachis hypogaea*, Hyderabad, A.S. Indulkar	1971

ITCC No.	Sources	Year of Deposition
1654	*Arachis hypogaea,* Hyderabad, A.S. Indulkar	1971
1655	*Arachis hypogaea,* Hyderabad, A.S. Indulkar	1971
1670	*Areca catechu,* New Delhi, J.N. Kapoor	1972
1717	*Arachis hypogaea,* Hyderabad, A.S. Indulkar	1972
1718	*Arachis hypogaea,* Hyderabad, A.S. Indulkar	1972
2392	*Glycine* sp. seeds, Nainital, Y.P.S. Pangtey	1977
5192	Groundnut, IARI, New Delhi, N. Mathur	2000
6284	Groundnut rhizosphere, Junagadh	2008
6347	Goundnut seed, IARI, A. Getachew	2008
6352	Soybean seed, IARI, A. Getachew	2008
6353	Goundnut seed, IARI, A. Getachew	2008
6720	Pea rhizosphere, IARI, New Delhi, S. Gaind	2011
6777	Groundnut, Imphal, G.S. Liamajum	2011
6972	Groundnut seed, Bachupally, Sasidharen	2011
	***ASPERGILLUS NIGER* van Tieghem**	
6117	Groundnut, NCIPM, New Delhi, S. Singh	2006
6359	*Dolichous lablab* pod, IARI, A. Getachew	2008
6438	Groundnut rhizosphere, Bikanare, V.K. Gaur	2009
6719	Soybean rhizosphere, IARI, New Delhi, S. Gaind	2011
6775	Groundnut, Imphal, G.S. Laimajum	2011
	***ASPERGILLUS OCHRACEUS* Wilhelm**	
1740	*Phaseolus vulgaris* seeds, New Delhi, S.P. Lal	1972
1840	New Delhi, N. Bahl	1974
3476	Soybean seeds, New Delhi, Vishwadhar	1985
6074	Groundnut, Junagadh, V. Kumar	2006
6350	Goundnut seed, IARI, New Delhi, A. Getachew	2008
6358	Soybean seed, IARI, New Delhi, A. Getachew	2008
6361	Black gram seed, IARI, New Delhi, A. Getachew	2008
6362	*Phaseolus vulgaris* pod, IARI, New Delhi, A. Getachew	2008
	***COLLETOTRICHUM CAPSICI* (Sydow) Butler and Bisby**	
6450	*Vigna mungo,* Navsari, Prof. and Head	2009
6870	Chickpeaa, Mahanadhi, Prof. and Head	2011
	***COLLETOTRICHUM TRUNCATUM* (Schweinitz) Andrus and Moore**	
1847*	*Cajanus cajan*, Wajid Khan	1974
7074*	Chick pea, Tirupati, Jayalakashmi	2012
	***CORTICIUM ROLFSII* (Sacc.) Curzi**	
2782*	*Vigna ungiculata,* P.N. Patel	1981
5518*	Gram, Junagarh, Head	2004
6415	*Glycine max*, Ahmed Nagar, Head	2009
6416	*Cicer arietinum,* Ahmed Nagar, Head	2009

ITCC No.	Sources	Year of Deposition
6421	*Beta vulgaris,* Ahmed Nagar, Head	2009
6866	Seed of groundnut, Navsari, B.P. Mehatha	2011
7069	Groundnut, Kadiri, Reddy	2012
7088	Cowpea, Navsari, B.P. Mahta	2012
DRECHSLERA AUSTRALIENSIS **(Bugnicourt) Subram and Jain**		
2562*	Ground nut, IARI, New Delhi, J.N. Kapoor	1978
3462*	Soybean seeds, Agartala, Vishwadhar	1985
7107	Cowpea, Allahabad, S. Simon	2012
DRECHSLERA BICOLOR **(Mitra) Subram. and Jain**		
4211*	Groundnut rhizosphere, Mohanpur, S.K. Roy	1991
FUSARIUM ACUMINATUM **Ellis and Everhart**		
4153*	Bean seeds, Solan, S.K. Gupta	1991
6867*	Cowpea seed, Navsari, B.P. Mehta	2011
FUSARIUM CHLAMYDOSPORUM **Wollenw. and Reink.**		
3355	Soil, Delhi, P.N. Chowdhary	1984
3451	Soybean seed, New Delhi, Vishwadhar	1985
3457*	Soybean New Delhi, Viswadhar	1985
4969*	Pigeonpea, Kanpur, Vishwadhar	1999
FUSARIUM OXYSPORUM **Schlecht**		
179*	CBS, Netherlands	1938
188*	CBS, Netherlands	1938
189*	CBS, Netherlands	1938
555*	*Cicer arietinum,* Udaipur, N. Prasad	1946
2389*	*Glycine* sp. seed, Nainital, Y.P.S. Pangtey	1977
3242*	*Cajanus cajan,* Mahendra Pal	1983
4807*	Pea, Nauni, Solan, N.P. Dohroo	1998
4918*	Lentil, Palampur, Sharma	2000
5085*	Gram roots, Kamal, A. Gupta	2000
5316	Gram root, Sriganganagar, R.B.Gaur	2002
6341	Chickpeaa, Bikaner, S. Gangopadhyay	2008
FUSARIUM OXYSPORUM **f. sp. CICERIS (PADWICK) Chattopadhyay and Sen Gupta**		
549*	*Cicer arietinum,* Udaipur, N. Prasad	1942
3636*	*Cicer arietinum,* New Delhi, Mahendra Pal	1986
FUSARIUMOXYSPORUM **f. sp. *LINI* (Bolley) Snyder and Hansen**		
197	CBS, Netherlands	1938
788*	K.V. Srinivasan	1950
FUSARIUM SOLANI **(Martius) Saco.**		
132	*Pisum sativum,* Rochester, G.W. Padwick	1936
136	*Pisum sativum,* G.W. Padwick	1936
1804	*Glycine max,* New Delhi, O.K. Agarwal	1973

ITCC No.	*Sources*	*Year of Deposition*
2751	*Cicer arietinum,* A. Mani	1981
3906	Pigeonpea Germany, H. Nirenberg	1989
4347	*Cajanus cajan,* Navsari, D.G.Vala	1993
5104*	Soil, Hyderabad, C. Manoharachary	2000
5142	Mungbean, BTC IARI, B.S. Bhan	2000
5472	Cowpea, Anand, J. Chouhan	2004
5585	*Glycine max,* Navsari, Head,	2005
6338	*Dolichous lablab,* Navsari, Prof. and Head	2008
6342	Chickpeaa, Bikaner, S. Gangopadhyay	2008
6633	Pea, Udaipur, R. Shah	2010
FUSARIUM UDUM **Butler**		
781*	*Cajanus cajan,* New Delhi, K.M. Dutt	1949
784	*Cajanus cajan,* New Delhi, K.G. Nema	1949
1050*	*Cajanus cajan,* New Delhi, K.M. Dutt	1957
3241*	*Cajanus cajan,* New Delhi, M. Pal	1983
3955*	*Cajanus cajan,* Vadodara, Neelima Nagar	1989
4873*	Chickpeaa, Junagarh, R. R. Khander	1999
4874	Chickpeaa, Ghed, R. R. Khander	1999
4968	Pigeonpea, Kanpur, Vishwa Dhar	2000
GILBERTELLA PERSICARIA **(Eddy) Hesseltine**		
4885	Bean pod, IARI, New Delhi, N. Mathur	1999
4887	Bean leaf, IARI, New Delhi, S.P. Lal	1999
MACROPHOMINA PHASEOLINA **(Tassi) Goidanieh**		
249*	*Glycine hispida*, New Delhi, G.W. Padwick and M. Mitra	1940
250*	*Phaseolus mungo* var. *radiatus*, New Delhi, M. Mitra	1940
480*	*Glycine max*, Lyallpur, B.B. Mundukur	1945
482*	*Arachis hypogea*, Lyallpur, B.B. Mundkur	1945
1567*	*Glycine max,* Kalyani, S.K. Ghosh	1970
1800*	*Glycine max*, New Delhi, D.K. Agarwal	1973
2411*	*Glycine* sp. Seed, Nanital, Y.P.S. Pangtey	1977
3134*	Urad seedlings, New Delhi, S.P. Lal	1982
3546 *	Mungbean, Assam, U.N. Saikia	1986
4253 *	Cowpea seed, Jobner	1992
5141 *	IARI, New Delhi, B.S. Bhan	2001
5143*	Mung bean, IARI, B. S. Bhan	2001
5156 *	Bean, Jalna, K.K. Zote	2001
5191*	Groundnut, IARI, New Delhi, N. Mathur	2001
5196*	Pea seed, Agra, R.M. Saxena	2001
5584*	Cowpea, Navsari	2005
6037*	Soybean, Navsari	2006

ITCC No.	Sources	Year of Deposition
6183*	*Raphanus sativus*, IARI, New Delhi, N. Mathur	2007
6375*	Root of chickpeaa, Anand, V.P. Gohil	2008
6486*	Wilted pigeonpea, Junagadh, R.N. Pandey	2009
6557*	Green gram leaf, ICAR, New Delhi, N.P. Singh	2010
6621*	Chickpeaa root, Anand, R.N. Pandey	2010
6630*	Chickpeaa, Navsari, K.U. Solunky	2010
6698*	Soybean seed, Jobner, S.C. Jain	2011
6749*	Soybean seed, Navsari, Prof. and Head	2011
7027	Beans, Ajmer, M. Gord	2012
7070	Groundnut, Kadiri, Reddy	2012
	***MYROTHECIUM* sp.**	
5126*	*Cajanus cajan*, IARI, New Delhi, S.P. Lal	2000
	***PHOMA CAJANICOLA* (Trottu) Aa and Boerema**	
7086	Pigeonpea, Navsari, Prof. and Head	2012
	***PHOMA* sp.**	
1763*	*Cyamopsis tetragonoloba,* New Delhi, S.P. Lal	1973
4155*	Bean, Solan, S.K. Gupta	1991
4967	French bean, IARI, New Delhi, S.P. Lal	2000
6196*	*Dolichous lablab*, Navsari, Patil V. Adhar	2007
	***PHOMOPSIS* sp.**	
4154*	Bean seeds, Solan, S.K. Gupta	1991
4631	New Delhi, S.P. Lal	1995
	***PYTHIUM* sp.**	
5125	*Cajanus cajan*, IARI, New Delhi, S.P. Lal	2000
	***RAMULARIA PHASEOLINA* Petr.**	
3961	*Phasiolus vulgaris*, R.P. Kaushal, Solan	1989
	***SCLEROTINIA SCLEROTIORUM* (Libert) de Bary**	
2774	Pea pod, J.N. Kapoor	1978
3292	New Delhi, M.L. Sahni	1986
6094	Pea pod, New Delhi, N. Mathur	2006
	***SETOSPHAERIA ROSTRATA* Leonard**	
1798	*Glycine max*, New Delhi, D.K. Agarwal	1973
3465*	*Glycine max* seeds, Beharampur, Vishwadhar	1985
4685	Cowpea, Durgapura, L.C. Sharma	1996
	***TRICHODERMA SPERELLUM* Rifai**	
7660	Groundnut rhizosphere,Head A.R.S.K., A.P, Anantapur	2014
7847	Groundnut, IARI, New Delhi, P. Sharma	2015
	***TRICHODERMA HARIZIANUM* Rifai**	
4532	Pea rhizosphere, Palampur, D. Singh	1995
7661	Groundnut rhizosphere, Head, A. R.S.K., A.P, Anantapur	2014
7823	Groundnut, IARI New Delhi, P. Sharma	2015

ITCC No.	Sources	Year of Deposition
***TRICHODERMA VIRENS* (Miller, Giddens and Foster) Arx**		
6461	IARI field,New Delhi, Rajesh A.N, and Prameela Devi	2009
6470	IARI field, IARI, New Delhi, Prameela Devi	2009
6983	Soil, Samastipur, N. Prabhakaran	2012
***TRICHODERMA LONGIBRCHITUM* Rifai**		
6999	Chickpea Bikaner, IARI, Prameela Devi	2012
7002	Chickpea, IARI, New Delhi, N. Prabhakaran	2012
7003	Chickpea, IARI, New Delhi, N. Prabhakaran	2012
7833	Groundnut, IARI, New Delhi, P. Sharma	2015
7834	Groundnut, IARI, New Delhi, P. Sharma	2015

Note: (*) = Sterile.

Utility of Fungal Diversity

Fungal diversity existing in culture collections has the following major applications:

Morphological and Molecular Identification

Conventional characterization of fungal species has been based mainly on morphological methods, which are the most routinely performed. Nevertheless, recognition by morphological characters sometimes is not enough for accurate identification of fungal isolates at the species level. Furthermore, morphological characterization is time consuming and requires considerable expertise in fungal taxonomy and physiology (Leslie and Summerell, 2006). As identification of fungal species is critical, there is a need for accurate and complementary tools which permit a rapid, sensitive and reliable specific diagnosis of fungal species.

The concerns indicated above may be overcome by appropriate DNA sequencing and species-specific PCR assays (Jurado *et al.*, 2006a and b). Various PCR assays have been developed for the identification of fungal species. Some of them are based on single copy of species-specific regions/genes (Jaén *et al.*, 2004; Mulé *et al.*, 2006) and some amplify multi copy target sequences, such as IGS or ITS regions (Intergenic spacer and internal transcribed spacer of rDNA units, respectively), which increases the sensitivity of the assay in comparison with PCR assays based on single copy sequences.

Furthermore, morphological and molecular characterization always needs a reference culture for accurate taxonomy. Therefore, cultures available with fungal resource center may help for correct identification.

Biogeochemical Role in Cycling of Matter

Soil acts as the source of nutrition for the growth of a spectrum of microorganisms which have remarkable ability to degrade a vast variety of complex organic compounds due to their metabolic versatility. Fungus carry out recycling of phosphorus, oxygen, carbon, nitrogen and sulphur elements and replenish the environment with these by degrading substrates obtained from dead and decaying plant and animal remains. There fungal consortium may be used for soil heath up-gradation (Bhardwaj and Garg, 2012). Here the repositories may play a vital role to conserve the fungi which having the ability to degrade the material and provide to researchers, students and industries.

Fungi for Biological Control

A variety of biological control agents are available at culture collection for use which provide outline of future directions that might lead to the development of more diverse and effective biological controls for plant diseases. Mungbean, Chickpeaa, cowpea, pea and pigeonpea are the important pulse crops gaining importance globally in the present day scenario. In India organic farming is becoming popular in pulse farming in large scale. The application of potential *Trichoderma* formulation was found effective to suppress all the fungal pathogens of pulse crops Khan *et al.*, 2014).

Fungi as Biotechnological Tools

From multiple uses of fungi in early historic times, "early biotechnological processes" developed. Fungi were used here already *e.g.* for production of cheese, bread, wine, beer and other foodstuff. Based on these processes, fungi today have become one of the most important groups of organisms in modern technology, where food, fodder and various metabolites such as antibiotics, enzymes, steroids *etc.* are produced on an industrial scale. An equally important role of fungi in biotechnology is also to be expected for the future where - in addition to existing uses - fungi will be increasingly used employing modern methods such as genetic engineering and will also new applications as in pollution control, biological control of pests, microbial leaching and even biotechnology (Scazzocchio, 2014).

Because fungi possess a relatively low level of organization and are easy to handle in the laboratory, fungi represent increasingly important tools in modern research, especially in genetics. The different types of life cycles, the different ways of genetic inheritance and *e.g.* the possibility to identify by tetratede analysis in certain ascomycetes the stepwise production of the meiosis products, are advantages of this group of microorganisms. These culture collections will give a window to take and deposit theses model fungi (Colot *et al.*, 2006).

Sustainable Land Use

Fungi play a key role in soil fertility including its maintenance. They can act as indicators for assessment of sustainable land use in space and time. A combination and improvement of several concepts of sustainability may help obtain an integrated signal of functional eco-physiology of microbiota. Fungi in the soil have the ability to increase nitrogen in the soil through decomposition of organic matter. They can also decrease the amount of nitrogen in the soil by consuming nitrogen and converting it into protein. There is also a rare, third function: a type of fungi known as arbuscular mycorrhizal fungi, or AMF, can form partnerships with plants, scouring the soil for nutrients in exchange for sugars from the plant. Soil microorganisms in agricultural soils seem to play an important role in nutrient cycling, but the body of literature is still relatively undeveloped. However, scientists still do not understand Soil microorganisms enough to make specific recommendations to improve soil fertility (Bender and Heijden, 2014).

Transgenic Crop Varieties Resistant to Pathogen

The most pathogen-resistance development programmes oriented towards the development of genetically modified (GM) plant varieties are concentrated on generating resistance to viruses and fungi, particularly in Asia and Latin America (FAO, 2004).

Transgene-mediated resistance strategies against fungi are still in their infancy. Although planting disease-resistant varieties is one of the better ways to develop resistance management strategies to limit the incidence of selection for resistance-breaking pathogen isolates. Here, the microbial resource Centre will provide a number of authentic and virulent isolate to develop the transgenic crop varieties.

REFERENCES

Bender, S.F. and Heijden, M.G. (2014). Soil Biota Enhance Agricultural Sustainability by Improving Crop Yield, Nutrient Uptake and Reducing Nitrogen Leaching Losses. *Journal of Applied Ecology*. DOI: 10.1111/1365-2664.12351

Bhardwaj, V. and Garg, N. (2012). Importance of Exploration of Microbial Biodiversity. *ISCA Journal of Biological Sciences*. **3:** 78-83.

Colot, H.V., Park, G., Turner, G.E., Ringelberg, C., Crew, C.M., Litvinkova, L. and Weiss, R.L. (2006). A high-throughput gene knockout procedure for *Neurospora* reveals functions for multiple transcription factors. *Proceedings of the National Academy of Sciences, USA* **103:** 10352 – 10357.

Deshpande, U.S. and Deshpande S.S. (1991). Legumes. In: Foods of Plant Origin: Production, Technology, and Human Nutrition (Eds. Salunkhe, D.K. and Deshpande, S.S.). Springer US, pp. 137-300.

FAO. (2004). The State of Food and Agriculture: Agricultural Biotechnology: Meeting the needs of the poor? http: //www.fao.org/DOCREP/006/Y5160E/ Y5160E00.HTM.

Jaén, M.T.G., Mirete, S., Patiño, B., López-Errasquín, E. and Vázquez, C. (2004). Genetic markers for the analysis of variability and for production of specific diagnostic sequences in fumonisin-producing strains of *F. verticillioides*. *Eur J Plant Pathol* **110:** 525–532

Jurado, M., Vázquez, C., Callejas, C., González-Jaén, M.T. (2006a). Occurrence and variability of mycotoxigenic *Fusarium* species associated to wheat and maize in the South West of Spain. *Mycotoxin Research*. **22:** 87–91.

Jurado, M., Vázquez, C., Marín, S., Sanchis, V., González-Jaén, M.T. (2006b). PCR-based strategy to detect contamination with mycotoxigenic *Fusarium* species in maize. *Systematic and Applied Microbiology*. **29:** 681–689.

Kannaiyan, J., Nene, Y.L., Raju, T.N. and Sheila, V.K. (1981). Screening for resistance to Phytophthora blight of Pigeonpea. *Plant Disease*. **65:** 61-62.

Khan, M.R., Ashraf, S., Rasool, F., Salati, K.M., Mohiddin, F.A. and Haque, Z. (2014). Field performance of Trichoderma species against wilt disease complex of chickpea caused by Fusarium oxysporum f. sp. ciceri and Rhizoctonia solani. Turkish Journal of Agriculture and Forestry. 38: 447-454.

Leslie, J.F. and Summerell, B.A. (2006). The Fusarium laboratory manual. Blackwell Professional, Ames, Iowa.

Mulè, G., Susca, A., Logrieco, A., Stea, G. and Visconti, A. (2006). Development of a quantitative real-time PCR assay for the detection of Aspergillus carbonarius in grapes. Int J Food Microbiol 111: 28–34.

Pal, M., Grewal, J.S. and Sarbhoy, A.K. (1970). A new stem rot of Arhar caused by Phytophthora. Indian Phytopath. 23: 583-587.

Prameeladevi, T. Mondal, K.K., Kamil, D., Mathur, N., Toppo, R.S. and Singh, O.P. (2012). Catalogue of Fungal and Bacterial Cultures 1936-2012. Division of Plant Pathology Indian Agricultural Research institute, New Delhi, pp. 1-115.

Scazzocchio, C. (2014). Fungal biology in the post-genomic era. Fungal Biology and Biotechnology 1: 7 DOI: 10.1186/s40694-014-0007-6.

Index

Color Plates

Figure 1.1. Chickpea Field Infected with Fusarium Wilt. p. 2

Figure 1.2. Fusarium Wilt Infected Chickpea Plants (*Cicer arietinum* L.). p. 5

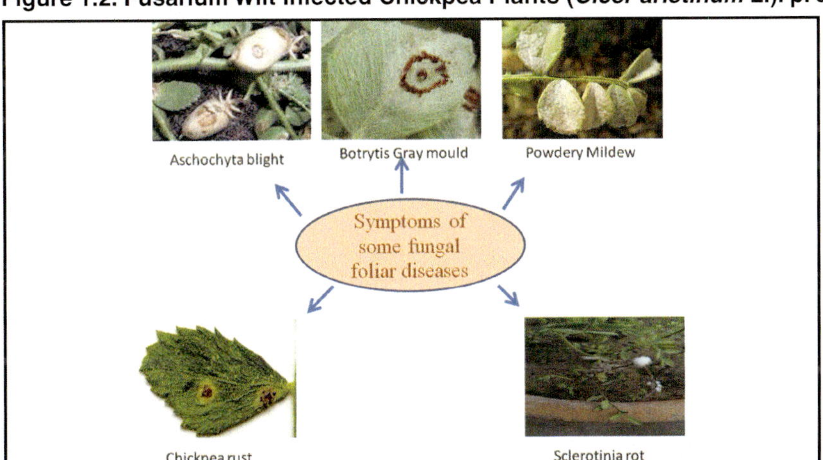

Figure 2.1. Foliar Diseases of Chickpea that Causes Severe Losses. p.30

Figure 3.1. Symptoms of Wilt on Pigeonpea Plants and Fields. p.30

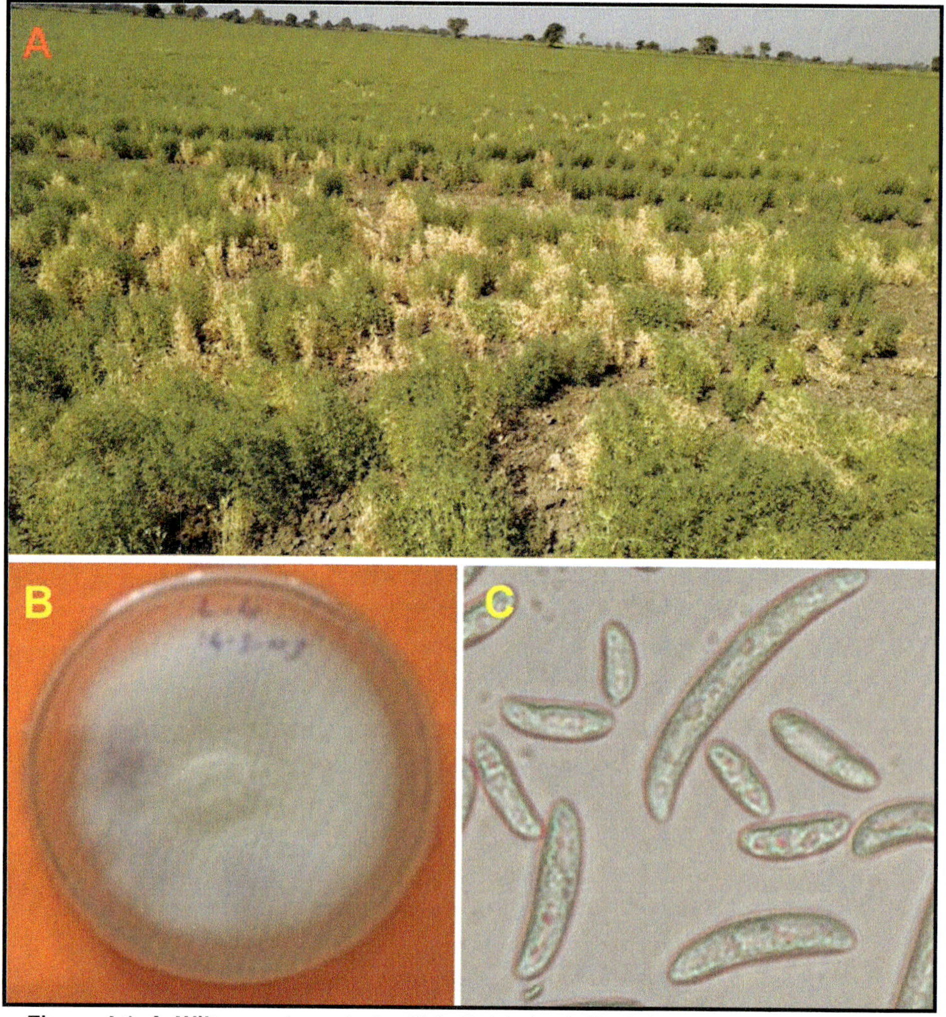

Figure 4.1. A: Wilt symptoms in lentil (a field view), B: Colony and growth pattern of *Fusarium oxysporium* f. sp. *lentis* and C: Typical macro and micro conidia of *Fusarium oxysporium* f. sp. *lentis. p. 68*

Figure 5.1: Infected Plant with Yellowish Green Leaves, a Bushy Appearance and without Flowers or Pods (Reddy *et al.*, 2012). p. 93

Figure 5.2: Mosaic Symptoms on Leaf (Reddy *et al.*, 2012). p.94

Figure 5.3: infected Leaf with Ring Spot Symptom (Reddy *et al.*, 2012). p.94

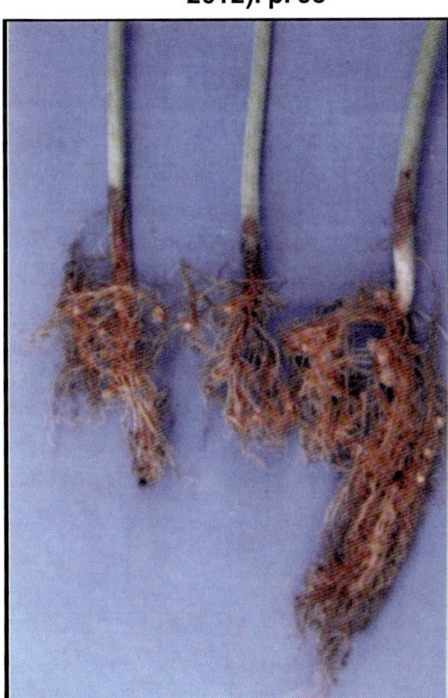

Figure 7.1a. Symptoms of Root Rot

Figure 7.1b. Symptoms of Web Blight. p.134

Figure 7.c. Symptoms on Pods. p.135

Figure 7.2a. Symptoms of Angular Leaf Spot on Leaves.

Figure 7.2b. Symptoms on Pods. p. 141

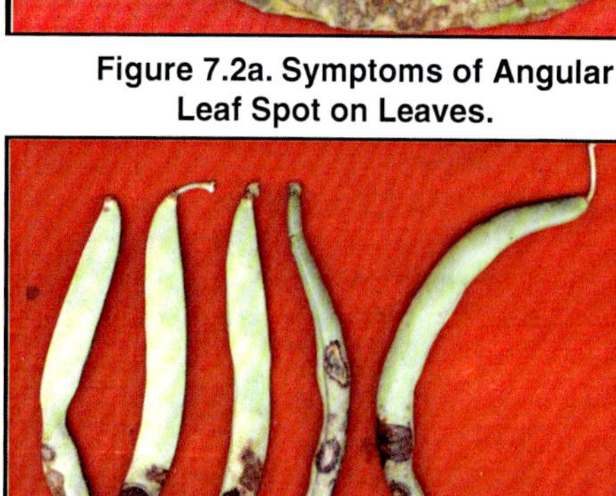

Figure 7.3. Symptoms of Anthracnose. p. 147

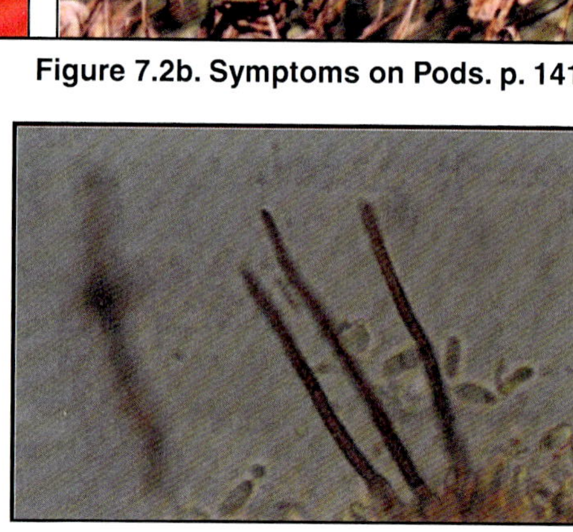

Figure 7.4. Acervulus of *Colletotrichum* Bearing Conidia and Conidiophores. p. 148

p. 165

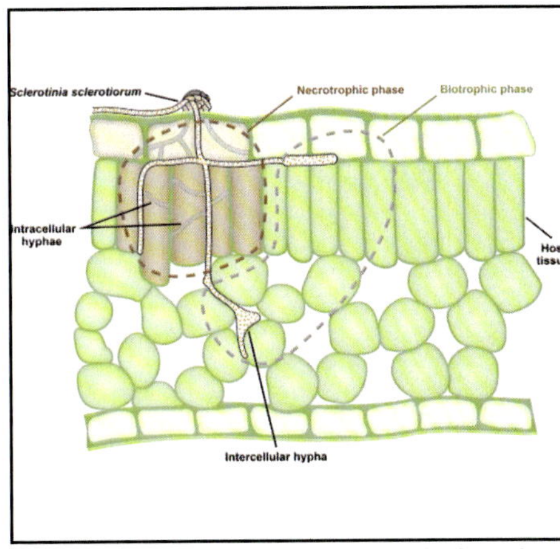

Figure 8.1. A New Model Depicting the Lifestyle Transition of *S. sclerotiorum*. p. 201

Figure 9.1. Rust Infected Fieldpea Leaves. p. 217

Figure 9.3. Powdery Mildew Infected Fieldpea Plants. p. 221

Figure 9.2. Greyish and Yellowish Symptoms on Fieldpea Leaves and Morphology of Pathogen. p. 220

Figure 9.4. Stem Rot Disease Infected Fieldpea Plants and Pods. p. 222

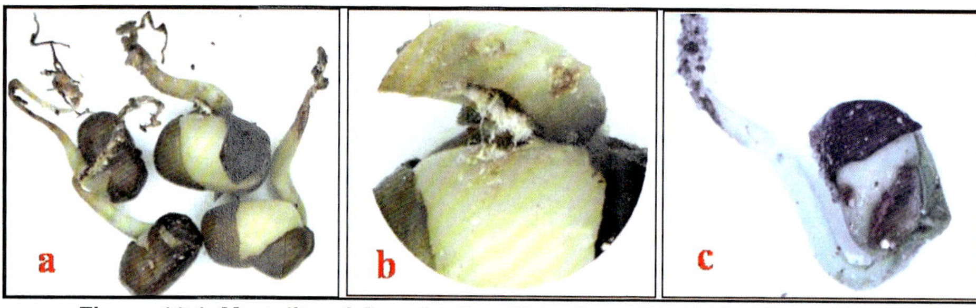

Figure 10.1. Mortality of Germinating Mungbean Seeds (cv. Pusa Vishal) Caused by *R. solani*. p. 231

Figure 10.2. Typical Aerial/Web Blight Symptoms on Mungbean Cultivar (Pusa Vishal). White mycelial growth on stems and leaves. p. 232

Figure 10.3. Sclerotia of *Rhizoctonia solani* on Soybean (a) Leaf and (b) Petiole. (Photograph: (a) Kharayat, B.S. (b) adapted from soybean diseases and courtesy of M. Emerson). p. 233

Figure 11.1. ***In vitro*** **Study of** ***Sclerotium rolfsii. p. 251***

Figure 11.2. Infection Cause by ***Sclerotium rolfsii. p. 251***

Figure 11.3. Seed Treatment and Soil Application of ***Trichoderma*** **for Management of** ***Sclerotium rolfsii. p. 252***

Figure 12.1. Field Symptoms of Chickpea Stem Sot, Distal Portions of the Plant Show Wilt and then become Necrotic. p. 278

Figure 12.2. Field Symptoms of Fieldpea Stem Rot. p. 278

Figure 12.3: Field Symptoms of Rajmash White Mold. p. 279

Figure 12.4. Infected Rajmash Plant Shows White Cottony Mycelium of the Pathogen that Grows on the Surfaces of Infected Flowers and Pod. p. 281

Figure 12.5. Infected Fieldpea Plant. p. 281

Figure 12.6. Pea-sized Clumps of Mycelium on Chickpea Diseased Plants. p. 282

Figure 12.7. Infected Lentil Plant with Mycelial Growth and Hard Black Sclerotia of the Fungus. Arrows indicate sclerotia. p. 282

Figure 12.8. Well Developed Sclerotia on Rajmash Seeds. p. 282

Figure 13.1. Symptoms of Stunt Disease in Chickpea. p. 302

Figure 13.1. Symptoms of Stunt Disease in Chickpea. p. 303

Figure 13.3. Symptoms of Leaf Curl/Necrosis Disease in Mungbean and Urdbean. p.302

Figure 13.4. Symptoms of Rosseting and Mosaic in Pea. p. 306

Figure 14.2. Yellow Mosaic Symptoms in Legume Crops.
A: Soybean, B: Rajmash, C: Mungbean, D: Horsegram, E: Urdbean, F: Dolichos, GvCowpea, H: Pigeonpea, I: Wild *Vigna. p. 317*